T0144224

Modeling the Power Consumption and Energy Efficiency of Telecommunications Networks

Modeling the Power Consumption and Energy Efficiency of Telecommunications Networks

Kerry James Hinton

Robert Ayre

Jeffrey Cheong

CRC Press
Taylor & Francis Group
Boca Raton London New York

CRC Press is an imprint of the
Taylor & Francis Group, an **informa** business

First edition published 2022
by CRC Press
6000 Broken Sound Parkway NW, Suite 300, Boca Raton, FL 33487-2742

and by CRC Press
2 Park Square, Milton Park, Abingdon, Oxon, OX14 4RN

© 2022 Taylor & Francis Group, LLC

First edition published by CRC Press 2022

CRC Press is an imprint of Taylor & Francis Group, LLC

ISBN: 978-0-367-25441-4 (hbk)
ISBN: 978-1-032-11596-2 (pbk)
ISBN: 978-0-429-28781-7 (ebk)

DOI: 10.1201/9780429287817

Typeset in Times
by SPi Technologies India Pvt Ltd (Straive)

Contents

Preface

There is a widely stated adage that to control a system you must be able to measure it. For many systems, measurement is a pretty straightforward process. However, for the Internet and large-scale telecommunications networks and services, measurement of their power or energy footprint is extremely challenging. This is due to many factors such as geographical spread, amount of equipment and equipment types, daily variation in network traffic, continual annual upgrading of equipment, sharing of common infrastructure across multiple services and introduction of new services, to name a few.

In addition to this, the paucity of up-to-date information on equipment and system parameters, especially power consumption and maximum throughput, makes estimating the power consumption and energy efficiency of these systems extremely challenging. Over recent years, as competition in the information sector has increased, equipment manufacturers and vendors appear to have taken an increasingly closed approach to disclosing these types of information. Typically, to secure such information, one requires either "inside knowledge" provided under some form of a non-disclosure agreement or on the basis of anonymised collected data. For some organisations, access to this type of information is not a problem because of their commercial arrangements with equipment vendors. However, for others it presents a major hurdle. Consequently, researchers, non-commercial organisations and policy makers may find it very difficult to construct estimates of current and future power consumption related to telecommunications services and networks.

Given the growing importance of Information and Communications Technologies (ICT), an understanding of their power consumption and energy efficiency is of immediate interest to many organisations and regulatory bodies. As ICT systems grow to meet increasing demand, the provision of power to their infrastructure is an issue, both on the global and local scales. Globally the growth in power consumption needs to be mitigated in order to ensure it does not become a burden on global electrical power production. Locally the use of renewable and stored energy sources requires energy-efficient equipment. In both cases, a good understanding of current and future power consumption is important.

This brings us back to the need for measurement of a system if we are to manage or control it. This principle applies both to telecommunications equipment, networks and services today and into the future. Because measurement of telecommunications network and systems is so difficult, in almost all cases physical measurement has to be replaced by partial measurement and mathematical modelling. This is where the material presented in this book comes into play. The focus of this book is on the modelling of telecommunications equipment, network and service "user phase" power consumption and energy efficiency. In other words, this book does not cover the life cycle stages of extraction, fabrication, deployment, removal and disposal/recycling in the overall lifecycle of equipment. The focus is on the power consumption during its use by customers and network operators.

This book presents a collection of mathematical methods and tools to enable the reader to construct power and energy efficiency models for telecommunications equipment, networks and services. In most cases, these models assume the reader does not have access to significant amounts of "insider knowledge" on the equipment, networks and services of interest. Indeed, in some cases the models presented in this book are expressly designed on the assumption of minimal knowledge and the principle that a "rough and ready" model is better than no model at all.

The early chapters can be taken as independent bodies of work. However, Chapter 6 assumes an understanding of the material presented in Chapters 4 and 5. Chapter 8 uses the results from Chapter 3 to Chapter 7 to provide mathematical tools to forecast future power consumption profiles of Edge and Core networks. Finally, Chapters 10 and 11 bring together material presented in previous chapters.

Referring to the chapters in order, Chapter 2 provides an overview of the motivation for the later chapters by looking at the reasons for constructing the models presented in this book. With telecommunications and the Internet becoming such an important contributor to economic growth and quality of life, their consumption of resources (particularly electrical power) needs to be understood and monitored. Telecommunications networks and the Internet have become too big to simply measure their power consumption. Consequentially the best we can do is to use a combination of measurement and modelling. This book provides an overview of modelling tools that can be used to construct appropriate power consumption and energy efficiency estimates.

Chapter 3 explains the standard approach to modelling the power consumption of a large-scale network. This is to separate out the segments of the network and construct models for each segment. After explaining segmentation this chapter focuses on wireline Access, Edge and Core networks. "Wireline" networks use technologies such as copper pair, coaxial cable and optical fibre to interconnect equipment. In the Edge and Core networks, wireline technologies are the most widely adopted technologies.

Chapters 4–6 focus on wireless technologies. Wireless Access networks are expanding and becoming the predominant access technology across the planet. Unfortunately, modelling wireless access is a particularly challenging task. To address this, Chapter 4 provides a brief introduction to 4th generation mobile LTE systems, to provide the reader with an overview of a common wireless access technology. The model described in Chapter 4 adopts a somewhat "blunt instrument" approach in that it does not describe any of the advanced Queueing Theory and numerical modelling approaches. Rather, it builds a model from the ground up by providing a (somewhat simplified) description of the physical aspects of LTE and their power consumption. The mobile network power model presented in Chapter 4 is based on one of the most simplified approaches published in the literature, the Full Buffer Model. However, this model is not capable of including some aspects of mobile network operation that significantly impacts the network performance and power consumption. In brief, the Static Model presented in Chapter 4 sacrifices accuracy for simplicity. However, it still can provide guidance for those who wish to understand the key factors that influence mobile network power consumption and energy efficiency.

Chapter 5 also presents a model for mobile access network performance and power consumption. This Advanced Model is somewhat more sophisticated (and consequentially difficult to construct) than the Static Model of Chapter 4. The Advanced Model (although still not a full "real world" representation of a mobile network) can provide information and understanding the Static Model cannot, such as dynamics of user network demand and network congestion behaviour of these networks.

In reality, both the Static and Advanced Models presented in this book are well away from a real-world representation of mobile network performance and power consumption. However, that is not their task. Rather they are designed to provide the modeller with understanding of the key aspect of mobile network power consumption so that forward strategies and service planning can be undertaken with some confidence.

If the reader seeks a real-world model of a mobile network, then they need to investigate the commercially mobile network planning platforms. Unfortunately, for many researchers these platforms are likely to be too expensive or require access to the appropriate equipment or network data to input into these highly sophisticated modelling tools. Further, they typically involve a serious learning commitment.

Chapter 6 uses the Advanced Model developed in Chapter 5 to construct network models for Heterogeneous mobile networks. These networks include a combination of large macro base stations, plus small low-power base stations to service users, particularly in areas of high user density. Heterogeneous networks are a part of the 4G/LTE and 5G mobile technologies being developed and deployed around the globe to meet increasing traffic demand.

Chapter 7 moves away from the physical layer to provide the reader with techniques for modelling trends for future network traffic diurnal cycles. To construct a model of possible future network power consumption and energy efficiency, we need the physical layer models (provided by Chapters 3–6) as well as potential future traffic levels to input into the physical layer models. Chapter 7 describes mathematical methods for constructing example future diurnal cycle profiles that accommodate the expected traffic growth trends for average and peak hour traffic. The chapter also exposes the issues of dimensioning a network to cover peak demand and consequences for the energy efficiency of running such a network over the low-demand periods during the day.

Chapter 8 brings together the physical equipment and traffic models to provide tools for forecasting power consumption over future years. These forecasts are based on future demand, technology evolution, and network upgrading strategies. With the increasing importance of data centres and content services, Chapter 9 provides an overview of power models for data centres. A key problem with constructing data centre models is again the lack of detailed data their power consumption and traffic levels. The models provided in Chapter 9 work from an assumption of minimal detailed information on the internals of data centres.

Chapter 10 shifts attention to modelling the power consumption of services. In reality, the experiences of most users of the Internet and telecommunications networks are the services they provide, not the details of the technologies underpinning those services. The purpose of Chapter 10 is to provide the modeller with tools that can address the power consumption of the Internet and telecommunications networks from that service perspective. In particular, the chapter addresses the issue of how the power consumption of shared network infrastructure can be apportioned across services, depending upon their level of usage of the shared infrastructure.

Chapter 11 brings together all of the previous chapters to provide the modeller with a range of energy efficiency metrics that can be used to quantify the energy efficiency of equipment, networks and services. This chapter includes a survey of existing metrics and provides a detailed description of several "energy per bit" and "power per user" metrics commonly used in the literature and industry.

Because this book describes methods for constructing mathematical models for telecommunications equipment, network and service power consumption and energy efficiency, it is highly mathematical. In some chapters, there are, unavoidably, several tens of symbols. To assist the reader, a Notation Index is provided at the end of the book. In this, we have tried to provide an exhaustive description of the symbols used in the equations throughout the book.

Although this book is targeted at technologists, some chapters also include an "Executive Summary" in which we have endeavoured to present a higher level overview of the contents of each chapter and why the results will be of interest to higher level management. A "Technical Summary" is also provided to provide the modeller an overview of the technical steps involved in modelling and how it could be applied.

The material presented in this book is based on the research undertaken by the authors when working in the Department of Electrical and Electronic Engineering at the University of Melbourne as members of the Centre for Energy Efficient Telecommunications. The Centre contributed to the international GreenTouch consortium which operated from 2010 to 2015 with the mission of constructing a "technology roadmap" to enable the energy efficiency of the internet to improve by a factor of 1000 between 2010 and 2020. The publications and details of this roadmap are identified in the text of this book.

The authors would like to acknowledge the significant assistance of Dr. Leith Campbell who provided both technical and editorial advice and also of Dr. Paul Fitzpatrick who provided expertise on the modelling of mobile networks.

Kerry James Hinton
Robert Ayre
Jeffrey Cheong

Authors

Kerry James Hinton, PhD, earned an Honours BE in 1978, an Honours BS in 1980 and an MS in mathematical sciences in 1982, all from the University of Adelaide. He earned a PhD in theoretical physics at the University of Newcastle upon Tyne, UK, and a diploma in industrial relations at the Newcastle upon Tyne Polytechnic, UK, in 1984.

In 1984, Dr. Hinton joined Telstra Research Laboratories (TRL), Victoria, Australia, where he worked on analytical and numerical modelling of optical systems and components. His work focused on optical communications devices and architectures, physical layer issues for automatically switched optical networks (ASONs) and monitoring in all-optical networks.

From 1998 until 2010, Dr. Hinton was a sessional lecturer for the Master of Telecommunications Engineering course at the University of Melbourne.

In 2006, Dr. Hinton commenced as a Senior Research Fellow in the Centre for Ultra Broadband Networks (CUBIN) at the University of Melbourne, Australia. In 2011, Dr. Hinton joined the Centre for Energy Efficient Telecommunications (CEET) as a Principal Research Fellow, researching the energy efficiency of the internet, communications technologies, and networks. In 2013 Dr. Hinton took on the role of Director of the CEET and continued in that role until July 2016 when CEET closed. Dr. Hinton has authored over 130 papers in peer reviewed journals and conferences.

Robert Ayre earned a BSc (with Distinction) at George Washington University in 1967 and a BE (with Honours) and an MESc at Monash University in 1970 and 1972, respectively.

In 1972, Mr. Ayre joined what was to become Telstra Research Laboratories (TRL). At TRL, he worked principally on optical transmission systems and networks, beginning with the earliest metropolitan multimode fibre systems through to long-haul optically amplified WDM systems. He was involved with Telstra's engineering teams on optical network planning and in the first deployments of each of the new generations of optical transmission technologies. He also represented Telstra in the International Telecommunications Union Study Group XV on standards for optical transmission systems and the International Electrotechnical Commission in developing standards for laser safety in optical communications systems. He became responsible for a number of TRL's infrastructure development teams, covering access networks, core transmission networks, data networks, and internet services.

Following the closure of TRL, in 2007, he joined the Centre for Ultra Broadband Networks (CUBIN), which later became the Centre for Energy Efficient Telecommunications (CEET) at the University of Melbourne as a Senior Research Fellow.

Jeffrey Cheong, PhD, earned a BE (Honours) and a PhD at Monash University in 1976 and 1983, respectively. He then joined the Telstra Research Laboratories (TRL) in 1983.

During his career with TRL (1983–2006), Dr. Cheong led TRL in the development of control and signalling systems to facilitate the introduction of advanced intelligent network (IN) PSTN services, service assurance systems for the rollout of VoIP, and performance and monitoring tools for Telstra online services. He was also involved in the standardisation of ITU-T Intelligent Network Capability Set 1 standards and the application of these standards for services rollout.

Dr. Cheong previously held appointments with the university as Senior Research Fellow in the Centre for Energy Efficient Telecommunications (CEET) and the Institute for a Broadband-Enabled Society (IBES) (2013–2014). In IBES, he developed models for costings of fibre and civil infrastructures to support a National Broadband Network based on FTTP and for also fixed wireless technology. His research in CEET involved the modelling of energy efficiency of 3GPP LTE networks, the development of the GreenTouch Consortium's Mobile Communications WG

Architecture, and development of models for quantifying energy savings of distributed vs. centralised mobile network architectures.

His research includes development of analytical models for quantifying the performance of broadband access networks and the dimensioning of these networks.

As members of CUBIN and CEET, Kerry James Hinton, Robert Ayre and Jeffrey Cheong participated in pioneering research into the energy efficiency of telecommunication systems, networks, and services. The activities of CEET covered a number of research areas, including the topology of core, metropolitan/edge and customer access networks with special attention to their energy consumption, the performance and power consumption of different network access technologies, the energy cost of cloud-based services, such as document storage and processing provided by data centres, and the potential impact of the evolving Internet of Things on energy consumption at both domestic and national levels, among others.

CEET was a foundation member of GreenTouch consortium, an international collaboration between industry, research organisations, and universities. As participants in GreenTouch, the authors were involved in the modelling of wireline and wireless network power consumption and energy efficiency.

GreenTouch was formed to address the need to substantially improve the energy efficiency of telecommunication networks as they continue to expand to meet the demand of new telecommunications services. Its mission was to provide a technology roadmap for 2010 to 2020 to provide a 1000 times improvement in the energy efficiency of the internet. GreenTouch delivered the roadmap in 2015.

Since the cessation of CEET in 2016, the authors have continued their research work as Honorary Research Fellows/Visitors at the University of Melbourne.

1 Introduction

Over recent years, it has become widely accepted that humanity has entered the "Anthropocene epoch" during which humankind has had a significant impact on the Earth, including its ecosystems [1]. One manifestation of this has been global climate change. This recognition has resulted in many changes in the way business and other economic activities are undertaken within society.

The growth in global use of Information and Communications Technologies (ICT), particularly the Internet, has resulted in ICT becoming an increasingly significant part of any organisation's operations. Today, most organisations use substantial "in-house" or outsourced ICT infrastructure. Either way, this infrastructure is becoming an important part of daily operations that need to be included in that organisation's carbon footprint.

For most organisations, energy consumption is an operational cost. As such, it needs to be controlled, if not minimised. Over recent years, it has been realised that energy consumption has another significant "cost": the environmental footprint, including production of greenhouse gases, arising from the generation of the electrical power an organisation consumes during its operation. Another cost is the "opportunity cost" associated with the use of fossil fuels to generate the electrical power consumed by ICT. Although originally considered plentiful, ongoing profligate burning of these materials to generate power removes the opportunities for other more long-term application of these materials in advanced technologies. For example, the use of plastics is an essential part of the electronics, automotive and aeronautical industries and much else.

As organisations increase their use of and reliance on ICT, the energy consumption costs of this aspect of their operations have become sufficiently large to require management and control. For some organisations, ICT energy costs dominate their power bill and there are reasons for expecting this will become increasingly common.

With the recognition of the importance of controlling our local and global carbon footprint, a variety of public policies have been developed to encourage corporations, governments and societies to reduce their footprint. In many cases, the "encouragement" process includes increasing the financial cost for operations that generate significant amounts of greenhouse gases. Consequentially, the adoption of stronger carbon control policies has encouraged organisations, for which energy consumption (such as in their ICT equipment) is a significant component of their operations, to control and/or reduce their energy consumption.

Today and in the coming years, governments and businesses are increasingly focusing on the energy usage and carbon footprint of their activities. This is reflected in the adoption of practices such as "triple bottom line" accounting, in which an organisation's environmental impact is included in its accounting system [2]. Another example is the development of the Greenhouse Gas Corporate Accounting and Reporting Standard [3]. In some jurisdictions, this accounting may be mandatory [4]. Consequently, assessment of current and estimation of future energy consumption or carbon footprint will become an increasingly important aspect of commercial practice.

This book was written to provide guidance and assistance to anyone with an interest or requirement to assess or estimate the energy consumption (and consequential carbon footprint) of telecommunications networks and services. This may include specialists such as business planners and policy developers with a need to provide quantitative estimates of potential power consumption of planned telecommunications networks and services.

The term "ICT" includes end-user hardware (home equipment, laptops, mobile phones, etc.), data centres (be they small and local or large and remote) and telecommunications networks connecting these centres to users. A significant contributor to overall ICT carbon footprint is the use-phase power consumption of this equipment [5, 6]. In this book, we are mainly focused on developing

DOI: 10.1201/9780429287817-1

models to estimate this use-phase component. Readers should understand that network power consumption is not the same as ICT power consumption; rather, it is a contributing component.

This book will provide guidance for constructing both total network (global) models and sub-network (local) models for telecommunications infrastructure and services. The scale of the model has an influence on how the model is constructed.

1.1 REASONS FOR CONSTRUCTING GLOBAL ENERGY CONSUMPTION MODELS

Since the mid-1990s, the Internet and related ICT services have grown exponentially. Over recent years, although the rate of growth has declined, the growth in traffic is still substantial and is expected to continue for the immediate future [7, 8]. As an illustration, energy consumption of ICT network operators was estimated to be 242 TWh in 2015 [9] and data centres 205 TWh in 2018 [10]. In comparison global electrical energy consumption in 2020 was 25,866 TWh [11]. These historical data suggest energy consumption of ICT network operations and data centres already represent a few percent of the global total and they will grow with increased data usage. This plus the forecasts of dramatic growth of Internet traffic has stimulated an interest in the resulting carbon footprint of the Internet and ICT.

Over recent years, many organisations and researchers have published estimates and articles on the potential growth of the energy consumption or carbon footprint of the Internet [12–16]. Although various models and forecasts are expected to provide different values for energy consumption [12, 17, 18], some recent publications have started to significantly diverge in their forecasts for trends of future energy consumption by ICT. Some forecast ongoing significant increases in annual energy consumption [19, 20], while others predict reductions [9].

Irrespective of the growth trends, global energy consumption models provide an important forewarning of future growth trends that may be potentially problematic. A specific example is given by the recently published concerns regarding the sustainability of cryptocurrencies such as Bitcoin [21, 22].

In the ICT industry, forecasts initially indicated that this sector of the economy is consuming an increasing proportion of the global electricity production [18]. Over recent years, some published models have predicted the annual growth rate in energy consumption by the ICT sector of 10% or more [18, 19], while the annual growth rate of electricity production continues to be around 3% [23]. If these trends were allowed to continue, the Internet would ultimately require an unsustainable proportion of global electricity production.

With annual global ICT power demand growing at 10% and global electricity production at 3%, even powering the Internet with renewable energy will not necessarily "decarbonise" the e-economy. If the global electrical power demand of the Internet grows much faster than global supply, it will likely shift the carbon footprint to users who will have to source their electrical power from fossil fuels as they compete against other operators in both ICT and non-ICT industries for a limited supply of renewables. Either that or the provision of Internet services will become highly congested due to a lack electrical power for the required infrastructure. This is where quantifying and improving energy efficiency will play a crucial role.

Similarly for ICT operators who have in-house renewable generation, quantifying and improving energy efficiency will reduce the cost of their renewable energy infrastructure as capacity requirements are optimised.

These forecasted trends in ICT energy consumption has stimulated a concerted effort to improve the energy efficiency in the ICT sector by a number of public and private organisations, such as GreenTouch [24], Green Grid [25], and GeSI [26]. Organisations such as GreenTouch and GeSI have developed and modelled energy-saving strategies that provide guidance on mitigating the growing global energy footprint of ICT [13, 24, 27].

As with other impacts of humankind on the global environment, forecasting the environmental footprint of a human activity that is growing as rapidly as the Internet and ICT is imperative to ensure humankind does not continue to over-burden the planet. This is where global energy consumption models play a crucial role.

1.2   REASONS FOR CONSTRUCTING LOCAL ENERGY CONSUMPTION MODELS

Global ICT energy consumption is the sum of the energy consumption in many thousands of local ICT facilities. Beyond this, there are also reasons for constructing local energy consumption models, in both a geographical and organisational context.

In some cases, the location of ICT infrastructure can significantly impact its environmental footprint. For example, a data centre located in a cold environment is likely to require less powered cooling by using the locally cool climate for passive cooling. The carbon footprint of an ICT facility using renewable electricity generation (e.g. hydropower, solar or wind) will be less than one that relies on coal-generated electricity. In cases such as these, the power consumption of a proposed ICT facility will need to be estimated to determine its potential environmental impact as well as to determine whether the local power supply has the capacity to supply the facility.

In other cases, there may be a limited choice for the source of electrical power. For example, network infrastructure needs to be close to the users to minimise service latency and data transport costs, as well as to connect to metro and distribution networks. In these circumstances, the network owner may have limited freedom regarding the choice of electrical power generation technology.

Beyond this, the adoption of carbon accounting and codes of conduct also require an ability to ascertain the energy consumption of ICT equipment on the local scale. For example, in an endeavour to improve energy efficiency, the European Union has developed a "Code of Conduct for ICT". This includes codes of conduct for data centres [28] and broadband equipment [29]. Independently of this, the Greenhouse Gas Protocol "provides standards, guidance, tools and training for business and government to measure and manage climate-warming emissions" [3, 30].

Some nations are now requiring carbon reporting [31]. This reporting requirement applies to organisations, most of whom will utilise ICT equipment and services. In many cases, direct item-by-item measurement of energy consumption of ICT equipment will be impractical. Therefore, to some extent, modelling will be required for both global and local assessments of network and service power consumption.

1.3   USING THIS BOOK

This book is primarily drafted for specialists, researchers and technologists. For these readers, this book will provide a set of principles and mathematical techniques for constructing estimates of power consumption. These principles and techniques will also be of interest to students in the environmental sciences who want a succinct, step-by-step introduction to developing or understanding how such estimates are made.

For policy managers, decision-makers and those with a more general interest in this field, each chapter of the book provides a high-level "Executive Summary" which gives an overview of how models for the energy consumption of telecommunications networks and services are constructed. For these readers, the Executive Summary provides a level of understanding that will give them enough insight to understand the overall ideas without delving into the technical aspects of these models.

An issue that needs to be appreciated when investigating or using the models presented in this book is the lack of publicly available detailed data on telecommunications equipment and networks. In regards to equipment, detailed power consumption data showing the dependence of power consumption on the equipment's traffic load is effectively unavailable in the public arena. In regard to networks, there is a significant paucity of detailed end-user and traffic data available in the public arena. This means we must develop models that are based on the small amounts of data that are available. Consequentially, some of the models contain a significant amount of approximation and extrapolation of available data. Although this means the numerical estimates from the models are only approximate, this situation is better than having no estimates at all. Any reduction in uncertainty of knowledge is of value [32].

1.4   NOTATION

This book contains a significant amount of mathematical derivations and expressions. Unfortunately, due to the wide range of models presented, the notation is sometimes complex and has multiple layers of arguments, subscripts and superscripts. To assist the reader, a Notation Index has been provided at the end of the book. This Index provides a chapter-by-chapter list of notation and corresponding description of the mathematical symbols.

Many of the mathematical models presented in this book utilise averaging over a variety of populations and in some cases there are multiple stages of averaging. An example is constructing a power model for a telecommunications network. In this case, the network consists of multiple network elements each transporting a number of services. Some of the quantities in the model may arise from averaging over the network elements, others from averaging over the services and yet others from averaging over both the network elements and services.

In many cases, it is important to distinguish between these different averages. Therefore, the following notation will be used to enable the reader to keep track of the population over which the averaging has been taken.

Let $X_j^{(k)}$ represent the value of X for the jth network element and the kth service. Let there be N_E network elements and $N^{(svc)}$ services. Therefore, $j = 1, 2, ..., N_E$ and $k = 1, 2, ..., N^{(svc)}$.

In almost all cases an index that covers services or users will be a superscript within parentheses, that is, $a^{(k)}$. For a network element or time step, the index will be a subscript without parentheses, that is, a_j.

If the $X_j^{(k)}$ are averaged over services, k, that average will be denoted as follows:

$$\left\langle X_j \right\rangle^{(svc)} = \frac{1}{N^{(usr)}} \sum_{k=1}^{N^{(usr)}} X_j^{(k)} \tag{1.1}$$

If the $X_j^{(k)}$ are averaged over the network elements, j, that average will be denoted by

$$\left\langle X^{(k)} \right\rangle_E = \frac{1}{N_E} \sum_{j=1}^{N_E} X_j^{(k)} \tag{1.2}$$

In this notation, the angular brackets $\langle . \rangle$ indicate an average has been taken. The quantity outside the angular brackets indicates which population has been averaged. Therefore, in (1.1) the superscript "(svc)" in $\langle \ \rangle^{(svc)}$ indicates the average was over the k-index for services or users. In (1.2) the subscript "E" in $\langle . \rangle_E$ indicates the average is over the j-index for network elements.

Also, note that the quantity that has not been averaged over remains inside the angular brackets. Thus, the j in $\langle X_j \rangle^{(svc)}$ indicates the j-index has not been averaged over and likewise in $\langle X^{(k)} \rangle_E$ the k-index has not been averaged over.

In some cases, multiple averages will be taken. For the case of the $X_j^{(k)}$, an average over both indices will be denoted as

$$\langle X \rangle_E^{(usr)} = \frac{1}{N_E N^{(usr)}} \sum_{j=1}^{N_E} \sum_{k=1}^{N^{(usr)}} X_j^{(k)} \tag{1.3}$$

There can also be the situation where the multiple averaging will be over indices that are subscripts. For example, in the mobile network power model we have a quantity of the form $X_{m,p,j}$ where $m = 0, 1, 2, \ldots, N_{BS}$, $p = 1, 2, N_{Sec}$ and $j = 1, 2, \ldots, N_{Tot}$. In this case, averaging over different indices follows the style in (1.2) in that

$$\langle X_{m,p} \rangle_{Tot} = \frac{1}{N_{Tot}} \sum_{j=1}^{N_{Tot}} X_{m,p,j}$$

$$\langle X_{p,j} \rangle_{BS} = \frac{1}{N_{BS}} \sum_{m=0}^{N_{BS}} X_{m,p,j} \tag{1.4}$$

$$\langle X_j \rangle_{BS,sec} = \frac{1}{N_{sec} N_{BS}} \sum_{p=1}^{N_{sec}} \sum_{m=0}^{N_{BS}} X_{m,p,j} = \frac{1}{N_{sec}} \sum_{p=1}^{N_{sec}} \langle X_{p,j} \rangle_{BS}$$

Because the amount of mathematical notation in some chapters is rather extensive, the Notation Index includes a table that defined the mathematical terms that appear in each chapter.

REFERENCES

[1] J. Zalasiewicz *et al.*, "The Anthropocene: A New Epoch of Geological Time?," *Philosophical Transactions of the Royal Society A*, vol. 369, p. 835, 2011.

[2] Atu and O.-E. Kingsley, "Triple Bottom Line Accounting: A Conceptual Expose," *IOSR Journal of Business and Management*, vol. 13, no. 4, p. 30, 2013.

[3] Carbon Trust, *The GHG Protocol Corporate Accounting and Reporting Standard*, Carbon Trust, 2017.

[4] N. Singh and L. Logendyke, "A Global Look at Mandatory Greenhouse Gas Reporting Programs," World Resource Institute, 27 May 2015. [Online]. Available: https://www.wri.org/blog/2015/05/global-look-mandatory-greenhouse-gas-reporting-programs. [Accessed 11 July 2020].

[5] W. Scharnhorst "Life Cycle Assessment in the Telecommunication Industry: A Review," *International Journal of Life Cycle Assessment*, vol. 13, no. 1, p. 75, 2008

[6] B. Raghavan and J. Ma, "*The Energy and Emergy of the Internet*," in *Proceedings of the 10th ACM Workshop on Hot Topics in Networks, Hotnets'11*, Cambridge, 2011.

[7] Cisco, "The Zettabyte Era: Trends and Analysis," Cisco White Paper, 2017.

[8] M. Meeker, *Internet Trends 2018*, Kleiner Perkins, 2018.

[9] J. Malmodin and D. Lunden, "The Energy and Carbon Footprint of the Global ICT and E&M Sectors 2010–2015," *Sustainability*, vol. 10, p. 3027, 2018.

[10] E. Masanet *et al.*, "Recalibrating Global Data Center Energy-Use Estimates," *Science*, vol. 367, no. 6481, p. 984.

[11] H. Ritchie, "Electricity Mix," *Our World In Data*. [Online]. Available: https://ourworldindata.org/electricity-mix [Accessed 9 June 2021].

[12] J. Baliga *et al.*, "Energy Consumption of Optical IP Networks," *Journal of Lightwave Technology*, vol. 27, no. 13, p. 2391, 2009.

[13] The Climate Group, "SMART 2020 Enabling the low-carbon economy in the information age," GeSI (Global eSustainability Initiative), 2008.

[14] D. Schien and C. Preist, "Approaches to the Energy Intensity of the Internet," *IEEE Communications Magazine*, vol. 52, no. 11, p. 130, 2014.

[15] R. Bolla *et al.*, "Energy Efficiency in the Future Internet: A Survey of Existing Approaches and Trends in Energy-Aware Fixed Network Infrastructures," *IEEE Communications Surveys and Tutorials*, vol. 13, no. 2, p. 223, 2011.

[16] V. Coroama *et al.*, "Assessing Internet Energy Intensity: A Review of Methods and Results," *Environmental Impact Assessment Review*, vol. 45, p. 63, 2014.

[17] D. Kilper *et al.*, "Power Trends in Communication Networks," *IEEE Journal of Selected Topics In Quantum Electronics*, vol. 17, no. 2, p. 275, 2011.

[18] S. Lambert *et al.*, "Worldwide Electricity Consumption of Communications Networks," *Optics Express*, vol. 20, no. 26, p. 513, 2012.

[19] A. Andrae and T. Edler, "On Global Electricity Usage of Communication Technology: Trends to 2030," *Challenges*, vol. 6, p. 117, 2015.

[20] P. Ryan, *et al.*, "Total Energy Model V2.0 for Connected Devices," International Energy Agency, 2021.

[21] C. Mora, *et al.*, "Bitcoin Emissions Alone Could Push Global Warming above 2° C," *Nature Climate Change*, vol. 8, p. 931, 2018.

[22] M. Krause, T. Talaymat, "Quantification of Energy and Carbon Costs for Mining Cryptocurrencies," *Nature Sustainability*, vol. 1, p. 711, 2018.

[23] International Energy Agency, "Electricity Statistics," International Energy Agency, 2020. [Online]. Available: www.iea.org/statistics/electricity/. [Accessed 11 July 2020].

[24] GreenTouch, *Reducing the Net Energy Consumption in Communications Networks by up to 98% by 2020*, GreenTouch 2015.

[25] The Green Grid, "The Green Grid," The Green Grid, 2020. [Online]. Available: https://www.thegreengrid.org/. [Accessed 11 July 2020].

[26] GeSI, "Global e-Sustainability Initiative," GeSI 2019. [Online]. Available: https://gesi.org/. [Accessed 11 July 2020]

[27] J. Elmirghani *et al.* "*Energy Efficiency Measures for Future Core Networks*," in *Optical Fiber Conference*, Los Angeles 2017.

[28] European Commission, "European Energy Efficiency Platform (E3P)," European Commission, 8 July 2020. [Online]. Available: https://e3p.jrc.ec.europa.eu/communities/data-centres-code-conduct. [Accessed 11 July 2020].

[29] European Commission "Code of Conducts for Broadband Communication Equipment," European Commission, 14 November 2016. [Online]. Available: https://ec.europa.eu/jrc/en/energy-efficiency/code-conduct/broadband. [Accessed 11 July 2020].

[30] Greenhouse Gas Protocol, "About the Greenhouse Gas Protocol," World Resource Institute [Online]. Available: https://ghgprotocol.org/. [Accessed 11 July 2020].

[31] Carbon Footprint, "Mandatory Greenhouse Gas (GHG) Reporting," Carbon Footprint Ltd. [Online]. Available: https://www.carbonfootprint.com/mandatorycarbonreporting.html. [Accessed 11 July 2020].

[32] D. Hubbard *How to Measure Anything* (2nd Edition) John Wiley & Sons 2020.

2 Why Model Network and Service Power Consumption and Energy Efficiency

A widely accepted management maxim is as follows: "You cannot control what you don't measure". This statement reflects the fact that if we wish to manage or improve some aspect of an activity or operation, we need a process by which we can ascertain (i.e. measure) the success of the strategies we implement to secure that change [1]. This applies to almost any strategy for change, including limiting or reducing the environmental footprint of an organisation, nation and humankind. In this book, we will follow this maxim and describe how to construct power consumption models that provide a "measure" of the energy footprint for telecommunications networks and services.

A process for securing ongoing control or directed change of a system is the "Shewhart Cycle" or PDCA (Plan, Do, Check, Act) Cycle [2], shown in Figure 2.1. Using this Cycle, those who are responsible for the process carry out the following steps:

Plan: Study the process and determine changes that might improve it. Determine how to measure "improvement".
Do: Carry out the changes, if possible on a small scale.
Check: Observe the outcomes of the Do phase. Observe the metrics adopted in the Plan phase.
Act: Utilise the knowledge acquired from the steps above.
Plan: Review plans as required to secure the desired improvements.
Repeat the PDCA cycle.

To undertake this process, we need to decide what will be measured and used as the metric in the "Plan" and "Check" steps. This choice is crucial to the process of control and management because, if the choice is inappropriate, we may find the strategies developed do not provide the desired outcomes. Our focus will be on quantities that are almost universally accepted as appropriate to reducing the environmental impact of the user phase of telecommunications networks and services by providing well-defined numerical values: Those quantities are based upon power and energy consumption of equipment that constitutes the network and provides the services.

The relationship between power and energy is well known. Energy is the power being used multiplied by the duration of its use. More precisely, energy is the integral of the instantaneous power used over the time duration of its use. Both energy and power are very important metrics. Energy use is directly relatable to the carbon footprint of the network or services. Power (more specifically, maximum power) is related to the maximum traffic load that a network can carry. Networks are designed to accommodate "peak load" when the traffic through the network is at its maximum. Typically, it is at peak load that the network power consumption is a maximum and it is essential that the power supply for the network has been designed to accommodate this peak. In the following sections, we will elaborate the role and importance of both energy and power metrics and how they can be defined for network equipment, total networks and for services provided via a telecommunications network.

DOI: 10.1201/9780429287817-2

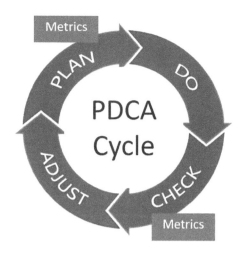

FIGURE 2.1 Shewhart Cycle: Plan, Do, Check, Act provides a process for ongoing control or change of a system. An important aspect of the process is the selection of metrics used in the process.

2.1 WHY POWER AND ENERGY CONSUMPTION MATTERS

Before discussing the reasons why energy and power consumption are issues of interest, it is instructive to briefly consider the difference between "power" and "energy" consumption and why both need to be considered. From any introductory physics book, we see that energy expenditure is the cumulative outcome of power consumption over time. Mathematically, we express this as energy expended, $E(T)$, over duration, T, as the time integral over power consumption, $P(t)$, over a duration, T:

$$E(T) = \int_0^T P(t)\,dt \tag{2.1}$$

Power is defined and can be measured at an instant in time, t. Energy is defined and can only be measured over a duration of time, T.

In many instances in this book, the terms "power consumption" and "energy consumption" will be used interchangeably without changing the intent of the text. However, it is important to appreciate the difference between the two.

Power consumption is important because power consumption and provisioning are limited by factors such as heat dissipation or the physical degradation of systems and sub-systems. Electrical generation systems are always limited in the amount of power they can generate at any specific time. Therefore, when designing and deploying telecommunications networks, due consideration of both the demand for power by the network equipment and the ability to provide this power is essential.

Energy consumption is different in that it can be arbitrarily large provided it is expended over a sufficiently long duration in time. However, for our purposes, energy consumption is important because, with fossil-fuel-based generation, it is directly related to the carbon footprint of telecommunications networks and services. Therefore, any carbon accounting system for these networks and services will require an estimation of energy consumption over some specified time period (e.g. annually).

The development and use of telecommunication networks and services has rapidly increased over recent decades as the myriad of Internet services have spread around the planet. Financial services, e-commerce, entertainment services (TV, music distribution), social media, social networks and e-health all involve the collection and transport of information, its analysis and storage and its distribution. All of these information-based activities involve the consumption of energy. Consequentially, society's demand for energy will continue to increase on the personal, organisational, national and global scales.

The dramatic growth of mobile services over recent years provides additional need for investigating network and service power consumption. In this context, "mobile services" refer to the expectation that services will be available anywhere and anytime on the planet. The provision of mobile connectivity to the Internet (and the services it provides) can be significantly more energy intensive than wired connectivity.

The spreading demand for connectivity to the Internet now includes areas where traditional "mains power", provided by an electricity grid, is not available. Mobile phones are the most ubiquitous example of this. Many mobile service base stations are not connected to mains power; rather, they rely on locally provided power via renewable energy generation (e.g. solar, wind) or fossil fuels (e.g. diesel-powered generators) [3].

Mobility of devices almost always requires mobility of its energy supply. For network elements (such as wireless base stations and remote regenerators), this typically includes batteries and/or portable (including renewable) power generation. Because these portable, energy-providing devices are almost always constrained in weight and size, the power they can provide is also constrained. Therefore, to maximise their use, we need to minimise the power consumption of the network element or mobile device they are energising. The need to minimise power consumption applies to all equipment that is powered by batteries or local non-mains-connected power supplies. In some telecommunications networks, this may include a significant number of devices and facilities.

From this, we see that power consumption matters on the global scale (carbon footprint and global electricity production) and on the local scale (equipment not attached to the power grid and mobile devices).

2.2   MEASUREMENT VERSUS MODELLING

Referring back to the management maxim, "You cannot control what you don't measure", the process of "measurement" requires some consideration. In some cases, it may not be at all obvious what needs to be measured in order to assess the success of an ongoing change strategy. In many cases, the selection of the quantities to be measured is, in itself, a sophisticated problem. For example, what if a numerical value is not easily applied to the quantities of interest? What if the most significant quantities are not even clearly measurable? A discussion of these issues will divert us from our main purpose in this book and the reader is referred to a book by Hubbard [1].

The actual determination of numerical values that provide a good quantitative representation of the power and/or energy consumption of a telecommunications network or service can be very difficult. A major reason for this is their geographical spread. In contrast to data centres and end-user equipment, telecommunications network power consumption is the cumulative consumption of many, widely spread devices and facilities. There has been a significant amount of work on the power consumption of data centres [4]. Likewise, manufacturers of end-user hardware (laptops, desktop computers, tablets, mobile phones, etc.) have demonstrated tremendous progress in energy use reduction [5, 6]. The power consumption of telecommunications networks – both fixed and mobile – however, is a complex problem and poorly understood. The challenge is that telecommunications networks are very complicated, geographically distributed collections of a very diverse range of equipment. Further, in many cases, the information required to construct an estimate of power consumption is extremely difficult to collect. Unlike measuring the power consumption of a fridge or TV, determining a numerical value for the power consumption of a telecommunications network or service is not just a simple task of getting out a power meter, plugging it into the network and looking at the display.

Most telecommunications networks are too large and complicated for measurement alone to ascertain their power consumption. When we consider the power consumption of a service, the challenge is even greater because services typically share infrastructure with many other services. The traffic through a router may consist of simultaneous data flows for many independent web searches, video streaming, social networking and much more. Therefore, a simple measurement of the power consumption of that router will not provide a direct value for the power consumption of any specific service that router is carrying.

This is where modelling has a crucially important role to play. Because direct measurement is not possible or feasible, estimates of the power consumption of telecommunications networks and services rely on a combination of measurements that are practical and (mathematical) models that are reliable. To date, two approaches to network power modelling are accepted as providing a realistic estimate of network power consumption. These approaches are referred to as "bottom-up" and "top-down" [7].

"Bottom-up" combines equipment power measurements with network design principles based on an understanding of the network architecture, equipment types, their traffic loads and how they are interconnected. Bottom-up models can be used to construct power consumption trend models based on traffic forecasts and equipment generations.

The alternate approach is "top-down" modelling based upon network equipment inventories and/or sales figures used in conjunction with typical power consumption values for the equipment. This approach may also include annual deployment trends to provide power consumption trends over several years.

Although neither of these approaches will provide a precise value for the power consumption of a telecommunications network or service, they do provide a significant reduction in the uncertainty of our knowledge of these values. This, in itself, is important because it provides improved understanding of power consumption and guidance as to how to reduce it [1].

Both these approaches will be discussed below. However, the models described in this work are primarily designed using the "bottom-up" approach.

2.3   WHAT IS "ENERGY EFFICIENCY"?

From the discussion above, the motivation for ascertaining the power consumption of a telecommunications network or service is relatively easy to understand. As a network grows in size (both geographically and in total traffic), its power consumption will also inevitably increase. Therefore, as we endeavour to develop strategies to control or minimise the power consumption of a network, a simple direct comparison of network power may not provide the best metric. For example, measuring the power consumption of different networks with different traffic throughputs or geographical spread will inevitably give different results. Unless we also account for network dimensions such as total network traffic and geographical extent, it will be difficult to ascertain which network best utilises its resources. This is where the concept of "energy efficiency" has a role to play.

Compared to network power consumption, network energy efficiency is a more subtle idea. The term "efficiency" is typically applied when considering minimising the wastage of limited resources when producing a good or service. Adopting this view to a telecommunications network, the "good or service" provided by a telecommunications network to its customers is information, which is typically measured in "bits". The limited resource in this case is energy consumed by the network to provide those bits. Therefore, it is intuitive to construct an energy efficiency metric for a telecommunications network based on "energy per bit". In this vein, the International Telecommunications Union (ITU) defines energy efficiency as "the relationship between the specific functional unit for a piece of equipment (i.e. the useful work of telecommunications) and the energy consumption of that equipment. For example, when transmission time and frequency bandwidth are fixed, a telecommunication system that can transport more data (in bits) with less energy (in Joules) is considered to be more energy efficient." [8]. This leads one to the following commonly used definition for an energy efficiency metric with dimensions of energy/bit [9]:

$$^2H_1(T) = \frac{\text{Total energy consumed over duration } T}{\text{Total bits transmitted over duration } T} \tag{2.2}$$

where the "bits transmitted" and "energy consumed" apply to the network element, total network, or service under consideration. (The notation $^2H_1(T)$ is for conformity with the notation adopted in, Chapter 8.)

This definition of "energy per bit" has been widely adopted and accepted as a useful measure. However, there are several subtleties that need to be considered. In particular, determining the number of bits to quantify the energy per bit requires some consideration.

In definition $^2H_1(T)$, the number of bits corresponds to the total over time duration T. However, many of those bits could be overhead (protocol, monitoring, encryption, retransmits, etc.) in addition to the customer data bits. For example, consider the case of an end user using a laptop at home to communicate with a service provider. That user will be connected to their service provider via the Access, Edge and Core networks. The (IP) packets generated by the user will most likely be encapsulated for transport by one or more protocols (e.g. Carrier Ethernet, DSL, SONET/SDH/OTN), each of which will add overhead bits to the user's IP packets. Although these overheads will be stripped off at the receive end of the connection, they still incur energy consumption during transport through the network.

Heuristically, we would expect a system that has a proliferation of overhead bits will be less energy efficient than one with fewer overhead bits. In that case, a more appropriate definition of energy efficiency may be

$$^2H_2(T) = \frac{\text{Total energy consumed over duration } T}{\text{Total customer bits transmitted over duration } T} \tag{2.3}$$

Because $^2H_2$ includes fewer bits than $^2H_1$, but the total energy consumed will be the same for both, $^2H_2 > {}^2H_1$ and may be significantly greater for networks and services that have significant overheads [10].

Alternatively, some networks (such as trans-oceanic and mobile) employ strong error-correction coding that increases the number of overhead bits, but reduces the number of transmission errors and the need to retransmit errored data traffic. This may increase the energy per bit for links with a high error rate.

Neither $^2H_1$ nor $^2H_2$ include the distance over which data is transmitted. For example, the energy efficiency of trans-oceanic systems has been quantified using energy/bit/km [11]. When applying the concept of "energy efficiency" to wireless networks, some authors use "energy/bit/area" [12, 13].

In this book, we adopt metrics with dimensions "energy per bit" to quantify energy efficiency and in so doing we well set the denominator, "Total customer bits transmitted during duration T", to be the end-user bits, stripped of the transport network overheads – no Carrier Ethernet, DSL, SDH/SONET/OTN, or other transport overhead bits. We include just the IP overhead that is sent or received by a customer router. This corresponds to a point of measurement at the customer side input of the CPE owned by the network provider, defined as the customer-end edge of the Access Network in Chapter 3 In some cases, the amount of overhead added to customer data is sufficiently small that ignoring it does not incur a significant error. However, with multiple protocol stacks (such as IP/TDM/WDM), the protocol overheads can have a significant impact on energy efficiency [14].

However, even with this specification, there are still issues of application. Across the telecommunications industry, another often-used metric for energy efficiency is "Power/data rate" [15]:

$$^1H(t) = \frac{\text{Power consumption at time } t}{\text{Data rate at time } t} \qquad (2.4)$$

The quantity $^1H(t)$ is measured in units of Watts/bit/s, which has dimensions of energy/bit similar to the $^2H(T)$ metrics. However, power/data rate is a distinctly different measure to a $^2H(T)$ "energy/bit" metric. We note that $^1H(t)$ is defined at an instant in time, t, whereas $^2H(T)$ is defined over a duration of time, T. This makes these two metrics distinctly different, as will be discussed in Chapter 11. Both have a role to play in understanding the energy efficiency of a network but they quantify distinctly different aspects of network and service energy efficiency.

2.4 NETWORK POWER AND SERVICE POWER

Focusing on power consumption, to calculate an estimate of network power consumption is conceptually straightforward. If we know or can calculate the number and power consumption of each network element (routers, switches, regenerators, etc.), then we merely sum up the individual power consumptions for each relevant element to get an overall value for network power consumption.

Although this is conceptually simple, the actual implementation of this approach can be very difficult. In particular, it is very rare for there to exist a comprehensive database of all network equipment in an organisation and their power consumptions. In many cases, there are multiple generations of equipment (old and new) operating in the network. Much older equipment was deployed before the issue of power consumption became significant. Consequently, the actual power consumption values (or power consumption profile) of such equipment were rarely included in equipment specifications or manuals.

The power consumption values provided in equipment specifications today typically present values for the capacity of equipment power supplies, not the actual consumption of the operating element. The power supply specification is greater (and in some cases significantly greater) than the equipment's actual consumption, because equipment vendors typically include a "safety margin" when specifying the power supply. This results in the specification giving a value above the device's actual operating consumption so that the power supply is operating safely below its operational limit. This means that merely using the stated power supply value in equipment specifications will most likely result in a significant over-estimate of overall power consumption.

Some modern telecommunications equipment may include either a power consumption reporting system in-built or its actual power draw made available to the purchaser; however, this is not universal.

The power consumption of many network devices is relatively constant over their full range of traffic throughput. For other equipment, the power consumption is somewhat dependent upon the amount of traffic it is processing. For example, many home modems have constant power consumption independent of their traffic load [16]. In contrast, mobile base-station power consumption shows a relatively strong dependence on its traffic load [17].

Where device power consumption is traffic dependent, it also becomes time dependent due to the traffic diurnal cycle observed in most networks. During peak traffic times, the power consumption will be greater than at the bottom of the diurnal traffic cycle.

From this, we can see that, although the idea of network power consumption is relatively intuitive, actually constructing a quantitative estimate of network power consumption is non-trivial. This is where network power modelling has an important role to play. The use of models significantly simplifies the task of estimating network power consumption. How this simplification comes about will be discussed in detail in later chapters.

As more and more organisations outsource their ICT requirements and for those organisations that provide or use Internet-based services (such as document preparation, social networking, cloud-based applications and data backup), there is an increasing need to construct estimates for the power consumption of services. For these organisations, the need to construct power/energy models for these services will become essential as mandatory carbon reporting requirements become more broadly adopted. The task of estimating service power/energy consumption is somewhat more complicated than for a network.

There are many conceptual and practical difficulties with service power estimation. As stated above, most services are one of many that are simultaneously being processed by a network element. This raises the question of how to apportion the power consumption of network equipment that simultaneously processes many services. The power consumption of all network elements includes an "idle" component, which represents the element's power consumption when it is not dealing with any throughput. Most network components also exhibit an additional traffic-level-dependent power consumption. Thus, in general, the power consumption of a network element has the form:

$$P(t) = P_{Idle} + \Delta P(R(t)) \tag{2.5}$$

where $R(t)$ is the element traffic throughput (in bit/s), P_{idle} is the power consumption of the network element independent of $R(t)$ and ΔP is the traffic-throughput dependent component of power consumption.

For a network element dealing with multiple services, we have that the total throughput, $R(t)$, is the sum of the traffic components of all the services the element is dealing with. That is:

$$R(t) = \sum_k R^{(k)}(t) \tag{2.6}$$

where $R^{(k)}(t)$ is the traffic component for the k th service.

The question then arises of how to fairly distribute the idle power consumption, P_{idle}, across the services. Further, ICT services typically have a diurnal cycle and so their traffic level varies with time. If the service of interest has a diurnal cycle that is out-of-phase[1] with the rest of the traffic diurnal cycle, the issue of apportioning power to that service increases the complexity of estimating the power consumption of that service. This requires careful consideration and is discussed in detail in Chapter 10.

From the discussion above, we can see there is a growing need to construct power consumption models for both networks and services. In the following chapters, we will step through how these models can be constructed to provide an overall estimate of the power consumption of telecommunications networks and services.

NOTE

1 By out-of-phase, we mean the diurnal-cycle peak traffic time for the service in question does not align with the peak traffic time of the other services being processed by that network element. For example, VOIP calls would be expected to peak during early evening to before midnight. Data backup services are likely to peak after midnight through to early morning.

REFERENCES

[1] D. Hubbard, *How to Measure Anything* (2nd Edition), John Wiley & Sons, 2020.
[2] M. Walton, *The Deming Management Method*, W. H. Allen & Co, 2010.
[3] M. H. Alsharif *et al.*, "Green and Sustainable Cellular Base Stations: An Overview and Future Research Directions," *Energies*, vol. 10, no. 5, p. 587, 2017.

[4] M. Dayarathna *et al.*, "Data Center Energy Consumption Modelling: A Survey," *IEEE Communications Surveys and Tutorials*, vol. 18, no. 1, p. 732, 2016.

[5] H. Kaeslin, "Semiconductor Technology and the Energy Efficiency of ICT," in *ICT Innovations for Sustainability, Advances in Intelligent Systems and Computing*, vol. 310, L. Hilty and B. Aebischer, Eds., 2015, p. 105.

[6] D. Careglio *et al.*, "Hardware Leverages for Energy Reduction in Large-Scale Distributed Systems," in *Large-Scale Distributed Systems and Energy Efficiency: A Holistic View*, J. Pierson, Ed., 2015, p. 17.

[7] K. Ishii *et al.*, "Unifying Top-Down and Bottom-Up Approaches to Evaluate Network Energy Consumption," *Journal of Lightwave Technology*, vol. 33, no. 21, p. 4395, 2015.

[8] ITU-T, *ITU-T L1310, Energy Efficiency Metrics and Measurement Methods for Telecommunications Equipment*, International Telecommunications Union, 2012.

[9] GreenTouch, *GreenTouch Green Meter Research Study: Reducing the Net Energy Consumption in Communications by up to 90% by 2020*, GreenTouch, 2015.

[10] C. Zhang *et al.*, "*Energy Efficiency of Optical IP Protocol Suites*," in *OFC/NFOEC Technical Digest*, 2012.

[11] R. S. Tucker, "Optical Communications—Part I: Energy Limitations in Transport," *IEEE Journal on Selected Topics in Communications*, vol. 17, no. 2, p. 245, 2011.

[12] M. Alsharif *et al.*, "Survey of Green Radio Communications Networks: Techniques and Recent Advances," *Journal of Computer Networks and Communications*, Vols. 2013, ID 453893, 2013.

[13] R. Mahapatra *et al.*, "Energy Efficiency Tradeoff Mechanism Towards Wireless Green Communication: A Survey," *IEEE Communications Surveys & Tutorials*, vol. 18, no. 1, p. 686, 2016.

[14] M. Fang *et al.*, "Network Energy Efficiency Gains Through Coordinated Cross-Layer Aggregation and Bypass," *Journal of Optical Communications Networks*, vol. 4, no. 11, p. 895, 2012.

[15] J. Baliga *et al.*, "Energy consumption of Optical IP Networks," *Journal of Lightwave Technology*, vol. 27, no. 13, p. 2391, 2009.

[16] J. Baliga *et al.*, "Energy Consumption in Wired and Wireless Access Networks," *IEEE Communications Magazine*, vol. 49, no. 6, p. 70, 2011.

[17] G. Auer *et al.*, "How Much Energy is Needed to Run a Wireless Network?," in *Green Radio Communication Networks*, E. Hossain *et al.*, Eds., Cambridge University Press, 2012, p. 360.

3 Network Segments and Wireline Equipment Power Models

Telecommunications networks cover very large geographical areas (some crossing oceans) and may include a few to many thousands of network devices. Constructing a mathematical model of an entire network may involve modelling many network elements interconnected over an area of many thousands of square kilometres. To construct a power consumption model of telecommunications network or service, we adopt the traditional approach of segmenting the total network into access network, edge network (also often referred to as the metro network) and core network [1]. In later chapters, we will construct additional models for home network equipment, data centres and content distribution networks.

This segmentation provides us with a natural way of breaking down the total power consumption model of a telecommunications network into a set of manageable tasks, because the segments often utilise different types of equipment and deal with different types of traffic flows. Access networks provide connections from many central "points of presence" of the network operator to individual customer premises and represent the first level of aggregation of traffic from many customers into a smaller number of traffic flows. The access network is the most cost-sensitive part of the network, which impacts the technologies deployed there. The edge network is the next level of traffic aggregation and typically interconnects the network operator's points of presence across multiple suburbs, towns, and minor regional cities. Therefore, edge network equipment deals with much more traffic than access network equipment. The core networks interconnect cities and nations and consequentially handle very large volumes of traffic.

Depending upon the requirements of the network power model, the modeller can choose between several different types of models. These range from very simple "first-cut" models to relatively complex throughput-based models.

This chapter focuses on wireline networks. That is, networks that use copper wire, coaxial cable or optical fibre to interconnect equipment. Power modelling of mobile networks requires a very different approach and so is discussed in subsequent chapters.

We start by describing a "first-cut" model of network power consumption for each segment. A first-cut model is the simplest model we can construct because it is merely a summation of different network element power consumption values. Despite its simplicity, providing numerical values for the model can still be a non-trivial exercise because just collecting the data can be challenging. A first-cut model does not provide us with enough information to investigate issues such as energy efficiency, diurnal traffic-cycle variation or any information about the relative peak and average network powers.

To study these issues, we then refine the first-cut models to include traffic load dependence. These refined models enable us to describe the power consumption of equipment over a diurnal (daily) traffic cycle. These models assume that, although the traffic will vary, the equipment is not changed over the time duration to which the model is applied. In later chapters, the impact of annual improvements in technology, the deployment of new generations of technologies and the annual growth in network traffic will be incorporated.

DOI: 10.1201/9780429287817-3

3.1 NETWORK SEGMENTATION

This segmentation has an intuitive fit with our purpose of constructing a power model because the equipment types also frequently align with the network segment in which they are deployed. For example, an edge router typically is located at the border between networks and implements functions different to those of core routers that are typically situated within the core network. An aggregation switch located in the access network is connected to many different user devices in the access network and aggregates their traffic. Each type of device will have different parameters and/ or parameter values that are appropriate for modelling their power consumption. For example, in simple terms we can characterise a core router as basically a very fast packet switch with minimal additional functionality (i.e. it is "dumb" and fast). In contrast, edge devices commonly also implement functions beyond connectivity into the core, such as authentication, traffic policing and security.

Figure 3.1 shows a representation of network segmentation. Using this segmentation, we can individually construct network power models for home networks, access networks, edge networks (sometimes referred to as the metro network), core networks and data centres. These models will then be combined to provide an overall network power model. For the purposes of modelling the power consumption of a service, we again use network segmentation and the refined equipment models to reflect the power consumption that can be assigned to a given service.

Simply put, the purpose of the Internet is to provide end-to-end connections for data flows. These flows provide all the data and information services for which we use the Internet. These flows typically start and finish in an access network or occur between an access network and a data centre. A flow may encounter multiple routers and switches on its path through the network. The number of such encounters can be parametrised by a "hop count", which represents the number of routers or switches through which the flow travels along its path.

It is important to note that a segmented network representation of the Internet is a somewhat simplified representation of these flows through the real Internet and does not include much of the fine detail of the Internet's true structure and topology. The networks that have been interconnected to create the Internet may have multiple "overlay" networks as newer technologies have been added to older network equipment. Also, special network connections are frequently added to provide

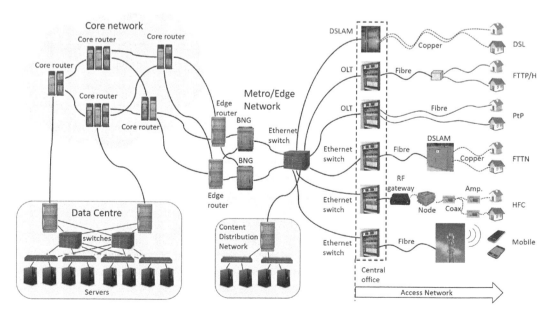

FIGURE 3.1 Network segmentation into an access network, edge network, core network and data centres.

services to specific customers or for particular traffic types. These add-ons and modifications result in a network structure that can be much more complex than that represented by the segmentation model. However, this model at least provides a starting point that is relatively generic across almost all parts of the Internet.

The segmented network representation does account for the typical hop count for packets that traverse the Internet [2] and provides a starting point for our analysis, which follows the approach of Baliga *et al.* [1]. However, it does not account for factors such as overlay networks and private enterprise networks.

3.2 FIRST-CUT (LOAD INDEPENDENT) POWER MODELS

The "first-cut" models described below are based on merely summing up the power supply values stated in the equipment specifications. The power supply values provided in equipment specifications are typically greater than the actual operational consumption values by about 20% to 25% [3] so that the power supplies are not overloaded. Therefore, using power supply values from equipment specifications will result in an overestimate of network power consumption, which can be considered an upper limit to network power consumption and so provide some guidance on overall power consumption. Although actual "real-life" consumption values are difficult to obtain, some manufacturers are becoming more open to providing these values. If they are not available, to get a tighter upper limit, we could discount the power supply values by 20%.

3.2.1 WIRED ACCESS NETWORKS POWER MODEL

Wired access networks are defined to be a network that provides a wired physical connection between the home or user's access device and local exchange/central office. We will define the customer-side boundary of the wired access network in the home/office to be the customer-facing ports of the service provider's customer premises equipment (CPE). Thus, in a home it will include the home modem/broadband gateway that the service provider supplies, but not any equipment on the customer's side of the modem/gateway.

The wired access technologies considered in this book are the following.

3.2.1.1 Digital Subscriber Line Technologies (DSL)

DSL operates over the copper pairs originally installed to provide traditional fixed-line telephone services (often referred to as PSTN). A DSL modem is located in the customer home and connects via a dedicated copper pair to a DSL Access Multiplexer (DSLAM) at the nearest central office/ telephone exchange.

3.2.1.2 Fibre to the Premises/Home (FTTP/H)

FTTP/H networks most commonly use a Passive Optical Network (PON) technology, in which a single fibre from the central office/exchange is connected to a passive optical splitter that enables the optical signal from an Optical Line Terminal (OLT) located in the central office to be shared by multiple customers. The OLT provides services to a number of home modems/gateways (also called optical network units (ONUs)), each located in a customer's premises. ONUs communicate with the OLT in a time-multiplexed order, with the OLT assigning time slots to each ONU based on its relative demand.

The number of users who share the OLT optical signal typically ranges between 24 and 128.

3.2.1.3 Point-to-Point Fibre (PtP)

A PtP access network provides a dedicated fibre between the customer premises and the central office/exchange. This provides the highest potential access speed of the access technologies currently available, wired or wireless. The modem/gateway located in the customer premises is often referred as an "optical media converter" (OMC) and is directly connected to the terminal unit in the exchange; there is no splitter in between. This means the customer does not share the terminal unit port and so has access to the full speed of that port.

3.2.1.4   Fibre to the Node (FTTN)

Fibre to the node (FTTN) uses existing copper pairs previously installed for a Public Switched Telephone Network (PSTN). A dedicated fibre is provided from a central office/exchange to a DSLAM in a remote node (typically located in a street cabinet) close to a cluster of customers. The PSTN copper pairs then carry the signal from the DSLAM to the customer premises. Between the DSLAM and the customer premises, a high-speed copper pair cable technology such as very high speed DSL (VDSL) or ADSL2+ is used. More recently, a technology to enable increased access speeds when using DSL over copper pairs, called vectoring (often referred to as "G.fast"), has been deployed in some FTTN networks [4, 5].

FTTN mitigates the distance limitations of VDSL-type technologies and enables high-speed broadband service delivery without having to re-lay cables for the entire access network. Variant of FTTN is "Fibre to the Kerb" (FTTK) or "Fibre to the Cabinet" (FTTC) in which the fibre is continued to a small DSLAM located very close to the customer premises.

3.2.1.5   Hybrid Fibre Coaxial (HFC)

Hybrid Fibre Coax (HFC) networks require more careful consideration than the wireline access network technologies discussed above. This is because HFC networks include a widely distributed collection of coaxial amplifiers whose power consumption must be included. The access networks discussed above almost always only include one or no power-consuming remote node between the central office terminal unit and the CPE. This is not the case for HFC access networks.

HFC distribution networks were initially deployed to deliver cable television services to homes. Over recent years, these networks have been adapted to also deliver Internet services. The service data are distributed on radio frequency (RF)-modulated optical carriers through an optical fibre to local distribution nodes, where the optical signal is converted into an electrical signal, which is then distributed to customers through a tree network of coaxial cables. The coaxial cable network includes electrical amplifiers placed as necessary in the network to maintain signal quality. Based on the neighbourhood layout, the cable may be branched, the signal re-amplified and signals for individual customers tapped off along the route. This architecture distinguishes HFC from the other wired access network technologies in that there may be multiple active network devices (i.e. the local distribution nodes and RF amplifiers) between the customer premises and the central office.

The electrical signal sent (downstream) to the customer on the coaxial cable includes a collection of modulated RF carriers. A reverse channel is also provided for customer data to be sent back (upstream) through the network.

Broadband Internet access is provided by using one or more of the downstream RF channels to deliver high-speed data to the customer, while one or more of the reverse channels is allocated to send data upstream from the customer into the network. The data/Internet customer has a cable modem to connect them to the network.

3.2.2   Remote Nodes

For FTTP/H, FTTN and HFC, the cable emanating from the central office will connect to at least one node (often referred to as a "remote node") that is geographically located between the central office and the home. For FTTN and HFC networks, the remote nodes require power because they contain electronic equipment that must be powered. FTTN typically has a single power-consuming remote node (containing the DSLAM) between the central office and the customer premises. HFC, however, will have multiple power-consuming devices (local distribution nodes and RF amplifiers) between the central office and customer premises. FTTN remote nodes are often located in street cabinets.

In FTTP/H access networks, the remote node is usually passive requiring no power supply. For PtP networks, each customer is allocated a separate fibre and no remote node is required.

The network side boundary of the access network (i.e. the boundary between the access network and the edge network) is taken to be the output ports of the first terminal unit inside the central office/local exchange. For example, for a DSL access connection the network side boundary is taken to be the edge network-facing ports of the DSLAM located in the central office. For an FTTP/H, the network side boundary is the edge network-facing ports of the OLT located in the central office. A diagrammatic representation of these access network technologies and their remote node is presented in Figure 3.2 and they are discussed in later chapters.

To calculate the power consumption of an access network, we include power consumption of the home modems/gateways, the remote nodes and the terminal equipment in the central office. For an access network that connects to N_{Access} CPEs, we can write the total power consumption as follows:

$$P_{Acess} = \sum_{j=1}^{N_{Access}} P_{CPE,j} + \sum_{k=1}^{N_{RN}} P_{RN,k} + \sum_{m=1}^{N_{TU}} P_{TU,m} \qquad (3.1)$$

where $P_{CPE,j}$ is the power consumption of the jth CPE, N_{RN} is the number of remote nodes, $P_{RN,k}$ is the power consumption of the kth remote node, N_{TU} is the number of terminal units in the central office that connect to the access network and $P_{TU,m}$ is the power consumption of the mth terminal unit. For the HFC technology, the k-sum includes the local distribution nodes and the RF amplifiers distributed along the HFC network coaxial cables.

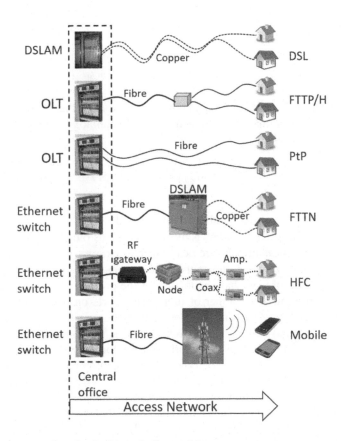

FIGURE 3.2　Access network technologies including mobile access.

For an access network that only uses a single technology throughout (e.g. a PON that only uses FTTP/H), this equation can be simplified because there will be only a single value for each of $P_{CPE,j}$, $P_{RN,k}$ and $P_{TU,m}$. In that case, (3.1) simplifies to:

$$P_{Acess} = N_{Access}P_{CPE} + N_{RN}P_{RN} + N_{TU}P_{TU} \tag{3.2}$$

For an HFC network, because of the two types of active remote nodes we have:

$$P_{HFC} = N_{Access}P_{CPE} + N_{RN}P_{RN} + N_{LA}P_{LA} + N_{TU}P_{TU} \tag{3.3}$$

where the subscript RN refers to the local distribution nodes and the subscript LA refers to the line amplifiers distributed throughout the network.

For an access network with a mixture of technologies, we can re-cast (3.1) in ways that may simplify the calculation of network power. For example, assume that an access network has proportions of X_{FTTP} FTTP connections, X_{FTTN} FTTN connections, X_{PtP} PtP connections and X_{DSL} DSL connections, where:

$$X_{FTTP} + X_{FTTN} + X_{PtP} + X_{DSL} + X_{HFC} = 1 \tag{3.4}$$

Then, we have for a wired access network:

$$P_{Wired\_Access} = N_{Access} \begin{pmatrix} X_{FTTP}\left(P_{CPE,FTTP} + W_{FTTP}\dfrac{P_{TU,FTTP}}{M_{TU,FTTP}}\right) + X_{FTTN}\left(P_{CPE,FTTN} + \dfrac{P_{RN,FTTN}}{M_{RN,FTTN}} + W_{FTTN}\dfrac{P_{TU,FTTN}}{M_{TU,FTTN}}\right) \\[2ex] +X_{PtP}\left(P_{CPE,PtP} + W_{PtP}\dfrac{P_{TU,PtP}}{M_{TU,PtP}}\right) + X_{DSL}\left(P_{CPE,DSL} + W_{DSL}\dfrac{P_{TU,DSL}}{M_{TU,DSL}}\right) \\[2ex] +X_{HFC}\left(P_{CPE,HFC} + \dfrac{P_{RN,HFC}}{M_{RN,HFC}} + \dfrac{P_{LA,HFC}}{M_{LA,HFC}} + W_{HFC}\dfrac{P_{TU,HFC}}{M_{TU,HFC}}\right) \end{pmatrix} \tag{3.5}$$

where $M_{Z,RN}$, $M_{Z,LA}$ and $M_{Z,TU}$ are the average number of customers that share the remote node, line amplifier (for HFC) and terminal unit, respectively, for technology Z (Z corresponding to DSL, FTTP/H, PtP, FTTN, HFC) given by

$$M_{Z,RN} = \frac{N_{Z,RN}}{N_Z}, \quad M_{Z,LA} = \frac{N_{Z,LA}}{N_Z}, \quad M_{Z,TU} = \frac{N_{Z,TU}}{N_Z} \tag{3.6}$$

where $N_{Z,RN}$, $N_{Z,LA}$ and $N_{Z,TU}$ are the number of remote nodes, line amplifiers (HFC) and terminal units for technology Z, respectively, and N_Z is the number of customers connected by technology Z.

The terms W_Z account for additional overheads such as external power supplies, electricity distribution losses, and cooling requirements in the building that houses the terminal equipment [6].

The form in (3.5) is useful if we have estimates for the proportion of users of each technology across the network and know the number of user access ports on the remote node and terminal equipment for each technology. (This can be accessed from the specification sheets of the equipment used.) Because FTTP, PtP and DSL don't have remote nodes, the power terms for the remote nodes for these technologies are zero. For HFC networks, the number of customers per terminal unit and number of line amplifiers will vary across customers in the neighbourhood serviced by the HFC. Therefore, the HFC contribution to total access network power given in (3.5) will be approximate.

From (3.5) we can construct a simple estimate of the energy efficiency metric of "average power per user" by dividing through by N_{Access}. It is important to note that the network power consumed by the connection to a given user may vary significantly from $P_{Wired\_Access}/N_{Access}$. This is because some technologies (such as DSL and HFC) utilise different modulation formats (e.g. multi-level QAM, QPSK) and different transmit powers depending upon the noise and distortion suffered by the signal to that customer. For high noise or distortion scenarios, a more robust (and hence greater power-consuming) signal may be implemented. Typically, the further the customer is located from the last active node, the greater the power allocation to the signal sent to that customer.

Calculating the energy efficiency (energy per bit) for the access network requires much more information and is discussed below. We will revisit access networks in a later chapter where we will add to the model above to develop a more detailed model that enables calculation of energy efficiency and power per service for an access network.

3.2.3 EDGE NETWORK POWER MODEL

The edge network serves as the interface between the access network and the core network. The edge network includes Ethernet switches that concentrate traffic from a large number of access nodes (such as DSLAMs) and typically connect to two broadband network gateways (BNGs) or broadband remote access server (BRAS) routers. The connection to two devices is to provide redundancy in case of failure. Switches in the edge network aggregate traffic and forward it to the customer's ISP. The ISP runs the BNG/BRAS to authenticate users, manage user sessions, maintain usage records, account for customer sessions, and address assignment, quality of service and security. These network elements provide the edge router connection to the core of the network. In general, edge routers accept inbound customer traffic into the network and forward it on into the core, and vice versa.

Edge network routers are also used between networks of different operators. This is typically done because each network operator needs to ensure the traffic they send into and receive from another operator is of acceptable quality (i.e. error free). If the service provided to a customer fails, each network operator needs to determine where the fault is. Therefore, ensuring adequate signal quality for data flows into and out of a network operator's network is an important function.

The first-cut power consumption model of the edge network is just the sum of the powers of the edge network elements:

$$P_{Edge} = \sum_{j}^{N_{Edge}} P_{Edge,j} \tag{3.7}$$

where N_{Edge} is the number of network elements (edge switches and routers) in the edge network and $P_{Edge,j}$ is the power consumption of the jth element.

3.2.4 CORE NETWORK POWER MODEL

The core network comprises a number of large routers in major population centres or interconnection hubs in a large (national or international) network. A core router forwards packets to other core routers within a carrier's network (or in other carrier networks). Core routers within a network carrier's or operator's network are typically highly meshed with other core routers in that operator's network. However, they will have only a few links to the networks of other network operators. High-capacity optical links interconnect core routers and connect to networks of other operators. Core routers offer packet forwarding between other core and edge routers and manage traffic to prevent congestion and packet loss.

As with the edge network, the power consumption of a core network is the sum of the power consumption of the core network elements. Because core networks may include very long (optical) links between core routers, the power consumption of the optical links between routers will also need to be included:

$$P_{Core} = \sum_{j}^{N_{Core}} P_{Core,j} \tag{3.8}$$

where N_{Core} is the number of network elements in the core network and $P_{Core,j}$ is the power consumption of the jth element. In this case, the jth element may be a router or a power-consuming component of a long-haul optical system.

3.2.5 TOTAL NETWORK POWER MODEL

Using the above equations, the total network power consumption will be given by

$$P_{Network} = P_{Wired\_Access} + P_{Wireless\_Acess} + P_{Edge} + P_{Core} \qquad (3.9)$$

With the above, a simple "first-cut" estimate of network power consumption can be constructed. Although writing down the equations is relatively straightforward, inserting values can be a major challenge. This is because, as stated previously, acquiring the actual power consumption values can be very difficult.

3.3 MEASURING AND MODELLING EQUIPMENT POWER CONSUMPTION

A key advantage of the models in Section 3.2 is their simplicity and minimal requirement for information about the network elements. Many network devices have a relatively constant power consumption over their full range of traffic load. However, a significant range of network equipment shows a relatively strong relationship between traffic throughput and power consumption.

In some cases, we may need to construct estimates of energy efficiency (such as power per user, energy per bit, etc.). In this case, the first-cut models above are too simple because we need to include any power dependence on traffic throughput or load. To include traffic load dependence of an element's power consumption in a network or service power model requires additional information on the network element, adding to the complexity of the model.

Another reason for including traffic load dependence is the widely accepted view that energy efficiency can be improved when equipment is designed so that its power consumption is "load proportional" [7, 8]. The power consumption of load-proportional equipment has a linear dependence on its traffic throughput and has almost zero power consumption when not processing any traffic. The desire for load dependence has resulted in a range of technologies and techniques to reduce power consumption at low and zero traffic loads.

We now discuss how to improve these first-cut models to include a power dependence on traffic load or throughput.

3.4 A GENERIC LOAD-DEPENDENT POWER CONSUMPTION MODEL

At an intuitive level, we can easily see why the power consumption of network equipment will be dependent upon its traffic load or throughput. Being machines that use electronic technologies (i.e. integrated circuits, hard disk drives, wireless transmitters, etc.), these devices require a continuous power supply for the circuits and, in many cases, cooling to remove the heat generated in electronic devices. Some of these functions will be independent of the traffic load, such as biasing and power supply. However, others will be dependent upon the number of packets arriving per second for processing. These machines operate digitally (i.e. they use two-valued digital logic to process packets), and the more packets arriving per second (i.e. the greater the traffic throughput) the more digital processing the electronics will have to undertake to process the packets. This increased amount of digital processing results in higher power consumption in the network elements' electronics. Consequentially, as equipment deals with more packets per second, the greater is its power consumption.

Those sub-systems of a network element that must remain on and powered, irrespective of the traffic throughput, will result in a power consumption that is independent of the traffic load even when there is no incoming traffic. We will refer to this power consumption as "idle power", P_{idle}.

The European Commission in its "Code of Conduct on Energy Consumption of Broadband Equipment" describes "idle state", in which a network element is not dealing with traffic, as [9]:

> In the idle-state, the device is idle, with all the components being in their individual idle states. In this state the device is not processing or transmitting a significant amount of traffic, but is ready to detect activity.

It is important to understand that, when equipment is in an idle state, it is not totally disabled and unable to respond to incoming traffic. Quoting the European Commission Code of Conduct [9]:

> When equipment is in an idle state, it needs to be able to provide services with the same quality as in the on-state, or to be able [to] transition to the on-state to deliver the service without introducing a significant additional delay from the user perspective.

For much equipment, its power consumption remains at, or very close to, P_{idle} irrespective of its traffic load. Such devices have a "load-independent" power consumption. Examples of this are home broadband gateways, optical link transmitters and receivers, and Time Division Multiplex equipment (SDH, SONET, OTN) [10].

For other types of equipment, their power consumption has a non-trivial dependence on traffic load. Examples include routers, Ethernet switches, some mobile base stations and Wi-Fi access points. For these types of equipment, the simplest power model has a linear dependence on throughput:

$$P(t) = P_{idle} + ER(t) \tag{3.10}$$

where $R(t)$ is the element traffic load or throughput (in bit/s) at time t and E is the "incremental energy per bit" which is a constant with dimensions of Joules/bit.

Studies have shown that Equation (3.10) provides a good approximation for router and switch power consumption [7, 8, 11, 12]. Under certain conditions, this linear model can also well approximate the power consumption of mobile base stations [13, 14] and Wi-Fi access points [15, 16], where the power consumption depends on throughput. However, for mobile base stations, in some cases the linear form in (3.10) needs to be modified. Wireless access technologies will be discussed in greater detail in the following chapters.

Although a linear model is not a perfect match for all cases, given the other approximations involved in estimating telecommunications network and service power consumption, in most cases a linear model is more than adequate to provide a good estimate for this purpose.

Given the linearity of (3.10), we can write it in the following form:

$$P(t) = P_{idle} + ER(t) = P_{idle} + \frac{\left(P_{max} - P_{idle}\right)}{C_{max}} R(t) \tag{3.11}$$

where $E = (P_{max} - P_{idle})/C_{max}$ and P_{max} is the equipment power consumption when $R(t) = C_{max}$, the maximum throughput the element is designed to handle.

There are several approaches to determining the values for C_{max}, E and P_{idle} for real equipment. We would expect C_{max} to be specified for the equipment, and hence available in the equipment data sheet or specifications. Given the linearity of (3.11), to determine E we only need two data points for power consumption and through traffic. If the power consumption of the element is P_1 with throughput R_1 and P_2 with throughput R_2, then we have:

$$E = \frac{P_1 - P_2}{R_1 - R_2} \tag{3.12}$$

If it is practical, we can set $R = 0$ in (3.11), which gives P_{idle}. If it is not practical to measure a second data point, it can be approximated using published information. For example, Jalali *et al.* contain maximum and idle power values for a range of network equipment types [17]. The data in that publication give a typical value of P_{idle}/P_{max} of around 80%–90%. That publication was in 2016 and since then there have been advances in reducing idle power. Unfortunately, there is a paucity of published information on the annual trend of idle power for network routers. Therefore, to get a handle on how to scale the ratio to make it more current, we refer to Niccolini *et al.* [7], which uses idle-power scaling trends of servers to represent the idle-power scaling for a router. Their justification comes from the fact that they are modelling software routers in which the CPU is the dominant power-consuming component. In a network router, the routing engine is likewise a dominant consumer of power [12] and so we can have some confidence that a similar approach can be applied. Adopting this approximation, Shehabi *et al.* [18] have published the yearly trend for the ratio of maximum-to-idle-power evolution up to 2020. Although this is not ideal, it will at least provide some degree of guidance for the element's load-dependent power consumption.

The next section shows that the linear form in (3.10) applies to equipment capacity upgrades. Then in Section 3.6 provides experimental justification of the linear form in (3.10).

3.5   EQUIPMENT CAPACITY UPGRADES

Equation (3.10) is appropriate for time scales corresponding to the diurnal (daily) traffic cycle. On longer time scales, such as months to years, the design of modern edge and core network switches and routers is based on the ability to increase switch or router throughput by adding cards and ports to an already operational machine. The network element generically consists, firstly, of a base configuration that provides the key network switching/routing functionality (this includes the chassis, core switching fabric, control plane, etc.). Connected to the base configuration, via a data-exchange backplane, are a series of line cards that provide ports that connect to the network and through which traffic ingresses and egresses the switch/router. Traffic flows from its source or a previous router/switch into a port on a line card, through the central component, and out of a port on another line card toward its destination. As traffic demand grows, the network operator can add additional cards and ports up to a maximum that is determined by the switching capacity of the central component. Some models provide for the addition of more switching/routing engines to the base configuration [11].

For some types of equipment, over a diurnal cycle, the power consumption of the line cards and central component is approximately constant (i.e. $P_{idle} >> EC_{max}$). However, as network traffic grows on the R_3 time scale of months to years, more line cards are typically added to the network element. The overall power consumption increases and can be expressed as follows [19]:

$$P = P_{base} + \sum_{v=1}^{V_{LC}} n_{LC,v} P_{LC,v} \tag{3.13}$$

where P_{base} is the power consumption of the base configuration, $n_{LC,v}$ is the number of line cards of type v, $P_{LC,v}$ is the power consumption of the type v line card and V_{LC} is the number of different line cards in the network element. The index v is used to account for the line-card port speeds, generation, port count, etc., because over the lifetime of the network element several types or generations of line cards may be added.

A line card is characterised by its power consumption, $P_{LC,v}$, and maximum throughput, $C_{LC,max,v}$. The maximum throughput of the whole network element will be:

$$C_{max} = \sum_{v=1}^{V_{LC}} n_{LC,v} C_{LC,max,v} \tag{3.14}$$

We consider a situation in which a network element has been deployed (hence P_{base} is set) and the network operator needs to accommodate an increase in traffic, ΔR. For simplicity, we assume the operator deploys a single type of (latest generation) line card for an upgrade. Then, the number of line cards the operator will add to the network element is:

$$n_{LC,v} = \left\lceil \frac{\Delta R}{C_{LC,max,v}} \right\rceil \tag{3.15}$$

in which $\lceil x \rceil$ is the smallest integer greater than x. If $\Delta R \gg C_{LC,max,v}$, then:

$$n_{LC,v} \approx \frac{\Delta R}{C_{LC,max,v}} \tag{3.16}$$

In this case, the increase in power consumption of the network element, ΔP, due to the increase in traffic, ΔR, can be written as:

$$\Delta P \approx \frac{P_{LC,v}}{C_{LC,max,v}} \Delta R = E_{LC,v} \Delta R \tag{3.17}$$

where $E_{LC,v}$ is the incremental power per throughput (or incremental energy per bit) of the type v line card.

We now consider a network element that has been subject to annual upgrades over N years via the addition of latest-generation line cards to accommodate traffic increases $\Delta R^{[n]}$, with $n = 1,...,N$. Let $\Delta R^{[0]}$ be the traffic in the year of initial deployment ($n = 0$). In year N, the power consumption will be:

$$P^{[N]} = P_{base} + \sum_{n=0}^{N} E_{LC,v^{[n]}} \Delta R^{[n]} \tag{3.18}$$

where $v^{[n]}$ represents the type of line card deployed in year n. We note that (3.18) has the same structure as (3.10), in that both have a constant part (P_{idle} and P_{base}) and a load-dependent part. In (3.10) there is a single load-dependent term, whereas in (3.18) there are multiple load-dependent contributions. This similarity in form for these power equations means we can use this basic form for the power consumption of a network element on time scales ranging from hours to years.

In the metro/edge and core networks, the routers and switches dominate the power consumption in each node [20]. The power dependence of lower-layer equipment is also of the form in (3.10) [21]; hence, their power consumption profile can be absorbed into that form by appropriately modifying the values of P_{idle} and E. With these considerations, the dependence of power consumption on traffic throughput of all equipment in the network nodes considered in this book will be assumed to be well approximated by a linear form such as (3.10).

3.6   LOAD-DEPENDENT POWER MODELS

The simple first-cut power models introduced in Section 3.2 above do not include a dependence on load or throughput, $R(t)$. Including this dependence improves the accuracy and broadens the possible uses of the model: for example, understanding of network power consumption at times of maximum, minimum and average traffic levels, ascertaining the variation of network power consumption over the daily traffic cycle, studying the increase in network power consumption as a network grows to accommodate increasing traffic demand. Also, the form in (3.11) is essential for us to develop and study energy efficiency metrics (energy per bit, power per data rate).

In the sections immediately below, we will provide references and measurement results that show the linear form (3.11) provides a good representation of the dependence of the power consumption of network equipment on their traffic load/throughput.

3.6.1   Wired Access Network Equipment

Most CPE types have a relatively constant power profile with E ≈ 0 [22, 23]. ONUs used in FTTH also have a constant power profile [23]. DSLAMs used in FTTN and DSL connect to a user via a port on a line card in the remote node/terminal unit. Once configured, and assuming network conditions do not dramatically change, the line cards have a relatively constant power consumption even if not all the ports are active [10]. The outcome of this is that, as the number of end users increases (which results in an increase in traffic), the power consumption of the DSLAM increases with a stepwise linear dependence on traffic [10]. This also applies to HFC networks, the line amplifiers, and the terminal unit (a Cable Modem Terminal System) [24].

DSL and HFC utilise technologies that transport data on multiple carrier frequencies to transport data to and from the customer premises [25, 26]. Each carrier transports data between a transmit/receive pair at each end of the access wireline, with each pair consuming power. Part of the initialisation process for the CPE and terminal equipment in DSL- and HFC-based networks is setting and fixing the number of carriers for the connection. There have also been proposals for "Dynamic Spectrum Management" in DSL and HFC access network equipment whereby the equipment dynamically reconfigures the number of carrier channels depending upon network conditions and traffic load. Depending upon the traffic load and link conditions, a number of channels can be turned off or reconfigured to optimise power consumption [27, 28]. The optimisation method applied to DSL in Tsiaflakis *et al.* [27] results in a linearly (or less) dependence of power consumption on data rate.

3.6.2   Edge and Core Network Equipment

A survey of power consumption by edge and core network equipment has been published by Van Heddeghem *et al.* [19, 29]. Over the daily (diurnal) traffic cycle, equipment such as optical transport systems (transceivers, optical amplifiers, transmitters and receivers) typically have a constant power profile with $E = 0$. In contrast, the power consumption of network elements such as routers and switches has a non-trivial traffic throughput dependence.

Power consumption measurements on routers and switches display a throughput dependence such as that shown in Figure 3.3. These measurements were undertaken on an online router dealing with real traffic over several days (diurnal cycles) showing an approximately linear average dependence of power consumption on traffic throughput, with a scatter. The grey line shows the linear best fit to the data.

To understand the scatter in Figure 3.3, a series of measurements were undertaken in the Centre for Energy Efficient Telecommunications on an enterprise switch (Huawei S5300), an edge switch (ALU 7450-ESS7), a metro router (Huawei CX600-X3) and an edge router (ALU 7550-SR7) [11, 12]. For each machine, a series of set-length packets (100, 576, 1000 and 1500 bytes) were input

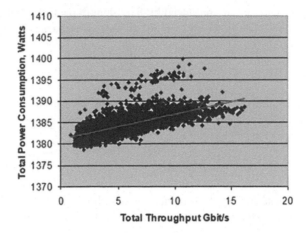

FIGURE 3.3 Throughput vs power measurements of an online router over several days (diurnal cycles). The line shows the best-fit linear approximation to the data.

into the device for a range of input packet rates, R_{pkt}. The results of these measurements are shown in Figure 3.4, which shows a linear relationship between the input bit rate, R_{bit}, and the device power consumption.

Comparing Figures 3.3 and 3.4, we can interpret the scatter in Figure 3.3 as a result of the power consumption being dependent not only on the traffic bit rate (bit/s) but also on the traffic packet rate (packets/s). There is a component of energy consumption involved in handling each packet, as well as the energy consumption involved in forwarding each bit. Because Figure 3.3 is based upon live traffic measurement, it shows power consumption when handling live traffic from hundreds of users, each generating different packet sizes and packet rates.

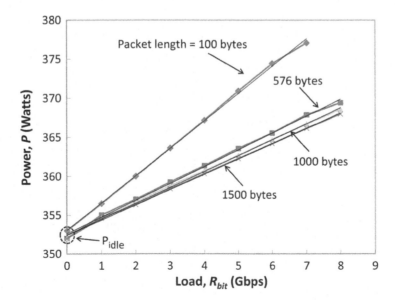

FIGURE 3.4 Router power as a function of bit/s load for different packet sizes. These results show a dependence not only on the data rate of bit/s, but also on the packet size.

The traces shown in Figure 3.4 are produced by fixed-length packets. From this we find the following relationships between the input bit rate, R_{bit} (bit/s), input packet rate, R_{pkt} (packets/s), and input byte rate, R_{byte} (byte/s):

$$R_{bit} = 8L_{pkt}R_{pkt} = 8R_{byte} \qquad (3.19)$$

where L_{pkt} is the length of the input packets in bytes. The measurements show that the power consumption of a router or switch, P, can be expressed in the form:

$$
\begin{aligned}
P &= P_{idle} + E_P R_{pkt} + E_{S\&F} R_{byte} \\
&= P_{idle} + E R_{bit}
\end{aligned}
\qquad (3.20)
$$

where P_{idle} is the power consumption of the device for zero throughput, $R_{bit} = 0$, E_P is the per-packet processing energy, and $E_{S\&F}$ is the per-byte store and forward energy. Also, we have:

$$E = \frac{1}{8}\left(\frac{E_P}{L_{pkt}} + E_{S\&F} \right) \qquad (3.21)$$

P_{idle} is the "idle power" because it is the power consumption when the network element is not processing traffic and therefore may be considered to be "idle".

To understand these results for the power consumption of a router/switch, we use a high-level schematic representation of how a router processes incoming packets, shown in Figure 3.5. The model is based on a generic "store and forward" description of a router or switch chassis, which consists of three sub-systems:

1. the data plane, which transports data (i.e. packets) through the element;
2. the control plane that controls the operation of the device, including directing packets to the appropriate output port; and
3. the environmental units, which include power provision and cooling.

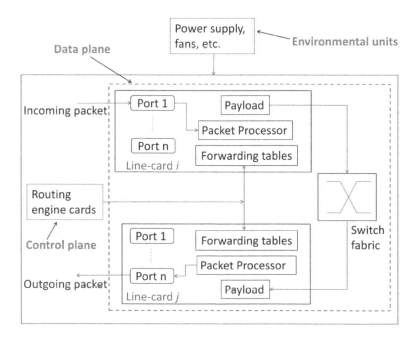

FIGURE 3.5 Schematic diagram of packet flow through a "store and forward" type router.

Line cards and switch fabric cards comprise key data plane elements; routing engine cards represent the control plane element; and the chassis power supply, fans, etc. constitute the environmental units. The power consumption of the environmental units is set to be constant, independent of traffic throughput.

Packet throughput occurs as follows: A packet arrives at an input port. Once all the bits of the packet are received, the packet processor performs the IP longest-prefix matching on the information contained in the packet header to identify the egress line card and port to which the packet must be switched. In a router, the control plane ensures the forwarding (IP and MAC address) tables are up to date. A switch operates below the IP layer and the lookups are based on the destination MAC address. The quantity E_P represents this header-processing energy per packet consumed by this process. Note that E_P is independent of the packet length.

While the lookup is occurring, the payload of the packet is stored in internal memory on the input line card. Once the lookup is completed, the payload is forwarded via the switch fabric to the appropriate destination output line card. On that line card, the packet is reconstructed and the appropriate headers are inserted. The packet is then sent to an output port for transmission on the network link to the next node. This storage and forwarding process consumes energy $E_{S\&F}$ per packet. The energy consumed by this store and forward process is dependent upon the length of the packet.

These measurement results show that, for a typical router/switch, Equation (3.20) provides a very good power consumption model for fixed-length input packets [11].

As mentioned above regarding the scatter around the linear model, we note that a router typically deals with packets of different length and ports that may have differing values for $E_{S\&F}$, E_P, and input/output bit rates, R_{bit}. To confirm this, Hinton *et al.* undertook a Monte Carlo simulation accounting for these factors and showed that the power consumption of a router/switch displays a scatter similar to that shown in Figure 3.3 [30].

3.6.3 SLEEP MODES

In recent years, proposals for the introduction of low-power "sleep" states in network equipment have been presented [31–35]. Sleep modes are designed to consume significantly less power than when a device is actively processing traffic or is idle. A low-power sleep state is typically implemented by ceasing a range of processes only required when the network element is dealing with traffic and powering-down the corresponding sub-systems. Such a state will consume power, but the expectation is that it will consume significantly less power than in a full-on state, when it is processing traffic, or in the equipment idle state.

With the introduction of sleep states, the power consumption profile of equipment that can implement these states becomes an $R(t)$-dependent two-step profile, which can be represented by

$$P(t) = \begin{cases} P_{sleep} & R(t) < R_{threshold} \\ P_{active} & R(t) \geq R_{threshold} \end{cases} \tag{3.22}$$

where P_{sleep} is the sleep-mode power consumption and P_{active} is the active-state power consumption. The value of $R_{threshold}$ is typically zero bit/s. In the active state ($R(t) \geq R_{threshold}$), the power consumption of the element is given by (3.10).

Low-power sleep modes are being researched for access network, edge and core network equipment. For multi-port and line-card equipment, the line cards and/or ports are placed into a low-power sleep state when its input traffic is below a threshold value [36]. Some proposals go further and optimise the overall network traffic by redirecting traffic to specific network elements to reduce the load on other elements to a level that enables those elements to be placed into a low-power state [37, 38].

From the measurement procedure that gives the results shown in Figure 3.4, we know that the power consumption arising from line cards has a linear form and hence multiple line cards will likewise give a linear form. Therefore, as line cards on a network element are stepped between active and sleep, the overall power consumption of the network element will have a stepwise dependence on throughput:

$$
\begin{aligned}
P(t) &= P_{idle} + \sum_{k=1}^{N_{sleep}} P_{sleep,k} + \sum_{k=1}^{N_{active}} P_{active,k}(t) \\
&= P_{idle} + \sum_{k=1}^{N_{sleep}} P_{sleep,k} + \sum_{k=1}^{N_{active}} P_{active\_idle,k} + E_{active,k} R_{active,k}(t)
\end{aligned}
\tag{3.23}
$$

This dependence is depicted in Figure 3.6. In (3.23), $P_{sleep,k}$ is the sleep state power consumption of the kth sub-system that is in sleep state, $P_{active,k}(t)$ is the active state power consumption of the kth sub-system that is active and has traffic throughput $R_{active,k}(t)$. $P_{active\_idle,k}$ is the load-independent component of the power consumption of the kth active sub-system.

As the load on the element increases toward the element's maximum capacity (C_{max}), more line cards/ports are taken out of a low-power sleep state and made active, resulting in an increase in total element power. As its traffic load decreases, these sub-systems are placed into a sleep mode, thereby reducing power consumption. To show the resulting power profile is approximately linear, we assume the total number of ports is large enough to apply averages to (3.23). This allows us to write:

$$
\begin{aligned}
P(t) &= P_{idle} + \sum_{k=1}^{N_{sleep}} P_{sleep,k} + \sum_{k=1}^{N_{active}} P_{active\_idle,k} + E_{active,j} R_{active,k}(t) \\
&\approx P_{idle} + N_{sleep}\langle P_{sleep}\rangle_{sleep} + N_{active}(t)\left(\langle P_{active\_idle}\rangle_{active} + \langle E_{active}\rangle_{active}\langle R_{active}\rangle_{active}\right) \\
&= P_{idle} + N_{total}\langle P_{sleep}\rangle_{sleep} + N_{active}(t)\left(\langle P_{active\_idle}\rangle_{active} - \langle P_{sleep}\rangle_{active} + \langle R_{active}\rangle_{active}\langle E_{active}\rangle_{active}\right)
\end{aligned}
\tag{3.24}
$$

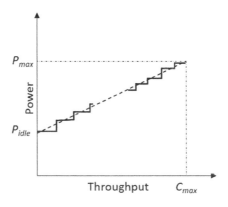

FIGURE 3.6 Stepwise power dependence of a network element in which ports and line cards can be independently placed into a low-power sleep state.

Here, the terms $\langle P_{sleep}\rangle_{sleep}$, $\langle P_{active\_idle}\rangle_{active}$, $\langle R_{active}\rangle_{active}$ and $\langle E_{active}\rangle_{active}$ are the mean values for sleep power, load-independent active power, throughput of active ports and energy per bit of the active ports, defined by

$$\langle X \rangle_{sleep} = \frac{1}{N_{sleep}} \sum_{k=1}^{N_{sleep}} X_{sleep,k}, \qquad \langle X \rangle_{active} = \frac{1}{N_{active}} \sum_{k=1}^{N_{active}} X_{active,k} \qquad (3.25)$$

Also, $N_{active} + N_{sleep} = N_{total}$ and we have treated $E_{active,k}$ and $R_{active,k}$ as independent random variables, which allows us to use $\langle E_{active}R_{active}\rangle = \langle E_{active}\rangle\langle R_{active}\rangle$. This expression is linear in the number of active ports, $N_{active}(t)$, which can be approximated by $N_{active}(t) \approx 2R(t)/\langle C_{port\_max}\rangle$, where $\langle C_{port\_max}\rangle$ is the mean maximum capacity of the ports and the two accounts for the fact that (3.24) includes both input and output ports.

We see that (3.24) has the same linear structure as (3.10), in which $P_{idle} + N_{total}\langle P_{sleep}\rangle$ is a constant and the remaining term is linearly dependent upon the element's traffic load.

3.7 TECHNICAL SUMMARY

A systematic approach to modelling telecommunications network power is to represent the network as consisting of "segments" as shown in Figure 3.1. With this approach, models can be developed for each segment with the resulting overall network power being the sum of the segmented network models.

With each segment, there are several types of equipment and technologies that may need to be modelled. For example, a typical access network will include several possible technologies including Fibre to the Home (FTTH), Fibre to the Node (FTTN) and Digital Subscriber Line (DSL) to name a few.

Within the edge and core networks, the range of technologies is not as broad thereby simplifying the modelling process.

If the modeller is seeking a very simple network power model, the model described in Section 3.2 may suffice. These constant power models assume no power dependence on the equipment load/throughput. Their advantage is simplicity in that they required a minimal amount of information. Also, some telecommunications equipment do have load-independent power consumption. Particularly in the access network.

Adopting a load-independent equipment power model, total access network power can be calculated using (3.5). For the edge network, the load-independent model power is given by (3.7) and for the core network (3.8). The total network power is then given by (3.9). The wireless component of this equation is the subject of the following three chapters.

For a load-dependent model, we can see that the linear form in (3.10) can be applied to almost all network equipment, including equipment that has sleep states. In all these cases, although (3.10) may not be precise, it is sufficiently accurate given the other approximations that will be applied to construct an estimate for network and service power consumption. A disadvantage of the load-dependent model is the need to collect more information on the equipment to be able to calculate the incremental energy per bit, E, in (3.10). Assuming that the equipment exhibits an approximately linear dependence on throughput load, $R(t)$, the value P_{idle} and E can be determined using several measurements and (3.12). This linear form can also be applied to equipment upgrades using (3.18). It can also be applied to sleep modes in equipment using (3.23) and (3.24).

For a "store and forward" router or switch, the incremental energy per bit, E, can be understood in as a combination of energy required for processing the packet headers (E_P) and storing the packet data while the header is being processed ($E_{S\&F}$) as shown in (3.21).

Using this linear form significantly simplifies power modelling in networks, which may include many hundreds, if not thousands, of network elements. This is primarily because the summation of the linear form over many devices, which will have different values for P_{idle}, E and C_{max}, will still be a linear function of overall network traffic. This is a major advantage of a linear model. Although it may not be perfect, it is sufficiently accurate so that, when applied to model networks, the results are meaningful and can be used for energy consumption estimation.

3.8 EXECUTIVE SUMMARY

To estimate the power consumption of a network, network "segmentation" provides a way of breaking up a large network into smaller segments. For example, a national network can be segmented into an access network, metro/edge network and a core network. Because the types of equipment used in each of these network segments are distinctly different, this approach provides an intuitive approach to developing an overall model.

Network segmentation can also be used in any network power modelling in which there are distinct groupings of equipment types.

There are several approaches to modelling network power consumption. As one would expect, the greater accuracy requires more information and hence a more complex model. For a simple rough approximation of network power, one can use simple constant power models that do not include the influence of the traffic throughput or load on the equipment. The advantage of simple constant power models is that they require a minimal amount of equipment and can provide an approximate "upper limit" to power consumption. Therefore, if one is seeking a simple "quick and rough" power consumption estimate of a network, a simple constant power model may be acceptable.

If the network power model is required to include the impact of traffic, then load-dependent equipment models will be required. The power consumption of most network equipment can be relatively accurately expressed as an idle power plus an incremental power that is linearly dependent on traffic throughput/load.

The linear load-dependent model can also be applied to network equipment as it is upgraded (which typically involves the addition of input/output ports and line cards) and the use of sleep modes in equipment.

REFERENCES

[1] J. Baliga *et al.*, "Energy consumption of Optical IP Networks," *Journal of Lightwave Technology*, vol. 27, no. 13, p. 2391, 2009.

[2] P. Mieghem, *Performance Analysis of Communications Networks and Systems*, Cambridge University Press, 2006.

[3] H. Mellah and B. Sanso, *"Routers vs Switches, How Much More Do They Really Consume? A Datasheet Analysis,"* in *2011 IEEE International Symposium on a World of Wireless, Mobile and Multimedia Networks*, 2011.

[4] ITU-T, *G.9700: Fast Access to Subscriber Terminals (G.fast) – Power Spectral Density Specification*, ITU-T, 2014.

[5] ITU-T, *G.9701: Fast Access to Subscriber Terminals (G.fast) – Physical Layer Specification*, ITU-T, 2014.

[6] U.S. Environmental Protection Agency, "EPA Report on Server and Data Center Energy Efficiency," August 2007.

[7] L. Niccolini *et al.*, *"Building a Power Proportional Software Router,"* in *Proceedings of the 2012 USENIX conference*, 2012.

[8] L. Barroso and U. Holze, "The Case for Energy-Proportional Computing," *IEEE Computer*, vol. 40, no. 12, p. 33, 2007.

[9] European Commission, "Code of Conducts for Broadband Communication Equipment," European Commission, 14 November 2016. [Online]. Available: https://ec.europa.eu/jrc/en/energy-efficiency/code-conduct/broadband. [Accessed 11 July 2020].

[10] R. Bolla *et al.* *"ECONET Deliverable D2.1: End User Requirements Technology Specifications and Benchmarking Methodologies,"* *ECONet Consortium* 2011

[11] A. Vishwanath *et al.*, "Modeling Energy Consumption in High-Capacity Routers and Switches," *Journal of Selected Areas in Communications*, vol. 32, no. 8, p. 1524, 2014.

[12] A. Vishwanath *et al.*, *"Estimating the Energy Consumption for Packet Processing, Storage and Switching in Optical IP Routers,"* in *Proceedings of OFC/NFOEC*, 2013.

[13] J. Gadze *et al.*, "Real Time Traffic Base Station Power Consumption Model for Telcos in Ghana," *International Journal of Computer Science and Telecommunications*, vol. 7, no. 5, p. 6, 2016.

[14] J. Lorincz et al., "Measurements and Modelling of Base Station Power Consumption under Real Traffic Loads," *Sensors*, vol. 12, p. 4281, 2012.

[15] A. Garcia-Saavedra *et al.*, *"Energy Consumption Anatomy of 802.11 Devices and its Implication on Modeling and Design,"* in *CoNEXT '12: Proceedings of the 8th International Conference on Emerging Networking Experiments and Technologies*, 2012.

[16] P. Serrano *et al.*, "Per-Frame Energy Consumption in 802.11 Devices and Its Implication on Modeling and Design," *IEEE/ACM Transactions on Networking*, vol. 23, no. 4, p. 1243, 2015.

[17] F. Jalali *et al.*, "Fog Computing May Help to Save Energy in Cloud Computing," *IEEE Journal on Selected Areas in Communications*, vol. 34, no. 5, p. 1728, 2016.

[18] A. Shehabi *et al.*, *United States Data Center Energy Usage Report*, Ernest Orlando Lawrence Berkeley National Laboratory, 2016.

[19] W. Van Heddeghem *et al.*, "Power Consumption Modeling in Optical Multilayer Networks," *Photonic Network Communications*, vol. 24, no. 2, p. 86, 2014.

[20] O. Tamm, "Eco-Sustainable System and Network Architectures for Future Transport Networks," *Bell Laboratories Technical Journal*, vol. 14, p. 311, 2011.

[21] R. Bolla *et al.*, "Energy Efficiency in the Future Internet: A Survey of Existing Approaches and Trends in Energy-Aware Fixed Network Infrastructures," *IEEE Communications Surveys and Tutorials*, vol. 13, no. 2, p. 223, 2011.

[22] N. Horowitz *et al.*, "Small Network Equipment Energy Consumption in U.S. Homes," *Natural Resources Defense Council (NRDC)*, June 2013.

[23] L. Valcarenghi *et al.*, *"How to Save Energy in Passive Optical Networks,"* in *13th International Conference on Transparent Optical Networks*, 2011.

[24] Z. Zhu, "Design of Energy-Saving Algorithms for Hybrid Fiber Coaxial Networks Based on the DOCSIS 3.0 Standard," *Journal of Optical Communications Networks*, vol. 4, no. 6, p. 449, 2012.

[25] E. Desurvire, *Wiley Survival Guide in Global Telecommunications: Broadband Access, Optical Components and Networks, and Cryptography*, John Wiley & Sons, 2004.

[26] J. Cioffi *et al.*, "Very-High-Speed Digital Subscriber Lines," *IEEE Communications Magazine*, vol. 37, no. 4, p. 72, 1999.

[27] P. Tsiaflakis, *"Green DSL: Energy-Efficient DSM,"* in *2009 IEEE International Conference on Communications*, 2009.

[28] K. Song *et al.*, "Dynamic Spectrum Management for Next-Generation DSL Systems," *IEEE Communications Magazine*, vol. 40, no. 10, p. 101, 2002.

[29] S. Chiaravalloti *et al.*, *Power consumption of WLAN network elements – TKN Technical Report TKN-11-002*, Technische Universitat Berlin, 2011.

[30] K. Hinton *et al.*, "Energy Consumption Modelling of Optical Networks," *Photonic Network Communications*, vol. 30, no. 1, p. 4, 2015.

[31] C. Gunaratne *et al.*, "Managing Energy Consumption Costs in Desktop PCs and LAN Switches with Proxying, Split TCP Connections, and Scaling of Link Speed," *International Journal of Network Management*, vol. 15, p. 297, 2005.

[32] K. Christensen *et al.*, "The next frontier for communications networks: power management," *Computer Communications*, vol. 27, p. 1758, 2004.

[33] S.-W. Wong *et al.*, *"Sleep Mode for Energy Saving PONs: Advantages and Drawbacks,"* in *IEEE Globecom Workshops*, 2009.

[34] I. Cerutti *et al.*, "Sleeping Link Selection for Energy-Efficient GMPLS Networks," *Journal of Lightwave Technology*, vol. 29, no. 15, p. 2292, 2011.

[35] R. Hirafuji *et al.*, *"Energy Efficiency Analysis of the Watchful Sleep Mode in Next-generation Passive Optical Networks,"* in *IEEE Symposium on Computers and Communication (ISCC)*, 2016.

[36] C. Lange *et al.*, "Energy Consumption of Telecommunication Networks and Related Improvement Options," *IEEE Journal of Selected Topics In Quantum Electronics*, vol. 17, no. 2, p. 285, 2011.

[37] Y. Zhang *et al.*, "Energy Optimization in IP-over-WDM Networks," *Optical Switching and Networking*, vol. 8, p. 171, 2011.

[38] F. Idzikowski *et al.*, "*Saving Energy in IP-Over-WDM Networks by Switching Off Line Cards in Low-demand Scenarios*," in *14th Conference on Optical Network Design and Modelling (ONDM)*, 2010.

4 Mobile Wireless

The power dependence of wireless access is somewhat different from that of wired access for several reasons. Firstly, wireless access is not a "guided media" system in that the signal from the base station is transmitted into the open atmosphere rather than into a wire or fibre. This means the "channel" that connects the transmitter to the end user is dramatically different for a wireless access network to that of a wired access network. To an extent, the signal within a fibre is effectively sealed off from its surrounding environment. Consequentially, provided the fibre is properly deployed, optical system designers need not worry about the introduction of noise or crosstalk to the signal from any aspect of the fibre's environment.

This is less so with copper cables. In this case, the designer must allow for crosstalk and other noise sources in DSL- and HFC-based access networks. A DSL service has its own wire pair, but receives interference from other wires in the cable. This is inherent, but can be partially compensated. Downstream channels in HFC networks are shared and can encounter interference, usually if the system is pushed too hard (into nonlinearity). However, wireless access stands apart from all guided media systems in that there are multiple base stations communicating with multiple other users in the area.

Unlike guided media systems, wireless signals naturally spread as they propagate. This has two important consequences: first, the spreading reduces the signal's power at the receiver much more dramatically than with guided media. Further, the frequencies used for wireless communications are subject to a diversity of additional losses, reflections and can display significant and rapid time variation on their loss profiles. This is especially so when a user is moving or has surroundings that are changing. Examples include a user in a moving vehicle or a standing in an area where there are vehicles and other objects moving around them.

The second consequence of wireless being an unguided media system is that in a mobile network, every user's receiver will experience interference due to signals transmitted from other base stations to service users in those base stations' coverage areas, or "cells". This interference is typically worse for users close to the edge of a cell or the boundary between two sectors of a cell. This interference is quantified by a "Signal to Interference and Noise Ratio" (SINR). This means that base stations must continually monitor the performance encountered by the user and, if need be, "hand-off" that user to another base station if it cannot maintain an acceptable performance. A diagrammatic representation of this is in Figure 4.1. This source of interference (from other base stations) is particularly troublesome in wireless access networks.

Users receiving poor signal quality will receive poorer download data rate and require more base station resources over a longer period to receive their service. This in turn means that delivery of the service consumes more energy. Consequentially, if we are to develop a "load-dependent" power model for a wireless base station, we need to consider models that can account for these circumstances when relating base station power consumption to user throughput (bit/s).

These factors mean that the construction of a mobile base station power, or energy efficiency, model is much more challenging than the relatively simple power models discussed above for wireline access networks. Unfortunately, the outcome is that even the "simple" mobile power models discussed below are rather complicated and may be a rather inexact representation of the operation of real-world mobile base stations.

For the wireless access network, the focus will be the wireless base stations that provide the wireless connection between the users' mobile devices (phone or tablet) and the network that interconnects the base stations. Further, we will focus on the downlink or downstream traffic flow. This is because the downlink from the base station to the user's mobile device dominates the power

DOI: 10.1201/9780429287817-4

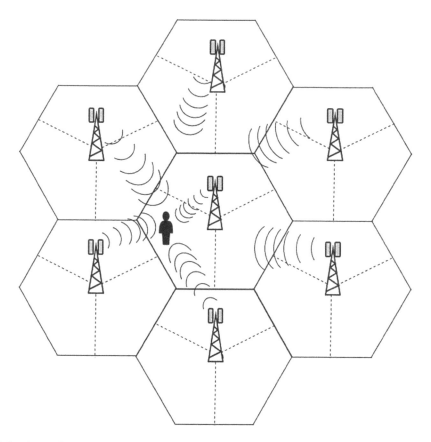

FIGURE 4.1   A user located in a service area that includes several mobile base station will receive signals from multiple antennas. These signals, although destined for other users, are the source of interference in the user's handset. This situation makes wireless access networks distinctly different to wireline access networks.

consumption of a mobile base station [1, 2]. The wireless access network does not include the users' devices.

We consider two types of wireless access technologies: mobile access networks and Wi-Fi access networks.

As with wireline access networks, in this chapter we will start by describing a "first-cut" model that does not include a dependence on utilisation. (A precise definition of the term "utilisation" will be developed below. For now, we can understand it to refer to the traffic load on the base station relative to its traffic maximum capacity.) We then construct a "static" mobile model that includes utilisation. The Static Model, however, does not include several key factors in the operation of a real mobile base station. An overview of an "Advanced" model that includes this greater detail is provided in the next Chapter. Such models are very sophisticated and require a substantial amount of computer programming beyond the Static Model described in this chapter.

The models presented in this book are not the only approach to modelling mobile networks. The models described below and in Chapters 5 and 6 could be colloquially described as "blunt instrument" models in that they are constructed in a direct physical "ground up" manner and do not use any advanced techniques such as signal processing and queueing theory. For an introduction to models based on more sophisticated mathematical techniques, the reader is referred to publications such as Bonald and Proutiere [3], Daigle [4], Chee-Hock and Boon-Hee [5] and Fowler *et al.* [6].

4.1 FIRST-CUT (LOAD-INDEPENDENT) POWER MODELS

4.1.1 MOBILE ACCESS NETWORK POWER MODEL

Mobile base stations are typically classified by intended coverage area as "macro", "micro", "pico" and "femto" base stations. In this context, a "macro" base station may cover an area of several square kilometres such as a metropolitan suburb or rural area or township. In contrast, a "micro" base station will cover tens to a few hundreds of square meters such as in a shopping centre or similar public area. Pico and femto base stations will cover even smaller areas and are typically deployed in areas where there the user density (per unit area) can be quite high. These small area base stations enable an acceptable service level for many users in a small area.

As would be expected, the maximum power consumption of a base station that covers a large area (macro) is much greater than that of a base station that covers a much smaller area. The specified power supply value can be used as a first-cut value to provide an upper limit to base station power consumption.

Many researchers use more detailed models such as in the EARTH project Deliverable D2.3 report "Energy efficiency analysis of the reference systems, areas of improvements and target breakdown" [1]. A "high level" power model of generic mobile base station is presented in Section 4.1 of this EARTH project report in which the power-consuming sub-systems of the base station are identified and their power consumption estimated. These sub-systems' power consumptions are then combined into an overall power consumption model. A similar, approach is described by Desset et al. [2], Deruyck, et al. [7, 8] Holtkamp et al. [9] and Lorincz and Matijevic [10], Liu et al. [11]. A block diagram of this model for a base station is shown in Figure 4.2.

As illustrated in Figure 4.2, a base station can have multiple transceivers and antennas. In particular, many base stations operate over the service area by splitting it into "sectors" surrounding the base station, with each sector covering an angular spread so that, together, they cover a full 360 degrees around the base station. A very common approach is to segment a station coverage area into three sectors each spanning approximately 120 degrees.

For modelling purposes, base station layout and coverage are split into adjacent hexagonal regions, with a base station at the centre of each hexagon as shown in Figure 4.1. In this case, each sector of a base station covers 120 degrees giving three sectors per base station. Over recent years

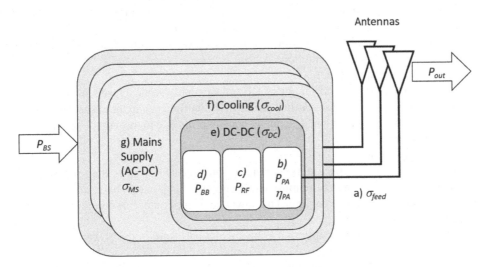

FIGURE 4.2 Base station sub-systems for base station power model. The labels a), b), etc. are referred to in the text to describe the function of each sub-system and how it contributes to overall base station power consumption.

consideration has been given to increasing the number of sectors to greater than three and coordinating transmission between sectors to improve energy efficiency. An example of this is "Large Scale Antenna Systems" [12, 13]. A power model for each transceiver is constructed and the overall base station power consumption is the sum of consumption across the transceivers. The power consumption of a sector of a base station in turn is the summation of the consumptions of its "components". These components are as follows:

(a) Antenna Interface

The antenna is represented as having a power loss that includes the feeder line to the antenna, band-pass filters, duplexers and matching components. For a macro station the feeder loss may be of order 50%, whereas for micro-base stations it is typically negligible. The feeder loss for new generation base stations is typically mitigated by introducing a "Remote Radio Head" (RRH) where the power amplifier is placed immediately next to the antenna, thereby removing the feeder line losses.

In the model, this component is represented by a loss factor, σ_{feed}, or efficiency, η_{feed} related by $\sigma_{feed} = 1 - \eta_{feed}$.

(b) Power Amplifier (PA)

In macro base stations, the design of the PA is a compromise between linearity (to minimise nonlinear distortions) and power consumption. Improving amplifier linearity typically incurs additional power consumption. In small (pico, femto) base stations, the base station power output is low and this trade-off is not so significant.

In the model, this power consumption is expressed in terms of the radio transmitter output power available to the base station antenna, P_{out}. This power is related to the PA power consumption, P_{PA} via an efficiency factor, η_{PA}: $P_{out} = P_{PA}\eta_{PA}$. (Note that in the case of systems where the transmitter PA is not adjacent to the antenna, the antenna radiated power is lower than P_{out} from the PA.)

(c) Radio Frequency small-signal transceiver (RF)

This component comprises the transmitter (for downlink communications to the user) and receiver (for uplink communications from the user). This component is represented by a power consumption, P_{RF}.

(d) Baseband interface including a receiver (uplink) and transmitter (downlink) sections (BB)

This interface performs digital processing of the signal including modulation/demodulation, channel estimation, equalisation, channel coding/decoding and more. This component also undertakes control of the station and media access control (MAC). This component is represented by a power consumption, P_{BB}.

(e) DC to DC power supply (DC)

This is required to provide a range of DC power supply voltage levels for the various electronic sub-systems in the base station. The intrinsic inefficiency in these supplies will incur an increase in power consumption. The power consumption due to this component is represented by a loss factor, σ_{DC}.

(f) Active cooling (cool)

In most cases, macro base stations require active cooling (rather than passive air flows). Smaller base stations may not require active cooling. Also, macro base stations that employ RRH will not require active cooling. The power consumption of this component is also represented by a loss factor, σ_{cool}.

(g) AC (mains) to DC power supply that connects the base station the power grid to power the station. (MS)

The inefficiency of the power supply that converts AC mains to DC for use in the base station also incurs an increase in overall consumption. This component is also represented by a loss factor, σ_{MS}.

The DC to DC power supply, cooling and AC to DC mains supply consumptions can be represented by loss factors because these represent a level of additional power consumption that is proportional to the power consumption of the other components.

Bringing all these factors together we get for a "first-cut" power model, the mains power input to the base station is as follows:

$$P_{BS} = N_{TRX} \frac{\dfrac{P_{out}}{\eta_{PA}\left(1-\sigma_{feed}\right)} + P_{RF} + P_{BB}}{\left(1-\sigma_{DC}\right)\left(1-\sigma_{MS}\right)\left(1-\sigma_{cool}\right)} \qquad (4.1)$$

where N_{TRX} is the number of transceivers (i.e. transmit/receive systems and antennas per base station).

4.1.2 BASE STATION TYPES

There is a much greater diversity of mobile base station types in mobile networks than for other types of access networks. A sparsely populated, low traffic area may be served by a single transmitter/ receiver and an omnidirectional antenna that covers the full 360 degrees around the base station. In areas with more traffic and a higher density of users, the base station would be equipped with a number of transmitter-receiver (TX-Rx) pairs each covering a "sector". Typically three sectors are deployed to 360 degrees, each with its own antenna(s) forming a directed radio signal beam, 120 degrees wide. There are no hard and fast rules for cell size. Rather it is a matter of adopting whatever is most functional. As an overall and very approximate characterisation "macro" base stations can cover a "cell" with a typical radius of 1 to as much as 10 km from the base station antenna [14, 15]. In yet higher traffic areas (urban and suburban areas) there will be many base stations, spaced to provide adequate coverage to users in the area, and with the beams oriented in such a way to minimise interference between base stations in areas of overlapping coverage. These "micro" base stations can have a cell radius ranging from 200 to 2000 m [14, 15]. In yet more population-dense areas, such as shopping centres, sporting grounds and entertainment complexes, many small cells may be used to cover the facility. In this case, the "pico" cells can have a radius from 10 to 200 m and may use omnidirectional antennas [15].

A mobile network covering a geographic region typically includes a number of base stations and supporting data switching equipment, to serve users in that region. In this chapter, we primarily focus on the energy consumption of the base stations themselves, whose power consumption dominates the overall network power.

Typical values for base station parameters for each base station type are given in Tables 4.1 and 4.2. This model for base station power consumption has been used by several researchers [14, 16, 17]. Note that some power contributions are directly quantified (P_{BB}, P_{RF}) and others are dependent on the transmitted power, P_{out}, via an efficiency coefficient (η_{PA}, σ_{feed}, σ_{DC}, σ_{MC}, σ_{cool}). It is important

TABLE 4.1

Example Values for the Parameters in (4.1) for a Single Antenna for Different Base Station Types Published in 2010 [1]

Base Station Type	N_{TRX}	η_{PA}	P_{BB} (W)	P_{RF} (W)	σ_{feed}	σ_{DC}	σ_{MS}	σ_{cool}
Macro	3	0.388	14.8	10.9	0.5	0.06	0.07	0.09
Macro (RRH)	3	0.388	14.8	10.9	0	0.06	0.07	0
Micro	1	0.285	13.6	5.4	0	0.064	0.072	0
Pico	1	0.080	1.5	0.7	0	0.080	0.10	0
Femto/Home	1	0.052	1.2	0.4	0	0.080	0.10	0

TABLE 4.2

Example Values for the Parameters in (4.1) for a Single Antenna for Different Base Station Types Published in 2013 [9]

Base Station Type	N_{TRX}	η_{PA}	P_{BB} (W)	P_{RF} (W)	σ_{feed}	σ_{DC}	σ_{MS}	σ_{cool}
Macro	3	0.36	29.4	12.9	0.5	0.075	0.09	0.1
Pico	1	0.08	4.0	1.2	0	0.09	0.11	0
Femto	1	0.05	2.5	0.6	0	0.09	0.11	0

to note that these parameter values can change significantly over technologies and the years as technologies develop [1]. Modellers should seek out the most relevant values for their model corresponding to the appropriate year.

For a small base station we have $\sigma_{feed} = \sigma_{cool} = 0$ giving:

$$P_{Small\_BS} = N_{TRX} \frac{\dfrac{P_{out}}{\eta_{PA}} + P_{RF} + P_{BB}}{\left(1 - \sigma_{DC}\right)\left(1 - \sigma_{MS}\right)} \tag{4.2}$$

The total power of a mobile wireless access network is the sum over the base stations in the network. Today, a mobile access network may include a combination of macro, micro and smaller base stations. Thus the total power consumption must account for these different configurations. Consequentially, we have for a mobile access network, the total power is given by:

$$P_{Mobile\_Access} = \sum_{j=1}^{N_{Mobile}} P_{BS,j} = \sum_{j=1}^{N_{Mobile}} N_{TRX,j} \frac{\dfrac{P_{out,j}}{\eta_{PA,j}\left(1 - \sigma_{feed,j}\right)} + P_{RF,j} + P_{BB,j}}{\left(1 - \sigma_{DC,j}\right)\left(1 - \sigma_{MS,j}\right)\left(1 - \sigma_{cool,j}\right)} \tag{4.3}$$

where N_{Mobile} is the total number of mobile base stations (of all sizes), $P_{BS,j}$ is the power consumption of the jth base station and the j-index represents the parameter values for the jth base station which may be a macro, micro, pico or femto base station.

This first-cut model does not account for any traffic load dependence on the base station power consumption. This will be discussed as we refine the model below.

4.2   WI-FI ACCESS NETWORKS POWER MODEL

Wi-Fi is a small-cell technology in that each Wi-Fi base station only services an area up to hundreds of square metres. Mobile macro-base stations can cover tens to hundreds of square kilometres. Because of its small-cell coverage, Wi-Fi power consumption is similar to wired access technologies because the open-air channels between the access point and the user are not as variable as with macro cell wireless technologies. Consequentially, some Wi-Fi access points show a relatively constant power profile [18], whereas the power consumption of other Wi-Fi access points has a non-trivial dependence on their through traffic [19, 20]. TKN Technical Report by Chiaravalloti *et al.* provides values for the idle receive and transmit powers for a wide variety of Wi-Fi equipment [21]. The paper by Serrano *et al.* demonstrates a linear dependence between traffic sending rate and power consumption [22].

A Wi-Fi access network consists of one or more Wi-Fi access points, each of which uses multi-antenna wireless technology. As a low-power small-cell technology, the PAs are is co-located next to the antennas. Wi-Fi access point maximum transmit power is dependent upon the regulatory rules for the location [23]. An example of local jurisdictional requirements is [24]. Being a low-power technology, Wi-Fi does not use active cooling. Therefore, one option for modelling Wi-Fi is to adopt the sub-system-based model introduced above for mobile base stations with both σ_{feed} and σ_{cool} set to zero giving (4.2) as a simple model for a Wi-Fi access point. In this case, the Wi-Fi access network power can be approximated using pico or femto cell values in Tables 4.1 or 4.2 which gives:

$$P_{WiFi\_Access} = \sum_{j=1}^{N_{WiFi}} P_{WiFi\_AP,j} = N_{TRX,j} \frac{\frac{P_{out,j}}{\eta_{PA,j}} + P_{RF,j} + P_{BB,j}}{\left(1-\sigma_{DC,j}\right)\left(1-\sigma_{MS,j}\right)} \tag{4.4}$$

where $N_{Wi\text{-}Fi}$ is the number of Wi-Fi access points and the j th index represents the parameter values for the j th access point. In many cases, it is likely the details of the sub-systems required for (4.4) will not be available to the modeller. In that case, it is even simpler is to adopt a single (measured) value for Wi-Fi access point power consumption [24, 25]. For 2020 technology systems, a constant power of around 2 W or less is typical. For example, the 2019 Code of Conduct on Energy Consumption of Broadband Equipment [26] lists "Idle-State" and "On-State" values for P_{WiFi} for a range of Wi-Fi equipment for 2020. In this document "Idle-State" for a Wi-Fi port is defined as [26]:

"Idle (link established and up, the interface is ready to transmit traffic but no user traffic is present)".

The "On-State" for a Wi-Fi port is defined as [26]:

"Active (link established and passing user traffic)".

In many cases, the power difference between idle and transmit modes is not significant [21]. Further, the definition of "On-State" in the Code of Conduct does not refer to utilisation, rather it only refers to "passing user traffic". Therefore, for the purposes of a "first-cut" power model, the modeller could use the "On-State" power recognising that this may result in an over-estimate of power consumption.

Enterprise and industrial Wi-Fi access points consume more power, up to around 10W [21]. Equipment combinations such as the inclusion of Wi-Fi with a VDSL or ADSL modem and router tend to flatten the power profile.

As wireless access evolves, there is consideration of unifying next generation mobile wireless and Wi-Fi in which the Wi-Fi component will cover small cells [24, 27]. For a network that includes both mobile and Wi-Fi access, the total wireless access network power is given by

$$P_{Wireless\_Access} = P_{Mobile\_Access} + P_{WiFi\_Access} \tag{4.5}$$

4.3 THROUGHPUT DEPENDENT POWER MODELS

Compared to the access network equipment discussed in Chapter 3, constructing a comprehensive power consumption model for mobile wireless access networks and equipment involves many more parameters. Consequentially, modelling the power consumption of a "real world" mobile access network is often iterative and not amenable to a closed-form or straightforward calculation. Network equipment manufacturers and network operators often use very complex and costly modelling software to assess their mobile network designs, which typically must be tailored to the geographic area to be covered by the network. Further, models that include both power and throughput, so as to calculate energy efficiency in the form of energy per bit, include sophisticated representations of the downlink traffic channels and data channel capacity. These tools are typically proprietary and well beyond the needs of expected readers of this book.

4.3.1 IDLE POWER AND DATA THROUGHPUT

Similar to other network equipment, the power consumption of each base station can be modelled as consisting of a constant "idle" component and a traffic-load-dependent component. When compared to most wireline network equipment, modern base station "idle" power consumption can be proportionately much less than its maximum power consumption when fully loaded [17, 28–30]. Most wireline access equipment has a relatively flat power-versus throughput profile (i.e. incremental energy per bit, $E \approx 0$). In contrast, macro-cell mobile base stations show a relatively strong correlation between power consumption and traffic throughput. Another difference is that, unlike the wireline network elements discussed in Chapter 3, a base station does not have a unique value for its maximum traffic capacity, C_{max}. The maximum bit-per-second capacity depends on many factors, principally the conditions encountered by the users' personal devices being served by the base station. For example, the maximum base station traffic capacity is dependent upon the amount of traffic being transmitted to other users by other base stations in the vicinity and local environmental conditions.

4.4 A STATIC MOBILE BASE STATION POWER MODEL

Because mobile networks are expensive to build and their performance can be highly dependent on the local geography in terms of coverage and throughput, many network operators use highly sophisticated, commercially available network simulators when designing and dimensioning new deployments or upgrades to existing deployments. These commercial simulators are well beyond the scope of this book and typically very expensive to acquire. The models discussed in this chapter are pretty much as simple as throughput-based models can be.

There are models published in the publicly available literature that are much simpler than the comprehensive proprietary models referred to above. Published models have adopted several different approaches for the three components of a base station model.

At a basic level, the development of a mobile base station model can be considered to have three components. The first component is the representation of the user traffic demand or service requests on the base station. In some simple models, the demand is set equal to the traffic download data rate (in bit/s). Such models do not intrinsically allow for the possibility of user demand being greater than the maximum download capacity, a situation that leads to congestion. Rather, the modeller may have to include or account for congestion manually. This is the approach adopted in the Static Model described in this chapter.

The second component is the modelling of downlink throughput. Downlink throughput is provided by allocating "radio resources" (or "Resource Elements" described in detail below) to satisfy user service requests. The number of resources allocated depends on the width of radio frequency (RF) spectrum available to the base station, and the number of users sharing those resources. The efficiency with which those resources can be used to transmit data is dependent on the user's location, and interference from other base stations creating interference as they service their users. This radio resources component thus relates the base station's "utilisation" (also described in detail below) of radio resources to provide downlink throughput from the base station to the user in bit/s. This relationship also depends upon a "link curve" which is also described in detail below. The allocation of resources also determines a base station's power consumption. The throughput model described below will also provide the reader with an understanding of how a mobile base station encodes user data onto the radio resources used to transmit data to users.

The third component is the model for the power consumption of the base station. A number of published power models express the base station power as a function of the "utilisation" of its "radio resources", which are often only vaguely defined [2, 31–33]. Such models have their use; however, to provide a quantitative value for mobile network energy efficiency, we need well-defined relationship between "utilisation" and data throughput. The power consumption model presented in this chapter is more detailed than some published models. The additional details in the model are included to provide the reader some insight into the relationship between base station utilisation and its radio resources.

The use of the term "simple" in describing the Static Model presented in this chapter is relative. Constructing a numerical simulation for base station power and throughput that provides a detailed, accurate representation of the operation of a base station but requires several thousands of lines of computer code. Such an "advanced" model has been developed by the authors, but is too long and complicated to be described in detail in this chapter. Therefore, a summary and flow chart of an advanced model is presented in Chapter 5. The Static Model described in this chapter side-steps much of the complexity of the advanced model. This Static Model should be adequate for a reasonable estimation of mobile base station power and energy efficiency values. A summary of the steps for constructing the Static Model are provided in Table 4.13.

4.4.1 OVERVIEW OF LONG-TERM EVOLUTION (LTE) MOBILE SYSTEMS

To assist with describing the fundamentals of constructing a mathematical model for base station throughput and power consumption, we will focus on the Frequency Division Duplex (FDD) "Long-Term Evolution" (LTE) wireless technology [34, 35]. LTE also can operate using Time Division Duplex (TDD) [34, 35]. LTE is a version of 4G mobile technology which is well established, and many of its technologies are included in 5G mobile networks now being developed and beginning to be deployed [34]. LTE was developed by 3GPP (the Third Generation Partnership Project consortium) [35] and although deployed since 2009, underwent several enhancements, leading to its being approved by the International Telecommunications Union (ITU-R) [36] as a standard for worldwide 4th-generation mobile networks in 2012. The ITU-R has agreed on a number of bands of frequencies, or channels, in which 4G (and now 5G) mobile systems can operate, although not all bands are available in all countries. All of these are in the Ultra High Frequency (UHF) frequency range.

This chapter is not designed to provide an extensive introduction to mobile network technologies and systems. There are many books, articles and industry white papers available to provide an overview or introduction to this technology [37–45]. These references range from an overview of mobile technology to a detailed background on the technology and evolution of LTE. Only a very cursory summary of details of this technology required to construct a power model will be provided here. The LTE-based model presented here applies to each sector of a base station. Here we assume three sectors per base station; however, there may be more or fewer in some deployments.

LTE was designed to deliver greatly increased transmission speeds over previous mobile systems, notably through more efficient and flexible use of the available frequency spectrum, and through the use of Multiple Input Multiple Output (MIMO) technology to increase the utilisation of spectrum and achieve more robust transmission.

A system that transmits a single data stream (or "layer) is referred to as a "Single Input Single Output" or SISO. Modern systems operate with multiple antennas transmitting 2 or 4 data streams (layers) requiring 2 or 4 antennas. This enables simultaneous transmission of multiple data streams to a user, or to multiple users within the coverage area. These systems are referred to as "Multiple Input Multiple Output" or MIMO [39, 40, 46–49].

Key to achieving higher downlink transmission is the implementation of Orthogonal Frequency Division Multiplexing Access (OFDMA) which enables flexible use and sharing of the available spectrum.

As implemented in LTE, OFDMA splits the Base Station allocated spectrum into blocks 200 kHz wide, with each block including 12 subcarriers. At the same time, the LTE signal is partitioned in time. The principal time partition is the Radio Frame, which has a duration of 10 ms. That is in turn divided into "sub-frames" of duration 1 msec, and "slots" of duration 0.5 ms. A block of

TABLE 4.3
LTE Downlink Physical Layer Parameters

Base Station Bandwidth (MHz)	1.4	3	5	10	15	20
Radio Frame duration (ms)				10		
Sub-frame duration (ms)				1		
Subcarrier spacing (kHz)				15		
Occupied subcarriers	72	180	300	600	900	1200
Number PRBs	6	15	25	50	75	100

12 subcarriers over a single slot duration forms a Resource Block (RB, often known as a Physical Resource Block PRB).

When allocating the transmission resources of a base station to its users, the minimum capacity that can be allocated is a pair of consecutive RBs, but the actual allocation is normally many times that. A base station with a 20 MHz spectrum allocation can thus apportion 100 RB pairs among its users each 1 ms interval, and can vary that allocation each successive ms.

As indicated, each 200 kHz frequency block includes 12 subcarriers spaced at 15 kHz intervals. There is a 10 kHz unused frequency space at the low-frequency end and another at the high-frequency end of that 200 kHz frequency block. These form a guard band between groups of subcarriers to avoid interference with adjacent blocks. Within each sub-frame, the carriers are modulated to form "symbols", with generally 14 symbols per sub-frame. One subcarrier spanning 1/14th of a sub-frame constitutes a Resource Element (RE) introduced earlier in this chapter. Thus in general, an RB (or PRB) carries 12x7 or 84REs. These REs carry all of the data and control signals to manage the LTE network.

Table 4.3 shows some of the downlink parameters for the LTE system bandwidths [47, 50–52].

4.5   FUNDAMENTALS OF MODELLING A MOBILE NETWORK

It is important to note that the model presented in this chapter is only one of multiple possible approaches to constructing a simplified model of a mobile base station. Each approach has its advantages and disadvantages and the modeller should review the literature before deciding on the model most appropriate for their requirements.

The basic classification of approaches to the components of a mobile base station model is listed in Table 4.4. All three components are required if we are to calculate mobile network energy efficiency in the form of energy per bit.

TABLE 4.4
An Elementary Classification of Common Approaches Taken to the Components of a Base Station Throughput and Power Model

Model Component	Common Approaches		
User/Traffic Demand	Utilisation Based	FTP Traffic Model	Full Buffer Model
Throughput	Shannon	Link curve	MCS mapping
Power	Utilisation based	3GPP Resource Element allocation	

Note: The details of each approach are in the text.

Some details of these model components of a base station include:

1. User/traffic demand-
   a. Full Buffer Model: This approach is presented in the ITU-R report "Guidelines for evaluation of radio interface technologies for IMT- Advanced" [53]. The demand is generated by placing a fixed number of users (typically 10 per sector [53]) at random locations in the service area of the base station. These users generate demand so that the base station is operating at 100% of its capacity all the time. This simplifies the calculation of signal interference between base stations, as well as data throughput. The simulation is run repeatedly with the users' locations changing for each run to get a statistically representative set of data [54–57].
   b. Utilisation based: This approach has similarities with the Full Buffer Model above in that it sets the utilisation of the base station. Whereas the Full Buffer Model has the base station operating at 100% of its capacity, this model provides for operation at utilisation of less than 100%. A mathematical definition of "utilisation" is given in (4.17). When modelling a mobile base station, the concept of utilisation is different to that when modelling wireline equipment. As discussed in detail below, a mobile base station services its users by broadcasting short duration pulses (called "Resource Elements") of RF power out to the users it is servicing. These pulses are encoded with user data. Each user is allocated a number of these pulses. The base station can simultaneously transmit up to a maximum number of these data-carrying Resource Elements. We define the "utilisation" as the ratio of broadcast user data carrying Resource Elements to the maximum number of these elements. In models that adopt a "Utilisation-based" approach to user demand, the modeller prescribes the value of the utilisation for each user and the base station [58–60]. Utilisation-based models typically do not involve repeated runs of the simulation to collect data.
   c. This approach is adopted in the Static Model described in this chapter.
   d. FTP Traffic Model: The "Utilisation-based" approach above does not properly reflect the actual operation of a base station. A closer representation of the real-world operation of a base station is that users in the service area make requests for a service at random intervals, albeit with a consistent longer term average demand. Further, the timing of any particular user's requests are typically uncorrelated with that of other users' requests. This enables the modeller to represent user requests using a Poisson distributed random process in time. User requests are taken to be a request to download data file of a prescribed size and users are typically assumed to be uniformly, but randomly distributed throughout the service area [1, 61]. This approach is used in the advanced model described in the next chapter. In this approach, the utilisation of the base station is determined by the time to download a file and the average rate at which user requests are made. If the user request rate is too large for the base station to accommodate all the requests, the base station can become congested. This approach enables the modeller to intrinsically include circumstances that lead to congestion [1, 12, 14, 62–64].
2. Power –
   a. Utilisation based: The simplest expression for base station power consumption includes typically an idle power, $P_{idle,BS}$, plus a contribution that is linearly dependent on the utilisation, μ_{BS}, having the form:

$$P_{BS}\left(\mu_{BS}\right) = P_{idle,BS} + \mu_{BS}\left(P_{max,BS} - P_{idle,BS}\right) \tag{4.6}$$

   where $P_{max,BS}$ is the base station power consumption at maximum utilisation, $\mu_{BS} = 1$.
   b. 3GPP Resource Element allocation: These models are based on the power consumption arising from the transmission of time and frequency Resource Elements (REs).

The model distinguishes between traffic-carrying REs and other REs used to monitor and manage the mobile network. For example, with the LTE (4G) wireless protocol (described in some detail below) downlink transmissions from the base station to the user consists of several "channels" only one of which carries user data. Although this data channel uses the most resources, it does not account for all 100% of the downlink traffic. Typically between 11% and 24% of the downlink traffic is "overhead" traffic used for management and monitoring of the network connections, the proportion depending upon the configuration of the base station (see for examples). Of the overhead, some is transmitted independent of the number of active users and some is transmitted dependent upon the number of users. The models based on this protocol follow the 3GPP standard for encoding information onto the downlink data channel and determine the downlink data rate based upon this standard [46, 60, 65–67].

3. Throughput-
    a. Shannon: Shannon's formula [48] quantifies the fundamental limits for the transmission of data as a function of channel capacity and noise, but full realisation of these limits is impractical. However modified versions of the Shannon formula can be used to calculate the achievable base station throughput, albeit without detailed consideration of base station downlink protocols [10, 68–72].
    b. Link curve: Models using a link curve relate SINR to spectral efficiency and hence link throughput. The link curve takes account of a range of other factors beyond the Shannon formula. Such curves commonly rely on the results of many very detailed simulations to provide a simplified performance correlation that relates the SINR to the throughput (in bit/s/Hz). Their use greatly reduces the computation effort involves in subsequent simulations. However, such conversion curves are usually proprietary [12, 14].
    c. Modulation and Coding Scheme (MCS) mapping: The MCS is a selection of a Modulation Format and Code Rate that is applied for downlink data transmission to users. Publications using this approach apply a table or mapping that relates the SINR at a user to a corresponding downlink data rate (in bit/s/Hz or bits/Resource Element) [54, 66, 73–76]. This approach ultimately relies on a modified Shannon formula or link curve (based on a set of assumptions) to provide the relationship between SINR and MCS level. This approach is used in the Static Model described in this chapter.

In the Static Model described in this chapter, the user demand in inserted by the modeller as described in approach 1)b. immediately above. In a model that includes multiple sectors of multiple base stations, this means the utilisation of each sector of each base station is set by the modeller. The base station throughput is essentially quantified by its utilisation, μ_{BS}, defined below. As with all communications networks, the quality of the received signal determines the data throughput, R. The quality of the received signal is quantified by the Signal to Noise Ratio (SNR). Typically, higher data rates are achieved by using a more efficient modulation and coding scheme, but require a higher SNR. To develop a model that provides a relationship between utilisation, μ_{BS}, and the base station data throughput, R, we need to account for the impact of any noise and interference power experienced by each user's receiver. This interference and noise power is quantified by the SINR at the user's receiver. We will now provide an overview of the various steps required to construct a base station power model.

4.6   MODELLING THE LTE NETWORK

From the discussion immediately above, in generating the data signal transmitted from the base station, the fundamental time step is a symbol duration, in the discussions below, when referring to the time, t, we will intrinsically assume this time corresponds to one of many sequential time steps. Therefore, when the time parameter, t, is used in an equation, it will be allocated a subscript indexing the time step that applies to that equation. Some of the entities defined in LTE provide different time scales. Entities such as symbol, PRB, sub-frame, TTI and Radio Frame each has an associated time scale. In each case, the appropriate time scale for that index will be provided.

The frequency span or bandwidth of a base station determines the number of PRBs it transmits in each time slot. For example, a 10 MHz bandwidth base station transmits 600 subcarriers in each time slot which corresponds to 50 PRBs in a time slot. (Recall, 12 subcarriers per PRB and each PRB is 200 kHz bandwidth.) A 20 MHz base station transmits 1200 subcarriers in each time slot corresponding to 100 PRBs in a time slot.

For a base station with bandwidth BW, the number of REs in a symbol duration, $N_{max,sd,RE}$ is:

$$N_{max,sd,RE} = 12BW / \Delta f \tag{4.7}$$

where Δf = spectral width of a PRB including guard band = 200 kHz for LTE and each PRB has 12 sub-carriers in a symbol duration. Using (4.7) for a 20 MHz base station this gives $12 \times 20 \times 10^6/200 \times 10^3 = 1200$ REs per symbol duration as stated above. For a 10 MHz base station this gives 600 REs per symbol duration.

For reasons that are discussed in detail in Section 4.6.4, with the Static Model described here we will use the average the power per RE, with the average taken over a Radio Frame. With 7 symbols per PRB and for 20 MHz bandwidth this corresponds to 16800 REs per 1 ms sub-frame (14 symbol durations) and 168000 REs per 10 ms Radio Frame. Similarly, for a 10 MHz base station, we get 8400 REs per 1ms sub-frame and 84000 REs per 10 ms Radio Frame. In general, the number of REs in a Radio Frame, $N_{max,RF,RE}$, given by:

$$N_{max,RF,RE} = 1680BW / \Delta f \tag{4.8}$$

The base station almost always transmits different numbers of REs to its various users in each symbol duration and sub-frame. The REs carry user data bits as well as a range of overhead data. The number of user data bits transported in an RE depends upon the modulation format and the "coding rate" used in that RE. For earlier versions of LTE, the modulation formats included QPSK, 16QAM or 64QAM. The latest release includes 256QAM. These modulation formats, QPSK, 16QAM, etc., determine the number of data bits that the RE can transport. For example, QPSK can carry 2 bits per RE, 16QAM can carry 4 bits per RE, 64QAM can carry 6 bits per RE and 256QAM can carry 8 bits per RE. A detailed introduction to these modulation formats is outside the scope of this book. For this, the reader is referred to references [39, 40, 44, 47, 50].

For any modulation format, one of several possible coding rates may apply. Blocks of data bits include both user data and sufficient extra data (or redundancy) to enable a level of error correction. The coding rate quantifies the amount of redundancy used in the RE to provide error correction of the data.

The choice of modulation and coding rate, and hence download data rate to a user, is dependent upon the SINR at the user's receiver. For a sufficiently high SINR, the REs will use 64QAM and minimal redundancy providing a higher data rate. With a low SINR, QPSK and a high redundancy code rate will be used, resulting in a lower data rate to the user. The amount of user data carried in an RE is quantified by the "Spectral Efficiency" which indicates the number of user data bit/s/Hz transported by that RE. The higher the spectral efficiency, the more user data is carried in an RE.

The appropriate modulation format and coding rate to reliably transmit data to a user results from a negotiation between the base station and the user's equipment. Using Reference Signal REs transmitted to the user's handset, the handset calculates a "Channel Quality Indicator" (CQI), estimating an MCS that would enable the highest throughput consistent with a downlink Block Error Rate of at most 10% [77]. This value is reported back to the base station. The base station then uses the reported value of CQI as well as its record of previous re-sends to the user to determine the Modulation Code Scheme (MCS) that is adopted for the user data carrying PDSCH REs for that user [46]. The user data and its MCS are then transmitted to the user device.

Since the transmission of each Resource Block requires RF power, various transmission schemes have been developed to maximise the overall efficiency of data transmission in PRBs. Examples of such schemes include Coordinated Multipoint (CoMP) and Large Scale Antenna Systems (LSAS), which aim to maximise users' SINR and hence minimises the number of PRBs (hence power) transmitted to satisfy user demand. It is beyond the scope of this text to provide an exposition of how technologies such as CoMP and LSAS schemes work. For further details of these schemes, interested readers could refer to [62, 78–80].

The PRB is often used as the basic unit in discussions on the physical layer of LTE [37, 42]. However, in line with the standards documentation for downlink power [77], the model presented here will use REs as the basic unit for transmitting data to users. In LTE multiple types of REs are transmitted in each PRB to provide a range of network signalling and management functions, some of which are transmitted irrespective of the traffic load. For example, some REs are used to enable users to initiate a call and to transfer calls between base stations. Others are required for network synchronisation with user equipment. These types of REs must be transmitted even when there is no user traffic; that is they are transmitted from the antenna irrespective of user traffic. Further not all REs are necessarily transmitted with the same power. PRBs can be transmitted in which some critical REs are transmitted at a higher power than others [44]. Finally most base stations have multiple transmit antennas covering each sector, and some RE are not transmitted during specified symbol durations to avoid interference with signals from other antennas.

In many cases, we are interested in both the energy efficiency of the transmissions to a specific user as well as the overall energy efficiency of all the transmissions from a base station. Therefore the model being developed here provides estimates for data download throughput and power for both individual users and all users. Constructing a model that tracks every RE transmitted to every user defeats the goal of simplicity, therefore we will use averaging the numbers of different RE types over a 10 ms LTE Radio Frame.

4.6.1 RESOURCE ELEMENTS AND DOWNLINK CHANNELS

The REs transmitted to the user are allocated to nine "downlink channels" each with a specific function for managing the connection and transferring data to the user. These channels are as follows:

1. "Reserved" The reserved REs are not transmitted in order to support the channel quality estimation process. When an antenna in a given sector of a base station is transmitting Reference Signals for the channel estimation process described above, adjacent antennae must not transmit REs in the same frequency band and symbol interval, otherwise they may corrupt the channel estimation process. For a four-port antenna system approximately 9.6% REs are reserved. As reserved REs are not transmitted, they do not contribute either to the base station power consumption or data throughout. This affects the overall spectral efficiency of a base station.
2. (RS): Reference Signals used by antenna ports for channel estimation.
3. (PCFICH): Control indicators in the Physical Control Format Indicator Channel.
4. (PBCH): Signals in the Physical Broadcast Channel.
5. (PSCH): Primary synchronisation signal in the Primary Synchronisation Channel (PSCH)
6. (SSCH): Secondary synchronisation signal in the Secondary Synchronisation Channel
7. (PHICH): ARQ (Automatic Repeat Request) signals in the Physical Hybrid ARQ Indicator Channel
8. Control information is sent in the Physical Downlink Control Channel (PDCCH). The allocation of REs to this channel is set by the "Control Format Indicator" (CFI = 1,2 or 3). Because

the control information is crucial to the operation of the downlink channels, the PDCCH is transmitted using QPSK modulation and its own Cyclic Redundancy Check (CRC) error checking to maximise its reliability.

9. User data is transported to the users in the Physical Downlink Shared Channel (PDSCH). This channel is shared across the active users, and its REs transport user data to users. The PDSCH channel can use one of several modulation formats (QPSK, 16QAM, 64QAM, 256QAM) and error correcting coding. The selection of modulation format and level of error correction is determined by the quality of the signal as reported by the receiver, and is principally based on SINR.

Depending upon the base station configuration, the number of REs used in each of these downlink channels varies. The details of the allocation of overhead REs is rather complicated and beyond the scope of discussion in this text [39, 45, 81]. Table 4.5 show examples of the number of REs allocated to the downlink channels in a Radio Frame for a 20 MHz base station. The number of REs of each downlink channel is not the same in every sub-frame. Also, the number of PDSCH REs used to download user data typically differs from sub-frame to sub-frame. As seen in Table 4.5, the number of REs in each downlink channel depends upon: Base station bandwidth, number of antenna ports (e.g. 1, 2 or 4) and the Control Format Indicator (1, 2 or 3).

The number of antenna ports depends on the number of antennas transmitting data to the user. MIMO operate with multiple antennas transmitting 2, 4 or 8 data streams (layers) requiring 2, 4 or 8 antennas. This enables simultaneous transmission of multiple data streams to a user, or to multiple users within the coverage area. The power consumption component of the Static Model being described here is applied to each layer. A discussion on MIMO is presented in Section 4.11.

The numbers and proportions of REs in each of the downlink channels are shown in Table 4.5 and Table 4.6, respectively, for 20 MHz LTE bandwidth. These numbers are totals or averages across the system bandwidth and over a Radio Frame. The majority of REs carry PDSCH (user data), with significant allocations to PDCCH, Reference Signals, or Reserved. However after accounting for the overheads, the REs available to the PDSCH channel amount to 70%–80% of theoretical capacity. This proportion becomes lower when the available bandwidth is reduced.

TABLE 4.5

The Number of REs Present in an LTE (10 ms) Radio Frame That Are Available for Transmitting Signalling Overheads and User Data for 20 MHz Base Station for Configurations with 1, 2 and 4 Antennas for CFI = 1, 2 and 3

Number of Ports		1			2			4		
CFI		1	2	3	1	2	3	1	2	3
Channel		REs	REs	REs	REs	REs	REs	REs	REs	REs
Index	Name									
1	Reserved	2076	2076	2076	8040	8040	8040	16040	16040	16040
2	RS	8000	8000	8000	8000	8000	8000	8000	8000	8000
3	PCFICH	160	160	160	160	160	160	160	160	160
4	PBCH	240	240	240	240	240	240	240	240	240
5	PSCH	124	124	124	124	124	124	124	124	124
6	SSCH	124	124	124	124	124	124	124	124	124
7	PHICH	1560	1560	1560	1560	1560	1560	1560	1560	1560
8	PDCCH	6280	18280	30280	6280	22280	34280	6280	14280	26280
9	PDSCH	149436	137436	125436	143472	127472	115472	135472	127472	115472
	Total	168000	168000	168000	168000	168000	168000	168000	168000	168000

Note: The values are for Ports 0 and 1.

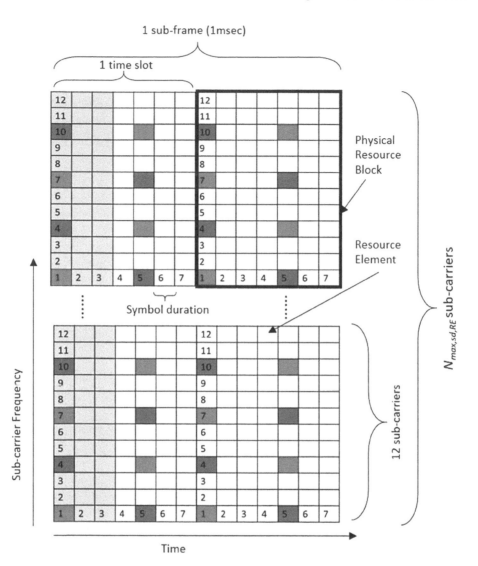

FIGURE 4.3   Structure of LTE downlink data flow for Port 0 of a 2 antenna base station with CFI = 3. The Reference Signal (RS) REs are indicated in red, the control REs (PDCCH) are indicated in light grey.(See below for details of these REs.) The REs in dark grey are not transmitted to avoid interference with Reference Signals transmitted from adjacent antennas in the base station. $N_{max,sd,RE}$ is given by (4.7) corresponding to the total number of REs transmitted in one symbol duration for a given base station bandwidth.

4.6.1.1   Example of the Impact of Overheads

To appreciate the impact of these overheads on base station throughput, we compare the theoretical maximum throughput of a base station assuming all REs in a Radio Frame are used for data download (i.e. no overheads) to that when the overheads accounted for. To make this comparison we consider an example where the download traffic stream SINR is high enough to use 64QAM with and the least amount of error correction coding [82]. This corresponds to a Channel Quality Index (CQI) of 15 (see Table 4.7). CQI of 15 transmits 5.55 bit/RE. The number of REs transmitted per 10 ms Radio Frame is 168000, giving the potential maximum throughput of the base station sector (with one antenna) as 5.55 bit/RE × 168000 (REs/Radio Frame) × 100 (Radio Frames/s) = 93.24 Mbit/s.

TABLE 4.6

The Percentage of REs Present in a LTE (10 ms) Radio Frame That Are Available for Transmitting Signalling Overheads and User Data for 20 MHz Base Station for Configurations with 1, 2 and 4 Antennas for CFI = 1, 2 and 3

Number of Ports		1			2			4		
CFI		1	2	3	1	2	3	1	2	3
Channel		% REs	% REs	% REs	% REs	% REs	% REs	% REs	% REs	% REs
Index	Name									
1	Reserved	1.24	1.24	1.24	4.79	4.79	4.79	9.55	9.55	9.55
2	RS	4.76	4.76	4.76	4.76	4.76	4.76	4.76	4.76	4.76
3	PCFICH	0.10	0.10	0.10	0.10	0.10	0.10	0.10	0.10	0.10
4	PBCH	0.14	0.14	0.14	0.14	0.14	0.14	0.14	0.14	0.14
5	PSCH	0.07	0.07	0.07	0.07	0.07	0.07	0.07	0.07	0.07
6	SSCH	0.07	0.07	0.07	0.07	0.07	0.07	0.07	0.07	0.07
7	PHICH	0.93	0.93	0.93	0.93	0.93	0.93	0.93	0.93	0.93
8	PDCCH	3.74	10.88	18.02	3.74	13.26	20.40	3.74	8.50	15.64
9	PDSCH	88.95	81.81	74.66	85.40	75.88	68.73	80.64	75.88	68.73
	Total	100	100	100	100	100	100	100	100	100

Note: The values are for Ports 0 and 1.

TABLE 4.7

Modulation Code Scheme (MCS) for LTE before Release 12

CQI Index	Modulation Format	Code Rate/1024	Efficiency (Bits/Symbol)	Spectral Efficiency (Bit/s/Hz)	SINR Threshold (dB)	SINR Threshold (dB) [54]	SINR Threshold (dB) [92]
1	QPSK	78	0.152	0.128	−6.91	−6.50	−6.936
2	QPSK	120	0.234	0.197	−5.077	−4.00	−5.14
3	QPSK	193	0.377	0.317	−3.106	−2.60	−3.18
4	QPSK	308	0.602	0.505	−1.19	−1.00	−1.253
5	QPSK	449	0.877	0.737	0.699	1.00	0.751
6	QPSK	602	1.176	0.988	2.643	3.00	2.699
7	16QAM	378	1.477	1.24	4.725	6.60	4.694
8	16QAM	490	1.914	1.608	6.558	10	6.525
9	16QAM	616	2.406	2.021	8.613	11.40	8.573
10	64QAM	466	2.73	2.294	10.39	11.80	10.366
11	64QAM	567	3.322	2.791	12.251	13.00	12.289
12	64QAM	666	3.902	3.278	14.167	13.80	14.173
13	64QAM	772	4.523	3.8	15.889	15.60	15.888
14	64QAM	873	5.115	4.297	17.833	16.80	17.814
15	64QAM	948	5.555	4.666	19.777	17.60	19.829

The modulation format and code rate are set by the SINR value. The spectral efficiency is determined by the Modulation Format and Code Rate. The three columns headed "SINR Threshold (dB)" are example values for the SINR threshold for each modulation format and code rate from three different sources [54, 92]. The first of the three SINR Threshold columns was derived by the authors. If the SINR value is greater the threshold value, the corresponding Modulation Format and Code Rate can be used.

An alternate approach is to calculate an "effective bandwidth" of the base station. This approach averages the reduction in throughput per Radio Frame (due to overhead) over the bandwidth of the base station.

With 5.55 bit/RE we get 466.20 bits/PRB (since there are 84 REs per PRB). After accounting for guard-bands, each PRB occupies a Resource Block of 200 kHz × 0.5 ms or 100 Hz-s. Hence its spectral efficiency is 466.20 bits/100 Hz-s or 4.662 bit/s/Hz. This gives a theoretical maximum throughput of a base station configured with 20 MHz bandwidth of 4.662 bit/s/Hz x 20 MHz = 93.24 Mbit/s (per sector), in agreement with the Radio Frame-based approach.

We now account for the overheads. Viewing and Table 4.6 we can see that the data-carrying PDSCH REs (Channel Index 9) make up 68% to 89% of total REs depending upon the CFI and number of transmitting antennas for 10 MHz and 20 MHz base stations. As an example, consider a 4 antenna, 20 MHz base station with CFI = 2, 76% of REs are user data carrying PDSCH REs. In this case with approximately 24% of REs in a Radio Frame being overhead and hence unavailable to carry user data. Keeping with CQI = 15, the spectral efficiency of the data-carrying REs remains the same. However only non-overhead REs are available for data transmission in each Radio Frame, therefore the number of REs available per Radio Frame is reduced by 24%. This gives the maximum throughput of the base station after accounting for overheads as 5.55 bits/RE × 0.76 × 168000 (REs/ Radio Frame) × 100 (Radio Frames/s) = 70.86Mbit/s. Note that this calculation is made for one base station sector, and applies to each of the (presumed) 4 antennas.

4.6.1.2 Spectral Efficiency, Modulation Coding Scheme (MCS) and Channel Quality Index (CQI)

The throughput of a base station (i.e. the amount of data it transmits to satisfy user demands) is set by the amount of data the base station can encode into the available PDSCH channel REs. This amount is quantified by the Spectral Efficiency of the Modulation Format and Coding Rate (also called the Modulation Coding Scheme or MCS) used to place data into the PDSCH REs. Although the other downlink channels are essential for base station operation, it is the PDSCH channel that transports data to users.

The relationship between SINR at a user's location and the resulting Spectral Efficiency for transmissions to a user is very complicated. The SINR is a ratio of the signal power to the combined received powers of interfering signals plus receiver noise. As such, the SINR does not account for any distortion or degradation internal to the signal channel. Degradation of the signal can take many forms, including issues well beyond just path loss. Examples include; broadband channel fading, multiple reflected signals causing selective fading over parts of the channel spectrum, additional losses due to stationary and moving obstacles in the vicinity, etc.

There is a variety of techniques available to overcome or mitigate these effects. However to model the relationship between the SINR at the user's receiver and the Spectral Efficiency accounting for all aspects of a user's local environment is a major task and well beyond the scope of this work. Such a model typically applies only to a specific network situation such as the geographical category, the motion of the user, multipath fading of the signal, specific details of the physical network layout and the user's surrounding environment. To avoid the complications of modelling the relationship between the SINR and Spectral Efficiency, in this book we shall use a "link curve" approach. Depending upon specific details of the model, a "link curve" will give a range of SINR to Spectral Efficiency profiles, examples of which are shown in Figure 4.4.

In its simplest form, the link curve relationship between the SINR and Spectral Efficiency can be based on the addition of "Additive White Gaussian Noise" (AWGN) to the desired signal. This allows for the use of Shannon's capacity formula [67]. However, with the use of technologies such as MIMO and inclusion of other signal degradation mechanisms, modifications of the Shannon formula can be applied [67] to develop a more realistic relationship between SINR and Spectral Efficiency. For example, LTE systems always include two transmitters and two receivers, which enable use of SIMO (Single In Multiple Out) and MIMO (Multiple In Multiple Out) techniques.

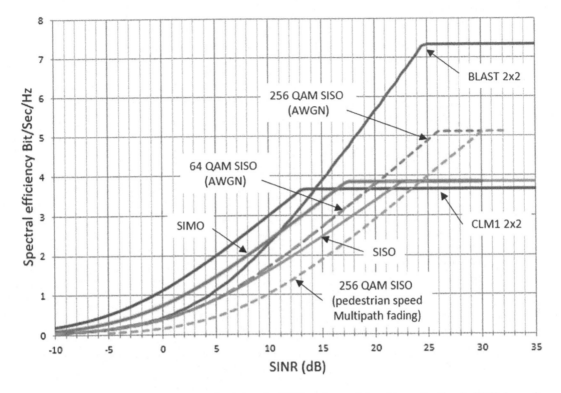

FIGURE 4.4 Examples of "link curves" relating the SINR at the user's hand set to the Spectral Efficiency for data downloading to the user. Depending upon the details included in the model and the technology used, the link curves provide a diversity of values.

MIMO sends the single data payload, split into two streams to the two receivers, whereas SIMO sends the same data payload via both transmitters [46, 50, 83]. The two SIMO receivers combine their respective detected signals to optimise performance.

The link curves shown in Figure 4.4 are labelled with the technology and model details [67, 84]. Morgensen *et al.* [67] introduce modified Shannon models to approximate the link curves derived by Wei *et al.* [84]. Beyond this, there are multiple publications that provide link curve models; [85–91] or the data needed to develop a simple AWGN channel model.

In particular, the link curves labelled BLAST and SIMO are modified versions of the Shannon model. (BLAST is a forerunner of the initial release of LTE, Release 8.) These curves were developed by Morgensen *et al.* [67] using the results of detailed link simulations by Wei *et al.* [84], which include several of the degradations mentioned above.

For the model described in this chapter, the link curve values are listed in Tables 4.7 and 4.8, which show an example of Modulation Format, Code Rates and Spectral Efficiency for the various CQI index values [77]. Also shown is an example of the SINR thresholds for each CQI index. The thresholds correspond to the SINR at which the simulation predicts a Block Error Ratio (BLER) of 0.1. If the BLER at a given MCS is greater than 0.1, then a lower MCS is used. All of these simulation results are for a link with a single transmitter and single receiver and for an Additive White Gaussian Noise (AWGN) channel model, but do include the error correction capability of the selected coding scheme (MCS). The SINR threshold values are not standardised, they may differ for different network models and different equipment types. For example, the mappings in Tables 4.7 and 4.8 are not included in Figure 4.4. The variation of threshold values across all the link curves in Figure 4.4, Tables 4.7 and 4.8 show the impact of adopting different approaches and approximations to determining the relationship between SINR and Spectral Efficiency.

TABLE 4.8

Modulation Code Scheme (MCS) for LTE Release 12 and Beyond

CQI Index	Modulation Format	Code Rate/1024	Efficiency (Bits/Symbol)	Spectral Efficiency (bit/s/Hz)	SINR Threshold (dB)
1	QPSK	78	0.152	0.128	−6.925
2	QPSK	193	0.377	0.317	−2.957
3	QPSK	449	0.877	0.737	0.859
4	16QAM	378	1.477	1.24	4.735
5	16QAM	490	1.914	1.608	6.552
6	16QAM	616	2.406	2.021	8.461
7	64QAM	466	2.73	2.294	10.368
8	64QAM	567	3.322	2.791	12.337
9	64QAM	666	3.902	3.278	14.184
10	64QAM	772	4.523	3.8	15.971
11	64QAM	873	5.115	4.297	17.849
12	256QAM	711	5.555	4.666	19.818
13	256QAM	797	6.227	5.23	21.514
14	256QAM	885	6.914	5.808	23.452
15	256QAM	948	7.406	6.221	25.451

The modulation format and code rate are set by the SINR value. The spectral efficiency is determined by the modulation format and code rate. [93]. The SINR Thresholds are based on "waterfall" diagrams for BLER such as in Figure 8 of Fuentes *et al.* [94].

Thus, if the modeller wishes to avoid the (considerable) task of constructing a model to generate their own link curve, they can seek to identify a link curve that corresponds closely to their situation and apply that to the SINR values derived by their model. This entails a greater degree of approximation than deriving a situation-specific link curve; however, it significantly reduces the work required to construct that relationship. References [67–91] above provide a variety of possible link curve options but is by no means exhaustive.

4.6.2 THROUGHPUT BASED POWER CONSUMPTION MODEL

Before undertaking a detailed description of how the relationship between downlink throughput and base station power consumption is modelled, we provide an overview as to the reasons why modelling the power and throughput of a mobile base station is much more challenging than wireline access network equipment.

Published measurements on some deployed base stations have shown that their throughput versus power consumption can be approximated by the linear form introduced in Chapter 3 used for wireline equipment. That is:

$$P(t) = P_{idle} + ER(t) \tag{4.9}$$

In this expression the time scale for t is typically much longer than a Radio Frame. Typically hours to days. However, these measurements show a significant scatter around the linear form [29, 30]. Deruyck, *et al.* [7] have shown the relationship between power per unit of coverage area and bit rate for a range of wireless technologies, which although increasing, is somewhat super-linear due to the interplay between the Signal to Noise Ratio (SNR) at the user location and the modulation format adopted for the connection. Looking at these results, we can see that, when considering mobile wireless access, there are circumstances in which the simple linear form (4.9) may need revision.

Power modelling of mobile base stations is rather complicated due to the fact that the base station downlink data capacity (i.e. total bit/s) must be shared across users whose individual circumstances (i.e. signal strength and interference power from other transmitters in their vicinity) may greatly differ. To address this, mobile base stations adopt physical layer protocols (which include the downlink channel MCS introduced in Section 4.6.1) that enable different modulation formats to be adopted depending upon the amount of interference noise experienced by the user's equipment. This means the base station must not only transmit user data, but must also exchange additional "overhead" data with the user equipment to enable the link between the two to be set up, monitored, modified as the user's circumstances change and possibly transferred to other portions of the base station's spectral range, or even to another base station if need be.

Further, the environment into which the base station transmits wireless signals can be very challenging and vary significantly from user to user. For example, with a user in a flat terrain and little vegetation, the transmission path between base station antenna and the user is relatively direct and can be modelled primarily as a loss. However, a user located in a central business district with many buildings, trees, reflecting surfaces and streetscapes typical of a central city presents a very complicated medium between the antenna and user, requiring a much more sophisticated transmission path model.

In addition a base station may be serving tens to hundreds of users (including some in motion), each experiencing differing circumstances. This means that a significant proportion of base station resources is consumed monitoring and accommodating the wireless path between each active user and the base station. Further, the base station also communicates with the hand-sets of inactive users so that, should new a call be required to be set up, it can be done with minimal delay. These resources consume power. Bringing all these factors together results in most detailed mobile base station power models being rather sophisticated, and constructed primarily as numerical models or total system simulations.

The static power model for a mobile base station described in this chapter is designed to enable a reader with access to an inexpensive mathematical software package to construct a model that will provide a "reasonable estimate" for base station power consumption. The model will also provide the reader with an introduction of the basics for constructing a mobile base station power model. Accurate power models require a significant amount of highly specialised information [95, 96] (e.g. detailed representation of the local terrain, building density, foliage type and density, individual base station antenna heights, user mobility, different user hand-set types, etc.) which are not included in the model described in this chapter. For readers who wish to consider developing a more sophisticated model, Chapter 5 provides an overview of such a model. However, it should be noted that almost all publicly available mobile base station models are somewhat idealised and well short of the commercially available and proprietary models.

User equipment (mobile handset or tablet) power consumption is not included in this model for several reasons. The sending of data to the user requires a high power transmitter that dominates the consumption of the base station, and of the whole mobile service [1, 2]. The high transmit power level simplifies the user's receiver. At the same time, uplink traffic from the user handset is sent at a much lower power level so as to preserve battery life, relying on the signal recovery through the use of multiple antennas and receivers at the base station. Finally, there is a very wide range of user handsets with consequent energy consumption that makes a handset model problematic. Therefore, this model focusses on the downlink communication from the base station to the user device. That is not to say that the power consumption modelling of uplink communication from the user to the base station is not important. It is important from the perspective of maximising the battery lifetime of the user's handset. However, this book's focus is on network power consumption and in this respect, it is the downlink that is dominant.

Unlike wireline access technologies, power consumption of a base station is principally dependent on the RF resources it uses to transmit data, rather than simply its bit/s throughput, which is why we reconsider the form in (4.9). With a mobile base station the throughput, $R(t)$, is determined by how efficiently data bits can be encoded by the base station onto the RF carriers used to transmit data to users, consistent with the ability of the users handset to accurately decode the transmitted data. That coding efficiency is determined by the quality of the radio link channels over which the data is transmitted. The quality of these channels is highly variable over time and location and differs from user to user. The channel quality is largely characterised by the SINR experienced at the receiver of the user's device.

4.6.2.1   Type A and Type B Resource Elements

This subsection is presented for completeness and may be omitted at first reading. The base station models presented in this work do not require details of the power values or the types of REs discussed in this subsection. However, it is highly likely the reader will come across these REs and so their relationship to base station transmit power is summarised here.

To enable determination of the SINR by the user equipment, the base station transmits Reference Signals, to which the user equipment responds. Reliable reception of the Reference Signal REs (RS REs) is essential for the successful operation of a mobile network. Therefore, the RS REs are modulated using QPSK, the most robust of the modulation formats used in LTE, and the standard allows for their transmit power can be enhanced relative to other REs. To do this, the 3GPP TS 36.213 V15.7.0 (2019–09) standard [77] provides for the network operator to specify transmit power value of the RS REs, which we denote as P_{RS} (W). This specification, referred to as the "Reference Signal Energy per RE" (RS-EPRE), is set via the "referenceSignalPower" parameter which is provided by higher layer network management parameter [77].

The value of RS RE power is set according to cell size, the higher the setting the larger the cell coverage on downlink [97]. The standard allows for the transmit power levels for all the other REs to be set relative to P_{RS}. Adopting this approach, we will also express the power of all REs as relative to P_{RS}.

The non-RS REs are categorised into Type A and Type B [44, 77]. The Type A REs are transmitted in symbol durations in which no RS REs are transmitted and can have their power adjusted relative to the power of the RS REs. This adjustment is given by a factor ρ_A that is defined in the 3GPP standard [77]. Type B REs are transmitted in symbol durations when the RS REs are transmitted. The power of Type B REs can also be adjusted relative to the RS RE power to ensure the RS REs are reliably detected [44], and allows higher power RS REs to be transmitted without increasing the overall total power transmitted in that symbol duration. This adjustment is given by a factor ρ_B also defined in the 3GPP standard [77]. Both ρ_A and ρ_B are user equipment specific.

The actual number and power allocated to the various channel REs is determined by a rather complicated set of factors specified in the standard [77]. These factors include; the number of antenna ports, the number of REs allocated to control channels and the base station bandwidth. The power of Type A and Type B REs are set by:

$$P_{TypeA,RE} = \rho_A P_{RS}$$

$$P_{TypeB,RE} = \rho_B P_{RS} = \frac{\rho_B}{\rho_A} P_{TypeA,RE}$$

(4.10)

Consequently, there are four possible transmit power levels for the REs within any symbol duration. These levels are as follows:

1. Reserved REs are not transmitted and hence have zero transmit power.
2. RS REs have a power set by the operator (see below).

3. Type A REs with power relative the RS REs set by the ρ_A parameter.

4. Type B REs with power relative to RS REs set by ρ_B parameter.

The 3GPP standard provides a choice of 8 values for ρ_A; -6dB, -4.77dB, -3dB, -1.77dB, 0dB, 1dB, 2dB, 3dB [77, 98]. With ρ_A set, the value of ρ_B is set by the ratio ρ_B/ρ_A. In the standard, the values ρ_A and ρ_B are specified using parameters P_A and P_B, respectively, defined in the standard [77]. The value of P_A sets ρ_A whereas the value of P_B sets the ratio ρ_B/ρ_A. For completeness, the possible values for this ratio are also specified in the standard [77] as listed in the Appendix in Table 4.14. These ranges of values for ρ_A and ρ_B/ρ_A result in the values for ρ_B given in the Appendix in Table 4.15 for a base station with a single antenna port, and Table 4.16 for a base station with two and four antenna ports. An example of relative RE power allocations is shown in Figure 4.5 for one specific base station configuration.

The overall goal is for the base station to adjust ρ_A and ρ_B to avoid power level variations of the signal at the user equipment receiver even when the PDSCH allocation is changed [44, 99]. The RS RE transmit power is fixed via the "referenceSignalPower" parameter, and the parameter P_B cannot be changed dynamically [44]. The PDSCH power depends upon the RE allocations which may change from sub-frame to sub-frame. Therefore to secure constant signal power at the user equipment receiver, the value of P_A can also from sub-frame to sub-frame [44].

To construct a transmit power model based upon the power contributions of REs in each downlink channel in each sub-frame it will need to include the power differences of Type A and Type B REs at the sub-frame level. This presents a significant problem. In particular, the modeller will need to know the values of ρ_A and ρ_B which typically change from sub-frame to sub-frame.

Even if we average over a Radio Frame to simplify the model, the values of ρ_A and ρ_B are also user equipment specific [77]. This means averaging of the ρ_A and ρ_B values over a Radio Frame which will require knowledge of numbers of each category of user equipment [40, 46] in the service

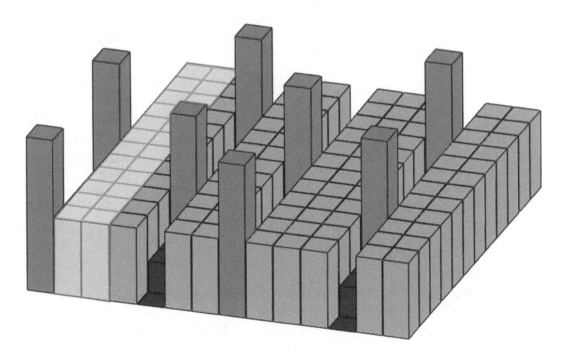

FIGURE 4.5 An example of the relative powers of RE types for the LTE PRB (downlink) for a 2 Antenna base station. The type of RE is indicated and the relative heights represent the relative powers. The Reserved REs have zero power because they are not transmitted.

area and their "transmission mode" [77]. This is clearly impractical and therefore the model presented here will not include this kind of detail.

4.6.3 UTILISATION AND POWER CONSUMPTION

A model based on calculating the transmit power in each symbol duration will have to deal with the variations in the number and power of PDSCH REs transmitted in each symbol duration as discussed in the preceding text. To avoid this complication, we adopt a different approach to modelling the power consumption of a base station by using the utilisation, as shown in (4.6), and averaging over a Radio Frame as described below. In the simple (load-independent) base station power model given in (4.1) the electrical power consumed by the base station, P_{BS}, is essentially dominated by a linear relationship with the radio base station antenna output (i.e. transmit) power, P_{out}, used for transmitting data to the user. Using this, some utilisation dependent models simply apply a linear relationship between the utilisation of RF resources used to transmit the data to users and the antenna output power P_{out} [2, 9, 100]. These models typically express the base station power in the form:

$$P_{BS}\left(\mu_{BS}\right) = N_{TRX}P_{idle} + \Delta_P P_{max,out} \sum_{p=1}^{N_{TRX}} \mu_{BS,p}$$

$$= P_{idle,BS} + N_{TRX}\mu_{BS}\Delta_P P_{max,out} \qquad (4.11)$$

where N_{TRX} is the number of transmit antennas on the base station, $P_{idle,BS}$ is the idle power consumption of each transmitter with zero traffic load (utilisation $\mu_{BS,p} = 0$), $P_{max,out}$ is the maximum antenna output power radiated by each antenna to users and Δ_P is a slope factor that relates the antenna output power to the base station incremental power consumption for each transmitter. Depending upon the technology, each sector within a cell may have multiple transmit/receive pairs of antennas. The total power consumption of the base station is the sum over all the N_{TRX} transmitters in the base station. All the transmitters are assumed to be identical thus $P_{max,out}$ is the same for each transmitter. The parameter $\mu_{BS,p}$ is the normalised load or utilisation of the p th transmitter ($0 \leq \mu_{BS,p} \leq 1$) and μ_{BS} is the utilisation averaged over all N_{TRX} transmitters

$$\mu_{BS} = \frac{1}{N_{TRX}} \sum_{p=1}^{N_{TRX}} \mu_{BS,p} \qquad (4.12)$$

As depicted in Figure 4.1, a base station independently serves several sectors within its coverage area, typically three sectors. In this model we assume three sectors for each base station and a single transmit antenna in each sector. This corresponds to a "Single Input Single Output" (SISO) system. There are also multiple transmit antenna technologies such as "Multiple Input Multiple Output" (MIMO). Irrespective of the technology, all transmitters must be included in the power consumption calculation.

Typical values for the parameters in (4.11) are given Table 4.9.

TABLE 4.9
Example Values for Parameters in (4.11) for Different Base Station Types Published in 2013 in Holtkamp, et al. [9]

Base Station Type	N_{TRX}	$P_{max,out}$ (W)	$P_{idle,BS}$ (W)	Δ_P (10 MHz)
Macro	3	40	460.4	4.2
Pico	1	0.25	17.4	4.0
Femto	1	0.10	12.0	4.0

Combining this with sub-component-based model in (4.1) and noting that the PA is the main influence on base station power consumption [100], we get for $P_{idle,BS}$ and Δ_P for each of the transmitters:

$$P_{idle,BS} = \frac{\dfrac{P_{idle,out}}{\eta_{PA}\left(1-\sigma_{feed}\right)}+P_{RF}+P_{BB}}{\left(1-\sigma_{DC}\right)\left(1-\sigma_{MS}\right)\left(1-\sigma_{cool}\right)}$$

$$\Delta_P = \frac{1}{\eta_{PA}\left(1-\sigma_{feed}\right)\left(1-\sigma_{DC}\right)\left(1-\sigma_{MS}\right)\left(1-\sigma_{cool}\right)}\frac{\left(P_{max,out}-P_{idle,out}\right)}{P_{max,out}} \tag{4.13}$$

In this equation, $P_{idle,out}$ is the antenna output power when the base station is not transmitting data to the users. In this model, this corresponds to the base station is transmitting all REs that are not PDSCH or PDCCH REs. Referring to Table 4.5 these REs correspond to RE Channel indices 2 to 7. (REs in Channel 1 are not transmitted.) PDSCH and PDCCH REs are transmitted when the base station is sending data to users. Typical values for the parameters in (4.13) are given in Table 4.1. As explained in detail below, even when there is no data to be transmitted to the users ($\mu_{BS} = 0$) the base station continues to transmit management and control signals.

Referring back to (4.6) we have:

$$P_{max,BS} - P_{idle,BS} = \Delta_P P_{max,out,BS}$$

$$= \frac{N_{TRX}\left(P_{max,out}-P_{idle,out}\right)}{\eta_{PA}\left(1-\sigma_{feed}\right)\left(1-\sigma_{DC}\right)\left(1-\sigma_{MS}\right)\left(1-\sigma_{cool}\right)} \tag{4.14}$$

The radio frequency power transmitted from a base station antenna is given by:

$$P_{out}\left(\mu_{BS}\right) = \mu_{BS}P_{max,out} \tag{4.15}$$

It is important to note the difference in the expressions for overall power consumption of the base staion, $P_{BS}(\mu_{BS})$ given in (4.6), and radio frequency power transmitted by the base station, $P_{out}(\mu_{BS})$ given in (4.15). Base station power consumption P_{BS} includes $P_{idle,BS}$ whereas bases station transmitted power, P_{out}, does not. This is because within the symbol durations when the base station is transmitting user data (i.e. PDSCH REs), its transmit power is overwhelmingly dominated by the PDSCH REs. Only a very small proportion of transmitted REs are overhead in those symbol durations. Therefore, to provide a correct expression for the SINR at the user, P_{out} is given by (4.15).

Although (4.11) to (4.15) provide a power consumption and transmission model in which the base station power is given as a function of its load or the utilisation of its radio resources, it does not provide a clear definition of utilisation, μ_{BS} nor does it provide a relationship between the base station throughput (bit/s) and its power consumption. Such a relationship is required if we wish to calculate the energy efficiency (energy per bit).

We will adopt the model given by (4.11) to (4.15). However, it is instructive to have a closer look at how the radio resources of a base station are utilised to transmit data to users and some details of the relative transmit power of REs of the downlink channels. This will also provide some preparation for the discussion of downlink data transmission.

4.6.3.1   Defining Utilisation $\mu_{BS,j}(t)$

As a first step, we need to refine the concept of "utilisation". As discussed in the sections above base station utilisation is an important parameter for calculating the power consumption of a base station. It also relates to the base station data throughput, $R(t)$, in bit/s enabling calculation of the base stations energy efficiency in Joules/bit.

The definition of utilisation adopted for wireline equipment in previous chapters is the ratio of current throughput, $R(t)$, to maximum possible throughput, C_{max}, both in bits per second. That is, wireline network element utilisation is defined by

$$\mu(t) = \frac{R(t)}{C_{max}} \qquad (4.16)$$

This definition means that when the network element is idle with effectively zero traffic user traffic, $R(t) = 0$, then $\mu(t) = 0$. Also, when the network element is operating at maximum throughput with no unused capacity, $R(t) = C_{max}$, then $\mu(t) = 1$.

The simple definition $\mu(t) = R(t)/C_{max}$ cannot be directly applied to mobile base stations because, as will be shown below, a base station does not have a fixed value of C_{max}. However, we still wish to construct a definition aligns with the notion that when base station utilisation is zero, $\mu_{BS}(t) = 0$, no user data is being throughput and when utilisation is unity, $\mu_{BS}(t) = 1$, the base station does not have any spare resources to transmit additional data to its users. The relevance of C_{max} will be explored in Section 4.7.

Because a mobile base station must keep track of all users in its service area, even when a user is not active, the base station exchanges information with that user's handset. This information exchange is essential for the network to provide an acceptable quality of service. The information exchanged with inactive users is minimal and is not the data a user seeks when active. Rather it is overhead data required for network monitoring and management. Consequentially we exclude this overhead data from our definition of "utilisation".

We base the definition of base station utilisation on the downlink channel that provides user data. That is, we use as a basis for "capacity" the tally of available PDSCH REs, which is the downlink channel that transports user data to users. Therefore, as a starting point, we define the utilisation of a base station as the ratio of number of PDSCH REs involved in transmitting user data to the maximum number of available PDSCH REs. That is:

$$\mu_{BS,p}(t) = \mu_{PDSCH,p}(t) = \frac{N_{PDSCH,p}(t)}{N_{max,PDSCH}} \qquad (4.17)$$

where the notation $\mu_{BS,p}$ refers to a single transmitter as given in (4.11) and (4.12), $N_{PDSCH,p}(t)$ is the number of PDSCH REs transmitted to users in sector p at time t and $N_{max,PDSCH}$ is the maximum number of available PDSCH REs. As discussed above, using the time scale of a symbol duration is impractical. Therefore the time scale represented by the parameter t is based on a Radio Frame as explained in Section 4.6.4.

In the following, we will undertake calculations on a "per transmitter" basis. Therefore, for MIMO and multi-antenna base stations, the power model is applied for each transmitter and summed over transmitters. For a three-sector base station with one antenna in each sector (SISO), we can include more detail for the p-index in (4.11). For SISO, the p-index covers the three sectors and the single antenna in each sector. Therefore, the power consumption of the m th base station in a mobile network will be:

$$P_{BS,m}(\mu_{BS,m}) = 3P_{idle,m} + \Delta_{P,m} P_{max,out,m} \sum_{p=1}^{3} \mu_{BS,m,p} \qquad (4.18)$$

where we have assumed all transmitters of the m th base station are identical except for their utilisation. For an MIMO system that has $N_{TRXSector}$ antennas in each sector the p-index in (4.11) covers each of the antennas in each sector. Therefore the power consumption of the m th base station will be

$$P_{BS,m}\left(\mu_{BS,m}\right) = 3N_{TRX,Sector}P_{idle,m} + \Delta_P P_{max,out,m} \sum_{p=1}^{3} \sum_{q=1}^{N_{TRX,Sector}} \mu_{BS,m,p,q} \qquad (4.19)$$

where $\mu_{BS,m,p,q}$ is the utilisation of the q th antenna in the p th sector of the m th base station. For example, 2x2 MIMO has $N_{TRX,Sector} = 2$ and 4x4 MIMO has $N_{TRX,Sector} = 4$.

Although the definition of $\mu_{BS,p}(t)$ in (4.17) aligns well with that in (4.16), there are several significant problems with (4.17) which we will now discuss. We also need to clarify the time scale for this definition.

4.6.4 Time Scales and Averaging

Although both utilisations are dimensionless, (4.16) is a quotient of instantaneous rates (bit/s), each having dimensions of (1/time). For wireline equipment the value of C_{max} is a constant over time and $R(t)$ will be constant in time for a fixed user traffic flow. In contrast, (4.17) is a ratio of dimensionless numbers of PDSCH REs and the values of $N_{max,PDSCH}$ and N_{PDSCH} depend on the time scale adopted for the simulation. For example, $N_{max,PDSCH}$ for a sub-frame is much less than that for a Radio Frame. (A Radio Frame consists of 10 contiguous sub-frames.) Consequentially, even if the traffic flow through the base station is constant, N_{PDSCH} and $N_{max,PDSCH}$ will both increase as the time scale under consideration increases from a sub-frame to a Radio Frame. It is important to keep in mind that these quantities are also set by the system bandwidth, which determines the number of subcarriers available

Next, on the time scale of one symbol duration, $N_{max,PDSCH}$ takes on different values depending upon the location of that symbol within its frame or sub-frame. For example, depending upon the CFI values, the first 1, 2 or 3 symbol durations of each sub-frame will be fully occupied by PDCCH and other overhead REs. The 6th and 7th symbol durations may contain PSCH and SSCH REs. This means, say for a 10 MHz base station, on the time scale of symbol durations, $N_{max,PDSCH}$ can take on values within the range zero to 600. These factors make the definition of utilisation as shown in (4.17) impractical.

There are several time scales involved in the LTE standard. From the summary of LTE in the sections above, there are three time scales of importance to constructing the model under consideration. These are as follows:

1. Symbol duration (71.43 microsecond): This is the useful symbol time duration of the REs in each sub-carrier frequency [37]. There are 14 symbol durations in each 1 ms sub-frame.
2. Sub-frame or Transmission Time Interval (TTI) (1 ms): When REs are allocated by the base station for transmission to its users, it does so on the basis of TTI's as the minimum unit of allocation.
3. Radio Frame (10 ms): This time scale corresponds to 10 sub-frames.

There is an additional time scale that is relevant to constructing a power model. This is the time scale of variations in the diurnal cycle. An example diurnal cycle is shown in Figure 4.6. Also shown is a stepwise approximation to the diurnal cycle. Stepwise approximations are often used to simplify the model [12, 14].

As noted just above, the value of $N_{max,PDSCH}$ varies dramatically at both the symbol and sub-frame time scales. However, for a given base station configuration (bandwidth, CFI and antenna count), the maximum number of PDSCH REs, $N_{max,RE,PDSCH}$ is constant from Radio Frame to Radio Frame. The values for $N_{max,RF,PDSCH}$ for a Radio Frame are listed for configurations of 10 MHz and 20 MHz base stations in Table 4.10. The value of $N_{max,RF,PDSCH}$ depends upon CFI value which can change from one sub-frame to the next [47, 101]. However, for higher bandwidth base stations (10 MHz and 20 MHz)

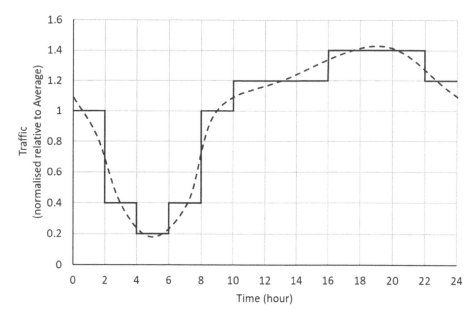

FIGURE 4.6 Example diurnal cycles for mobile network power modelling shown by the dashed line. This cycle is based upon a Dense Urban area with the traffic scaled relative to the mean traffic over the diurnal cycle. Also shown is a step-wise approximation of the measured cycle (solid line). The step-wise approximation can be used to simplify the time scales used in the model.

TABLE 4.10

Values for the Maximum Number of PDSCH REs in a Radio Frame, $N_{max,RF,PDSCH}$, for Given Mobile Base Station Configurations

10 MHz								
Antennas	**1**			**2**			**4**		
CFI	1	2	3	1	2	3	1	2	3
$N_{max,RF,PDSCH}$	74436	68436	62436	71472	65472	59472	67472	63472	58672
% Error	8.8	0.0	8.8	9.2	0.0	9.2	6.3	0.0	7.6

20 MHz								
Antennas	**1**			**2**			**4**		
CFI	1	2	3	1	2	3	1	2	3
$N_{max,RF,PDSCH}$	149436	137436	125436	139472	127472	115472	135472	127472	115472
% Error	8.7	0.0	8.7	9.4	0.0	9.4	6.3	0.0	9.4

The "% Error" shows the maximum percentage error incurred in the value of $N_{max,RF,PDSCH}$ by assuming CFI = 2 for PRBs in a Radio Frame.

adopting the values of CFI = 2 results in the variation of $N_{max,RF,PDSCH}$ away from the CFI = 2 value is 10% or less as shown in the "% Error" row in Table 4.10. Thus we shall adopt $N_{RF,PDSCH}$ for CFI = 2 as a representative average and set $N_{max,PDSCH}$ in (4.17) to be

$$N_{max,PDSCH} = N_{max,RF,PDSCH} \qquad (4.20)$$

where $N_{max,RF,PDSCH}$ is for a single transmitter in the base station. For a base station with multiple transmitters (multiple sectors and/or MIMO), we assume all transmitters are identical hence

$N_{max,RF,PDSCH}$ is the same for all transmitters in the base station. Adopting (4.20) will give a constant value for the denominator of (4.17) for a given base station bandwidth and antenna count.

With the denominator based on a Radio Frame, the numerator, $N_{PDSCH}(t)$, in (4.17) is the total number of PDSCH REs allocated in the Radio Frame centred on time t_u where the u indexes the Radio Frames. By "centred on the sub-frame at time t_u" we mean the values of the t_u correspond to the times of the centre of consecutive Radio Frames. Therefore, $t_{u+1} - t_u = 10$msec. Next, we index the sub-frames within a Radio Frame with index $j = 1,. 2,.. . 10$. With this, we set $N_{PDSCH}(t_u)$ in (4.17) to be $N_{RF,PDSCH}(t_u)$ which is given by

$$N_{PDSCH}\left(t_u\right) = N_{RF,PDSCH}\left(t_u\right) = \sum_{j=1}^{10} N_{sf,PDSCH}\left(t_u + \left(j-5\right)\delta t_{sf}\right) \tag{4.21}$$

where $N_{sf,PDSCH}(t_u)$ is the number of PDSCH REs in the sub-frame being transmitted at time t_u and δt_{sf} is the time duration of a sub-frame. Therefore, $N_{RF,PDSCH}(t_u)$ is the total number of PDSCH REs transmitted in the previous 5 sub-frames and the subsequent 5 sub-frames around time t_u. This is shown in Figure 4.7

With this we have effectively defined the utilisation as average utilisation over a Radio Frame. However, we note that because unallocated REs not transmitted [98, 102], power and throughput variations occur on the time scale of symbol durations. That is, we have averaged out the base station utilisation and throughput variations that may occur on the time scale of symbol durations.

With this, we have determined the tally of PDSCH REs available for carriage of user data for a given system bandwidth and over a Radio Frame accounting for all the REs allocated to overhead channels across the system bandwidth, in a Radio Frame. Utilisation then becomes the quotient of total PDSCH REs used to total available PDSCH REs.

(An important side note here: A well-documented disadvantage of the use of OFDM for LTE transmission is its tendency to have a high "Peak to Average Power Ratio" (PAPR) of the analogue signal that results from modulating the multiple independent sub-carriers used to transmit the REs [40, 47, 48, 103]. A large PAPR and the requirement for linearity places strong requirements on the PA that feeds the antenna. The static power model being developed here has its focus on the transmit power allocated to REs, not on how these powers combine across sub-carriers to give the overall instantaneous total overall transmit power.)

Therefore, we replace (4.17) with the following definition of base station utilisation:

$$\mu_{BS,p}\left(t_u\right) = \frac{N_{RF,PDSCH,p}\left(t_u\right)}{N_{max,RF,PDSCH}} \tag{4.22}$$

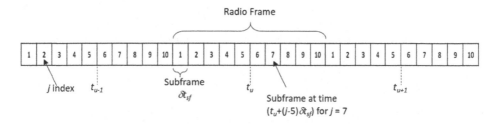

FIGURE 4.7 Schematic for calculating the number of allocated PDSCH REs in a Radio Frame as given by (4.21). For the sub-frame indicated ($j = 7$) $N_{sf,PDSCH}(t_u+(j-5)\delta t_{sf})$, is the number of allocated PDSCH REs in the sub-frame transmitted at time $t_u+2\delta t_{sf}$. The value of $N_{RF,PDSCH}(t_u)$ is the total number of allocated PDSCH REs in the sub-frame centred on the time t_u which is given by the summation in (4.21).

where t_u is the time index for the Radio Frames and the number of PDSCH REs is tallied over the bandwidth of the base station and a Radio Frame. Again, we remind the user that the presence of the p subscript indicates that this is utilisation per transmitter. For multi-transmitter technologies, these parameters are modelled per transmitter. With this understanding, we will drop the p index from now on unless it is explicitly required.

Adopting this approach means the utilisation, $\mu_{BS}(t_u)$, is effectively the running average of the utilisation over a Radio Frame. The duration of a Radio Frame being 10 msec is significantly less than the typical time periods over which we are interested in the power consumption of a base station and far less than the time scale of the diurnal cycle variations shown in Figure 4.6.

This brings to the fore an important aspect of the Static Model being described in this chapter and a difference between this model and the Advanced Models described in Chapters 5 and 6. The Static Model does not include any time dynamics in that there are no intrinsic time delays in the model's response to changes in demand traffic load, such as shown in Figure 4.6. In contrast, the Advanced Model includes some (but certainly not all) dynamics that occur in real mobile base stations.

Real mobile base stations operate with multiple feedback loops and system delays that introduce numerous dynamic response time scales. Modelling all these is somewhat impractical, particularly with a model as simplified as the Static Model described in this chapter and the Advanced Model described in the following chapters. In fact, the types of models presented in this book (and other similar published models [1, 12, 14, 58, 63]) are not expected to replicate the behaviour of real base stations. Rather, they are designed to provide reasonable guidance on issues such as power consumption and energy efficiency.

The lack of time dynamics in the Static Model means the model's response to a change in traffic is effectively instantaneous which is not representative of real mobile base stations. Consequently, the Static Model is best considered as essentially a "steady-state" model in which the number of users, their locations and base station utilisation are fixed and unchanging. What this means, in respect of the traffic diurnal cycle, is that the Static Model results are based on the assumption that the change in user demand and their environment is slow enough for the Static Model base station to have attained a steady state and the model's results correspond to that steady state.

This then raises the question of "To what does the time argument t_u correspond?" In the discussion above, t_u was defined as the time at the centre of the u th Radio Frame, as shown in Figure 4.7. Given the context described immediately above, the Static Model describes a period of "steady-state" operation in which the time t_u is located. The purpose t_u is therefore not to provide a time dependence of the model's output, rather it is to locate the model within the diurnal cycle. By this we mean the role of t_u is to provide the modeller a value user demand, $R(t)$, that corresponds to a time within the diurnal cycle.

This is also the case with the Advanced Model described in Chapters 5 and 6. The Advanced Model does include some time dynamics, represented by a Poisson process describing the occurrence of new user service requests. Despite this, the Advanced Model is also largely a steady-state model in that is assumes the user requests although arriving at random do follow a consistent average rate for the duration of the model's simulation. Referring to Figure 4.6, the reader will note a "step-wise" approximation to the diurnal cycle. This approximation is applied to provide values of the Poisson process describing the service request rate. The Advanced Model is then applied to describe the network's response, in the steady state, to a service request rate at the corresponding fixed value. The details of this are provided in Chapter 5.

4.6.5 OVERHEAD RESOURCE ELEMENTS

The idle output power, $P_{idle,out}$, in (4.11) is determined by the number and power consumption due to the overhead REs transmitted irrespective of base station utilisation. The incremental output power, $\mu_{BS}\Delta_P P_{max,out}$ in (4.11) is determined by $N_{RF,PDSCH}(t_u)$, the number of PDSCH REs transmitted in Radio Frame.

The downlink channel that carries user data is the PDSCH. The other channels carry overhead information to provide network management and monitoring functions. Some of the overhead channels must be transmitted irrespective of whether or not any user data is being downloaded. These are referred to as "Load Independent Overhead" REs. Other overhead REs, in particular the PDCCH REs, are required if user data being transmitted to active users [104]. These REs are referred to as "Load Dependent Overhead" RE's. The number of PDCCH REs transmitted is determined by a number factors and requirements for serving active users [55].

To include the overheads in the base station power model, we assume all the downlink channels apart from the PDCCH and PDSCH are Load Independent Overhead and therefore transmitted irrespective of base station traffic utilisation [102]. Thus these REs contribute to the idle power component, $P_{idle,out}$, of the transmitted power from the base station

The incremental transmitter output power, $\mu_{BS}\Delta_P P_{max,out}$, is determined by the number of transmitted PDSCH and PDCCH REs, which in turn depends on the amount of user data traffic transmitted to the users. As noted above, the number of PDCCH REs transmitted in a sub-frame is determined by the CFI value. In modern LTE systems, this value can be dynamically set for each sub-frame [105]. The number of PDCCH REs transmitted has a complicated relationship to the number of users, the type of user equipment, the signal quality at the users' handsets and the radio link technology being used by the base station [44, 105–107]. We define the utilisation of PDCCH REs in the same manner as the utilisation of PDSCH REs in (4.17) and (4.22). That is:

$$\mu_{PDCCH}\left(t_u\right) = \frac{N_{RF,PDCCH}\left(t_u\right)}{N_{max,RF,PDCCH}} \tag{4.23}$$

where $N_{RF,PDCCH}(t_u)$ is the number of PDCCH REs transmitted to users across the base station bandwidth in the Radio Frame centred on t_u and $N_{max,RF,PDCCH}$ is the average maximum number of PDCCH REs across the base station bandwidth in a Radio Frame.

In this Static Model we shall set the number of PDCCH REs transmitted, when averaged over a Radio Frame, so that the utilisation of PDCCH REs, $\mu_{PDCCH}(t_u)$ is equal to that of the of PDSCH REs, $\mu_{PDSCH}(t_u)$ which, from (4.17), is defined to be the utilisation of the base station, $\mu_{BS}(t_u)$. That is, we set $\mu_{PDCCH}(t_u)$ such that

$$\mu_{PDCCH}\left(t_u\right) = \mu_{PDSCH}\left(t_u\right) = \mu_{BS}\left(t_u\right) \tag{4.24}$$

Adopting the same approach of averaging over a Radio Frame as with PDSCH REs, from (4.23), the average number of PDCCH REs transmitted in a Radio Frame is:

$$N_{RF,PDCCH}\left(t_u\right) = \mu_{BS}\left(t_u\right) N_{max,RF,PDCCH} \tag{4.25}$$

This is a simplification of the real situation because although (4.25) is unlikely to be exactly correct, we would expect it to be a reasonable estimation. This is because the number of PDCCH REs sent to a user is dependent upon the signal quality of the transmission at the user's location. When the signal quality is high at a user's location, the number of PDSCH and PDCCH REs sent to that user (in order to transmit a given amount of data) is less than when the signal quality at the user's location is poor. This approximation has been used in advanced mobile network power modelling [12].

Using (4.24) and (4.25), the power model retains the form given in (4.11) to (4.14) for a base station.

4.7   RESOURCE ELEMENTS AND DATA THROUGHPUT

The sections above have provided an overview of LTE technology and base station power consumption. The purpose of this chapter is to construct a generic model that relates base station throughput to power consumption. We now bring together the factors discussed above to complete this endeavour.

To calculate the relationship between base station power and energy efficiency (W/bit/s), we also need to model the base station data throughput in bit/s. Therefore we now turn our attention to modelling the data transmitted from the base station to users. The base station allocates resources on the basis of pairs of PRBs (i.e. two contiguous PRBs) that correspond to the TTI (sub-frame) time scale [97, 108] and within that time scale it allocates REs within the pairs of PRBs [47]. The base station does not transmit unallocated PDSCH REs [98, 102]. Having adopted the Radio Frame as the time scale for modelling base station power consumption, we will retain this for modelling base station throughput.

Despite adopting the Radio Frame time scale, we need to start at the time scales of a TTI and symbol duration. During each TTI, the base station will allocate a number of PDSCH REs which may range from 0 (if the base station is idle and $\mu_{BS} = 0$) to its full complement of PDSCH REs within that TTI. Because the base station is transmitting to multiple users and the SINR inevitably varies for user to user, the modulation and coding rate will differ across the PDSCH REs within a symbol duration of the TTI. This means spectral efficiency will differ across transmitted PDSCH REs within the allocated pairs of PRBs.

To move our focus to a Radio Frame we assume the circumstances (e.g. user location, propagation loss) for each active user remains approximately constant over time scales much longer than a Radio Frame. Without this assumption, the model becomes extremely complicated, as noted above, the power variations from sub-frame to sub-frame need to be included and the definition of utilisation runs into difficulties.

To do a quick sanity check on this assumption, consider the following scenario: In one Radio Frame for a 20 MHz system with CFI = 2, there are 127472 PDSCH REs. With MCS 15, we have 5.55 bits per RE, therefore a Radio Frame will deliver 5.55 × 127472 = 707 kbit (≈ 71Mbit/s). But if it uses MCS 4, it will deliver only 0.6 × 127472 = 76 kbit (≈ 7.6Mbit/s). Therefore, with 10 simultaneous users each downloading a file size of around 1Mbit and equally sharing the PDSCH REs will require around 14 Radio Frames for MCS 15 and 130 Radio Frames for MCS 4. Based on this, the time duration we assume "circumstances" remain approximately constant over several 10's to possibly 100 Radio Frames each of 10ms which corresponds to around a second. This is a reasonable time duration based on our assumption the all users are stationary.

If the number of PDSCH REs transmitted in Radio Frame at time t_u is $N_{RF,PDSCH}(t_u)$, then the total number of user data bits transmitted in that Radio Frame is given by

$$B_{RF}\left(t_u\right) = \sum_{i=1}^{N_{RF,PDSCH}\left(t_u\right)} B_i\left(t_u\right) = \sum_{i=1}^{N_{RF,PDSCH}\left(t_u\right)} SE_i\left(t_u\right)\delta t_{sd}\delta f \qquad (4.26)$$

where $B_i(t_u)$ is the number of bits contained in the i th PDSCH RE transmitted in the u th Radio Frame ($i = 1,2,...N_{RF,PDSCH}(t_u)$), $SE_i(t_u)$ is the spectral efficiency the i th PDSCH RE transmitted in the u th Radio Frame $\delta t_{sd} = 71.4$ microseconds (1/14 ms = one 14th of a TTI) and $\delta f = \Delta f/12$ is the spectral width available for a modulated sub-carrier.

The number of bits given in (4.26) is for the data stream from a single antenna. This corresponds to Single Input Single Output (SISO) [40, 46] system in which the base station transmits user data from a single antenna. (Also referred to as a single "layer" system.) Most modern systems use a multiple layers (Multiple Input Multiple Output, MIMO) [39, 40, 46–49] in which the base station transmits multiple data streams (or "layers") [46, 48, 50, 109] to users from multiple antennas. In the case of MIMO, the data in each Radio Frame transmitted to the user will be given by (4.26). The user's handset applies algorithms to maximise the quality of the decoded signal from the multiple incoming data streams it receives from the base station. In this section, we focus on a single antenna which may be one of several antennas in an MIMO system. MIMO will be discussed in Section 4.11 and in Chapter 6. Because of the diversity in the number of transmitters per sector, the model presented below will relate to a single transmitter. The modeller will need to modify the model to account for MIMO or other multi-antennae systems.

Starting with (4.26), the total data download from a single transmitter, $B_{RF}(t_u)$, in the Radio Frame at t_u and can be re-written in the form:

$$B_{RF}\left(t_u\right) = \sum_{i=1}^{N_{RF,PDSCH}\left(t_u\right)} SE_i\left(t_u\right)\delta t_{sd}\delta f = N_{RF,PDSCH}\left(t_u\right)\left\langle SE\left(t_u\right)\right\rangle_{RF}\delta t_{sd}\delta f \qquad (4.27)$$

In this we have introduced the mean spectral efficiency, $\langle SE(t_u)\rangle_{RF}$, averaged over all PDSCH REs transmitted in the u th Radio Frame.

$$\left\langle SE\left(t_u\right)\right\rangle_{RF} = \frac{1}{N_{RF,PDSCH}\left(t_u\right)}\sum_{i=i}^{N_{RF,PDSCH}\left(t_u\right)} SE_i\left(t_u\right) \qquad (4.28)$$

We want to express the number of PDSCH REs transmitted in the Radio Frame, $N_{RF,PDSCH}(t_u)$, in terms of the utilisation, $\mu_{BS}(t_u)$, and total transmitter bandwidth, BW. To do this we use maximum number of available PDSCH REs in a Radio Frame, $N_{max,RF,PDSCH}$, and the number of REs in a Radio Frame, $N_{RF,RE}$. Also we introduce the duration of a Radio Frame, T_{RF}:

$$\begin{aligned} B_{RF}\left(t_u\right) &= N_{RF,PDSCH}\left(t_u\right)\left\langle SE\left(t_u\right)\right\rangle_{RF}\delta t_{sd}\delta f \\ &= \left\langle SE\left(t_u\right)\right\rangle_{RF}\mu_{BS}\left(t_u\right)\frac{N_{max,RF,PDSCH}}{N_{RF,RE}}N_{RF,RE}\delta t_{sd}\delta f \\ &= \left\langle SE\left(t_u\right)\right\rangle_{RF}\mu_{BS}\left(t_u\right)\sigma_{PDSCH}N_{max,RF,RE}\delta t_{sd}\delta f \\ &= \left\langle SE\left(t_u\right)\right\rangle_{RF}\mu_{BS}\left(t_u\right)\sigma_{PDSCH}T_{RF}BW \end{aligned} \qquad (4.29)$$

The utilisation $\mu_{BS}(t_u)$ is defined in (4.22), σ_{PDSCH} is the ratio $N_{max,RF,PDSCH}/N_{RF,RE}$ where $N_{RF,RE}$ is the total number of REs in the Radio Frame. The value of σ_{PDSCH} is given as a percentage with Channel Index 9 in Table 4.6 for a 20 MHz transmitter. $T_{RF} = 10$ ms, the duration of a Radio Frame and BW is the transmitter bandwidth (in Hz).

Using (4.26) the throughput (bit/s) $R(t_u) = B_{RF}(t_u)/T_{RF}$, of a single transmitter at time t_u is:

$$R\left(t_u\right) = B_{RF}\left(t_u\right)/T_{RF} = \mu_{BS}\left(t_u\right)SE\left(t_u\right)_{RF}\sigma_{PDSCH}BW = \mu_{BS}\left(t_u\right)C_1\left(t_u\right) \qquad (4.30)$$

where C_1 is the throughput when the utilisation $\mu_{BS}(t_u) = 1$:

$$C_1\left(t_u\right) = \left\langle SE\left(t_u\right)\right\rangle_{RF}\sigma_{PDSCH}BW \qquad (4.31)$$

The purpose of introducing $C_1(t_u)$ is to provide a comparison with (4.16), the definition of utilisation of wireline equipment. $C_1(t_u)$ also relates to the discussion in Section 4.6.1 relating to the impact of overheads on the maximum possible throughput of a transmitter. It is important to note that $\langle SE\rangle_{RF}$ is dependent upon the locations of the users being served by the antenna because, as discussed in Sections 4.7.1 and 4.7.2, their locations will determine the spectral efficiency values for the data transmitted to them.

Referring back to Section 4.6.1.1 in which the impact of overhead REs on the download data rate was shown and viewing (4.31) we see that the effective bandwidth in Section 4.6.1.1 was given by $\sigma_{PDSCH}BW$. That comparison used MCS15 which gives $\langle SE(t_u)\rangle_{RF} = 4.662$ bit/s/Hz and when combined with $BW = 20$ MHz and $\sigma_{PDSCH} = 0.76$, the throughput was 70.86 Mbit/s.

In relation to the definition in (4.16), $C_1(t_u)$ for a base station is a very different quantity to C_{max} for a wireline network element. The quantity $C_1(t_u)$ is defined over a Radio Frame that includes both frequency range (BW) and time duration (T_{RF}) but its value is not unique; it is contingent on many factors. In contrast, C_{max} for a wireline network element is a definite and instantaneous quantity.

Another important perspective $C_1(t_u)$ is the throughput a transmitter for maximum utilisation $(\mu_{BS} = 1)$ and is dependent upon all the factors that feed into $\langle SE(t_u)\rangle_{RF}$. Because the quantity $C_1(t_u)$ is dependent upon the average of the spectral efficiency over all transmitted PDSCH REs, it will be dependent upon the circumstances of the all users to whom the PDSCH REs are transmitted. Their circumstances will depend upon their location, environment and the activity of other users in their vicinity. Further, $C_1(t_u)$ will likely differ between each of the transmitters of a base station, and across base stations within a network. From this we see that the concept of "maximum throughput" for a mobile network is significantly more nuanced than that of a wireline network element.

Another detail the modeller must be aware of is that the equation for $R(t_u)$ given above is for a single layer (i.e. single transmitter and receiver). For multiple-layer systems (MIMO) there are multiple transmissions to a user and the user's equipment applies signal processing methods to each received signal to decide the data stream being provided to the user. These methods may include proprietary algorithms which are beyond the scope of the Static Model presented this chapter. A brief discussion of applying this Static Model to MIMO systems is provided below.

Reviewing the definition of $\mu_{BS}(t_u)$ in (4.17) and its appearance in (4.30) the transmitter utilisation can be expressed in several forms:

$$\mu_{BS}(t_u) = \frac{N_{RF,PDSCH}(t_u)}{N_{max,RF,PDSCH}} = \frac{R(t_u)}{\langle SE(t_u)\rangle_{RF} \, \sigma_{PDSCH}BW} \tag{4.32}$$

The downlink traffic $R(t_u)$ is in response to user demand. In turn, user demand is dependent upon the user density in the geographic area served by the base station and download requirements of the users.

4.7.1 SIGNAL TO INTERFERENCE AND NOISE RATIO (SINR)

The spectral efficiency values used in (4.28) to calculate the average $\langle SE(t_u)\rangle_{RF}$ are determined by the SINR at the receivers of the active users during the time duration of the Radio Frame at t_u. Therefore, to calculate the average spectral efficiency, we need the model to evaluate the SINR at the locations of the users being serviced at time t_u.

The SINR at a user location is dependent on many factors including the transmit signal power from the transmitter serving them, the propagation and other loss properties of the path from the base station to the user and the interference power from adjacent base stations as they serve their users. The interference power will depend upon the utilisation of those base stations. Consequentially the SINR at different locations within the service area will vary as conditions change with location and time over a diurnal cycle. This means each location will be characterised by its own SINR which may also vary over time and hence be described by a probability density at that location.

Deriving an accurate closed-form analytical expression for SINR at user locations around a base station is generally extremely very difficult if not impossible. Consequently, to accurately determine the SINR across a service area, situation-specific simulation models are almost always the preferred approach [1, 12, 95, 96]. There are also commercial simulators that typically include proprietary statistical models and algorithms, hence the details of their modelling methods are not publicly available. Commercial simulators are generally expensive and marketed to network operators rather than individuals.

To reduce the amount of computational power required to construct a power model for a mobile base station, a range of simplifying assumptions are needed. Such simplifications may include: the location of users and that their surrounding environment is static (unchanging) in time; users are assumed to be evenly distributed in the geographical area of interest; use of streamlined propagation loss model (simplified local terrain and environment models); the resource scheduling algorithm that allocates PDSCH REs to users assumes an equal share of the REs at each location when down-loading (i.e. no prioritisation of resources allocated); user downloads are assumed identical (all users download a single-file size) and much more. Changing any these assumptions can affect the calculated SINR probability density function at a user's locations.

We now describe a somewhat simplified model for base station data download to users. The model is based on a highly symmetric mobile network consisting of multiple hexagonal cells each serviced by a base station in the centre of the hexagon. Each hexagonal cell is split into three sectors each being serviced by a transmit/receive pair; these pairs being located in the base station. Such models are common, for example [7, 28, 32, 76]. The model is a single layer (i.e. single antenna port) SISO model. Therefore, each base station consists of three sets of transmitters/receivers, one set for each sector. The model can be applied to each layer of more advanced (MIMO) systems. Combining the multiple layers is discussed below.

To construct the model, we consider a user located in a sector of base station BS_0 as shown in Figure 4.8. We refer to this user as the "k th user" who is located as position $z^{(k)}$, $k = 1,2,...,N^{(usr)}(t_u)$. User k is one of many active users being serviced by BS_0 at time t_u. We wish to determine the SINR for user k. In Figure 4.8 we include transmit powers of the closest 18 other base stations, BS_1 to BS_{18}, in the calculation of the SINR at the location of the user k. A modeller may choose to include more or fewer base stations; however, a reasonable number of nearby stations must be included to provide a reasonably good approximation to the interference noise experienced by the k th user from other cells. We now provide an overview of calculating the powers incident on the k th user from all the base stations that are required to calculate the SINR.

The SINR at the location of user k, at time t_u of the u th Radio Frame, $SINR(z^{(k)},t_u)$, can be expressed as the received signal power for the user k from the chosen antenna, $P_{Rx,Sig}(z^{(k)},t_u)$ divided by the total power of signals from the other base stations serving other users and incident on the receiver of user k, $P_{interf}(z^{(k)},t_u)$, plus the users handset receiver noise, $P_{RxN}^{(k)}$. That is:

$$SINR\left(z^{(k)},t_u\right) = \frac{P_{Rx,Sig}\left(z^{(k)},t_u\right)}{P_{interf}\left(z^{(k)},t_u\right) + P_{RxN}^{(k)}} \qquad (4.33)$$

The receiver noise $P_{RxN}^{(k)}$ is determined by thermal noise power spectral density of -174dBm/Hz, the bandwidth of the received signal and a noise figure to account for imperfect receiver. Typical value for the receiver noise figure is 9dB. For the k th user, the proportion of PDSCH REs allocated to that user in a symbol duration, averaged over Radio Frame, is $\mu_{BS}^{(k)}(t_u)$. This corresponds to a receiver bandwidth of $\mu_{BS}^{(k)}(t_u)BW$ where BW is the total bandwidth of the base station. Therefore, for the k th user:

$$P_{RxN}^{(k)} = 0.001 \times 10^{-17.4}10^{0.9}\mu_{BS}^{(k)}\left(t_u\right)BW \ \left(\text{Watts}\right) \qquad (4.34)$$

For the full bandwidth of a 20 MHz system this gives $P_{RxN} = 6.32 \times 10^{-10}$W.

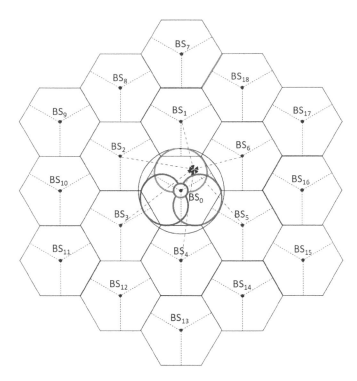

FIGURE 4.8 Mobile network physical layout for the simple simulation. The k th user is located in a sector of base station BS_0. The user's equipment will also receive signal power from the surrounding base stations which are transmitting to service their users. The user's equipment will determine the best transmitter from which to receive its signal. The signal from that base station will be the "signal power". The power received by the k th user's equipment from other transmitters constitute "interference power". The paths from BS_1 to BS_6 to the user are indicated by the dot-dash arrows. The paths for BS_7 to BS_{18} are not shown to avoid cluttering the diagram. Using these powers in (4.33) we can calculate the SINR for at the user's location. Also shown are the horizontal radiation pattern for the three sectors of BS_0. The pattern for sector 1 of BS_0 is a solid line. The corresponding patterns for sectors 2 and 3 are shown with dashed lines.

We expect that each base station is continually setting up and tearing down connections to their users as those users request or complete their data downloads. Consequentially, the interference power contributions from the various neighbouring base stations will vary across base stations and symbol durations. Therefore, keeping with the simplification adopted in (4.22) for the definition of utilisation, we also take P_{interf} as the average interfering power over a Radio Frame.

The received power (W) at location $z^{(k)}$ from the antenna for sector p of base station m can be expressed as follows:

$$P_{Rx,m,p}\left(z^{(k)},t_u\right) = P_{out}\left(t_u\right) G_{Tx} G_{Rx} \Pi_{m,p}\left(z^{(k)}\right) L\left(r_m^{(k)}, f, h_b, h_r, F, \upsilon\right) \qquad (4.35)$$

where $P_{out}(t_u)$ is the power provided to the antenna given by (4.15), G_{Tx} the base station transmission antenna gain factor and G_{Rx} the user handset receiver antenna gain factor. We assume these parameters are have the same value for all the base stations and handsets.

The term $\Pi_{m,p}(z^{(k)})$ describes the angular dependence of the transmission pattern from the base station antenna to the location $z^{(k)}$, given by:

$$\Pi_{m,p}\left(z^{(k)}\right)(dB) = \Pi_{m,p}\left(\theta,\varphi\right)(dB) = \max\left\{\Theta\left(\theta\right) + \Phi\left(\varphi\right), -25\right\} \qquad (4.36)$$

where the quantity $\Theta_{m,p}(z^{(k)})$ is the radiation pattern value for the angle subtended by the position $z^{(k)}$ relative to the horizontal antenna pattern peak of sector p of base station m and $\Phi_m(z^{(k)})$ the radiation pattern value for the position $z^{(k)}$ relative to the vertical radiation pattern peak of the m th base station. In some cases, it is more convenient to express the location vector z in radial coordinates, $z = (r, \theta)$ relative to the base station location and sector transmission pattern. In (4.36) θ and φ are the horizontal and vertical angles subtended by the user at $z^{(k)}$ relative to the transmitter in section p of base station m.

When calculating $\Pi_{m,p}$, it is important to note that both terms in the brackets {.} are negative. Therefore, taking the maximum corresponds to taking the value closest to zero dB. The specific forms for $\Theta_{m,p}(z^{(k)})$ and $\Phi_m(z^{(k)})$ are given in the Appendix.

The path loss function $L\left(r_m^{(k)}, f, h_b, h_r, F, \upsilon\right)$ is dependent upon the distance, $r_m^{(k)}$, from the m th base station antennas to the k th user and f, the frequency used to transmit user data. The height of the base station antenna above the ground is h_b and the height of the user equipment antenna above the ground is h_r. The parameter F accounts for the type of local environment; central business district, dense urban, suburban, rural, etc.

The parameter υ represents "fading loss" which covers a variety of process that impact the signal power and quality at the receiver [40, 48, 110]. To keep the model simple as practical, in this work we focus on "slow" or "shadow" fading and "fast" or "multipath" fading.

Shadow fading loss represents processes that cause significant fluctuations of the signal power. These processes include objects, with sizes greater than the carrier wavelength, which obstruct the signal propagation paths from the transmitter and the user device. For example, buildings and other large-scale terrain undulations. The shadowing object attenuates the transmitted signal and causes signal power variations relative to the path loss without shadow fading. Because of the arbitrary nature of the objects that may induce shadow fading, a statistical approach is usually adopted for modelling the effect of shadow fading. Models for shadow fading are frequently based upon measurement to determine the range of values and statistical variation. Most shadow fading models adopt a log-normal distribution for the resulting loss. Because users located in a given vicinity will have similar environments, shadow fading models often include a location correlation so that the shadow fading losses experience by two users is correlated relative to their distance apart [111]. The shadow fading loss may also include an additional penetration loss allowance to account for the possibility that a receiver may be indoors, or in a vehicle.

Multipath fading is a consequence of multiple reflections causing multiple copies of the transmitted signal to arrive at the user's receiver with small delays between them. The delays between these multiple copies can result in destructive interference causing the signal power at the user's receiver to be significantly reduced. Multipath fading be accounted for with a Rayleigh distributed fading margin, but is more typically accounted for in the link curve derived from a link-level simulation in the relevant faded channel, as discussed in Section 4.6.1.2.

Although the base station serving user k is also transmitting to the other users it is serving, it will manage those transmissions so that their signals do not interfere with its transmissions to the user k. Therefore, the interference power will be dominated by transmissions from other base stations to their users. In this model, we include transmissions from base stations surrounding the cell of BS_0, thus $N_{BS} = 18$. The interference power is the sum of the transmitted signal powers from all antennas, as received at the location of user k, other than the signal from the antenna serving user k.

The transmit power values used in this model are the instantaneous transmit powers averaged over a Radio Frame. Therefore, we calculate the interference noise in a symbol duration (i.e. "instantaneous power") averaged over Radio Frame assuming only PDSCH REs from other base stations contribute to the interference power. In reality this model is overly simplistic; the process of transmitting PDSCH REs is more complicated, involving transportation blocks of multiple PDSCH REs and network operator policies on how resources are distributed across users. However, to include these details will make the model too complicated, so we have adopted a "simplest available model" approach to provide a generic result.

We express the interference power using the average power in a symbol duration (with the average taken over that Radio Frame) in terms of the probability PDSCH REs transmitted by the signal base station antenna (sector 1 of BS_0, hence $m = 1$ and $p = 0$) suffering interference from PDSCH REs simultaneously received from other base station antennas (sectors 2 and 3 of BS_0 and all sectors of all other base stations). Therefore, with user k being served by sector 1 of BS_0, the expression for interference power, P_{interf}, has the form:

$$P_{interf}\left(z^{(k)}, t_u\right) = \sum_{m=1}^{N_{BS}} \sum_{p=1}^{3} \left(p_{interf,m,p}\left(t_u\right) P_{Rx,m,p}\left(z^{(k)}, t_u\right) \right.$$
$$\left. + p_{interf,0,2}\left(t_u\right) P_{Rx,0,2}\left(z^{(k)}, t_u\right) + p_{interf,0,3}\left(t_u\right) P_{Rx,0,3}\left(z^{(k)}, t_u\right) \right) \quad (4.37)$$

where $p_{interf,m,p}(t_u)$ is the probability PDSCH REs are transmitted from the p th sector antenna of the m th base station are also being used by BS_0 to service user k for over the Radio Frame centred on time t_u. That is, $p_{interf,m,p}(t_u)$ is the probability PDSCH REs being simultaneously used by both BS_0 and sector p of BS_m. This means those PDSCH RE transmissions from sector p of BS_m will contribute to interference noise for that PDSCH RE.

This is a point of divergence between the Full Buffer Model described in 1)a. of Section 4.4 and this utilisation-based model (see 1)b. of Section 4.4). The Full Buffer Model operates at 100% utilisation, hence every base station has $\mu_{BS,m,p}(t_u) = 1$ for all index values u. With 100% utilisation every signal PDSCH RE suffers interference from PDSCH REs simultaneously transmitted by all other base stations. In contrast, the utilisation base model has base stations operating at less than 100% utilisation and therefore not every signal PDSCH RE suffers interference noise power. To determine the SINR using a utilisation-based model requires calculating the probability of transmitted PDSCH REs being simultaneously used for both the user signal and other users' signals.

To calculate (4.37) we need $p_{interf,m,p}(t_u)$. To do this we consider the probability of allocating PDSCH REs by different transmitters into the same RE frequency/time slot. This probability will be determined by the operational policies of the network operator. Data allocated to users in whole PRBs and transmitted in transport blocks of multiple PRBs and the system uses frequency selective scheduling [39, 47]. Users are allocated PRBs according to the highest SINRs and a scheduling algorithm. For example, with round-robin scheduling the next user scheduled to be allocated PRBs is the user with the highest SINR, then if there are any leftover PRBs they are allocated to the next user again based on the user with the next highest SINR and so on.

Unfortunately, this depth of detail makes the model much too complicated. Therefore, to keep the model simple, we assume the selection of energised PDSCH REs by each base station can be approximated as distributed uniformly randomly across the total of $N_{PDSCH,max}$ PDSCH REs available to transmit user data during a Radio Frame. With this, if sector 1 of BS_0 has energised $N_{PDSCH}^{(k)}\left(t_u\right)$ PDSCH REs to service user k during the Radio Frame, then the probability an arbitrarily chosen, PDSCH RE is used to service user k is:

$$p_{BS,0,1}\left(t_u\right) = \frac{N_{PDSCH}^{(k)}\left(t_u\right)}{N_{PDSCH,max}} = \mu_{BS,0,1}^{(k)}\left(t_u\right) \quad (4.38)$$

where $\mu_{BS,0,1}^{(k)}\left(t_u\right)$, is the utilisation of sector 1 of BS_0 to service the k th user. Because we have assumed the energising of PDSCH REs is uniformly random (4.38) is constant across all the PDSCH REs in the Radio Frame. The propagation delay across the service area is much less than the symbol duration transmissions from other base stations, therefore the probability of an arbitrarily chosen PDSCH RE being energised in sector p of base station m is:

$$p_{BS,p,m}\left(t\right) = \frac{N_{PDSCH,p,m}\left(t\right)}{N_{PDSCH,max}} = \mu_{BS,p,m}\left(t\right) \quad (4.39)$$

The allocation of PDSCH REs within in a symbol period by different base stations is assumed to be independent. Therefore, using these results we can write for $p_{interf,m,p}(t)$:

$$p_{interf,m,p}\left(t_u\right) = p_{BS,0,1}^{(k)}\left(t_u\right)p_{BS,m,p}\left(t_u\right) = \mu_{BS,0,1}^{(k)}\left(t_u\right)\mu_{BS,m,p}\left(t_u\right) \qquad (4.40)$$

The transmit powers are per symbol duration, averaged over a Radio Frame. Therefore, the transmitted interference power per symbol duration, averaged over a Radio Frame is:

$$p_{interf,m,p}\left(t_u\right)P_{out,max} = \mu_{BS,0,1}^{(k)}\left(t\right)\mu_{BS,m,p}\left(t\right)P_{out,max} \qquad (4.41)$$

Using (4.35), (4.37) and (4.41) this the interference power in a symbol duration, averaged over a Radio Frame is:

$$P_{interf}\left(z^{(k)},t\right) = P_{out,max}G_{Tx}G_{Rx}\mu_{BS,0,1}^{(k)}\left(t\right)\left(\sum_{m=0}^{N_{BS}}\sum_{p=1}^{3}\Pi_{m,p}\left(z^{(k)}\right)\mu_{BS,m,p}\left(t\right)L\left(r_m^{(k)},f,h_b,h_r,F,\upsilon\right)\right.$$
$$\left. - \mu_{BS,0,1}^{(k)}\left(t\right)\Pi_{0,1}\left(z^{(k)}\right)L\left(r_0^{(k)},f,h_b,h_r,F,\upsilon\right)\right) \qquad (4.42)$$

In this we have summed over all transmitters and subtracted out the contribution of sector 1 of BS_0. This form applies to users located in sectors 2 and 3 of BS_0 with the appropriate changes to the indices.

An important detail that must be included in the model is that even if user k is located within the expected coverage area of BS_0, it may not be served by an antenna at that base station. That is, $P_{Rx,Sig}(z^{(k)},t_u)$ may not originate from the antenna that is geographically closest to the k th user. For example, viewing Figure 4.8 the horizontal radiation pattern for each sector of that base station is shown with solid and dashed lines around BS_0. We can see that the patterns have peak transmission at 120-degree separation around the circle of directions. For a user located close to the border between two sectors, the received power for those two sectors may be less than that from an antenna at the adjoining base station, which has its peak horizontal radiation pattern directed at the user. In this case, the k th user may be serviced by the adjoining cell base station. For example, in Figure 4.8 a user located at the border between the sectors with the purple and red horizontal radiation patterns may be served by BS_6.

This is accounted for comparing the SINR for all signals from all the antennas around the user and selecting the antenna for which the receiver SINR is greatest. This antenna then becomes the signal power transmitter, at the k th user's location, $z^{(k)}$ at time t_u, $P_{Rx,sig}(z^{(k)},t_u)$. That is:

$$P_{Rx,Sig}\left(z^{(k)},t_u\right) = P_{Rx,m',p'}\left(z^{(k)},t_u\right)$$

such that

$$\frac{P_{Rx,m',p'}\left(z^{(k)},t_u\right)}{P_{interf}\left(z^{(k)},t_u\right)+P_{RxN}^{(k)}} \geq \frac{P_{Rx,m,p}\left(z^{(k)},t_u\right)}{P_{interf}\left(z^{(k)},t_u\right)+P_{RxN}^{(k)}} \qquad (4.43)$$

for all $m = 0,1,2,..N_{BS}$ *and* $p = 1,2,3$

where the notation "Sig" represents the tuplet "m',p'" where m' and p' correspond to the base and sector antenna for which the SINR at the k th user is greatest. In other words, we identify as the signal power for the k th user $P_{Rx,Sig}(z^{(k)},t_u)$ which is set to be $P_{RX,m',p'}(z^{(k)},t_u)$ where m' and p' correspond to the base station and sector that transmit the signal that provides the highest SINR at $z^{(k)}$, the location of the k th user. We adopt this notation so that it is easily seen that $P_{RX,m',p'}(z^{(k)},t_u)$ is the signal power in the equations below.

To calculate the power at location $z^{(k)}$ at time t_u from sector p of base station BS_m, $P_{Rx,m,p}(z^{(k)},t)$, we require the transmit power from the base station and the path loss between the base station and the location of the user of interest. To keep the model simple, we assume that the physical configuration of the network being modelled is static and the only parameter that will be time dependent is the utilisation of the base station sectors, $\mu_{BS,m,p}(t_u)$. This means the all users are stationary and the local environment is also unchanging. As discussed in detail in Section 4.6.4, the time dependence t_u is introduced merely to indicate the different utilisations over a diurnal cycle. For any specific time, the utilisation will be assumed constant on time scales much longer than a Radio Frame.

4.7.2 PROPAGATION

As a radio signal propagates from the base station to the user's handset, it encounters a propagation loss represented by the $L\left(r_m^{(k)},f,h_b,h_r,F,\upsilon\right)$ term in (4.35). The signal power at the receiver needs to be sufficiently greater than the inherent receiver noise, plus any interference noise from other sources, to achieve an acceptably low error rate of the received signal. In areas covered by more than one base station, the user's receiver will receive signals from a dominant base station, usually the closest, plus weaker signals from a number of other, more distant, base stations. These lesser signals constitute interference at the receiver, degrading its performance. The SINR, given by the signal power divided by the sum of interference power and internal receiver noise power, determines data rates that can be achieved by the receiver to download user data.

In contrast to wired access network technologies, the power transmitted to users in a mobile network is not confined to a guided medium (e.g. copper wire, coaxial cable, optical fibre). Rather it radiates into free space. This means the power at the user's handset is only a small fraction of the transmitted radio power. There are many models used to determine the signal propagation loss. The simplest is the classic inverse-square law for (zero attenuation) free-space transmission, which represents the reduction in power per unit area as the signal diverges as it propagates away from the transmitter:

$$L_{free\_space} = \left(\frac{4\pi rf}{c}\right)^2$$

$$L_{free\_space}(dB) = 20\left(\log(r) + \log(f)\right) + 32.44$$

(4.44)

where r is the distance from the antenna in metres, f is the frequency and c is the speed of light in free space in metres/s. While well understood, the propagation loss calculated using (4.44) is invariably unrealistically optimistic. In practice, mobile radio propagation rarely satisfies the requirement for free space propagation and is affected by many factors including reflections from buildings and from the ground, by attenuation due to local foliage, antenna heights, and a number of other factors. The simple form in (4.44) has been replaced by a diverse range of models developed for various situations. These loss formulae are generally empirically determined from very extensive measurement programmes in various types of locations, after curve fitting an equation to provide their final form. Many of these have been published by 3GPP and ITU-R among others, for use in modelling to compare the performance of different technology developments as well as being used for more general assessments [48, 51, 112–119]. These models typically share a generic form:

$$L\left(r,f,h_b,h_r,F,\upsilon\right)dB = -\left(A\left(f\right)+B\left(f\right)\log\left(\frac{r}{r_{norm}}\right)+C\left(h_b,h_r\right)+F+\upsilon\right) \qquad (4.45)$$

where A, B, C, D, $r_{norm} > 0$. Parameter A is dependent on the signal frequency, f, B is the loss coefficient over distance, r, from the base station to the user (note that B may be dependent on f and possibly other parameters) and r_{norm} is a normalising distance parameter. Parameter $C(h_b,h_r)$ encompasses factors such as the vertical height of the base station antenna above ground level, h_b, and the height above ground of the user hand-set, h_r. The factor F includes factors to account for the type of local environment; e.g. central business district, dense urban, suburban, rural, etc. The factor υ represents a statistical "shadow or fading loss" that is modelled by a random parameter that follows a log-normal distribution with set mean and variance. It is also common to include an additional penetration loss allowance, recognising the possibility that a receiver may be indoors, or in a vehicle, etc. The forms for A through F may vary depending upon the environment of the area being serviced by the base station. In natural units (4.45) has the form:

$$10^{0.1L(r,f,h_b,h_r,F,\upsilon)} = H\left(f,h_b,h_r,F,\upsilon\right)r^{-B(f)} \qquad (4.46)$$

To simplify the model, for a given local environment and base station we will assume h_b, h_r and F are fixed and we will assume the propagation loss is dependent only on distance, r, and carrier frequency, f. An additional path loss component arises because the radiation pattern of the base station antennas is directional, shaped in both horizontal and vertical planes to best serve their coverage areas. There will be loss components relating to the offset angles between the antenna's directivity patterns (horizontal and vertical) and the direction of the user from the base station (θ and φ, respectively).

For the model described here, we will use the "SUI" propagation model developed by Stanford University, which has the advantage of being applicable to some of the evolving 5G systems being planned. Details of this model are provided in the Appendix to this chapter. The reader should select a model that is most suited for their purposes.

Using form of (4.42) and subscript $_{Sig}$ to represent the quantities related to the signal transmitted to the k th user gives for the interference power

$$P_{interf}\left(z^{(k)},t_u\right) =$$

$$P_{out,max}G_{Tx}G_{Rx}\mu^{(k)}_{BS,Sig}\left(t_u\right)\left(\sum_{m=0}^{N_{BS}}\sum_{p=1}^{3}\mu_{BS,m,p}\left(t_u\right)\Pi_{m,p}\left(z^{(k)}\right)H\left(f,h_b,h_r,F,\upsilon\right)\left(r^{(k)}_m\right)^{-B}\right. \qquad (4.47)$$

$$\left. -\mu^{(k)}_{BS,Sig}\left(t_u\right)\Pi_{Sig}\left(z^{(k)}\right)H_{Sig}\left(f,h_b,h_r,F,\upsilon\right)\left(r^{(k)}_{Sig}\right)^{-B}\right)$$

where for sector p of BS_m, $\Pi_{m,p}(z^{(k)})$ is the value of the antenna's gain pattern in the direction towards user k.

The signal power in (4.33) and (4.43) for user k, $P_{Rx,Sig}(z^{(k)},t)$ is given by

$$P_{Rx,Sig}\left(z^{(k)},t_u\right)$$
$$= \mu^{(k)}_{BS,Sig}\left(t_u\right)P_{out,max}\Pi_{Sig}\left(z^{(k)}\right)H_{Sig}\left(f,h_b,h_r,F,\upsilon\right)\left(r^{(k)}_{Sig}\right)^{-B_i} \qquad (4.48)$$

where $\Pi_{Sig}(z^{(k)}) = \Pi_{m',p'}(z^{(k)})$, $\mu^{(k)}_{BS,Sig}(t_u) = \mu^{(k)}_{BS,m',p'}(t_u)$, $H_{Sig}(f,h_b,h_r,F,\upsilon)$ and $r^{(k)}_{Sig} = r^{(k)}_{Sig,m',p'}$ are, respectively, the three-dimensional antenna pattern value, utilisation, H function from (4.45) and distance corresponding to the base station and antenna that is transmitting to serve the k th user. The values m' and p' are defined in (4.43)

As explained immediately after (4.15), the transmitted power from the base stations, $P_{out,m,p}$, is given by:

$$P_{out,m,p}\left(\mu_{BS,m,p}\right) = \mu_{BS,m,p}P_{max,out} \tag{4.49}$$

This form assumes that base stations in the vicinity of the user are coordinated such that the Radio Frames are aligned in time. This avoids the idle power transmissions interfering with the data-carrying REs within a Radio Frame.

If the modeller assumes the base stations are not coordinated then there can be interference due to transmission of overhead REs and the interference transmitter output power has the form:

$$P_{out,m,p}\left(\mu_{BS,m,p}\right) = P_{idle,out,m,p} + \mu_{BS,m,p}\left(P_{max,out,m,p} - P_{idle,out,m,p}\right) \tag{4.50}$$

From (4.47) we see that the higher the utilisation $\mu_{BS,m,p}$, of the surrounding base stations, the lower will be the SINR at locations in the service area. Using (4.34), (4.47) (4.48) and (4.49) in (4.33) we can use a numerical computational package to calculate the SINR at any location $z^{(k)}$ in the service area at time t_u.

Using a fixed grid of coordinate points covering the hexagonal cell (such as step values of $z = (r,\theta)$ where $r = v\Delta r$ and $\theta = w\Delta\theta$ for integers v, $w = 0,1,2,....$, with fixed Δr and $\Delta\theta$ fixed, around the base station BS_0 located at $(r,\theta) = (0,0)$) (4.33) provides contour plot of the SINR values around the base station. In calculating (4.33) the vertical angle φ is calculated using the radial distance, r, from the base station the transmit antenna height above ground, h_b, and user handset height above ground, h_r.

Such a plot is shown in Figure 4.9. The shade/colour at a location indicates the value of the SINR (within pre-set ranges) We can see that the SINR reduces close to the boundaries of the cells as the interference power from other cell antennas becomes more significant relative to the signal. The "valleys" in the SINR value at the borders between sectors of a base station due to the directional profile of the transmit antennas.

Note: if radial (r,θ) coordinates are adopted, one must be aware that for a radial coordinate, r, the angular spatial distance between points is $r\Delta\theta$. This reduces the number of points per unit area as the radial distance increases which will impact spatial averages across the service area. To provide spatial averaging that is not biased in this manner, the modeller should use rectilinear coordinates (x,y) with $x = v\Delta x$ and $y = w\Delta y$ for integer ranges v, w, $= ...,-3,-2,-1,0,1,2,3,..$ enough to cover the service area.

The plots in Figure 4.9 are for a simulation with minimal variation in the shadowing loss, υ. As the variance of the shadowing loss increases, the clean contours seen in Figure 4.9 start to break up as the randomness of the shadowing loss increases. Examples of SINR and Spectral Efficiency contour plots with significant shadowing loss are shown in Figure 4.10.

Figure 4.11 shows the SINR averaged over all the grid locations for values of network-wide base station utilisation for a Suburban geographical category. This shows the typical profile of a declining SINR as base station utilisation increases causing greater interference for signal transmissions.

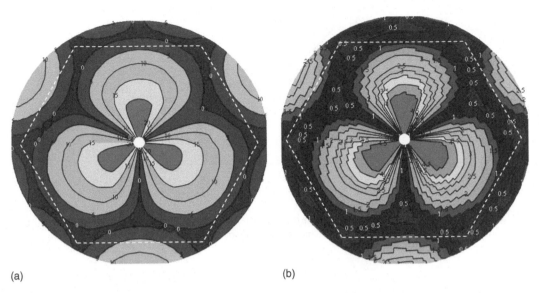

(a)   (b)

FIGURE 4.9 a) Contour plots of SINR for locations around BS_0 for a multiple base station mobile network with inter base station distance 1000m. The SINR includes the impact of interference power from base stations BS_1 to BS_{18}. The plot shows the cell in the mobile network as a dashed white hexagon. The contour shades/colours are set by the range of SINR values for each shaded/coloured region with the displayed values in dB. b) The corresponding contour plot for the Spectral Efficiency. In this case, the steps due to the thresholds in the SINR to spectral efficiency mapping (see Table 4.7) produce the jagged contours.

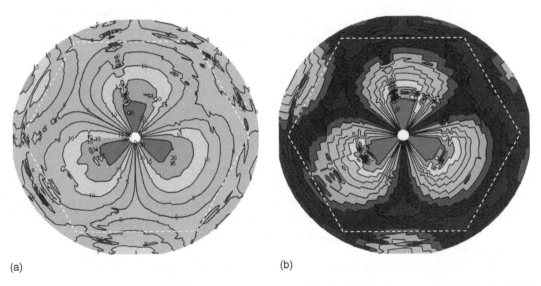

(a)   (b)

FIGURE 4.10 Contour plot of SINR, a), and Spectral Efficiency, b), for locations around BS_0 for a multiple base station mobile network with inter base station distance 1000m. The plots include the impact of interference power from base stations BS_1 to BS_{18} and significant shadowing loss variation.

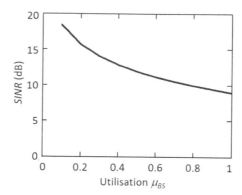

FIGURE 4.11 Average SINR dependence on base station utilisation for a Suburban geography. All base station utilisations are equal resulting in more interference power as the network becomes more active.

4.8   USER UTILISATION AND THROUGHPUT

It is important to note a fixed grid of the form used to create Figure 4.9 does not provide the SNIR values required to calculate the downlink throughput, $R(t_u)$. To do this we calculate the SINR value only at the locations of the users, $z^{(k)}$, for $k = 1,2,....N^{(usr)}(t_u)$, where $N^{(usr)}(t_u)$ is the number of active users receiving data in the Radio Frame at time t_u.

As stated above, the appropriate modulation format and coding rate to reliably transmit data to a user results from a negotiation between the base station and the user's equipment. Using the received RS REs, the user's handset calculates its SINR to establish a "Channel Quality Indicator" (CQI). Broadly, the CQI that the handset reports to the base station indicates a modulation and coding scheme that would incur an "acceptable" a downlink Data Block Error Rate of at most 10% [77]. The base station then uses that reported value of CQI, as well as its record of previous re-send requests from the user, to determine the MCS that is adopted for the PDSCH REs for that user [46]. The user data is then transmitted to the user device.

The MCS sets the spectral efficiency of the PDSCH REs. The mapping from the SINR to the MCS is not unique. Different handset makers use additional data and different algorithms to decide a CQI value, as do network equipment manufacturers. Different network operators may also adopt different threshold values for the SINR to MCS mapping. Thus it is only possible in system modelling to approximate this mapping process using the link curves such as shown in Figure 4.4 or Tables 4.7 and 4.8.

Using the MCS that gives the highest throughput is important, and typically the SINR threshold for an MCS is chosen to allow a higher throughput rate consistent with a block error rate below 10%; that level of errors being generally recoverable. Example SINR threshold values are shown in the right-most columns of Table 4.7 [54, 73, 75, 92]. Using this type of table or an appropriate link curve, the SINR at the location of any user can be mapped into a spectral efficiency of the PDSCH REs transmitted to that user.

The PDSCH REs transmitted from a base station within a given Radio Frame are very likely to be destined to multiple users. The details of the process for determining the number of PDSCH REs allocated to each user are beyond the scope of this book. The reader can find these details in 3GPP TS 36.213 [77] as well as texts [46, 50, 51, 97]. In most cases there will be multiple active users geographically spread throughput a base station or sector service area, with each being the recipient of a number of PDSCH REs transmitted within a Radio Frame. As discussed above, the location of user k in the service area has a significant impact on the SINR experienced by that user and therefore on the spectral efficiency of the PDSCH REs transmitted that user.

The average spectral efficiency, $\langle SE(t_u)\rangle_{RF}$ defined in (4.28) is an average over all PDSCH REs transmitted to all active users in the Radio Frame at time t_u. The spectral efficiencies, $SE^{(k)}(t_u)$ are mapped from the SINR values, $SINR(z^{(k)},t_u)$, at the locations of all the active users, $z^{(k)}$, serviced at time t_u. An example of this mapping is given by the link curves in Figure 4.7 and the SINR threshold values in Table 4.7. The quantity $\langle SE(t)\rangle_{RF}$ is useful for calculating the total base station throughput, $R(t_u)$ which is required to calculate the base station energy per bit. However, in some cases we are interested in the power consumption and energy per bit for a specific user. For example, Chapter 10 focuses on power consumption modelling of services. In this case, we need to consider those users who are using the service of interest. Also, Chapter 11 includes methods for estimating the energy efficiency of services. Again, this requires focussing on those users of the service of interest.

With this move to a focus on the individual users, a significant simplification that has been applied in constructing this model has to be emphasised. As shown in Table 4.4, there are multiple approaches that can be adopted when constructing a power model of a mobile base station. In particular, there are three common approaches for modelling the user/traffic demand. The approach adopted for this model is "Utilisation based". An advantage of this approach is its simplicity. However, in the "FTP Traffic Model" is a better representation of how a base station deals with traffic demand. Keeping with the "Utilisation-based" approach, it is important the modeller appreciates the shortcomings of this approach. These are discussed in greater detail at the end of this chapter and in Chapter 5.

Moving our attention to the individual users, the k th user will receive $N_{RF,PDSCH}^{(k)}(t_u)$ PDSCH REs with a spectral efficiency $SE(z^{(k)},t_u)$ in the Radio Frame at time t_u. Different users may receive different number of PDSCH REs with different SINR and, hence, different spectral efficiency. If we wish to determine the data throughput for the k th user (to calculate the energy efficiency for a service used by that user), we need to evaluate $SE(z^{(k)},t)$.

Knowing the locations of users ($z^{(k)}$ for $k = 1, .., N^{(usr)}$) and their $SE(z^{(k)},t_u)$, we can calculate the average of the spectral efficiency over the number of users; $\langle SE(t_u)\rangle^{(usr)}$:

$$\left\langle SE\left(t_u\right)\right\rangle^{(usr)} = \frac{1}{N^{(usr)}\left(t_u\right)} \sum_{k=1}^{N^{(usr)}(t_u)} SE\left(z^{(k)},t_u\right) \tag{4.51}$$

Given all the transmitted REs are received by the users, we expect there to be a relationship between the spectral efficiency averaged over PDSCHs, $\langle SE(t_u)\rangle_{RF}$, and spectral efficiency averaged over users, $\langle SE(t_u)\rangle^{(usr)}$.

To understand this relationship, we define base station utilisation by the k th user as:

$$\mu_{BS}^{(k)}\left(t_u\right) = \frac{N_{RF,PDSCH}^{(k)}\left(t_u\right)}{N_{max,RF,PDSCH}} \tag{4.52}$$

where $N_{max,RF,PDSCH}$ is assumed the same across all transmitters in the base station serving user k. Recall that the utilisation is defined on a "per transmitter" basis. Because the sum of PDSCH REs allocated to all active users at time t_u equals the total number, $N_{RF,PDSCH}(t_u)$, of allocated PDSCH REs at that time, from (4.22) we have:

$$\mu_{BS}\left(t_u\right) = \frac{N_{RF,PDSCH}\left(t_u\right)}{N_{max,RF,PDSCH}} = \sum_{k=1}^{N^{(usr)}(t_u)} \mu_{BS}^{(k)}\left(t_u\right) = N^{(usr)}\left(t_u\right)\left\langle\mu_{BS}\left(t_u\right)\right\rangle^{(usr)} \tag{4.53}$$

where $\langle\mu_{BS}(t)\rangle^{(usr)}$ is the average of $\mu_{BS}^{(k)}(t)$ taken over active users serviced by the Radio Frame at time t_u, i.e.

$$\left\langle\mu_{BS}\left(t\right)\right\rangle^{(usr)} = \frac{1}{N^{(usr)}\left(t\right)} \sum_{k=1}^{N^{(usr)}(t)} \mu_{BS}^{(k)}\left(t\right) \tag{4.54}$$

(**Note:** the notation $\langle X \rangle_{RF}$ denotes an average of X over PDSCH REs in a Radio Frame whereas $\langle X \rangle^{(usr)}$ denotes an average of X over users served by that Radio Frame.)

Referring back to (4.26), the data per PDSCH for the k th user will depend upon the spectral efficiency at the location of that user. The data bits delivered to the k th user can be expressed as:

$$B^{(k)}\left(t_u\right) = N_{RF,PDSCH}^{(k)}\left(t_u\right) SE\left(z^{(k)},t_u\right)\delta t_{sd}\delta f \tag{4.55}$$

Therefore the data rate to the k th user will be:

$$R^{(k)}\left(t_u\right) = N_{RF,PDSCH}^{(k)}\left(t_u\right) SE\left(z^{(k)},t_u\right)\delta f = \mu_{BS}^{(k)}\left(t_u\right) SE\left(z^{(k)},t_u\right)\sigma_{PDSCH}BW \tag{4.56}$$

where we have applied the same approach as when deriving (4.30). Summing over all active users, $N^{(usr)}(t_u)$ and using (4.52) we get for the total base station throughput:

$$\begin{aligned} R\left(t_u\right) &= \sum_{k=1}^{N^{(usr)}\left(t_u\right)} R^{(k)}\left(t_u\right) = N^{(usr)}\left(t_u\right)\left\langle \mu_{BS}\left(t_u\right) SE\left(t_u\right)\right\rangle^{(usr)} \sigma_{PDSCH}BW \\ &= N^{(usr)}\left(t_u\right)\left\langle R\left(t_u\right)\right\rangle^{(usr)} \end{aligned} \tag{4.57}$$

where $\langle R(t_u)\rangle^{(usr)}$ is the average download data rate per user at time t_u. Using this with (4.30) we get:

$$N^{(usr)}\left(t_u\right)\left\langle \mu_{BS}\left(t_u\right) SE\left(t_u\right)\right\rangle^{(usr)} = \mu_{BS}\left(t_u\right)\left\langle SE\left(t_u\right)\right\rangle_{RF} \tag{4.58}$$

In this equation on the left-hand side the spectral efficiency is averaged over users whereas the right-hand side the spectral efficiency is averaged over PDSCH REs transmitted from the base station. Note that the average $\langle \mu(t)SE(t)\rangle^{(usr)}$ is over the product of utilisation and spectral efficiency of the users. Because these two terms are not independent the mean of their product may not equal the product of their means. In fact, in most cases the number of PDSCH REs allocated to a user will be dependent upon the spectral efficiency of the signal used to transmit to that user. This relationship is often determined by network operator policy as the operator endeavours to provide an acceptable quality of service to all users in the base station service area.

Using (4.53) in (4.58) and cancelling the common terms, we get the relationship between the spectral efficiency averaged over PDSCH REs and that averaged over users:

$$\left\langle SE\left(t_u\right)\right\rangle_{RF} = \frac{\left\langle \mu_{BS}\left(t_u\right) SE\left(t_u\right)\right\rangle^{(usr)}}{\left\langle \mu_{BS}\left(t_u\right)\right\rangle^{(usr)}} \tag{4.59}$$

In this equality, we note that changing the relative utilisations of users will change the Spectral Efficiency averages because the number of PDSCH REs transmitted to each user will change.

In the special case of $\mu_{BS}(t_u) = 1/N^{(usr)}(t_u)$, we have $\mu_{BS}(t_u) = 1$ and hence $\langle SE(t_u)\rangle^{(usr)} = \langle SE(t_u)\rangle_{RF}$. In fact this equality also holds if the utilisation of all users is the same, $\mu_{BS}^{(k)} = $ constant for all k. A

Also note, an important outcome of (4.59) is that even if we have fixed the utilisation rate of a base station, we have not fixed its throughput. Its throughput won't be fixed until we have also determined how the resources are allocated to each user, i.e. the $\mu^{(k)}$'s. Therefore, by "clever" scheduling of its resources, a base station can minimise the utilisation of its resources to meet a given total user traffic demand.

We also have the condition that the throughput of the base station cannot be more than its capacity with 100% utilisation; that is, $R(t_u) \leq C_1(t_u)$. Using (4.31) and (4.57) this gives the requirement

$$N^{(usr)}\left(t_u\right)\left\langle \mu_{BS}\left(t_u\right)SE\left(t_u\right)\right\rangle^{(usr)} \leq \left\langle SE\left(t_u\right)\right\rangle_{RF} \tag{4.60}$$

This places a condition on the allocation of resources to users, $\mu_{BS}^{(k)}\left(t_u\right)$. Because the Static Model allocates resources to the users by hand (i.e. the modeller sets the value of $\mu_{BS}^{(k)}\left(t_u\right)$), care must be taken to not "overload" the base station by violating the inequality. Because averages involving the spectral efficiency appear in multiple terms in this inequality, the modeller must be aware of the SINR conditions experienced by users when setting the values of $\mu_{BS}^{(k)}\left(t_u\right)$.

From (4.56) and (4.57) we see that the data download rate for users, $R^{(k)}(t)$ and hence for the base station, $R(t)$, is dependent upon the location, $z^{(k)}(t)$, of the users. Therefore, to calculate these quantities, we have to input into the model the location of all the users being serviced by the base station and their utilisations. The simplest case is a uniform random distribution of users over the base station cell.

To construct a uniform, random distribution of users in a cell within a network with inter-site distance ISD, we use a uniform random number generator to generate a number in the range $-ISD/\sqrt{3}$ to $ISD/\sqrt{3}$. This is taken as the x-coordinate of the user. We repeat this process to provide the y-coordinate. This will give a candidate location $z^{(k)} = (x^{(k)},y^{(k)})$ for the k th user. This will produce a square area in which users are randomly located. This square is shown in Figure 4.12.

We then reduce the area in which the users are located to a circle centred on the base station (shown as the central black dot in Figure 4.12) by removing users further out than a radial distance of $ISD/\sqrt{3}$. We also remove users who are too close to the base station, say less than 30m from the base station. This leaves a uniformly distributed population of users located within the shaded area of Figure 4.12. An example of such a uniform distribution is shown in Figure 4.13a).

More generally, user locations can be set up using a probability distribution. With $N_{BS}^{(usr)}$ users in the base station service area, applying a probability density function, $f^{(usr)}(z)$ the number of users in an area dA centred on position z will be given by

$$N^{(usr)}\left(z,t_u\right) = N_{BS}^{(usr)}\left(t_u\right)f^{(usr)}\left(z\right)dA \tag{4.61}$$

In the simplest form, this will be uniform random distribution as shown in Figure 4.13a) for which $f^{(usr)}(z) = 1/A_{BS}$, where A_{BS} is the area serviced by the base station. For the simulation, the area dA can be set up as a small square $dA = \Delta x \Delta y$, centred at location z where (x,y) are rectilinear coordinates covering the service area and $z = (x,y)$.

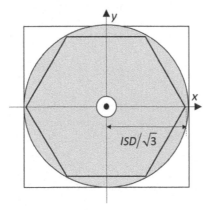

FIGURE 4.12   Area over which users can be randomly located around a base station located at the dot in the centre. Users are distributed within the shaded area.

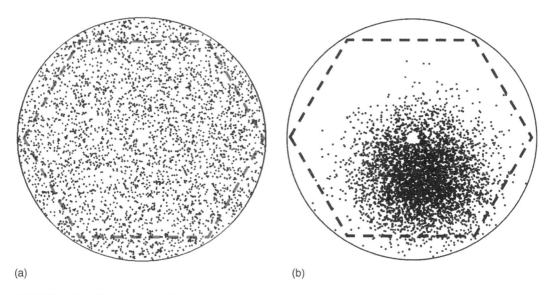

(a)   (b)

FIGURE 4.13 Simulated user distributions, $z^{(k)}(t)$, around base station BS_0. Both distributions consist of 3500 users. a) Uniform distribution throughout the service area, b) uneven distribution with the majority of users in one sector of the base station located in the centre of the hexagonal region. Note that not all these users are simultaneously active. Rather, the model consists of multiple simulations each with a sub-population of active users. The results are then averaged over all users to provide an indicative outcome.

Depending upon the situation being modelled, a non-uniform distribution may be more appropriate. For example, the distribution shown in Figure 4.13b) has the majority of users in one sector of the cell with the $f^{(usr)}(z)$ is a Gaussian centred on that sector of BS_0. This may correspond to the base station being located at the edge of a populated area with the expectation that future growth will extend the user distribution into the sparsely populated area.

For the purposes of modelling the base station power consumption and throughput we are only interested in the locations of the, $z^{(k)}$, of the users such as depicted in Figure 4.13. Therefore we focus on the active users, $k = 1,2,\ldots N^{(usr)}(t_u)$ at time t_u. The network operator policy determines the number of PDSCH REs to be allocated to each user. This policy can be expressed via the utilisation of the k th user, $\mu_{BS}^{(k)}(t_u)$. For real networks, the operator policy will be based upon issues such as fairness for access across users, the depth of the queue of user requests awaiting service, the CQI reported by user equipment and more. However, in simple models such as this, these details are too sophisticated to be included. Therefore, very simple approaches are typically taken to the allocation of resources to users. For example, a policy of equal allocation of PDSCH REs across the $N_{BS}^{(usr)}(t_u)$ users, irrespective of the spectral efficiency at their location, will give for all k:

$$\mu_{BS}^{(k)}(t_u) = \mu_{BS}(t_u)/N_{BS}^{(usr)}(t_u) \tag{4.62}$$

This policy for allocating PDSCH REs has been adopted in other models [14, 70, 120]. However, is unlikely to provide users with equal share of the available base station total data rate, $R(t_u)$. Using (4.56) the download data rate experienced by the k th user will be

$$R^{(k)}(t_u) = \frac{\mu_{BS}(t_u)\sigma_{PDSCH}BW}{N_{BS}^{(usr)}(t_u)} SE\left(z^{(k)},t_u\right) \tag{4.63}$$

In this we see that the download data rate of the k th user is set by the spectral efficiency at their location. This means users at the edge of a cell or the boundary between two sectors in a cell who typically have a lower spectral efficiency than those closer to the base station and/or in the centre of a sector, will experience a slower download speed.

The allocation of resources can also be used to ascertain the range possible user data rates to estimate upper and lower bounds for power consumption per user. One approach for making this estimation is to assume an even allocation the available download throughput across all users. That is, the download throughput per user is set to $R^{(k)}(t_u) = R(t_u)/N^{(usr)}(t_u)$ which will require different utilisations, $\mu_{BS}^{(k)}(t_u)$, for users to account for the variation of the spectral efficiency across users' locations. Using (4.56) and (4.30) such an allocation requires:

$$\mu_{BS}^{(k)}(t_u) = \frac{R(t_u)}{N_{BS}^{(usr)}(t_u) SE(z^{(k)},t_u)\sigma_{PDSCH}BW} = \frac{\mu_{BS}(t_u)\langle SE(t_u)\rangle_{RF}}{N_{BS}^{(usr)}(t_u) SE(z^{(k)},t_u)} \qquad (4.64)$$

Using this in (4.56), we get:

$$R^{(k)}(t_u) = \frac{\mu_{BS}(t_u)\langle SE(t_u)\rangle_{RF}\sigma_{PDSCH}BW}{N_{BS}^{(usr)}(t_u)} \qquad (4.65)$$

This data rate is the same for all users. Setting the base station utilisation to unity, there is an upper and lower limit to the value which $R^{(k)}(t_u)$ can take. From Tables 4.7 and 4.8 we see that the range of spectral efficiency values that can be applied in this equation is constrained by the MCS levels. For example, applying Table 4.7 the spectral efficiency has minimum value (MCS = 1) of 0.128 bit/s/Hz and the maximum (MCS = 15) of 4.666 bit/s/Hz. The value of $\langle SE(t_u)\rangle_{RF}$ will have to lie within this range. Consequentially, for a given number of users a maximum and minimum per user download throughput, $R^{(k)}(t_u)$, can be calculated. This, in turn, can be used to calculate the bounds on the average user energy consumption.

Using the user distributions displayed in Figure 4.13, and resource allocation based on (4.62) a "heat map" of the spectral efficiency at each user location is shown in Figure 4.14.

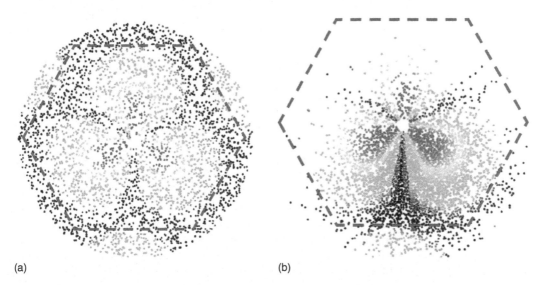

(a)   (b)

FIGURE 4.14 Heat map for the spectral efficiency at the location of users displayed in Figure 4.13. The shade/colour of each dot represents the value of the spectral efficiency experienced by the user at that location. In a) the users are uniformly distributed over the entire cell. In b) the same number of users are more localised to the lower region of the cell.

Using (4.30) the total download traffic for the uniformly distributed users shown in Figure 4.14a) is $R(t)$ = 23 Mbit/s, and for the non-uniformly distributed users shown in Figure 4.14b) is $R(t)$ = 35 Mbit/s. The reason the non-uniform user distribution gives a higher total traffic is that more users are located in areas of higher spectral efficiency, as can be seen comparing the two heat maps. The improvement in spectral density in the more populated areas in Figure 4.14b) has more than compensated the reduction in resources allocated in each user in these areas.

This is a facet of the Static Model. The modeller can adjust $\mu^{(k)}{}_{BS}(t_u)$ to set the load on the network subject to the condition they satisfy (4.53) with $\mu_{BS}(t_u) \leq 1$. In a real network, at high utilisation the probability of the network becoming congested and providing degraded service is significant. When using the Static Model in this regime, the modeller will need to be careful to appropriately interpret the results. For example it is highly likely that, rather than having the higher throughput indicated in Figure 4.13b), the base station will become congested with this densely packed user distribution.

A real mobile network has protocols and processes designed to enable the base station to respond to high traffic loads. These processes include continuing reports on service quality from the user handset to the base station, decisions by the base station on resource allocation between users who attempt to optimise the quality of service across all users. The Static Model does not include any of these processes but assumes the network has settled these issues and is in a resulting steady state.

Recall that apart from (4.30) the total throughput of the base station can also be calculated using (4.57). Assuming the utilisation of each user is given by (4.62), the equality of (4.30) and (4.57) requires $\langle SE(t_u)\rangle_{RF} = \langle SE(t_u)\rangle^{(usr)}$. This makes sense on the basis that the location of the users is the same for both calculations.

The heat plots in Figure 4.14 are for a three-sector base station with a single transmitter per sector each of bandwidth 20 MHz, an inter base station distance of 1km and all sectors operating at $\mu_{BS}(t_u)$ = 1 with 3500 users in the service area around BS_0. Because of the model includes allocation of users to the base station and sector that provides the highest SNIR, which is done using (4.43), the number of users allocated to be served by BS_0 is, in fact, 2370 for both heat plots in Figure 4.14.

This means that in Figure 4.14a) where the users are evenly distributed around the base station cell, $N^{(usr)} \approx 790$ users in each sector of the base station. This gives utilisation of each user of $\mu_{BS}^{(k)}\left(t_u\right) \approx 1/790$. Using Table 4.5 for CFI = 2 and a single antenna, and using (4.62) each user is allocated approximately 137 PDSCH REs in each Radio Frame.

Although the average user download data rate is 30 kb/s, the download data rate ranges from a low of 3.8 kbit/s to a high of 90 kbit/s. Users close to the antenna experience a higher spectral efficiency than those further away, as shown in Figure 4.15. In this figure each dot is the spectral efficiency for a user for the given distance $r^{(k)}$ (Figure 4.15a)) and angle $\theta^{(k)}$ (Figure 4.15b)) for the k th user. We see that the spectral efficiency is higher for users close to the base station and not located at angles corresponding to the borders between sectors. For users further away Figure 4.15a) shows the spectral efficiency declines until distances getting close to the edge of the cell after which it increases. This increase is because those users are being served by the surrounding base stations.

Looking at Figure 4.15b) we see that users at angles corresponding to the centre of each sector have can have a high spectral efficiency. Those close to the border between two sectors and those further away (irrespective of their angle) experience a lower spectral efficiency.

Turning our attention to the non-uniform distribution of users shown in Figure 4.14b), the application of this Static Model for such non-uniform user population requires some care and consideration. The Static Model simulation based on the user distribution shown in Figure 4.14b) assumes all base stations are operating at utilisation $\mu_{BS,m,p}$ = 1 and the allocation of PDSCH REs for all users is given by (4.62). Given the asymmetry of the user locations, this resource allocation is unlikely to be realistic. Adding to the Static Model a more sophisticated resource allocation method to provide a fairer allocation of resources is beyond the scope of the model. The allocation of resources to users for a mobile network with non-uniform demand across the service area requires advanced methods [121].

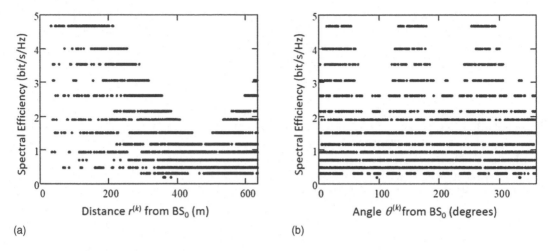

FIGURE 4.15 Spectral Efficiency of users distributed uniformly around BS_0 showing a) dependence on distance, $r^{(k)}$ from the base station and b) dependence on angle, $\theta^{(k)}$.

4.9 BASE STATION POWER CONSUMPTION AND THROUGHPUT

In the simulations that produced Figure 4.13 through Figure 4.15, the utilisation of all the base stations was set to unity, that is, $\mu_{BS,m,p} = 1$. In many cases, it is quite reasonable to adopt this as standard practice because it ensures all active users receive the maximum number of resources available thereby maximising their download speed. However, for the purposes of improving our understanding of the relationship between traffic and power consumption, it can be informative to set the utilisation of the base station below unity to determine the base station power consumption profile. Also, it is important to leave some headroom in base station capacity (in REs) to meet surges in demand. Therefore, setting $\mu_{BS} < 1$ is a method to learn the energy cost of keeping (say) 50% headroom compared with (say) 30% headroom in capacity to deal with traffic surges. However, it would most likely be better to use an FTP-type model (such as described in Chapter 5) or a more sophisticated one to also explore how often that headroom was needed.

The utilisation of the sector of a base station, $\mu_{BS,m,p}$ is set by the sum of the utilisations of the users it is serving, as given in (4.53). To operate the base station at utilisation less than unity requires the total number of PDSCH REs allocated in a Radio Frame to be less than $N_{max,RF,PDSCH}$. On a per user basis, this corresponds to setting the average utilisation for each user at less than $1/N^{(usr)}$. For the two resource allocations examples given above (i.e. equal allocation of PDSCH REs as in (4.62) and equal allocation of available throughput as in (4.64)) if we wish to represent less than unity utilisation in terms of user download data rate, then we have to set $R^{(k)}(t_u)$ to be less than the right-hand sides of (4.63) or (4.65), respectively.

The power consumption profile of a mobile base station expressed in terms of its utilisation is given in (4.18) for a single antenna (SISO) system and (4.19) for a multi-antenna (MIMO) system. From these expressions, we can see that the base station power can be modelled as linear in utilisation. However, often it is more instructive to express the power consumption of a base station in terms of its throughput, $R(t_u)$. To do this we can use (4.30) to substitute for $\mu_{BS,m,p}(t_u)$ in (4.18) which gives, for a single antenna per sector system,

$$P_{BS}(R) = 3P_{idle} + \frac{\Delta_P P_{max,out}}{\sigma_{PDSCH} BW} \sum_{p=1}^{3} \frac{R_p}{\langle SE_p \rangle_{RF}} \qquad (4.66)$$

In this we have dropped the time dependence, t_u, and the base station index, m, for convenience. The p-index covers the three sectors and R_p and $\langle SE_p \rangle_{RF}$ are the download data rate and mean spectral efficiency (over a Radio Frame) for sector p, respectively. We have assumed the equipment for all sectors is identical.

Viewing (4.66) we note the following: a) The term $R_p/\sigma_{PDSCH} BW \langle SE \rangle_{RF,p}$ ranges between zero and unity. For a given base station the slope of the R_p dependent term is set by the average spectral efficiency in each sector, $\langle SE_p \rangle_{RF}$. And b) the range of values attainable by R_p depends upon the average spectral efficiency. In other words, the maximum possible throughput, in bit/s, is set by the average spectral efficiency of the PDSCH REs transmitted to users.

In the simulation above, the mean spectral efficiency was 1.469 bit/s/Hz. However, from Table 4.7, we see that the spectral efficiency can range from a lower limit of $\langle SE \rangle_{RF,l} = 0.178$ bit/s/Hz to an upper limit of $\langle SE \rangle_{RF,u} = 4.666$ bit/s/Hz. This means the power consumption, as a function of base station throughput, $R_{BS} = \sum_{p=1}^{3} R_p$, of the base station will be within the shaded region in Figure 4.16

The end goal of the model is to calculate the energy efficiency, H_{BS}, of a mobile base station. This is given by the base station power, (4.66), divided by its total throughput:

$$H_{BS}\left(R_{BS}\right) = \frac{P_{BS}\left(R_{BS}\right)}{R_{BS}} = \left(3P_{idle} + \frac{\Delta_P P_{max,out}}{\sigma_{PDSCH} BW} \sum_{p=1}^{3} \frac{R_p}{\langle SE_p \rangle_{RF}}\right) \bigg/ \sum_{p=1}^{3} R_p \qquad (4.67)$$

Viewing Figure 4.16, we can see that the energy efficiency of the base station will also have a range of values for a given total throughput due to the possible range of values of $\langle SE_p \rangle_{RF}$.

In some cases, we are interested in the power consumption and energy efficiency for specific users (or the services they are using). This requires allocation of the base station idle power to users and/or the services they are accessing. Allocating the base station power to users requires careful consideration. Although allocating the incremental power, $\Delta_P P_{max,out}$ in (4.11) to users is relatively straightforward, allocating idle power requires some thought. Chapter 10 provides a methodology for allocating a share of the idle power to each user. The reader is referred to Chapter 10 to see how this can be done.

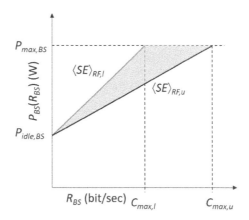

FIGURE 4.16 Power consumption of mobile base station as a function of throughput R_{BS}, in bit/s. The power profile will lie in the shaded area which is bounded by (4.66) with the upper bound corresponding to the lower limit of spectral efficiency, $\langle SE \rangle_{RF,l}$ and the lower bound corresponding the upper limit of spectral efficiency $\langle SE \rangle_{RF,u}$. The maximum throughput of the base station also depends upon the spectral efficiency. The values of $C_{1,l}$ and $C_{1,u}$ are given by (4.31) with $\langle SE \rangle_{RF,l}$ and $\langle SE \rangle_{RF,u}$, respectively.

4.9.1 DATA COLLECTION

The Static Model implements the "utilisation-based" user/traffic demand model component of the model listed in Table 4.4. With this, the utilisation of users and the base stations are set by the modeller. The model results focus on the users located around base station BS_0 and are derived by analysing their download data rates and the power consumption of BS_0. In this model, once the utilisations of the other base stations, BS_1 to BS_{18}, are set, the interference noise power, $P_{inter}(z,t_u)$, at any location z in the service area is independent of the number of users, $N^{(usr)}$. Likewise, the signal strength $P_{Rx,Sig}(z,t_u)$ and hence the SINR, $SINR(z,t_u)$ and spectral efficiency, $SE(z,t_u)$, are also independent of $N^{(usr)}$. By this we mean that the values of these quantities is determined by the location z, whether or not a user is located at that position.

The value $N^{(usr)}$ only comes into play when determining details such as the allocation of PDSCH REs, $N_{RF,PDSCH}^{(k)}$ to the k th user and hence user utilisation, $\mu_{BS}^{(k)}$ via allocation policies such as (4.62) and (4.64). The total number of users and their locations come into play when calculating results such as averages over the user population. For example, download data rate per user, $\langle R(t_u)\rangle^{(usr)}$ and average utilisation per user, $\langle\mu_{BS}(t_u)\rangle^{(usr)}$. This provides an opportunity to simplify the modelling process.

As an example, we consider a simulation in which we wish to calculate the average download throughput data rate per user, $\langle R(t_u)\rangle^{(usr)}$, across a population of $N^{(usr)}$ users. From (4.56) this has the form:

$$\langle R\rangle^{(usr)} = \frac{\sigma_{PDSCH}BW}{N^{(usr)}}\sum_{k=1}^{N^{(usr)}}\mu_{BS}^{(k)}SE\left(z^{(k)}\right) = \sigma_{PDSCH}BW\left\langle\mu_{BS}SE\left(z\right)\right\rangle^{(usr)} \tag{4.68}$$

where we have dropped the time coordinate, t_u, for convenience and the average $\langle\mu_{BS}SE(z)\rangle$ is over all users in the same manner as in (4.57).

The Full Buffer Model prescribed in ITU-R M.2135-1 (12/2009) [53] constructs average performance metrics (such as user download speed) by repeating the simulation multiple times, each time with 10 users at randomly distributed locations around the service area. The simulation is repeated until the averaged performance metrics converge. Quoting from that document:

> For a given drop the simulation is run and then the process is repeated with the users dropped at new random locations. A sufficient number of drops are simulated to ensure convergence in the user and system performance metrics. The proponent should provide information on the width of confidence intervals of user and system performance metrics of corresponding mean values, and evaluation groups are encouraged to provide this information

Referring to (4.68), if we repeat the simulation N_{sim} times calculating $\langle R\rangle^{(usr)}$ each time, we get values $q = 1, 2,.., N_{sim}$ values, $\langle R\rangle_q^{(usr)}$. Averaging these values gives an overall average:

$$\langle R\rangle^{(usr)} = \frac{1}{N_{sim}}\sum_{q=1}^{N_{sim}}\langle R\rangle_q^{(usr)} = \frac{\sigma_{PDSCH}BW}{N_{sim}N_{BS}^{(usr)}}\sum_{q=1}^{N_{sim}}\sum_{k=1}^{N_{BS}^{(usr)}}\mu_{BS}^{(k)}SE\left(z_q^{(k)}\right)$$

$$= \frac{\sigma_{PDSCH}BW}{N_{sim}N_{BS}^{(usr)}}\sum_{k=1}^{N_{BS}^{(usr)}}\mu_{BS}^{(k)}\sum_{q=1}^{N_{sim}}SE\left(z_q^{(k)}\right) \tag{4.69}$$

where the $z_q^{(k)}$ represents the location of the k th user in the qth run of the simulation. Being randomly generated, this location changes from run to run. In the second line, the double sum can be interpreted as $N_{BS}^{(usr)}$ groups of users each group consisting of N_{sim} users and all of those N_{sim} users have the same utilisation of $\mu_{BS}^{(k)}$. Over the N_{sim} simulations a total of $N_{sim}N_{BS}^{(usr)}$ locations are generated using the probability density function, $f^{(usr)}(z)$ in (4.61).

With this interpretation of that double sum, rather than running the simulation of $N_{BS}^{(usr)}$ users N_{sim} times, an alternate approach is to run a single simulation with a total of $N_{sim}N_{BS}^{(usr)}$ users all located using the same probability density function $f^{(usr)}(z)$ and grouping them into representative sub-populations of $N_{BS}^{(usr)}$ users and applying $\mu_{BS}^{(k)}$ to the users in each sub-population. On the basis that these sub-populations are all representative of the population distribution $f^{(usr)}(z)$ across the service area, we can then use (4.69) to calculate the simulation results.

For the case of PDSCH REs being evenly distributed across all users, $\mu_{BS}^{(k)} = 1 / N_{BS}^{(usr)}$ we get:

$$\langle R \rangle^{(usr)} = \frac{\sigma_{PDSCH} BW}{N_{sim}\left(N_{BS}^{(usr)}\right)^2} \sum_{k=1}^{N_{sim}N_{BS}^{(usr)}} SE\left(z^{(k)}\right) \tag{4.70}$$

This approach will be beneficial if there are operational advantages to running one large population simulation rather than multiple smaller population simulations.

4.10   CONSTRUCTING DIURNAL CYCLE RESULTS

The results derived so far in this chapter relate to a specific base station utilisation, μ_{BS}. Typically, the modeller is also interested in the base station's performance over a diurnal cycle. We can use the results above to construct simulation results over a diurnal cycle. An important detail to note is that the model presented in this chapter does not include any time dynamics. That is, it does not account for delays in the system's response to changes in demand or user locations. The model is effectively a steady-state model which assumes the base station instantaneously adjusts to changes in demand, user location and other potential variations.

Therefore, in the simplest terms, constructing results for a diurnal cycle merely requires the modeller to input the base station utilisation, $\mu_{BS}(t)$, for various times, t, over the diurnal cycle. In many cases, the diurnal cycle is split into durations each of fixed demand, such as shown in Figure 4.6. In that Figure, the diurnal cycle is approximated with five durations of fixed demand. The modeller may choose a different number, even to the extent of using hour-by-hour values of utilisation. For example, let the diurnal cycle be split into N_{stp} equal time steps. Let the utilisation for each step be $\mu_{BS,S}$ where S = 1, 2, ..., N_{stp}. We have $0 \le \mu_{BS,S} \le 1$ for all values of S. Also, let the model output of interest be $Y(\mu_{BS,S})$ which is provided by the simulation with utilisation $\mu_{BS,S}$. The average value of Y over the diurnal cycle, $\langle Y \rangle_D$, will be:

$$\langle Y \rangle_D = \frac{1}{N_{stp}} \sum_{S=1}^{N_{stp}} Y\left(\mu_{BS,S}\right) \tag{4.71}$$

Although this is relatively straightforward for results such as average base station power consumption, $\langle P_{BS} \rangle_D$, or average base station throughput, $\langle R \rangle_D$, in some cases the results required are per user (such as user average SINR, $\langle SINR \rangle^{(usr)}$). In other cases the results are related to user data throughput (such as user data rate, $\langle R \rangle^{(usr)}$) or base station power dependence on throughput, $P_{BS}(R)$). In these cases we require a relationship between base station utilisation, μ_{BS}, and user throughput such as that given in (4.56), (4.57) and (4.68).

The complication arises from the fact that a given base station utilisation can correspond to many values of user utilisations, $\mu_{BS}^{(k)}$ (though (4.53)) and user throughputs, $R^{(k)}$ (through (4.57)). We can simplify this by selecting a policy for allocation of resources to users such as that in (4.62) or (4.64). However, this still does not "lock down" the values of $R^{(k)}$ because the download data rate experienced by users is dependent upon their the spectral efficiency at their location as shown in (4.63). It is these dependencies that result in the relationship between base station throughput and power consumption shown in Figure 4.16.

Therefore to determine a dependence of a simulation output over a diurnal cycle, the modeller needs to specify details such as user locations and user utilisations over the diurnal cycle. User locations can be specified using a pre-determined location probability density function (pdf), using (4.61), examples of which are shown in Figure 4.13. Assuming users do not move about, the modeller can use the same location pdf for each stage of the diurnal cycle.

Another issue the modeller needs to be aware of is the dependence of C_1 upon user locations. From (4.31) we see that the throughput of a transmitter for 100% utilisation, C_1, is dependent upon the average spectral efficiency of the PDSCH REs transmitted. This, via (4.59), is dependent the average SINR over users which, in turn, is dependent upon their locations. This means the modeller must be careful to use the same user locations for each run over the diurnal cycle, or the results will be confounded by the different location patterns. That issue can be overcome by re-running many more times or using a sufficiently large number of users so that the average spectral efficiency $\langle SINR \rangle^{(usr)}$ and hence $\langle SINR \rangle_{RF}$ are approximately constant giving an approximately constant value of C_1. In general, is a better option to model the system over a diurnal cycle by fixing the number of users and varying the utilisation per user either by fixing $\mu_{BS}^{(k)}$ user-by-user or by setting the overall base station utilisation, μ_{BS}, and using a policy such as in (4.62) to set the per user utilisation. It is also preferred to use a large number of users (or large number of runs) to ensure the randomness of user location does not cause significant variation on the results.

As can be seen, there are multiple approaches the modeller can adopt to generate diurnal cycle results and the modeller needs to carefully consider possible unintended consequences of these various approaches.

4.11   MULTI-INPUT MULTI-OUTPUT (MIMO)

The Static Model presented above is based on transmission from a single transmit antenna to users. This of referred to as Single-Input Single-Output (SISO). Today the technology uses multiple transmit antennas at the base station transmitting to multiple receive antennas in the user handset. This technology is referred to as Multiple Input Multiple Output (MIMO) [49].

There are several MIMO technologies each labelled by the number of transmitters and number of receivers. Two transmitters in the base station transmitting to two receivers in the handset is referred to as 2×2 MIMO. Four transmitters to two receivers is 4×2 MIMO. Typically these systems use 2, 4 or 8 transmit antennas per sector. MIMO requires physical separation of antennas by at least ½ wavelength at the transmitter and receiver for efficient separation of the data streams. The compact size of handsets makes even two receive antennas difficult, and more antennas not feasible at today's operating frequencies hence current generation systems are typically N×2 MIMO indicating two received antennas in the user's handset.

An MIMO system with N transmitters and M receivers (N×M) can serve the smaller of N and M different traffic streams to a user. In the N×2 case this means 2 streams, which provides a possible doubling of the throughput. MIMO is commonly used in LTE-4G to provide independent streams of data to the user, also in modern Wi-Fi systems and in other applications. Tables 4.5 and 4.6 include the numbers and percentages of REs for different channels for 1, 2 and 4 transmit antenna systems.

To calculate the power consumption of an MIMO base station, $P_{BS}(\mu_{BS})$, the Static Model, which is effectively a SISO model, can be easily modified to deal with MIMO by applying the Static

TABLE 4.11

Example Values for Parameters in (4.11) for Different Base Station Types (Published in 2011) in NT-DOCOMO, et al. [64]

Base Station Type	Sectors	$P_{max,out}$ (Watts)	$P_{idle,BS}$ (Watts)	Δ_P
Macro (with 3dB feeder loss)	3	40	260.0	4.7
Remote Radio Head	3	40	168.0	2.8
	1	5.0	103.0	6.5
	1	1.0	96.2	1.5
	1	0.25	13.6	4.0
Pico	1	5.6	103.0	6.5
	1	1.0	96.2	1.5
	1	0.25	13.6	4.0
Femto	1	0.1	9.6	8.0

The parameters are based on 10 MHz bandwidth and 2×2 MIMO.

TABLE 4.12

Example Values for Parameters in (4.11) for Different Base Station Types (Published in 2012) in Auer, et al. [1]

Base Station Type	N_{TRX}	$P_{max,out}$ (Watts)	$P_{idle,BS}$ (Watts)	Δ_P
Macro single cabinet	6	20	130.0 (per sector)	4.7
Macro Remote Radio Head	6	20	84.0 (per sector)	2.8
Micro	2	6.3	56.0	2.6
Pico	2	0.13	6.8	4.0
Femto	2	0.05	4.8	8.0

The parameter values given here are for 10 MHz, 2×2 MIMO which has two transmitters for each sector. The Macro base stations have three sectors (each covering an angular span of 120 degrees) whereas the Micro, Pico and Femto base stations have a single sector covering all 360 degrees. For the Maco base stations the values in this table are per sector.

Model to each transmitter. For example, using the parameter sets in Tables 4.11 and 4.12 correspond to, 2×2 MIMO base stations. Calculating the throughput $R(t_u)$ and throughput for a user, $R^{(k)}(t_u)$, is more complicated. It is important to note that these parameter values can change significantly over technologies and the years as technologies develop [1]. Modellers should seek out the most relevant values for their model corresponding to the appropriate year.

Modelling received signal detection and processing in MIMO technologies to provide the user data stream $R^{(k)}(t_u)$ can be very sophisticated [49]. A detailed calculation of throughput $R^{(k)}(t_u)$ in an MIMO system and is well beyond the scope of the Static Model presented here. The Advanced Model described in the next chapter uses the SIMO and BLAST link curves show in Figure 4.4, [67] to model an implementation of SIMO and 2 × 2 MIMO, respectively. This includes the issues with separation and resulting inefficiency, and even slow signal frequency-selective fading. More advanced applications of MIMO technology will require yet more sophisticated models to determine user and base station throughput.

4.12 ADVANCED MODELLING OF MOBILE NETWORKS

It is important to note that the Static Model described in this chapter assumes the mobile network is in steady state, has a population of users that is uniformly distributed over the service area, is not congested and provides no dynamic description of the operation of the network. By this, we mean the model does not consider the impact on the network of details such as; new users becoming active (i.e. requesting a download), current users becoming inactive, whether or not user download requests are fulfilled and what happens if the download requests are not fulfilled. This Static Model is not designed to represent the dynamics of a real-world network, particularly when a mobile network becomes congested with increasing traffic loads. Its primary advantage is its simplicity relative to more representative models and can provide a "ball-park" estimate for the power consumption and energy per bit.

In a real network, even if the number of users, $N_{BS}^{(usr)}(t)$ and utilisation $\mu_{BS}(t)$ are approximately constant over time, there will be an ongoing stochastic process of new users becoming active and current users leaving becoming inactive. Therefore, in a real network, the utilisation of individual users, $\mu_{BS}^{(k)}(t)$ can change significantly as they or other nearby users become active and go inactive once their request has been serviced. The model described above does not account for this. In particular, because the process of users becoming active is stochastic, at high values of $\mu_{BS}(t)$, the random nature of the stochastic process can cause the network to move into congestion. Once this move into congestion starts, even if the overall number of users remains constant, the network remains in a state of congestion, resulting in many users failing to have their service request fulfilled. The model presented in this chapter does not intrinsically include this behaviour. Rather, it is up to the modeller to interpret the results appropriately.

To construct a model that can accommodate more real-world operation we must include details such as network congestion, the statistics and impact of new users coming online and then dropping out, the download file size and duration, failed downloads and more. To incorporate details such as these requires us to move from the numerical calculation described in this Static Model to a detailed numerical simulation that includes details missing from this model. An appropriately designed simulation-based model can include a more realistic representation of the actions of a mobile base station in response to dynamic user. Such a simulation can include time dependences and provide information not generated by the Static Model described above. However, this is at the price of much greater effort being required to construct the simulation.

Chapter 5 provides a description of such a mobile network simulation. The description in Chapter 5 has been drafted to provide a reader who is so motivated, some guidance on how to construct a detailed mobile network simulation.

4.13 TECHNICAL SUMMARY

Modelling the power consumption of a mobile network or of a mobile base station is substantially more complex than for other access network technologies. Despite its mathematical complexity the base station power Static Model described in this chapter is a somewhat simplified approach describing a relatively idealised, steady-state network. The transmit power, P_{out}, and total base station consumption, P_{BS}, are provided by (4.11) to (4.15) using the appropriate parameters from Table 4.9. To calculate the power consumption of an MIMO base station, $P_{BS}(\mu_{BS})$, the Static Model, which is effectively a SISO model, can be easily modified to deal with MIMO by applying the Static Model to each transmitter. For example, using the parameter sets in Tables 4.11 and 4.12 correspond to, 2×2 MIMO base stations. Calculating the throughput $R(t_u)$ and throughput for a user, $R^{(k)}(t_u)$, is more complicated. The base station throughput, $R(t)$ is given by (4.30) or (4.57).

A straightforward way for describing the model presented in this chapter is using a flow chart that summarizes of the steps for constructing the model, as shown in Table 4.13.

TABLE 4.13

Step-by-Step Summary for the Static Model of the Power and Throughput of a Mobile Base Station

Step	Model Component
1)	Set up network of cells, base station array and coordinate system. Typically three sectors per cell, each with $2\pi/3$ angular span. (Figure 4.8) Set inter-base station distance
2)	Setup coordinate system for user locations, $z^{(k)}$, of the $N_{BS}^{(usr)}$ users relative to the base station BS_0 in the simulation. Typically these locations are uniformly, randomly distributed throughout the coverage area of BS_0.
3)	Set utilisation, $\mu^{(k)}{}_{BS}(t_u)$, for users according to (4.52). Utilisation of that transmitter, $\mu_{BS}(t_u)$, is given by (4.53).
4)	Set utilisation for other sectors of each base station, $\mu_{BS,m,p}(t_u)$.
5)	Calculate transmission powers for all base stations using (4.15)
6)	Calculate the power consumption of the base station for the given utilisation $\mu_{BS}(t_u)$ using (4.11)
7)	Set transmit antenna gain G_{Tx}, radiation pattern, $\Pi(\theta,\varphi)$(dB), given in (4.72) for each base station sector antenna
8)	Select propagation path loss model and corresponding parameters for use in (4.47) and (4.48)
9)	Determine the distance $r_m^{(k)}$ and radiation pattern $\Pi_{m,p}(\theta^{(k)})$, from each sector antenna of each base station to each of $N_{BS}^{(usr)}$ users to use in steps 7) and 8).
10)	Calculate the received power, $P_{RX,m,p}(z^{(k)},t_u)$ from each antenna in the network at the location, $z^{(k)}$, of the $N_{BS}(usr)$ users' locations using (4.35)
11)	Calculate $P_{interf}(z^{(k)},t_u)$ for $N_{BS}^{(usr)}$, users from every sector of transmitter using (4.47)
12)	Select signal power, $P_{Rx,Sig}(z^{(k)},t_u)$ for each of the $N_{BS}^{(usr)}$ users based on the antenna signal that provides the highest SINR by applying (4.43)
13)	Track which users are allocated to be served by BS_0
14)	Calculate SINR for the $N_{BS}^{(usr)}$ users with (4.33)
15)	Determine Spectral Efficiency, $SE^{(k)}(t_u)$, for $N_{BS}^{(usr)}$ users. E.g. using Table 4.7 or Table 4.8.
16)	Calculate the power consumption of BS_0 using (4.18) or (4.19)
17)	Calculate $R^{(k)}(t_u)$ using (4.56) and $R(t_u)$ using (4.57).

4.14 EXECUTIVE SUMMARY

The prevalence and importance of mobile networks and services is expected to continue to increase over coming years. The network power consumption of mobile networks is an increasing proportion network operator operational expenditure. Further, as traffic demands increase, the scaling of mobile networks and their power consumption is becoming an issue, to the point that techniques to mitigate this by offloading traffic onto small cell or fixed networks are evolving. Thus, an understanding of current and future network power consumption and energy efficiency is an important business parameter.

Highly detailed mobile network power models are commercially available, but they are typically quite expensive. Constructing a detailed "in-house" mobile network power model can require a significant commitment in terms of staff hours and resources. A potential compromise is to build a simple approximate model that provides overall guidance on power consumption. The Static Model described in this chapter provides such an alternative. This model is simplified by assuming the network is operating in a "steady state" in which the number of users is fixed as is their data allocations.

This Static Model can be constructed using an inexpensive numerical computer package and even a spreadsheet with statistical functions.

4.15 APPENDIX

4.15.1 Values of ρ_A and ρ_B

TABLE 4.14
Ratio ρ_B/ρ_A for 1, 2 or 4 Antenna Ports

	ρ_B/ρ_A	
P_B	One Antenna Port	Two and Four Antenna Ports
0	1	5/4
1	4/5	1
2	3/5	2/4
3	2/5	1/2

TABLE 4.15
Values of ρ_B for Given ρ_A and Ratio ρ_B/ρ_A for a 1 Antenna Port

ρ_A(dB)	3	2	1	0	−1.77	−3	−4.77	−6
ρ_B/ρ_A	2.00	1.58	1.26	1.00	0.67	0.50	0.33	0.25
1.00	2.00	1.58	1.26	1.00	0.67	0.50	0.33	0.25
0.80	1.60	1.27	1.01	0.80	0.53	0.40	0.27	0.20
0.60	1.20	0.95	0.76	0.60	0.40	0.30	0.20	0.15
0.40	0.80	0.63	0.50	0.40	0.27	0.20	0.13	0.10

Note: The ρ_A values are given in dB and natural units. The values of ρ_B/ρ_A are in natural units.

TABLE 4.16
Values of ρ_B for Given ρ_A and Ratio ρ_B/ρ_A for 2- and 4-Antenna Ports

ρ_A(dB)	3	2	1	0	−1.77	−3	−4.77	−6
ρ_B/ρ_A	2.00	1.58	1.26	1.00	0.67	0.50	0.33	0.25
1.25	2.49	1.98	1.57	1.25	0.83	0.63	0.42	0.31
1.00	2.00	1.58	1.26	1.00	0.67	0.50	0.33	0.25
0.75	1.50	1.19	0.94	0.75	0.50	0.38	0.25	0.19
0.50	1.00	0.79	0.63	0.50	0.33	0.25	0.17	0.13

Note: The ρ_A values are given in dB and natural units. The values of ρ_B/ρ_A are in natural units.

4.15.2 Transmission and Propagation Models

The mobile base station power model requires several inputs that are provided by 3GPP documentation [122] and industry-accepted formulations. The mathematical representation of the horizontal and vertical transmit antenna patterns are examples. The form adopted for the horizontal antenna transmit pattern in dB is taken from Table A.2.1.1–2 of 3GPP TR 36.814 V9.2.0 (2017-03) [61]:

$$\Theta(\theta)(dB) = -12\left(\frac{\theta}{\theta_{3dB}}\right)^2$$

$$\Phi(\varphi)(dB) = -12\left(\frac{\varphi - \varphi_{tilt}}{\varphi_{3dB}}\right)^2$$

(4.72)

where θ and φ are the horizontal and vertical angles subtended by the user relative to the base station transmitter under consideration, respectively. Also, θ_{3dB} is the half-power angle typically set to $\theta_{3dB} = 70$ degrees, φ_{3dB} is the half-power angle, typically set to $\varphi_{3dB} = 10$ degrees and φ_{tilt} is the angle of the downward tilt of the peak power direction, typically set to around $\varphi_{tilt} = 10$ degrees. The overall three-dimensional transmission pattern is given by:

$$\Pi_{m,p}\left(z^{(k)}\right)(dB) = \Pi_{m,p}(\theta,\varphi)(dB) = \max\left\{\Theta(\theta) + \Phi(\varphi), -25\right\} \qquad (4.73)$$

Note: when calculating $\Pi_{m,p}$, both terms in the brackets {.} are negative. Therefore, taking the maximum corresponds to taking the value closest to zero dB.

The value of transmitter antenna gain G_{TX} in (4.35) can be set to 8dBi.

There are a variety of path loss models [112, 114, 115, 117, 118, 123–125]. For this work, we have adopted the Stanford University Interim (SUI) loss model [112, 125]. The model is based on multiple empirical measurements and provides a mean path loss by categorising the local terrains into three different types:

Category	Terrain Description
A	Hilly terrain with moderate-to-heavy tree densities. (maximum path loss)
B	Either mostly flat terrain with moderate-to-heavy tree densities, or hilly terrain with light tree densities. (moderate path loss)
C	Mostly flat terrain with light tree densities. (minimum path loss)

Typically for these three terrain categories, the general scenario this model covers is as follows:

- Inter Base Station distance (IBS) < 2000 m
- Receiver antenna height above ground is between 2 m and 10 m
- Base station antenna height above ground is between 15 m and 40 m
- High cell coverage is required, around 80% to 90%.

The SUI model basic path loss equation is applicable for frequencies around 2 GHz, receiver antenna height around 2 m and is suitable for suburban environments. The mean path loss is given by:

$$L_{path}(dB) = A + 10\gamma \log\left(\frac{r}{r_{norm}}\right) + \upsilon \qquad (4.74)$$

where

$$r > r_{norm}$$

$$A = 20\log\left(\frac{4\pi r_{norm}}{\lambda}\right) \qquad (4.75)$$

$$\gamma = a - bh_b + \frac{c}{h_b}$$

with

$$r_{norm} = 100m$$
$$10m < h_b < 80m \qquad (4.76)$$

The parameter υ represents the "shadowing fading". This is a random process included to account for the possible occurrence of local environmental features that will induce a random additional loss in the path. For example, vegetation, nearby objects, reflective surfaces and the like may be present causing additional losses that are very difficult it include in a deterministic manner. Hence, the shadowing loss/effect is typically represented as a log-normal random variable (i.e. Gaussian distribution in dB). The probability density function of υ has the form:

$$P_s(\upsilon) = \frac{1}{\sigma_s \upsilon \sqrt{2\pi}} \exp\left(-\frac{\left(\log\upsilon - \mu_s\right)^2}{2\sigma_s^2}\right) \tag{4.77}$$

where $0 < x < \infty$. In this case we set the, $\mu_s = 0$, and $\sigma_s = 4$ dB to 8 dB depending upon the environment [124].

Parameter	Description		
r	Distance from base station to receiver (m)		
λ	Sub-carrier wavelength (m) for frequency range ≤ 2 GHz		
γ	Path loss exponent		
h_b	Base station height above ground (m)		
h_r	User receiver antenna height above ground (m)		
υ	Shadowing fading, probability distribution given by (4.77)		
	Terrain category A	Terrain category B	Terrain category C
a	4.6	4	3.6
b	0.0075	0.0065	0.005
c	12.6	17.1	20

For frequencies above 2 GHz and receiver antennas between, h_r, 2 m and 10 m, the following correction factors are introduced:

$$L_{path}(dB) = A + 10\gamma \log\left(\frac{r}{r_{norm}}\right) + \upsilon + \Delta L_f + \Delta L_{rh} \tag{4.78}$$

where ΔL_f is a frequency correction factor:

$$\Delta L_f = 6.0 \log\left(\frac{f}{2000}\right) \tag{4.79}$$

And ΔL_{rh} is a receiver antenna height correction factor:

$$\Delta L_{rh} = \begin{cases} -10.8 \log\left(\dfrac{h_r}{2}\right) & terrain\ category\ A\ and\ B \\ -20 \log\left(\dfrac{h_r}{2}\right) & terrain\ category\ C \end{cases} \tag{4.80}$$

Other detailed path loss models for a range of geographical categories are provided by 3GPP TR 38.901 V16.1.0 (2019-12), "Technical Specification Group Radio Access Network; Study on channel model for frequencies from 0.5 to 100 GHz (Release 16) [124].

4.15.3 Power Consumption Model Values

Tables 4.1, 4.2, 4.9, 4.11, and 4.12 have introduced values for many of the base station performance and power consumption parameters, in particular those relevant to the Static Model. A careful reader may note that some of the values in Table 4.17 may not directly agree with similar tables in the cited references. The reader needs to take careful note of additional factors stated in those references that lead to these different values. In the course of working with the Advanced Model, additional parameters will be needed. Table 4.17 summarises a number of published consumption values for different base station types. For macro base stations, the values are for three sectors and 2×2 MIMO (or in one instance, 8×2 MIMO). The transmitter Power Output column lists the power output of

TABLE 4.17
Published Base Station Power Consumption Values

Reference	BS Configuration	MIMO	Transmitter Power Output, (W)	Sectors per Base Station	RF Chains per Sector	Max Power (W)	Idle Power (W)
	Macro Base Station with Cabinet Baseband Equipment					**2010 Vintage Performance**	
1	Earth D2.3 (Auer *et al*)	2×2	40	3	2	1350	329
14	Blume et al	2×2	40	3	2	1394	712
9	Holtkamp	2×2	40	3	2	1370	335
	Macro Base Station with Remote Radio Heads					**2010 Vintage Performance**	
1	Earth D2,3 (Auer *et al*)	2×2	20	3	2	755	299
	Small Cell Base Stations					**2010 Vintage Performance**	
1	Earth D2.3 Micro	2×2	6.3	1	2	145	79
1	Earth D2.3 Pico	2×2	0.13	1	2	14.7	9.4
						2012 Performance (Estimated in Reference)	
	Macro Base Station with Cabinet Baseband Equipment						
1	Earth D2.3 2012 Base	2×2	40	3	2	965	191
						2012 Performance (Estimated in Reference)	
	Macro Base Station with Remote Radio Heads						
1	Earth D2.3 2012 RRH	2×2	20	3	2	528	175
						2012 Performance (Estimated in Reference)	
	Small Cell Base Stations						
1	Earth D2.3 2012 MICRO	2×2	6.3	1	2	94.00	42
1	Earth D2.3 2012 Pico	2×2	0.13	1	2	9.00	5.2
	Macro base station with Remote Radio Heads					**2020 Performance (Estimated in Reference)**	
14	Blume et al 2020 RRH	8×2	10	3	8	665	189
12	Blume et al 2020 RRH	2×2	40	3	2	638	132
126	Debaille et al 2020 RRH	2×2	40	3	2	702	115

(Continued)

TABLE 4.17 (*Continued*)
Published Base Station Power Consumption Values

Reference	BS Configuration	MIMO	Transmitter Power Output, (W)	Sectors per Base Station	RF Chains per Sector	Max Power (W)	Idle Power (W)
	Small Cell Base Stations					2020 Performance (Estimated in Reference)	
14	Blume et al 2020 Small	2×2	0.5	1	2	11	4
126	Debaille et al 2020 Pico	2×2	0.5	1	2	6.9	2.3

the transmitter power amplifier. For a system where the power amplifier is located at the base of the antenna, a cable loss of 3dB means that the power radiated by the antenna, as required for signal strength calculations, is 3 dB or a factor of two lower than this. The "Number of RF Chains" column lists the number of power amplifiers, baseband signal processors, radio signal generators and receivers used in each base station sector. The maximum power consumption of a fully-loaded base station, and the idle power consumption, make up the last two columns. The idle power consumption is the sum of the power consumption of the RF chains, less the consumption of the power amplifiers.

For small cells, the power values apply for a single 2×2 MIMO installation. Small cells generally have just one 2×2 MIMO system, although cells with single output SISO systems are also deployed.

The values quoted in this table use the power values for the various components of the base station directly, where these are available, [1, 12, 14, 126], or developed from other data in the quoted reference [9].

It has been mentioned that much of this type of information is regarded as commercial and confidential by equipment manufacturers, thus difficult to find and much of this information is dated. A useful source of assistance in this area is an online base station power consumption modelling tool made available by IMEC [127]. IMEC has separated the various components of a base station, and used models for the trends of performance in each component to build an estimate of likely future energy consumption performance.

REFERENCES

[1] G. Auer *et al.*, "INFSO-ICT-247733 EARTH Deliverable D 2. 3 Energy Efficiency Analysis of the Reference Systems, Areas of Improvements and Target Breakdown," 2010.

[2] C. Desset *et al.*, "*Flexible Power Modeling of LTE Base Stations,*" in *2012 IEEE Wireless Communications and Networking Conference (WCNC)*, 2012.

[3] T. Bonald and A. Proutiere, "*Wireless Downlink Data Channels: User Performance and Cell Dimensioning,*" in *MobiCom'03, Proceedings of the 9th Annual International Conference on Mobile Computing and Networking*, 2003.

[4] J. Daigle, *Queueing Theory with Applications to Packet Telecommunications*, Springer Science, 2006.

[5] N. Chee-Hock and S. Boon-Hee, *Queueing Modelling Fundamentals with Applications in Communications Networks*, John Wiley & Sons, 2008.

[6] S. Fowler *et al.*, "*Analysis of Vehicular Wireless Channel Communication Via Queueing Theory Model,*" in *2014 IEEE International Conference on Communications (ICC)*, 2014.

[7] M. Deruyck *et al.*, "Modelling and Optimization of Power Consumption in Wireless Access Networks," *Computer Communications*, vol. 34, p. 2036, 2011.

[8] M. Deruyck *et al.*, "Reducing The Power Consumption In Wireless Access Networks: Overview And Recommendations," *Progress In Electromagnetics Research*, vol. 132, p. 255, 2012.

[9] H. Holtkamp *et al.*, "A Parameterized Base Station Power Model," *IEEE Communications Letters*, vol. 17, no. 11, p. 2033, 2013.

[10] J. Lorincz and T. Matijevic, "Energy-Efficiency Analyses of Heterogeneous Macro and Micro Base Station Sites," *Computers and Electrical Engineering*, vol. 40, no. 2, p. 330, 2014.

[11] K. Liu *et al.*, *"Base Station Power Model and Application for Energy Efficient LTE,"* in *15th IEEE International Conference on Communication Technology*, 2013.

[12] GreenTouch, "GreenTouch Final Results from Green Meter Research Study: Reducing the Net Energy Consumption in Communications Networks by up to 98% by 2020," GreenTouch, 2015.

[13] H. Yang and T. Marzetta, *"Total Energy Efficiency of Cellular Large Scale Antenna System Multiple Access Mobile Networks,"* in *2013 IEEE Online Conference on Green Communications (OnlineGreenComm)*, 2013.

[14] O. Blume *et al.*, *"Energy Efficiency of LTE Networks under Traffic Loads of 2020,"* in *The Tenth International Symposium on Wireless Communication Systems*, 2013.

[15] J.-C. Nanan and B. Stern, "Small Cells Call for Scalable Architecture," Freescale, 2012.

[16] M. Alsharif *et al.*, "Green and Sustainable Cellular Base Stations: An Overview and Future Research Directions," *Energies*, vol. 10, no. 5, p. 587, 2017.

[17] G. Auer *et al.*, "How Much Energy is Needed to Run a Wireless Network?," in *Green Radio Communication Networks*, E. Hossain *et al.*, Eds., Cambridge University Press, 2012, p. 360.

[18] D. Halperin *et al.*, *"Demystifying 802.11n Power Consumption,"* in *Proceedings of the 2010 International Conference on Power aware computing and systems (HotPower'10)*, 2010.

[19] G. Palem and S. Tozlu, "On Energy Consumption of Wi-Fi Access Points," in *2nd IEEE International Workshop on Densely Connected Networks*, 2012.

[20] A. Garcia-Saavedra *et al.*, *"Energy Consumption Anatomy of 802.11 Devices and its Implication on Modeling and Design,"* in *CoNEXT '12: Proceedings of the 8th International Conference on Emerging Networking Experiments and Technologies*, 2012.

[21] S. Chiaravalloti *et al.*, "Power Consumption of WLAN Network Elements – TKN Technical Report TKN-11-002," Technische Universitat Berlin, 2011.

[22] P. Serrano *et al.*, "Per-Frame Energy Consumption in 802.11 Devices and Its Implication on Modeling and Design," *IEEE/ACM Transactions on Networking*, vol. 23, no. 4, p. 1243, 2015.

[23] National Instruments, "Introduction to Wireless LAN Measurements – From 802.11a to 802.11ac," National Instruments.

[24] Eidgenössisches Departement für Umwelt, Verkehr, Energie und Kommunikation UVEK, "WLAN Factsheet – Wireless Local Area Networks," 2017.

[25] S. Andrade-Morelli *et al.*, "Energy Consumption of Wireless Network Access Points," in *Green Communication and Networking. GreeNets 2012*, Lecture Notes of the Institute for Computer Sciences, Social Informatics and Telecommunications Engineering, vol. 113, J. Mauri and J. Rodrigues, Eds., Springer, 2013, p. 81.

[26] P. Bertold, "Code of Conduct on Energy Consmumption of Broadband Equipment," European Comission, 2019.

[27] AT&T, "Will 5G replace WiFi?," AT&T, 2019.

[28] M. Mourato *et al.*, "A Novel and Realistic Power Consumption Model for Multi-Technology Radio Networks," *URSI Radio Science Bulletin*, vol. 2018, no. 364, p. 20, 2018.

[29] J. Lorincz *et al.*, "Measurements and Modelling of Base Station Power Consumption under Real Traffic Loads," *Sensors*, vol. 12, p. 4281, 2012.

[30] J. Gadze *et al.*, "Real Time Traffic Base Station Power Consumption Model for Telcos in Ghana," *International Journal of Computer Science and Telecommunications*, vol. 7, no. 5, p. 6, 2016.

[31] A. Ambrosy *et al.*, *"Dynamic Bandwidth Management for Energy Savings in Wireless Base Stations,"* in *Globecom 2012 – Symposium on Selected Areas in Communications*, 2012.

[32] B. Delaillie *et al.*, *"A Flexible and Future-Proof Power Model for Cellular Base Stations,"* in *IEEE 81st Vehicular Technology Conference (VTC Spring)*, 2015.

[33] Y. Zhang *et al.*, *"An Overview of Energy-Efficient Base Station Management Techniques,"* in *24th Tyrrhenian International Workshop on Digital Communications – Green ICT (TIWDC)*, 2013.

[34] GSMA, "Road to 5G: Introduction and Migration," 23 April 2018. [Online]. Available: https://www.gsma.com/futurenetworks/wp-content/uploads/2018/04/Road-to-5G-Introduction-and-Migration_FINAL.pdf. [Accessed 15 August 2020].

[35]   Third Generation Participation Project (3GPP), "Third Generation Participation Project," 3GPP, 2020. [Online]. Available: https://www.3gpp.org/. [Accessed 15 August 2020].

[36]   ITU-T, "International Telecommunications Union Radiocommunication Sector," 2020. [Online]. Available: https://www.itu.int/en/ITU-R/Pages/default.aspx. [Accessed 15 August 2020].

[37]   Telesystem Innovations Inc., "LTE in a Nutshell: The Physical Layer," Telesystem Innovations Inc., 2010.

[38]   J. Zyren and W. McCoy, "Overview of the 3GPP Long Term Evolution Physical Layer," Freescale Semiconductor, 2007.

[39]   T. Ali-Yahiya, *Understanding LTE and its Performance*, Springer, 2011.

[40]   S. Sesia *et al.*, Eds., *LTE – The UMTS Long Term Evolution: From Theory to Practice*, John Wiley & Sons, 2009.

[41]   D. Astely *et al.*, "LTE Release 12 and Beyond" *IEEE Communications Magazine*, vol. 51, no. 7, p. 154, 2013.

[42]   Ericsson, "Long Term Evolution (LTE): An Introduction," Ericsson, 2007.

[43]   M. Sukar *et al.*, "SC-FDMA & OFDMA in LTE physical layer," *International Journal of Engineering Trends and Technology (IJETT)*, vol. 12, no. 2, p. 74, 2014.

[44]   S. Ahmadi, *LTE-Advanced: A Practical Systems Approach to Understanding 3GPP LTE Releases 10 and 11 Radio Access Technologies*, Elsevier Science & Technology, 2013.

[45]   E. Dahlman *et al.*, *4G: LTE/LTE-Advanced for Mobile Broadband*, 2nd Edition, Academic Press, 2013.

[46]   H. Holma and A. Toskala, Eds., *LTE for UMTS – OFDMA and SC-FDMA Based Radio Access*, John Wiley & Sons, 2009.

[47]   C. Cox, *An Introduction to LTE, LTE-Advanced, SAE, VoLTE and 4G Mobile Communications* (2nd Edition), John Wiley & Sons, 2014.

[48]   Huawei Technologies Co. Ltd., Long Term Evolution (LTE) Radio Access Network Planning Guide, Huawei Technologies Co. Ltd., 2011.

[49]   A. Mohammadi and F. Ghannouchi, *RF Transceiver Design for MIMO Wireless Communications, Lecture Notes in Electrical Engineering* Vol. 145, Springer-Verlag, 2012.

[50]   A. Ghosh and R. Ratasuk, *Essentials of LTE and LTE-A*, Cambridge University Press, 2011.

[51]   F. Khan, *LTE for 4G Mobile Broadband: Air Interface Technologies and Performance*, Cambridge University Press, 2009.

[52]   Third Generation Participation Project (3GPP), "TS 36.101 V14.3.0 (2017-04) LTE; Evolved Universal Terrestrial Radio Access (E-UTRA) User Equipment (UE) radio transmission and reception," Third Generation Participation Project, 2017.

[53]   ITU-R, " Report ITU-R M.2135-1, Guidelines for evaluation of radio interface technologies for IMT-Advanced," International Telecommunications Union, 2009.

[54]   D. Lopez-Perez *et al.*, "*Optimization Method for the Joint Allocation of Modulation Schemes, Coding Rates, Resource Blocks and Power in Self-Organizing LTE Networks*," in *30th IEEE International Conference on Computer Communications (IEEE INFOCOM 2011)*, 2011.

[55]   J. Liu *et al.*, "*Design and Analysis of LTE Physical Downlink Control Channel*," in *IEEE 69th Vehicular Technology Conference – VTC Spring 2009*, 2009.

[56]   D. Lee *et al.*, "*Inter-Cell Interference Coordination for LTE Systems*," in *IEEE Global Communications Conference (GLOBECOM)*, 2012.

[57]   O. Ramos-Cantor *et al.*, "Centralized Coordinated Scheduling in LTE-Advanced Networks," *EURASIP Journal on Wireless Communications and Networking*, vol. 2017, p. 122, 2017.

[58]   C. Chan *et al.*, "Assessing Network Energy Consumption of Mobile Applications," *EEE Communications Magazine*, vol. 53, no. 11, p. 182, 2015.

[59]   P. Frenger *et al.*, "*Reducing Energy Consumption in LTE with Cell DTX*," in *2011 IEEE 73rd Vehicular Technology Conference (VTC Spring)*, 2011.

[60]   M. Alsharif *et al.*, "Energy Efficiency and Coverage Trade-Off in 5G for Eco-Friendly and Sustainable Cellular Networks," *Symmetry*, vol. 11, p. 408, 2019.

[61]   Third Generation Participation Project (3GPP), "TR 36.814 V9.2.0 (2017-03) Technical Specification Group Radio Access Network; Evolved Universal Terrestrial Radio Access (E-UTRA); Further advancements for E-UTRA physical layer aspects (Release 9)," Third Generation Participation Project, 2017.

[62]   F. Salem *et al.*, "*Energy Consumption Optimization in 5G Networks Using Multilevel Beamforming and Large Scale Antenna Systems*," in *IEEE Wireless Communications and Networking Conference (WCNC)*, 2016.

[63] M. Ismail *et al.*, "A Survey on Green Mobile Networking: From The Perspectives of Network Operators and Mobile Users," *EEE Communications Surveys & Tutorials*, vol. 17, no. 3, p. 1535, 2015.

[64] NTT DOCOMO, Alcatel-Lucent, Alcatel-Lucent Shanghai Bell, "R1-11 3495 "Base Station Power Model"," 2011.

[65] A. Ambrosy *et al.*, "*Energy Savings in LTE Macro Base Stations*," in *7th IFIP Wireless and Mobile Networking Conference (WMNC)*, 2014.

[66] B. Bossy *et al.*, "*Optimization of Energy Efficiency in the Downlink LTE Transmision*," in *IEEE ICC 2017 Green Communications Systems and Networks Symposium*, 2017.

[67] P. Morgensen *et al.*, "*LTE Capacity Compared to the Shannon Bound*," in *2007 IEEE 65th Vehicular Technology Conference – VTC2007-Spring*, 2007.

[68] G. Li *et al.*, "Energy-Efficient Wireless Communications: Tutorial, Survey, and Open Issues," *IEEE Wireless Communications*, vol. 18, no. 6, p. 28, 2011.

[69] S. Buzzi *et al.*, "A Survey of Energy-Efficient Techniques for 5G Networks and Challenges Ahead," *IEEE Journal on Selected Areas in Communications*, vol. 34, no. 4, p. 697, 2016.

[70] R. Litjens *et al.*, "*Assessment of the Energy Efficiency Enhancement of Future Mobile Networks*," in *IEEE WCNC'14 Track 3 (Mobile and Wireless Networks*, 2014.

[71] A. Abdulkafi *et al.*, "*Energy Efficiency of LTE Macro Base Station*," in *1st IEEE International Symposium on Telecommunication Technologies*, 2012.

[72] P. Mogensen *et al.*, "*LTE Capacity Compared to the Shannon Bound*," in *IEEE 65th Vehicular Technology Conference (VTC2007 Spring)*, 2007.

[73] J. Fan *et al.*, "*MCS Selection for Throughput Improvement in Downlink LTE Systems*," in *Proceedings of 20th International Conference on Computer Communications and Networks*, 2011.

[74] M. Salman *et al.*, "*CQI-MCS Mapping for Green LTE Downlink Transmission*," in *Proceedings of the Asia-Pacific Advanced Network Conference*, 2013.

[75] P. Vieira *et al.*, "*LTE Spectral Efficiency Using Spatial Multiplexing MIMO for Macro-cells*," in *2nd International Conference on Signal Processing and Communication Systems*, 2008.

[76] S. Kitanov, "*Simulator for the LTE Link Level Performance Evaluation*," in *ICT Innovations 2011 Web Proceedings*, 2011.

[77] Third Generation Participation Project (3GPP), "3GPP TS 36.213 V15.7.0 (2019-09), Technical Specification Group Radio Access Network Evolved Universal Terrestrial Radio Access (E-UTRA); Physical layer procedures" (Release 15)," European Telecommunications Standards Institute, 2019.

[78] S. Ali, "On the Evolution of Coordinated Multi-Point (CoMP) Transmission in LTE-Advanced," *International Journal of Future Generation Communication and Networking*, vol. 7, no. 4, p. 91, 2014.

[79] D. Lee, "Coordinated Multipoint Transmission and Reception in LTE-Advanced: Deployment Scenarios and Operational Challenges," *IEEE Communications Magazine*, vol. 50, no. 2, p. 148, 2012.

[80] F. Qamar *et al.*, "A Comprehensive Review on Coordinated Multi-point Operation for LTE-A," *Computer Networks*, vol. 123, p. 19, 2017.

[81] M. Rumney, Ed., *LTE and the Evolution to 4G Wireless: Design and Measurement Challenges*, John Wiley & Sons, 2013.

[82] B. Classon *et al.*, "Channel Coding and Link Adaptation," in *LTE – The UMTS Long Term Evolution*, S. Sesia, *et al.*, Eds., John Wiley & Sons, 2009, p. 207.

[83] A. Sibille *et al.*, Eds., *MIMO From Theory to Implementation*, Elsevier, 2011.

[84] N. Wei *et al.*, "*Baseline E-UTRA Downlink Spectral Efficiency Evaluation*," in *IEEE Vehicular Technology Conference (VTCF 2006)*, 2006.

[85] R. Srinivasan, *et al.*, "IEEE 802.16m Evaluation Methodology Document; IEEE 802.16m-08/004r2," 2008.

[86] W.-B. Yang and M. Souryal, "LTE Physical Layer Performance Analysis; NISTIR 7986," National Institute of Standards and Technology, 2014.

[87] A. Sassan, *LTE-Advanced: A Practical Systems Approach to Understanding 3GPP LTE Releases 10 and 11 Radio Access Technologies*, Elsevier Science and Technology, 2013.

[88] I. Latif *et al.*, "*Link Abstraction for Multi-User MIMO in LTE using Interference-Aware Receiver*," in *012 IEEE Wireless Communications and Networking Conference (WCNC)*, 2012.

[89] J. Wang *et al.*, "Interference Coordination for Millimeter Wave Communications in 5G for Performance Optimisation," *EURASIP Journal on Wireless Communications and Networking*, vol. 2019, p. 46, 2019.

[90] Ericsson, "Coverage and Capacity Dimensioning – Recommendation," Ericsson, 2009.

[91] C. Mehlführer *et al.*, "The Vienna LTE Simulators – Enabling Reproducibility in Wireless Communications Research," *EURASIP Journal on Advances in Signal Processing*, vol. 2011, p. 29, 2011.

[92] G. Basilashvili, "Study of Spectral Efficiency for LTE Network," *American Scientific Research Journal for Engineering, Technology, and Sciences*, vol. 29, no. 1, p. 21, 2017.

[93] G. Göktepe *et al.*, *"Reduced CBG HARQ Feedback for Efficient Multimedia Transmissions in 5G for Coexistence with URLLC Traffic,"* in *2018 IEEE International Symposium on Broadband Multimedia Systems and Broadcasting (BMSB)*, 2018.

[94] M. Fuentes *et al.*, "5G New Radio Evaluation Against IMT-2020 Key Performance Indicators," *IEEE Access*, vol. 8, p. 110880, 2020.

[95] M. Rupp *et al.*, *The Vienna LTE-Advanced Simulators: Up and Downlink, Link and System Level Simulation*, Springer, 2016.

[96] Y. Yang *et al.*, *5G Wireless Systems: Simulation and Evaluation Techniques*, Springer International Publishing, 2018.

[97] X. Zhang, *LTE Optimization Engineering Handbook*, John Wiley & Sons, 2018.

[98] P. Carro, "Evaluation of Radio Resource Management Impact on RoF Signal Transmission for Downlink LTE," *Journal of Lightwave Technology*, vol. 36, no. 9, p. 1591, 2018.

[99] A. Adouane, Dynamic management of spectral resources in LTE networks, Université de Versailles-Saint Quentin en Yvelines, 2015.

[100] A. Conte, "Energy-Efficient Base Stations," in *Green Communications: Principles, Concepts and Practice*, K. Samdanis *et al.*, Eds., John Wiley & Sons, 2015, p. 73.

[101] S. Adegbite *et al.*, *"Improved PCFICH decoding in LTE systems,"* in *The 21st IEEE International Workshop on Local and Metropolitan Area Networks*, 2015.

[102] A. Bürgi *et al.*, "Time Averaged Transmitter Power and Exposure to Electromagnetic Fields from Mobile Phone Base Stations," *International Journal of Environmental Research and Public Health*, vol. 11, p. 8025, 2014.

[103] M. Rana *et al.*, *"Peak to Average Power Ratio Analysis for LTE Systems,"* in *2nd International Conference on Communication Software and Networks*, 2010.

[104] Third Generation Participation Project (3GPP), "Third Generation Participation Project (3GPP); 3GPP TS 136 211 V10.0.0 (2011-01) LTE; Evolved Universal Terrestrial Radio Access (E-UYTRA); Physical channels and modulation (Release 10)," 3GPP Organizational Partners, 2011.

[105] P. Rajesh *et al.*, *"Impact of Dynamic Control Format Indicator on Downlink Throughput Performance in LTE System,"* in *2018 International Conference on Electronics, Information, and Communication (ICEIC)*, 2018.

[106] M. Baker and T. Moulsley, "Downlink Physical Data and Control Channels," in *LTE – The UMTS Long Term Evolution: From Theory to Practice*, S. Sesia, *et al.*, Eds., John Wiley & Sons, 2011, p. 189.

[107] J. Liu *et al.*, *"Design and Analysis of LTE Physical Downlink Control Channel,"* in *VTC Spring 2009 - IEEE 69th Vehicular Technology Conference*, 2009.

[108] M. Kottkamp *et al.*, "LTE-Advanced Technololgy Introduction White Paper," Rhode & Schwarz, 2012.

[109] B. Schulz, "LTE Transmission Modes and Beamforming White Paper," Rhode and Schwarz, 2015.

[110] S. Sun *et al.*, *"Path Loss, Shadow Fading, and Line-Of-Sight Probability Models for 5G Urband Macro-Cellular Scenarios,"* in *2015 IEEE Global Communications Conference Workshop (GC Wkshps)*, 2015.

[111] K. Kumaran *et al.*, "Correlated Shadow-Fading in Wireless Networks and its Effect on Call Dropping," *Wireless Networks*, vol. 8, p. 61, 2002.

[112] V. Erceg *et al.*, "Channel Models for Fixed Wireless Applications," IEEE 802.16 Broadband Wireless Access Working Group, 2001.

[113] Z. Naseem *et al.*, "Propagation Models for Wireless Communication System," *International Research Journal of Engineering and Technology (IRJET)*, vol. 5, no. 1, p. 237, 2018.

[114] R. Jain, "Channel Models: A Tutorial, V1," 21 February 2007. [Online]. Available: https://www.cse.wustl.edu/~jain/cse574-08/ftp/channel_model_tutorial.pdf. [Accessed 15 August 2020].

[115] T. Chrysikos and S. Kotsopoulos, *"Site-specific Validation of Path Loss Models and Large-scale Fading Characterization for a Complex Urban Propagation Topology at 2.4 GHz,"* in *Proceedings of the International Multi-Conference of Engineers and Computer Scientists 2013 (IMECS 2013)*, 2013.

[116] V. Abhayawardhana *et al.*, *"Comparison of Empirical Propagation Path Loss Models for Fixed Wireless Access Systems,"* in *IEEE 61st Vehicular Technology Conference*, 2005.

[117]   A. Zreikat and M. Djordjevic, *"Performance Analysis of Path loss Prediction Models in Wireless Mobile Networks in Different Propagation Environments,"* in *Proceedings of the 3rd World Congress on Electrical Engineering and Computer Systems and Science (EECSS'17)*, 2017.

[118]   I. Khan and S. Kamboh, "Performance Analysis of Various Path Loss Models for Wireless Network in Different Environments," *International Journal of Engineering and Advanced Technology (IJEAT)*, vol. 2, no. 1, p. 161, 2012.

[119]   M. Hata, "Empirical Formula for Propagation Loss in Land Mobile Radio Services," *IEEE Transactions on Vehicular Technology*, vol. 29, no. 3, p. 317, 1980.

[120]   J. Salo and E. Zacarfas, "Analysis of LTE Radio Load and User Throughput," *International Journal of Computer Networks & Communications*, vol. 9, no. 6, p. 33, 2017.

[121]   T. Lin *et al.*, *"A Distributed Multi-objective Optimisation Framework for Energy Efficiency in Mobile Backhaul Networks,"* in *IEEE Conference on Control Applications (CCA)*, 2015.

[122]   Third Generation Participation Project, "3GPP TR37.840 V12.1.0 Technical Specification Group Radio Access Network; Study of Radio Frequency (RF) and Electromagnetic Compatibility (EMC) requirements for Active Antenna Array System (AAS) base station (Release 12)," 3GPP, 2013.

[123]   P. Sharma and R. Singh, "Comparative Analysis of Propagation Path loss Models with Field Measured Data," *International Journal of Engineering Science and Technology*, vol. 2, no. 6, p. 2008, 2010.

[124]   Third Generation Participation Project, "3GPP TR 38.901 V16.1.0 Technical Specification Group Radio Access Network;Study on channel model for frequencies from 0.5 to 100 GHz (Release 16)," 3GPP, 2019.

[125]   V. Erceg *et al.*, "An Empirically Based Path Loss Model for Wireless Channels in Suburban Environments," *IEEE Journal on Selected Areas in Communications*, vol. 17, no. 7, p. 1205, 1999.

[126]   B. Debaillie *et al*, *"A Flexible and Future-Proof Power Model for Cellular Base Stations,"* *IEEE Vehicular Technology Conference (VTC-spring)*, May 2015.

[127]   IMEC, "Power Model for Wireless Base Stations," https://www.imec-int.com/en/powermodel, Accessed 28 June 2021.

5 Advanced Modelling of Mobile Networks

From the discussion in Chapter 4, it is apparent that estimating the power consumption of a mobile network or base station is substantially more complicated than for routers and other access networks. Despite its mathematical complexity the base station power model described in Chapter 4 adopted a somewhat simplified approach that describes a relatively unsophisticated steady-state network scenario. That approach assumes the mobile base station can accommodate all the traffic demands made of it by the users it is servicing. It does not account for the network response to new users' requests for service from the network and how this may impact the users already on line. A dynamic model that accommodates more realistic situations such as network congestion, new users coming on line and completed or failed downloads leaving the network requires us to use a numerical simulation rather than just a model as in Chapter 4; one that more closely characterises the operation of a real mobile base station. Such a simulation can include the feedback loops that are part of the operation of a mobile base station in response to user demands. It can also include the random nature of user traffic demands not manifest in the Static Model in Chapter 4.

For those who wish to undertake the task of constructing a more representative model than the Static Model described in Chapter 4, this chapter provides an overview of a mobile network simulation that includes such factors. This model provides much more data on the performance of the network, thereby providing the modeller with a deeper understanding of the energy efficiency of the network and its dependencies.

A mobile network covering a geographic region typically includes a number of base stations and the supporting back-haul, data switching, and transport equipment interconnecting the base stations. These interconnections are required for the mobile network to serve customers in the region. In the model presented below, we focus on the energy consumption of the base stations themselves, whose power consumption dominates the overall network power.

We continue our focus on 4th-generation mobile networks and equipment using LTE (Long Term Evolution). This is a system developed by 3GPP (Third Generation Partnership Project consortium) [1] and approved by the International Telecommunications Union (ITU) as the standard for worldwide 4th-generation mobile networks [2]. Much of the technology of 4G/LTE is being carried through into the coming 5G networks.

As with the objectives of Chapter 4, this chapter will cover techniques to model the energy consumption of a small network of mobile base stations, taking account of networks in differing demographic regions and different user traffic demand levels. These techniques can enable calculation by an end user of the energy consumed by a mobile service. Complementing this type of work, the ITU and the European Telecommunications Standards Institute (ETSI) are developing standards for the measurement and reporting of mobile network performance and energy consumption, which will be introduced in Appendix 2.

We start by making some comparisons with the Static Model of Chapter 4. The Static Model assumes a static steady-state situation in which the user demand is within the downlink capacity provided by the network. It does not include any dynamic or stochastic aspects in the network, e.g. details such as new users becoming active and active users completing their task and leaving the network. The Static Model uses a time, t_u, based upon Radio Frames for which the time step t_u to t_{u+1} is 10 ms, corresponding to a single Radio Frame. For convenience, we can take the t_u to be the time corresponding to the mid-point of the u th Radio Frame as depicted in Figure 4.7. The number

DOI: 10.1201/9780429287817-5

of users served by the base station at time t_u, is fixed at $N_{BS}^{(usr)}(t_u)$, and likewise each user's base station resource utilisation, $\mu_{BS}^{(k)}(t_u)$ is fixed, corresponding to the proportion of available base station PDSCH REs allocated to that user. These two parameters are effectively constant, with the time parameter t_u primarily used to indicate the stage of the diurnal cycle that is being simulated.

Although the modeller can adjust $\mu_{BS}^{(k)}(t_u)$ to represent high and low network load, in each case the network is in a steady state. In circumstances of high network load, with $\mu_{BS}(t_u)$ close to unity, the modeller has to check, by hand, that the corresponding throughput is less than the 100% utilisation throughput of the base station in accordance with (4.31). In a real network, at high utilisation the probability of the network becoming congested and providing a degraded service is significant. When using the Static Model in this regime, the modeller will need to be careful to appropriately interpret the results.

In contrast, a real mobile network has built-in protocols and processes designed to enable the base station to respond to high traffic loads. These processes include continuing reports on service quality from the user handset to the base station, decisions by the base station on resource allocation between users who attempt to optimise the quality of service across all users. The Static Model of Chapter 4 does not include any of these processes but assumes the network has settled these issues and is in a resulting steady state.

One strategy for dealing with new users coming on line is for the base station to always operate at its maximum possible utilisation. This is done so that users will have their requests fulfilled as quickly as possible so that the allocated resources are freed up as soon as possible, and thus made available for other users. As total user demand increases (around peak traffic times) user demands on the base station may start to overburden the resources available. Consequentially when a new user comes on line, the resources allocated to that user are likely to be taken from others. Thus the download data rate achieved by each user decreases, and users will be on line for longer. If the rate of new users coming on line does not decrease, the longer duration required to service current users may result in the network moving into a situation where some users experience a download data rate so low that they cannot have their service requests fulfilled within an acceptable time. In the ordinary course of network design, network operators provision their networks such that an acceptable level of service quality is maintained over the daily "busy hours"; however, the concept of "acceptable" is at the whim of the operator. Users receiving poor performance may "try again later", thereby inadvertently but temporarily improving the outcomes for other users, or in some cases, poorly performing user downloads may be terminated. Such outcomes are not provided for in the Static Model of Chapter 4.

In the Static Model, the k th user is described by their location, $z^{(k)}$ and their "user demand" set by their utilisation, $\mu_{BS}^{(k)}$, as discussed in Table 4.4. In contrast, the advanced model adopts the FTP Traffic Model, [3, 4] in which user demand is derived from requests to download a file (see Table 4.4). Specifically, a user enters the network, is allocated a proportion of the serving base station traffic capacity, downloads a file, and leaves the network. Depending on the user's locations, each will generally encounter different signal levels and interference power levels, leading to a different SINR and hence a different download data rate for each user. Users in poorer reception areas encounter lower data rates, require longer service times, and thus their serving base station may operate with an extended duty cycle and thus create more interference to users in adjacent cell areas. The stochastic nature of the FTP model can simulate performance over demand troughs and peaks when traffic levels are high, and a proportion of users may receive unacceptable service quality. Any factor influencing service times has an effect on base station energy consumption.

In simulations it is common to remove poorly performing downloads before they are completed; these instances would rarely arise in real, well-designed networks [4]. The Static Model described in Chapter 4 is not able to incorporate this type of interactive dynamic behaviour. To do so requires a much more sophisticated network simulation that includes the random processes of customer demand and random customer locations, traffic types and levels.

5.1   OVERVIEW OF THE ADVANCED MODEL

This section is intended to provide an overview the structure and operational phases of an Advanced Model simulation. This model was developed in the Centre for Energy Efficient Telecommunications (CEET) that operated in the University of Melbourne from 2010 to 2016. The CEET modelling was undertaken as part of the GreenTouch international consortium [5, 6]. A step-by-step summary of the Advanced Model is presented in Table 5.8.

Each operational phase of the Advanced Model and related issues in the simulation will be discussed in greater detail in the following sections. The end objective in this modelling is to estimate the energy consumption of a service being delivered by a mobile network, both from the network operator perspective as a function of the network traffic, and from the perspective of an interested network user as to the impact of a task requested by the user. This model is considerably more involved than the Static Model of Chapter 4. It can yield reasonable results for energy consumption, but still omits much of the capability of more complex commercial and proprietary models. Broadly, the model follows the modelling methods used in early releases of LTE [3, 7, 8].

To begin such a model, we set up a service area populated by a number of users, and set up a network of serving base stations. In modelling undertaken at CEET we chose to model one central base station, BS_0, surrounded by a hexagonal array of base stations, BS_1 to BS_6, as shown in Figure 5.1. That cluster of base stations is surrounded by a second hexagonal ring, BS_7 to BS_{18}, and potentially a third outer ring, BS_{19} to BS_{36}. Performance parameters of download services provided to users by the central core of BS_0 to BS_6 are recorded for model output data collection. The sole purpose of the outer ring, or rings, is to serve their own users with the same services, so that the interference seen by all of the central core users is consistent. Data is not collected for users serviced by these outer rings.

The antennae of the base stations are set in the same orientation, spaced at 120 degrees, and with – the main lobes of each base station antenna facing the troughs between lobes of its adjacent base station antennae, as can be seen in Figure 5.1. Hexagonal shapes show approximate coverage areas of each base station, and colour codes indicate contours of SINR in a similar manner to Figure 4.9a. Such

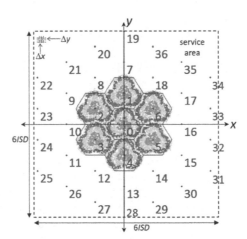

FIGURE 5.1   CEET Advanced Model mobile network layout of base stations. The service area is a square of sides *6ISD* centred on BS_0. Results were collected for users located in the central base station, BS_0, as well as the inner-most ring BS_1 to BS_6. The remaining base stations, BS_7 to BS_{36} are included in the model to ensure the results collected include the effect of a consistent level of interference from surrounding base stations as they serve their users. A colour heat plot of the SINR for the inner cluster of base stations is shown. The colour scale of red, pink, light blue, dark blue and green correspond to decreasing SINR with red for the highest SINR and green the lowest. From the orientations of the SINR heat plots, we can see that the lobes of each base station are directed towards the troughs between the lobes of the adjacent base stations.

a base station arrangement is generally used in simulations to enable cross-model comparison, and is broadly in line with network deployment practices; although the latter do vary to take account of terrain and local variations in user population distribution.

The spacing between base stations, and the density of the user population depend on the type of service area. This model follows the approach of Blume *et al.* [7] adopting four representative service area or geographical categories: Dense Urban, Urban, Suburban, and Rural areas, together with a representative base station spacing and user density for each. The details of these categories of areas are expanded in Section 5.4.

5.2   PLACING USERS IN THE SERVICE AREA

The shape and size of the overall service area, (sometimes referred to as a "playground") is set as a square with at least 5 or 6 times the inter-base station distance (*ISD*) per side, and centred on one base station BS_0. Figure 5.1 shows an example service area that uses six times the *ISD* per side as indicated by the dashed line surrounding all the base stations. This size of service area ensures that all of the likely coverage area of the inner cluster of base stations, BS_0 to BS_6, and first interfering ring of base stations, BS_7 to BS_{18}, is included, plus most of the coverage area of the outermost ring, BS_{19} to BS_{36}. Larger service areas can lead to longer simulation times.

The simulation is based on introducing randomly located users each with a service request. The new users are introduced at time intervals using a Poisson process with a set average service request rate. To simplify the positioning of new users across the service area it is convenient to define a number of pre-set grid locations covering the service area that are potential locations for the introduction of users during the simulation. This can be done by setting up a square rectilinear (x,y) grid centred with BS_0 at its origin, and sides of length N_{ISD} multiples of the inter-site distance, *ISD*, enough to cover the service area. For example, a square grid with sides $N_{ISD} = 6$, as shown in Figure 5.1. The grid is evenly spaced using spacing Δx in the x-direction and Δy in the y-direction with $\Delta x = \Delta y = \delta ISD$ with δ set to a few percent (say 3%) of the inter-site distance, *ISD*. The grid of user locations extends over the whole service area. This provides grid points at locations (x,y) with $x = a\Delta x$ and $y = b\Delta y$ for integers a and b ranging from $-N_{ISD}/2\delta$ to $N_{ISD}/2\delta$. For example, adopting 6*ISD* as the length of the service area and $\Delta x = \Delta y = 0.03 ISD$, we get $6/2\delta = 3/0.03 = 100$. Therefore, (x,y) point on the grid is identified with an index doublet (a,b) where $(x,y) = (a\Delta x, b\Delta y)$ with a and b ranging from -100 to 100. A diagrammatic representation of the grid points is shown in the upper left corner of the service area in Figure 5.1.

The number of grid points available for the location of users, $N_{grid}^{(use)}$, will be:

$$N_{grid}^{(usr)} = \frac{N_{ISD}^2}{\delta^2} \tag{5.1}$$

Each location on the grid is given an index number, w, between 1 and $N_{grid}^{(use)}$. That is, each of the $N_{grid}^{(use)}$, locations $(x,y) = (a\Delta x, b\Delta y)$ will have a unique index number w.

For example, using geographical category Urban from Table 5.1, we have *ISD* = 1000 m. For a square service area with length six times *ISD* on each side as in Figure 5.1, and adopting a value of $\delta = 0.03$, we would get $N_{grid}^{(use)} \approx 40,000$ potential user locations each spaced 30 m apart.

Throughout the simulation, at intervals as described in Section 5.4 users at randomly selected locations in the service area make service requests. When the k th user makes their service request, that user is assigned a location in the grid by allocating a random integer, w, between 1 and $N_{grid}^{(use)}$. Using that random integer the location of the k th user, $(x^{(k)}, y^{(k)})$, is set to be the grid point with index number w. With this approach, for a uniform random distribution of user requests, we set the probability of the location of the k th user being at (x,y) as:

$$\text{Prob}\left(\left(x^{(k)}, y^{(k)}\right) = (x,y)\right) = \frac{1}{N_{grid}^{(usr)}} \tag{5.2}$$

TABLE 5.1

Representative Values of these Modelling Parameters for the Four Geographical Categories [7]

Geographical Category	Dense Urban (DU)	Urban (U)	Suburban (SU)	Rural (R)
Population density $U^{(usr)}$ (users/ km^2),	10000	1000	300	30
Data Download Volume per user (GB/month/user)	16.30	16.30	16.30	16.30
Base Station Inter-site Distance (*ISD*) (km)	0.50	1.0	1.73	4.33
Average download data rate per user (based on above Data Download Volume), $D^{(usr)}$ (kbit/s)	49.50	49.50	49.50	49.50
Approximate service area of base station (based on above ISD) $A_{BS} = \frac{1}{2}(\sqrt{3})(ISD)^2$ (km^2)	0.22	0.87	2.60	16.24
Total service area $A_{SA} = (6ISD)^2$ (km^2)	6.50	26.00	78.00	487.47
Average user request rate over service area, A_{SA}, based on a file download size of 2 MByte, λ (files/s)	201.09	80.44	72.39	45.24

The user file size, $B^{(usr)}$ is set to 2 Mbyte for all users. The total service area in a simulation, A_{SA}, is chosen by the modeller to cover N_{BS} base stations. In the table, a square area six ISDs per side is assumed. The user request rate, λ, is given by (5.11).

Upon the k th user' service request, the signal level received from every sector of every base station at the k th user's location $(x^{(k)}, y^{(k)})$ is calculated. This calculation includes the signal propagation loss, the beam pattern of each base station transmitting antenna in both vertical and horizontal planes, a building penetration loss, plus a random additional "shadowing" loss. Following Blume *et al.* [7], shadowing loss is calculated as the sum of two equal log-normal components with mean value zero and standard deviation 10 dB. One component, v_{BS}, is specific to each base station, and the second, $v(z^{(k)})$, specific to the user location. For the download to the k th user, the shadow loss for signals reaching the user from each base station is given by:

$$v_{shdo}^{(k)} = \sqrt{v_{BS}^2 + v\left(z^{(k)}\right)^2} \Big/ \sqrt{2} \tag{5.3}$$

In addition to the signal power that would be received by the user at location $(x^{(k)}, y^{(k)})$ from each sector of every base station, on the presumption that this station were to be selected to serve the user, the interference power that would be received from each base station is calculated based on the station's current activity. Together, these received power levels enable calculation of the SINR that would be presented to the user for every possible serving station option. The k th user's serving base station and sector are then selected on the basis of giving the best SINR to the new user. As an alternative, a simpler and faster approach sometimes used would be to simply assign the user to the BS and sector that give the highest signal strength. Note, this simplified approach may not always achieve the best user throughput. This is because the base station that supports the best SINR may already be supporting many user downloads and is unable to (re)allocate adequate resources to serve the new user.

With the servicing base station and sector selected for the new user, the SINR is used to select an appropriate modulation and coding format, which will determine the number of effective bits the user will receive per RE or per TTI. This in term leads to a spectral efficiency value. The details of this process are expanded in Section 5.5.

Upon entry, each new user begins to receive a share of the resources transmitted by their servicing base station sector. At regular time intervals (referred to as "epochs"), each active user receives a number of data bits. The number is determined by the spectral efficiency attributed to the user, the number of other users sharing a base station sector and hence the proportion of the base station

Resource Blocks (RBs) allocated to the user, and the length of the epoch. Typically the epoch duration is one or a small number of TTIs or subframe intervals, but at times of high traffic the simulation will adapt to shorter epoch durations; a fraction of a TTI. At each epoch, the simulation checks to identify whether it is time to add a new user, whether a user has reached its target download volume, and whether a user has exceeded its allowed download time. These details are discussed in Sections 5.4 and 5.6. If a user has reached their target download volume or has exceeded the allowed download time, that user is removed from the service area, the volume of data received and the time taken to receive that data is recorded. Upon the departure of a user the transmission resources that user had received are reallocated among continuing users. If the base station sector that has been serving that user has no other current users, its transmitted power level is reduced to zero.

During each of the epochs, a running record of the activity of each base station sector status (i.e. serving a user or not) is maintained so as to determine the average system utilisation, Likewise for each user, a running record of the number of other users sharing the base station sector is maintained.

The simulation is allowed to run until a target number of file downloads has been completed. The simulation then continues to run, with new users being added so as to maintain stability in interference levels, until the last of the user downloads within the target number has been completed, or terminated as running over time. In either case, the file volume and service time of these user downloads are recorded, but the results of any new user downloads above the target number are not.

Upon termination of the simulation, all the collected data is stored for analysis and display.

5.3 BASE STATION POWER CONSUMPTION

As with the Static Model of Chapter 4, there are several time scales intrinsic to the Advanced Model. A detailed discussion of the time scale to be adopted is given Section 5.4. For now, we shall consider the time variable, t, introduced in this section to be generic in that its precise definition of the time scale it refers to will be made clear in Section 5.4.

The base station operation in this model assumes a base station to be transmitting at the maximum possible utilisation, $\mu_{BS,m,p}(t) = 1$, at all times when it is transmitting data to users in that sector. (Recall $\mu_{BS,m,p}$ is the utilisation of the p th sector of base station m). As described above, this is done to fulfil user requests as quickly as possible so as to free resources up as soon as possible and make them available for new users' requests. When there is no data to be transmitted to users in a particular sector, the base station transmitters facing that sector are modelled as not transmitting any power. Therefore, we have for a TTI at time t:

$$\mu_{BS,m,p}\left(t\right) = \begin{cases} 1 & \text{at times } t \text{ when transmitting data to users} \\ 0 & \text{at times } t \text{ when not transmitting data to users} \end{cases} \tag{5.4}$$

where the p-index is generic and covers the different base stations as well as the sectors within a base station and the transmitters within each sector. With this, the average utilisation of a base station sector $\langle\mu_{BS,m,p}\rangle_T$ over time period T (which can be the duration of the simulation) will be:

$$\left\langle\mu_{BS,m,p}\right\rangle_T = \frac{\text{total base station transmitting time}}{T} \tag{5.5}$$

where the "total base station transmitting time" $\leq T$. The details of this are provided below. For a base station we define its utilisation as the average utilisation of its sectors:

$$\left\langle\mu_{BS,m}\right\rangle_T = \frac{1}{N_{sec}}\sum_{p=1}^{N_{sec}}\left\langle\mu_{BS,m,p}\right\rangle_T \tag{5.6}$$

where N_{sec} is the number of sectors of the base station. The number of sectors for the various base station types is shown in Table 4.11. Also note that the number of transmitters, N_{TRX}, equals the number of sectors for SISO systems, but will be greater for MIMO systems as shown in Table 4.12.

The transmit power is set to zero at times when the transmitter is idle; that is, $P_{idle,out,m,p} = 0$ W. This is because it is assumed that the transmissions from all the base stations in the vicinity of BS_0 are coordinated at the frame level; therefore all base stations are transmitting PDSCH REs simultaneously and no base station will be transmitting PDCCH or other overhead REs whenever a nearby base station is transmitting PDSCH REs. This means the interference power from a nearby base station will either be at a level corresponding to its full power, $P_{max,out}$, (because it is transmitting all the available PDSCH REs) or zero in any particular subframe interval. Therefore, in so far as simulating the transmit power from a base station antenna we write:

$$P_{out,m,p}\left(\mu_{BS,m,p}\right) = \mu_{BS,m,p}P_{max,out} \qquad (5.7)$$

where $\mu_{BS,m,p} = 0$ if the base station is not transmitting PDSCH REs, $\mu_{BS,m,p} = 1$ if it is transmitting PDSCH REs. In this simulation we adopt a value $P_{max,out} = 40$ Watts (46 dBm) and $P_{idle,out} = 0$ Watts for all antennas in all base stations.

Note that even if the base station idle transmit power is set to zero ($P_{idle,out} = 0$), the total base station power consumption, when $\mu_{BS} = 0$, i.e. $P_{BS}(\mu_{BS} = 0)$, will not be zero due to the base station operational powers detailed in (4.1). For a 3 sector base station with N_{ant} antennas in each sector, and retaining the forms used in (4.11) and (4.13), the base station power consumption for the m th base station in this simulation is given by:

$$P_{BS,m}\left(\mu_{BS,m}\right) = 3N_{ant}\frac{P_{RF} + P_{BB}}{\left(1-\sigma_{DC}\right)\left(1-\sigma_{MS}\right)\left(1-\sigma_{cool}\right)} +$$

$$+ \frac{3N_{ant}P_{idle,out} + P_{max,out}\sum\limits_{p=1}^{3N_{ant}}\mu_{BS,m,p}}{\eta_{PA}\left(1-\sigma_{feed}\right)\left(1-\sigma_{DC}\right)\left(1-\sigma_{MS}\right)\left(1-\sigma_{cool}\right)} \qquad (5.8)$$

$$= 3N_{ant}P_{idle} + \Delta_P P_{max,out}\sum\limits_{p=1}^{3N_{ant}}\mu_{BS,m,p}$$

For SISO, $N_{ant} = 1$, for 2 × MIMO, $N_{ant} = 2$, 4 × MIMO, $N_{ant} = 4$ and 8 × MIMO, $N_{ant} = 8$. Also, the σ_{feed} term does not apply if the base station uses a "Remote Radio Head" amplifier immediately adjacent to the antenna.

As noted above, in (5.8) $\mu_{BS,m,p}(t) = 0$ or 1 and we have assumed all the transmitters in the base station are identical. Example values for these parameters are given in Tables 4.1, 4.2, 4.11 and 4.12. Also, we have:

$$P_{idle} = \frac{\dfrac{P_{idle,out}}{\eta_{PA}\left(1-\sigma_{feed}\right)} + P_{RF} + P_{BB}}{\left(1-\sigma_{DC}\right)\left(1-\sigma_{MS}\right)\left(1-\sigma_{cool}\right)} \qquad (5.9)$$

$$\Delta_P = \frac{1}{\eta_{PA}\left(1-\sigma_{feed}\right)\left(1-\sigma_{DC}\right)\left(1-\sigma_{MS}\right)\left(1-\sigma_{cool}\right)}\frac{\left(P_{max,out} - P_{idle,out}\right)}{P_{max,out}}$$

5.4   TRAFFIC DEMAND AND THROUGHPUT

Before starting a detailed discussion of how user demand and network throughput is simulated, we need to have a clear idea of what these terms mean. The term "demand" refers to user download requests. These requests can be in the form of a "traffic demand" in bit/s, or "service request rate" in requests/s. The term "throughput" refers to the network's download data rate in response to the user demand. The throughput is in bit/s and can be applied on either a "per user" basis, for a base station or the entire network. The throughput may be moderated by the ability of the network to support the user demand. A given level of user demand may or may not be fully satisfied by the network throughput.

Constructing a realistic simulation of the traffic demand for a mobile network base station is complicated. The current traffic level depends on the recent history of demands and the base station's ability to service those demands. This simulation uses the FTP traffic model introduced in Section 4.5. The FTP traffic model is based on representing service requests by users (which is to download a file) as a stochastic process, parametrised by the average number of requests per unit time, λ, and then tracking how the base station handles those requests as new customer requests occur and current customers go off line. Current customers may go off line either because their request has been fulfilled or because their request "times out" before it is fulfilled and their session is terminated by the base station [4, 7]. A session times out if it takes longer than a pre-set limit, $\Delta t_{max,dl}$, to complete the download. The value of $\Delta t_{max,dl}$ is set by the modeller.

As discussed in Chapter 4, there are several time scales that can be defined in the operation of a base station. These are as follows:

1. Subframe duration or Transport Time Interval (TTI) (1 ms): When REs are allocated by the base station for transmission to its users, it does so on the basis of TTI's as the minimum time unit of allocation.
2. Symbol duration (71.43 ms): This is the time duration of the REs in each sub-carrier frequency. There are 14 symbol durations in each 1 ms subframe.
3. Radio Frame (10 ms): This time scale corresponds to 10 subframes.

In the simulation described in this chapter an additional time scale, an "epoch", is introduced. The simulation is based on equal time steps of Δt, each step being an epoch. The total duration of the simulation, T, is the time over all N_{Tot} epoch steps:

$$T = N_{Tot}\Delta t \qquad (5.10)$$

The time of the j th epoch is given by $t_j = j\Delta t$ where the j indexes the steps, $j = 1,2,\ldots,N_{Tot}$. In the equations below, the epoch step index, j, is an un-bracketed subscript, $_j$. The simulation also tracks users in the service area and the user index, k, will be a bracketed superscript, $^{(k)}$.

At the start of each epoch, new user download requests may enter the simulation. Each current download receives a number of bits, dependent upon on the spectral efficiency at that user's location, and the number of PDSCH RE they receive which form their share of the resources of in their serving sector of the base station. All current user downloads are checked to see if the download has reached the volume target, $B^{(usr)}$, or whether the download has exceeded a set maximum allowed download time, $\Delta t_{max,dl}$.

The use of an epoch as the simulation's time step allows its length to be set for a simulation run depending upon the level of demand. At low demand levels an epoch duration, Δt, of one TTI is used, or at very low rates an epoch may have a duration of several TTIs to speed up the simulation. In special cases, such as very large service areas and high demand levels, it is sometimes necessary to adopt a shorter epoch duration for the simulation to ensure that all incoming user requests during any epoch can be recorded. This aspect is discussed in detail below.

The user demand is implemented using the FTP traffic model in which users request a file be download [3, 5, 7–10]. The file to be downloaded is set to be the same size, $B^{(usr)}$ bits, for all users. The user demand is represented using a stochastic process parameter of user requests per second, λ. This parameter is set by the total service area, A_{SA}, that includes all N_{BS} base station cells, user population density, $U^{(usr)}$, (users/km^2), and the average traffic demand per-user, $D^{(usr)}$ as described below. To set the parameter values A_{SA}, $U^{(usr)}$, $D^{(usr)}$ for the service area, four different "geographical categories" are modelled each represented by a generic population density and inter-base station distance [7]:

1. Dense Urban (DU),
2. Urban (U),
3. Suburban (SU) and
4. Rural (R).

Each geographical category will be characterised by its population density (i.e. number of users/km^2), user demand (GByte download/month/user) and an inter-site distance (ISD) between base stations (which indirectly sets base station cell area, A_{BS}). Table 5.1 shows the representative values of these modelling parameters for countries in the "mature market" segment for 2020 (such as North America, Western Europe, Japan) [7]. These values are used to provide an estimate of typical values of parameters needed to determine base station downlink traffic. In this table the average user request rate, λ, is given for a user file size, $B^{(usr)} = 2$ Mbyte.

For each geographical category, we calculate the request rate, λ, across the whole service area, A_{SA}. The total service request rate across the entire service area, λ is given by:

$$\lambda = A_{SA}U^{(usr)}D^{(usr)}\big/B^{(usr)} \tag{5.11}$$

After each new user is added to the service area, they are assigned to a base station and sector based upon the SINR at their location as described below.

5.4.1 Deriving $D^{(usr)}$

The basis of this model is the generation of user requests, quantified by the parameter λ defined in (5.11) and the model (described below in this chapter) then represents how the network responds to these requests. To simulate a range of possible mobile network configurations, the model uses the "geographical categories" listed in Table 5.1.

From observation, the random nature of user requests is well represented by a Poisson process [11]. To directly measure a value for λ, would require observation of user behaviour or detailed monitoring of service request arrivals at the base station. Neither of these options is easily implemented. Therefore, we need a method to determine a value for λ that makes intuitive sense.

A statistic that can be easily determined from base station logs is the downloaded traffic over multiple diurnal cycles [12]. With knowledge of the number of users of the network, the "download per user per month" (GByte/month/user) can be calculated. Denote the GByte/month/user with $GPM^{(usr)}$. This quantity can be converted to an equivalent average user download traffic demand per user, $D^{(usr)}$, bit/s/user using:

$$D^{(usr)} = \frac{GPM^{(usr)}}{30.5\times24\times3600}8\times10^9 \tag{5.12}$$

where the 8×10^9 converts Gbytes to bits, we have set an average of 30.5 days in a month with 24 hours in a day and 3600 seconds in an hour.

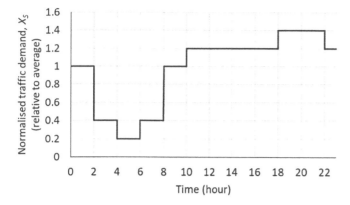

FIGURE 5.2 Diurnal cycle, normalised relative to average traffic, used to represent traffic demand $D^{(usr)}$ over a diurnal cycle [7]. The traffic demand has 5 levels given by $D_S^{(usr)} = X_S^{(usr)} D^{(usr)}$.

The value of $D^{(usr)}$ is an average over a month of diurnal cycles. To represent the variations within a diurnal cycle, this average multiplied by a factor, $X(t)$, such that at time t in the diurnal cycle, the average user download traffic is $X(t)D^{(usr)}$ and we have the average of the $X(t)$ over a diurnal cycle is unity. To simplify the model, we represent $X(t)$ with a stepwise approximation X_S as shown in Figure 5.2 in which the diurnal cycle is split into five stages each with duration T_S and demand traffic $X_S D^{(usr)}$.

In reality, λ is the result of a diverse range of user request rates and $D^{(usr)}$ is an average over many downloaded data file sizes. It is possible to model multiple user categories and multiple file sizes, resulting in a model with multiple simulations running separate λ values simultaneously. This level of complexity is not warranted here, nor used in model simulations [4]. It is sufficient to base the calculation on averages and adopt a fixed file size of $B^{(usr)}$.

Although the $X_S \lambda$ values determined using (5.11) and a diurnal cycle approximation X_S given in Figure 5.2 may not precisely replicate a real-world service request rate, the purpose of these simplified models is not to precisely replicate base station behaviour. Rather the purpose of this model is to provide guidance and understanding of (the power consumption of) mobile networks.

5.4.2 Representing Traffic over a Diurnal Cycle

User demand varies in accord with a diurnal cycle such as that shown in Figure 5.2. This cycle is a step-wise approximation to the diurnal cycle in Figure 4.6. Using such an approximation, several simulation runs with different λ values are required to build up a diurnal cycle of utilisation and hence power consumption and energy efficiency profile. To represent the diurnal cycle of traffic, the average download data rate per user, $D^{(usr)}$, is scaled. An example of such scaling is given in Figure 5.2 in which the peak traffic is scaled up by a factor of 1.4 and the minimum traffic is scaled down by a factor of 0.2. The values for the steps, normalised relative the diurnal cycle average, $D^{(usr)}$, are the X_S values also shown in Figure 5.2. The stepwise approximation to a diurnal cycle is adopted to simplify the model by reducing the number of simulations required to cover the full diurnal cycle. The profile in Figure 5.2 is an approximation of a measured diurnal cycle [7]. The value of the request rate, λ, is constant for a given geographical category and stage of the diurnal cycle.

The simulation collects data by tracking the download details of all users in the service area for each epoch. The choice of the epoch duration is a balance between a short duration, which leads to an extended simulation run time, or a longer epoch duration which increases the probability of multiple user requests being made within an epoch, necessitating increased software complexity.

As an example, the simulation described allows for up to two user requests in each epoch, but sets the time step of the j th epoch such that the probability of no more than two new users being added to the service area within an epoch is close to unity (>0.95). That is, Δt is set by the probability, *Prob*, requirement:

$$Prob\left(\Delta N_j^{(usr)+} \leq 2\right) > 0.95$$

$$= e^{-\lambda\Delta t} \sum_{n=0}^{2} \frac{(\lambda\Delta t)^n}{n!} = e^{-\lambda\Delta t}\left(1 + \lambda\Delta t + \frac{(\lambda\Delta t)^2}{2}\right) \qquad (5.13)$$

where $N_j^{(usr)+}$ is the number of users added to the service area in the j th epoch.

This approach is adopted because keeping track of adding many new users within an epoch can be quite demanding on the simulation. The simulation needs to be run for several different user demand levels, representing user behaviour at different times of day. The requirement that the duration of each epoch at least satisfies condition (5.13) in each case means that the epoch duration, Δt, will vary over this diurnal cycle as $D^{(usr)}$ is scaled accordingly. The simulation is run over a time period in which $D^{(usr)}$ is effectively constant. Note that although $D^{(usr)}$ is scaled to represent the stages of the diurnal cycle, there is actually no morning or evening "time of-day" knowledge in the simulation, only a demand level awareness provided by $D^{(usr)}$.

All activities are timed relative to the epoch duration; thus entry of a new user, updating of the downloaded data tally for each active user, retirement of users completing their task and removal of under-performing users are done at the start of an epoch. Because the epoch quantisation of time does not correspond with the duration of a TTI, this leads to a different delay in commencement and cessation of a download relative to the "next TTI" timing used in LTE.

A user may start earlier if the epoch time is less than a TTI, or later if the epoch duration is greater. Likewise for cessation, there is the possibility of delay in retiring a user by up to 1 epoch could occur, with the possibility of unused REs being allocated. An epoch time Δt just satisfying (5.13) sets an upper bound to this unused allocation of $\approx 1.2\%$ at the highest demand level, or $\approx 4\%$ at the lowest. A shorter epoch than required by (5.13) means a shorter delay and lower excess. In the simulation, we allow data download to continue until the end of the epoch, and the total volume of data that the user would receive (including the overshoot) is recorded.

For example, with a 9 km² service area with 10,000 users/km², peak user demand of $X_S D^{(usr)} = 70$ kbit/s, and $B^{(usr)} = 16$ Mbit and epoch duration of $\Delta t = 1$ ms epoch (i.e. one TTI) we find greater than 99% of epochs include 2 or fewer requests, and the maximum download over-run is approximately 0.6%. However with smaller file sizes, larger service areas or greater average user demand, a larger over-run is possible; hence shorter epoch lengths are necessary.

The scaling to provide a diurnal cycle is used in conjunction with the average download rates shown in Table 5.1. For example, for the busy hour the average download rate is scaled 1.4 times the average for the corresponding geographical category. Focussing on Dense Urban the busy hour traffic demand per base station would be $X_S D^{(usr)} U^{(usr)} A_{BS} = 1.4 \times 49.50$ (kbit/s) $\times 10000$ (users/km²) $\times 0.22$(km²) $= 152.5$ Mbit/s For a file size of $B^{(usr)} = 2$ MByte this corresponds to a service request rate of 152.5(Mbit/s)/16(Mbit) = 9.53 service requests per second per base station. For the minimum during the diurnal cycle, the traffic demand per base station will be $X_l D^{(usr)} U^{(usr)} A_B = 21.78$ Mbit/s corresponding to a service request rate 1.36 service requests per second per base station. Using this example, for the busy hour, (5.13) predicts $Prob\left(N_j^{(usr)+} \leq 2\right)$ user drops in a 1 ms epoch is ≈ 0.96, which is acceptable. But, if a smaller file size, such as $B^{(usr)} = 0.5$ Mbyte which is also part of 3GPP simulation options, the epoch duration, Δt, must be significantly reduced to keep the number of additional users per epoch at 2 or less.

The values in Table 5.1 are only meant to be a representative example of "typical" LTE deployments. In practical implementations, factors such as the non-homogeneity of traffic demand, variations in population density, download behaviour and topology of the service area will all influence how base stations are configured and deployed. It may also require the selection of more appropriate *ISD* values, antenna beam orientation and scheduling of resources to suit the deployment terrain. It is beyond the scope of this book to delve into these details.

The choice of file size to be downloaded (2 Mbyte) can significantly impact the feasibility of the simulation methodology. The value of 2 Mbyte has been adopted by the 3GPP for network simulation comparisons [4]. Selecting a smaller $B^{(usr)}$ file size for a given traffic demand, $D^{(usr)}$, will increase the request rate, λ. From (5.13) this, in turn, will require a reduction in the epoch duration or require the simulation to track the higher numbers of users that would come on line during an epoch. In either case, the simulation run time would increase.

5.4.3 RUNNING THE SIMULATION

With a service area category selected, the simulation follows the approach adopted by GreenTouch [5] and Salem *et al.* [9]. With this approach, the simulation is started with no active users, and new users are added in accordance with Section 5.4.2. For each epoch, $j = 1, 2, \ldots$, any new user is placed at a random location somewhere in the service area. Figure 5.3 is a re-draft of Figure 5.1 with the

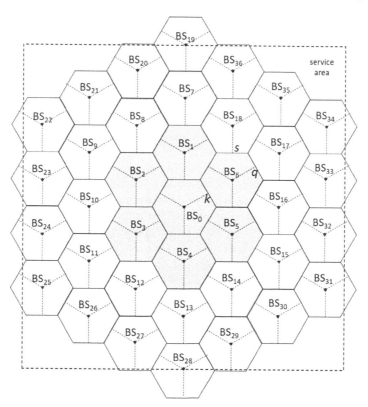

FIGURE 5.3 Mobile network layout for the simulation consisting of identical hexagonal cells with a base station at the centre of each cell. The boundary of each sector within each base station cell is indicated with the dashed lines from the base station in the centre of each cell. Only users who are serviced by base stations BS_0 to BS_6 (shaded) are included in the data set used to analyse the data download and power consumption of the network. Base stations BS_7 to BS_{36} are included to provide an accurate evaluation of the interference powers experienced at the locations of users serviced by the inner (shaded) base stations. Users are served by the base station and sector that provides the highest SINR. This can result in user q not being included in the data set because it is served by BS_{17} whereas users being included because it is served by BS_6.

central cells, BS_0 to BS_6 shaded and the perimeter of the service area corresponding to the dashed line. In this simulation, the number of surrounding active base stations is $N_{BS} = 36$; however the area for which user data is collected is the shaded inner group of base stations, BS_0 to BS_6.

Each new user is allocated an identification index, $k = 1, 2,\ldots, N^{(usr)}$. Each user remains stationary at their location, $(x^{(k)}, y^{(k)})$ in rectangular coordinates.

During the initial epochs, the network is progressively "loaded up" with users who request a file download within the epochs in accordance with the Poisson process. The request rate, λ, is independent of a user's location in the service area.

As each service request by users occurs, the most appropriate base station and sector is selected, and that base station sector's resources are reallocated among the new user and other currently active users. Although in a real network resources are allocated on the basis of a Transmission Time Interval, $\delta_{TTI} = 1$ ms, the simulation uses an epoch as that basis; the epoch duration is scaled from a few TTIs to (in some cases) a fraction of a TTI to enable the simulation to run more efficiently.

To do this the allocation procedure is as follows: As stated above, in this simulation the base station utilisation is assumed to be $\mu_{BS} = 1$ whenever data is downloaded to users. Therefore, when the base station is transmitting PDSCH REs to users, it transmits using all available REs within a TTI. To adjust the parameters from a TTI time scale of δt_{TTI} to the epoch time scale Δt, set by (5.13), the simulation scales the number of PDSCH REs transmitted to users by the ratio $\Delta t / \delta t_{TTI}$. If $N_{TTI,PDSCH}$ is the average number of PDSCH REs in a TTI then in the epoch time scale the number of available PDSCH REs is $\Delta t N_{TTI,PDSCH}/\delta t_{TTI}$.

We have $N_{TTI,PDSCH} = \sigma_{PDSCH} N_{TTI,RE}$ where σ_{PDSCH} was defined in (4.29) and $N_{TTI,RE}$ is the total number of REs in a TTI. The σ_{PDSCH} parameter accounts for the average number of overhead REs (which do not carry user data) in a TTI (or subframe) with the average taken over a Radio Frame.

The number of available PDSCH REs within an epoch is equally shared over all active users for each sector in a base station at the end of the previous epoch. Therefore, if there are $N^{(usr)}_{j,m,p}$ active users being served by the sector p of base station m at the end of the j th epoch, each user in the $(j+1)$ th epoch is allocated $N^{(usr)}_{j+1,m,p,PDSCH}$ PDSCH REs where:

$$N^{(usr)}_{j+1,m,p,PDSCH} = \left\lfloor \frac{\Delta t N_{TTI,PDSCH}}{\delta t_{TTI} N^{(usr)}_{j,m,p}} \right\rfloor = \left\lfloor \frac{\Delta t \sigma_{PDSCH} N_{TTI,RE}}{\delta t_{TTI} N^{(usr)}_{j,m,p}} \right\rfloor \qquad (5.14)$$

where $\lfloor x \rfloor$ means the largest integer less than x. This means if the k th user is already active, their allocation of REs will change from the j th epoch to the $(j+1)$ th epoch if the number of active users changes.

In order to use (5.14) we need to know $N^{(usr)}_{j,m,p}$, the number of active users of the p th sector in the m th base station in the j th epoch. During each epoch, a number of users are added to the service area, as determined by (5.13). Also a number of users are removed from the service area because either (a) the download of their $B^{(usr)}$ bit-sized file is successfully completed before their session times out or (b) their session times out and the base station terminates their download even though it is not completed.

The time duration for which the k th user has been on line is $\Delta t^{(k)}_{on,j}$ given by the number of epochs from the epoch when user k was added to the service area, $j^{(k)}_{on}$, until the current j th epoch. That is:

$$\Delta t^{(k)}_{on,j} = \left(j - j^{(k)}_{on} \right) \Delta t \qquad (5.15)$$

With this the two criteria for a user's session being terminated and their going off line are as follows:

(a) User k successfully finalises their download before their session times out. In this case the epoch during which they complete their download and leave the service area is given by:

$$j_{off}^{(k)} \text{ such that } \sum_{j=j_{on}^{(k)}}^{j_{off}^{(k)}} \Delta B_j^{(k)} = B^{(usr)}$$

$$t_{off,j}^{(k)} = j_{off}^{(k)} \Delta t$$

(5.16)

(b) User k's download is unsuccessful in that their download is not completed within the pre-set time limit, $\Delta t_{max,dl} = j_{max,dl} \Delta t$. In this case the epoch, $j_{off}^{(k)}$ at time, $t_{off,j}^{(k)}$, in which they leave the service area is given by:

$$j_{off}^{(k)} = j_{on}^{(k)} + j_{max,dl}$$

$$t_{off,j}^{(k)} = t_{on,j}^{(k)} + \Delta t_{max,dl}$$

(5.17)

The number of users whose session is terminated, hence go off line, in the j th epoch is $N_j^{(usr)-}$ is given by the sum of users going off line in that epoch due to criteria a) above, $^{a)}N_j^{(usr)-}$ and criteria b) above, $^{b)}N_j^{(usr)-}$. That is

$$N_j^{(usr)-} = {}^{a)}N_j^{(usr)-} + {}^{b)}N_j^{(usr)-}$$

(5.18)

We now have the number of users added in the j th epoch given by (5.13) and the number of users removed in the j th epoch given by (5.18). Each of these equations applies to a sector, p, of a base station, m. Therefore, the number of active users in the j th epoch, $N_{j,m,p}^{(usr)}$ is given by the cumulative sum of added users less the cumulative sum of removed users. That is:

$$N_{j,m,p}^{(usr)} = \sum_{i=1}^{j} N_{i,m,p}^{(usr)+} - N_{i,m,p}^{(usr)-}$$

(5.19)

In this simulation, as discussed in Section 5.4.2, at high traffic levels an epoch duration can be set as low as $0.2\delta t_{TTI}$; however, this is not a hard rule. The model developed in conjunction with GreenTouch [5] uses a value of $0.5\delta t_{TTI}$ with a maximum of two new users per epoch. At low traffic levels, an epoch duration can be multiple TTI durations. Taking the greatest integer less than the quotient in (5.15) may result in some PDSCH REs not being allocated in some epochs. The number of unallocated PDSCH REs is typically very small and does not impact the overall results of the simulation.

Although the allocation given in (5.14) equally shares PDSCH REs across active users, because each different user's handset may receive a different SINR, the spectral efficiency, and hence achieved data download rate (bit/s) will differ. This means that the RE allocation algorithm in (5.14) will result in a longer time to download the file to some users relative to others.

5.5 SIGNAL TO INTERFERENCE AND NOISE RATIO (SINR)

As with the Static Model in Chapter 4, we assess the influence of antenna radiation pattern and path loss on the signal received at a user site. We represent the impact of the antenna pattern and the path from the p th sector of the m th base station antenna to the location, $z^{(k)}$ of user k as the product of transmit and receive antenna gains, $G_{Tx}G_{Rx}$, angular transmission factors, Θ and Φ and path loss, L. We introduce a channel function, $W(m,p,z^{(k)})$ defined by:

$$P_{Rx,m,p}\left(z^{(k)}\right) = \mu_{BS,m,p}P_{max,out}W\left(m,p,z^{(k)}\right) = \mu_{BS,m,p}P_{max,out}G_{Tx}G_{Rx}\Pi_{m,p}\left(z^{(k)}\right)L\left(r_m^{(k)},f_i,h_b,h_r,F,\upsilon\right) \quad (5.20)$$

where the generic form for the path loss L is given in (4.45) and we assume all the terms are constant over the time of the download. There are multiple path loss models published in the literature, [13–19].

In contrast to the Static Model in Chapter 4, active users are distributed over the entire simulated service area, rather than just limited to the sectors of the central base station, BS_0. As with the Static Model, the simulation allows users to be serviced by a base station outside the nominal coverage area of their closest base station when appropriate. A user is served by the base station and sector that offers the highest SINR. Users close to the notional boundary between adjacent base stations or sectors may be served by a more distant base station if that delivers a higher SINR.

Only users serviced from the inner ring of cells (BS_0 to BS_6) are included in the data collection by the simulation. This means users k and s in Figure 5.3 will be included (both being served by one of BS_0 to BS_6) in the simulation data collection but user q will not (being served by BS_{17}).

We need the SINR at the location, $z^{(k)}$, of user k to determine the spectral efficiency of the data transfer to user k. Therefore, we need the interference power experienced by k's handset at the location $z^{(k)}$. The interference power experienced by user k is the sum of the power received at k's location from all other active base stations, (**NOTE**: The Advanced Model being described includes 2x2 MIMO, SISO and SIMO systems. In the notation $\mu_{BS,m,p,j}$ used in this chapter, the subscript j-index refers to the epoch index number. This is different to Chapter 4 in which in the term $\mu_{BS,m,p,q}$, the q-index referred to the antennas in the transmitters of base station m, sector p of the MIMO system).

With the assumption that base station transmissions are time coordinated in the service area, the only interference to k's PDSCH REs is due to PDSCH REs being transmitted from the base stations servicing their users in other cells near-by to user k. Thus the interference power from other base stations is to the sum of their $\mu_{BS,p,m}P_{max,out}$ signals as attenuated by distance, antenna pattern, and shadowing. The overhead REs transmitted as part of idle power, $P_{idle,out}$, from other base stations in the service area are transmitted during different symbol intervals. Further, because propagation delay from near-by base stations to user k's location is very short relative to the symbol duration, δt_{TTI}, there will be minimal overlap of idle power and other overhead REs transmitted from those stations with the PDSCH REs destined for user k. This means the only non-negligible interference power comes from base stations transmitting PDSCH REs located in cells near-by to user k.

Bringing these details together, we can set the contribution to interference power as zero when a base station is not transmitting PDSCH REs (i.e. $\mu_{BS,m,p,j} = 0$). Therefore, for the j th epoch the interference power at location $z^{(k)}$ will be

$$P_{Interf,j}^{(k)} = \left(\sum_{m=0}^{N_{BS}}\sum_{p=1}^{3}\mu_{BS,m,p,j}W\left(m,p,z^{(k)}\right) - W\left(m^{(k)},p^{(k)},z^{(k)}\right)\right)P_{max,out} \quad (5.21)$$

In this equation all transmitting antennas when active are transmitting power $P_{max,out}$ and $W(m,p,z^{(k)})$ is the channel function defined in (5.20) for the path from sector p of base station m to user k's location. The sum is over all sectors and base stations and the active base stations will have

$\mu_{BS,m,p,j} = 1$. The term $W(m^{(k)}, p^{(k)}, z^{(k)})$ represents the channel function from the base station and sector serving the k th user, denoted $m^{(k)}$ and $p^{(k)}$, respectively.

The simulation needs to identify the base station, sector that serves each user. The approach taken to make this identification is the same as with the Static Model in Chapter 4. That is, the base station, m, and sector, p, chosen to serve the k th user, at location $z^{(k)} = (x^{(k)}, y^{(k)})$, is the base station and sector that provides the maximum SINR for that user. Therefore, the SINR for the k th user is:

$$SINR^{(k)} = \frac{W\left(m^{(k)}, p^{(k)}, z^{(k)}\right) P_{max,out}}{P_{Interf}^{(k)} + P_{RxN}^{(k)}}$$

$$= \max_{m,p} \left\{ \frac{W\left(m, p, z^{(k)}\right) P_{max,out}}{P_{Interf}^{(k)} + P_{RxN}^{(k)}} \right\} \qquad (5.22)$$

where the base station index, $m^{(k)}$, and sector index, $p^{(k)}$, correspond to the base station for which $SINR^{(k)}$ is a maximum at location $z^{(k)}$ and $P_{RxN}^{(k)}$ is the receiver noise power over the bandwidth of the k th user's handset.

The receiver noise $P_{RxN}^{(k)}$ is determined by thermal noise power spectral density of -174 dBm/Hz and a noise figure to account for imperfect receiver. Typical value for the receiver noise figure is 7 dB to 9 dB. For the k th user, the proportion of PDSCH REs allocated to that user in a symbol duration, averaged over Radio Frame, is $\mu_{BS}^{(k)}(t_u)$. This corresponds to a receiver bandwidth of $\mu_{BS}^{(k)}(t_u)BW$ where BW is the total bandwidth of the base station. Therefore, for the k th user:

$$P_{RxN}^{(k)} = 0.001 \times 10^{-17.4} 10^{0.9} \mu_{BS}^{(k)} BW \quad \text{(Watts)} \qquad (5.23)$$

For a 20 MHz system this gives $P_{Rx,N} = 6.32 \times 10^{-10}$W. In this model, the full base station bandwidth is typically allocated to each user in any given TTI.

The value of $SINR^{(k)}$ is mapped to $SE^{(k)}$ using an appropriate link curve as discussed in Section 4.6.1.2. Depending upon the technology (e.g. SISO, SIMO, MIMO) and the user's local environment under consideration the modeller may be able to select one of the link curve examples presented in Figure 4.4 or seek a more appropriate alternative from the references listed in that section.

An example of a link curve mapping between SINR and spectral efficiency, SE, such as shown in Table 4.7 for pre-Release 12 LTE (up to 64QAM based modulation and coding formats) and Table 4.8 for Release 12 LTE and beyond (up to 256QAM). Using the appropriate link curve, the value of $SE^{(k)}$ is then used to calculate the number of data bits downloaded to the k th user, $\Delta B_j^{(k)}$ and the resulting data throughputs for the simulation (see Equation 5.32 and 5.33).

Although this provides a pathway to constructing the simulation, there are a few details that need to be addressed. It was stated in Section 5.4 that it is assumed $SINR^{(k)}$ and hence $SE^{(k)}$ are constant for the k th user for the duration of their session from the time their download starts, $t_{on,j}^{(k)}$ to the time their session finishes, $t_{off,j}^{(k)}$. This means the $SINR^{(k)}$ and $SE^{(k)}$ are constant with respect to the epoch index, j, for a given k th user. This is done to simplify the model because keeping track of the SINR for every active user for every epoch significantly increases the demand on the computer running the simulation. This assumption is justified by noting that users do not move about and the number of new users who can come on line in any epoch, given by (5.13), is generally zero but at most 1 or 2. In steady state, the number of users going off line in any epoch is also of this order. Although the SINR will change as users come on line and go off line, in steady-state upward changes and downward changes of the SINR for different users are equally likely. Hence, averaging over many users these variations will balance these out.

5.6   DATA COLLECTION

The simulation starts with zero active users, then goes through a start-up phase as – users are added (i.e. go "on line" by making a service request to download a file) at time intervals according to the Poisson process. Each added user downloads their file and, when the download is completed, goes off line. Users whose download fails to be completed within the $\Delta t_{max,dl}$ time limit have their download terminated and are sent off line by the base station. This process of users coming on line and going off line continues, with increasing numbers of active users until the system reaches a steady state where the average number of new users coming on line over multiple epochs approximately equals the average number of users going off line. Due to the random nature of the Poisson process, there can still be significant fluctuations of the steady-state values epoch to epoch. The steady state may involve a significant proportion of users having their download terminated by the network before completion because the user demand is too high for the network to satisfy. This indicates the network is congested.

Once the network is in steady state the data collection commences at a time designated the "start-time", t_{start}. The epoch index, j, is set to zero at time t_{start}. The start time is not dependent on the number of users over which the simulation is run and typically occurs when between 1000 and 5000 users have come on line, after which the performance of subsequent users is recorded. It is while the network is in this steady state that the data for analysis of the network is collected. Hence, t_{start} is the time at which data generated by the simulation starts to be collected. The simulation runs until a pre-set total number of users, $N^{(usr)}$, across the entire service area, have completed their download (or had their download terminated).

Of this total number of users, data is recorded only for users who are served by those base stations for which data is to be collected. In the example shown in Figure 5.3, the data is collected for users served by the inner cluster of base stations, BS_0 to BS_6. The simulation ends when all users $k = 1, 2, ..., N^{(usr)}$ across the entire service area finish their session.

The random process used to add new users distributes them uniformly across the entire playground. Therefore, only a portion of the $N^{(usr)}$ users will be located in the inner group of base stations for which data is collected (the 7 base stations in the shaded area in Figure 5.3). This means, the number of users added to the service area over the duration of the simulation, $N^{(usr)}$, will be greater than the number of users for whom data is recorded. The area of the total service area is A_{SA} (the square area bounded by the dashed line in Figure 5.3) and the area covered by the inner ring of base station cells is $7A_{BS}$. Therefore, $N^{(usr)}$ is related to the number of users for whom data is recorded, $N_{Tot}^{(usr)}$, by:

$$N_{Tot}^{(usr)} \approx N^{(usr)} \frac{7 A_{BS}}{A_{AS}} \tag{5.24}$$

Once the required tally of $N^{(usr)}$ users has either completed their download, or been terminated, the simulation is brought to an end. A typical value for $N^{(usr)}$ is around 60,000 users which, depending upon the geographical category will give $N_{Tot}^{(usr)} > 10,000$.

Note that the last user to be active is not necessarily the $k = 60,000$ th user. This can arise because of a user's location or base station loading, the 59,999th or an earlier user may finish their download session after the 60,000th user. Once that final user's session is completed, the simulation is ended. Therefore, the total number of epoch steps in the simulation, N_{Tot}, is given by:

$$N_{Tot}\Delta t = T = \text{time to serve } N^{(usr)} \text{ users} \tag{5.25}$$

This process is depicted in Figure 5.4.

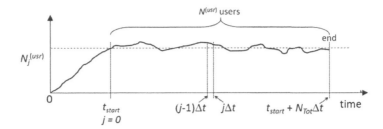

FIGURE 5.4 Simulation run time includes a lead time during which the number of users increases to a steady state. Once the number of users reaches a steady-state value, at time t_{start}, the data collection commences setting $j = 0$ at time t_{start}. The simulation is run until $N^{(usr)}$ (typically 60,000) users have been served (either fully downloading a file, or terminated through failing to download within a time limit), at which time the simulation is ended. During the data collection phase, the simulation data on every user served by the inner cluster of base stations are collected in every epoch. After attaining steady state, the simulation maintains the steady state until termination.

Referring back to Section 5.2, an example calculation of the number of grid points, $N_{grid}^{(usr)}$, for potential user location gave $N_{grid}^{(usr)} \approx 40,000$. Note that with $N^{(usr)} = 60,000$, we have $N^{(usr)} > N_{grid}^{(usr)}$. This is not a problem. Indeed, it can occur that not all of the 40,000 locations are used because, over the duration of the simulation, the random process used to locate users can locate more than one user to any given site on the location grid. Further, because users enter and leave the network, any specific grid location is typically not occupied for a significant portion of the simulation duration.

The data collected for analysis focuses on users served by the inner base stations, BS_0 to BS_6 inclusive. Any user who is served by one of these base stations is included in the analysed data. Users served by BS_7 to BS_{36} are part of the simulation but not included in the data collection. Their role is to provide a consistent level of interference power, representative of the surrounding network. It can occur that users in one of the notional coverage areas of the outer cells BS_7 to BS_{36} are actually serviced by one of BS_0 to BS_6 (e.g. a user in cell of BS_{16} close to the boundary of BS_6). These users are included in the data analysis. Likewise, it can also occur that a user in the ring of cells BS_1 to BS_6 will be serviced by one of the outer cells BS_7 to BS_{18}. Those users are not included in the data set used for analysis.

New users are introduced at time intervals generated by the Poisson process in (5.13) and the users are randomly placed into the service area. When the k th user is added to the service area, the epoch at which they come on line, $j_{on}^{(k)}$, their location, $z^{(k)}$, the $SINR^{(k)}$ at that location and the identification of the base station servicing that user are recorded. The simulation generates an output table of users in which each row corresponds to a user and the results for that user are listed in the columns of that row. An example of such a data table is shown in Table 5.2. This table shows several but not all of the data values that can be recorded for each user in the simulation. For those users served by BS_0 to BS_6, while they are on line the data values listed in Table 5.3 are recorded for each epoch, j, they are active. When the user connection finishes, user data such as their total download, $B_{Tot}^{(k)}$ as well as the epoch in which their session finished and they go off line, $j_{off}^{(k)}$, are recorded.

The simulation continues until all users $k = 1$ to $N^{(usr)}$ all finish their session (whether successful or terminated). Because users are randomly located around the entire service area, the SINR for some users can be much lower than for others. As a consequence of this, it can occur that over the total of $N^{(usr)}$ users, a user other than the $N^{(usr)}$ th user will be the last to finish their session. For example, assume user number $N^{(usr)}$-2 has not completed their session before user number $N^{(usr)}$. The simulation continues to run until user number $N^{(usr)}$-2 finishes their session so that all users $k = 1$ to $N^{(usr)}$ have completed their sessions.

Further, the simulation continues to add new users to the network according to the Poisson process with rate λ given by (5.13) after user $N^{(usr)}$ has come on line to ensure the environment in which

TABLE 5.2

Example of an User Download Outcomes Data Collection

User Number, k	$z^{(k)}$	$B^{(k)}$	$SINR^{(k)}$	...	...	...	On line epoch index $j^{(k)}_{on}$	Off-line epoch index $j^{(k)}_{off}$
1	$z^{(1)}$	$B^{(1)}_{Tot}$	$SINR^{(1)}$	...	...	...	$j^{(1)}_{on}$	$j^{(1)}_{off}$
2	$z^{(2)}$	$B^{(2)}_{Tot}$	...	...	...	...	$j^{(2)}_{on}$	$j^{(2)}_{off}$
..	...	...	...	...	...	...	...	...
...	...	...	...	...	...	...	...	...
..	...	...	...	...	...	...	...	...
$N^{(usr)}-1$	$z^{\left(N^{(usr)}-1\right)}$	$B^{\left(N^{(usr)}-1\right)}_{Tot}$	$SINR^{(N^{(usr)}-1)}$	...	...	...	$j^{\left(N^{(usr)}-1\right)}_{on}$	$j^{\left(N^{(usr)}-1\right)}_{off}$
$N^{(usr)}$	$z^{\left(N^{(usr)}\right)}$	$B^{(N^{(usr)})}_{Tot}$	$SINR^{(N^{(usr)})}$	...	...	...	$j^{\left(N^{(usr)}\right)}_{on}$	$j^{\left(N^{(usr)}\right)}_{off}$
$N^{(usr)}+1$	$z^{\left(N^{(usr)}+1\right)}$	$B^{(N^{(usr)}+1)}_{Tot}$	$SINR^{(N^{(usr)}+1)}$	...	...	...	$j^{\left(N^{(usr)}+1\right)}_{on}$	$j^{\left(N^{(usr)}+1\right)}_{off}$

In this example the nominal number of users is $N^{(usr)}$; however, the $(N^{(usr)}+1)$ th is also included because that user's download commenced within the time $N^{(usr)}$ was downloading and finished before the simulation was terminated. The modeller can choose to either include the data of user number $N^{(usr)}+1$ or not.

users are downloading continues to remain in the same steady state, while data continue to be collected. These new users will have an index $k > N^{(usr)}$ and may well either still be downloading or have completed their download when user $N^{(usr)}$ goes off line. For example, consider the situation depicted in Figure 5.5. We assume all users in this figure have their data collected. User number $N^{(usr)}+1$ has come on line and completed their simulation while user $N^{(usr)}$ is still on line. Because that user commenced their download before $N^{(usr)}$ finished, their data will be included. User number $N^{(usr)}+2$ also comes on line while $N^{(usr)}$ is still downloading. However, user $N^{(usr)}+2$ does not finish their download

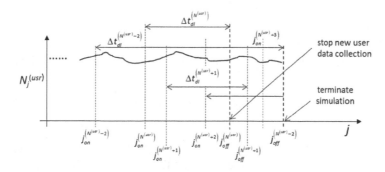

FIGURE 5.5 Detail of termination of the simulation. In this example user $N^{(usr)}-2$ is the last of the $k = 1$ to $N^{(usr)}$ users to finish their download. Data is collected for all the users who commence their download before user $N^{(usr)}-2$ finishes. However, the data is not collected for users added to the network after user $N^{(usr)}$ has finished their download. That user may have their session terminated by the end of the simulation. In this example, the $(N^{(usr)}+1)$ th user commences and finishes their download before user $N^{(usr)}-2$ finishes. Their data will be recorded. User $N^{(usr)}+2$ also commences downloading before user $N^{(usr)}-2$ goes off line. However, user $N^{(usr)}+2$ is still on line when the simulation finishes. That user is terminated when the simulation finishes and data for the partially complete download collected. In contrast, user $N^{(usr)}+3$ commences their download after user $N^{(usr)}-2$ finishes. User $N^{(usr)}+3$ and all subsequent new users do not have their data collected. New users continue to be added (and their data not collected) solely to ensure steady-state network conditions do not change for the full duration of the simulation.

before the simulation ends. Their download will terminate with the end of the simulation and their data will be recorded. In contrast, user $N^{(usr)}+3$ comes on line after $N^{(usr)}$ has finished their download but before the simulation is ended. Their appearance is solely to keep the simulation in steady state until it is ended when user $N^{(usr)}-2$ goes off line. The data for users who come on line after epoch $j_{off}^{\left(N^{(usr)}\right)}$ is discarded irrespective of which base station serves them.

The user count $N^{(usr)}$ is for all users across the entire service area. However, data is collected only for users served by BS_0 to BS_6 which is a portion of $N^{(usr)}$, as shown in (5.24). The precise number of users, $N_{Tot}^{(usr)}$, for which data is collected will be vary from run to run of the simulation because of the random timing and location procedure applied to add users to the network.

Although $N_{Tot}^{(usr)}$ may vary over multiple runs of the simulation, because the data is collected only while the simulation is in steady state, the variation in $N_{Tot}^{(usr)}$ between runs will not impact the results. The reason for this is presented in the Appendix. There is normally a variation between runs, hence the need to run a simulation covering many user drops. Indeed, provided the simulation is ended while it is in a steady state and the termination process does not upset this steady state, the precise details of how the simulation is terminated will have a negligible impact on the results. Provided the steady-state condition is satisfied, the modeller may choose a simulation termination procedure that differs from the above.

The total data download recorded by the simulation, B_{Tot}, include all downloads of users served by inner cluster BS_0 to BS_6, whether the download was completed or terminated. For all users included in the data collection, their downloaded data, $B_{Tot}^{(k)}$, is included in the analysis. That is:

$$B_{Tot} = \sum_{k=1}^{N_{Tot}^{(usr)}} \sum_{j=1}^{N_{Tot}} \Delta B_j^{(k)} = \sum_{k=1}^{N_{Tot}^{(usr)}} B_{Tot}^{(k)} \tag{5.26}$$

where $B_{Tot}^{(k)}$ is the total amount of data downloaded to the k th user whether or not their session terminated before their download was completed:

$$B_{Tot}^{(k)} = \sum_{j=1}^{N_{Tot}} \Delta B_j^{(k)} = \sum_{j=j_{on}^{(k)}}^{j_{off}^{(k)}} \Delta B_j^{(k)} \tag{5.27}$$

The total data downloaded to active users in the service area during the j th epoch is given by:

$$\Delta B_j = \sum_{k=1}^{N_j^{(usr)}} \Delta B_j^{(k)} \tag{5.28}$$

The $\Delta B^{(k)}_j$ are zero for all j when the k th user is not on line. While user k is on line, the value of $\Delta B^{(k)}j$ is determined by the SINR at $z^{(k)}$, $SINR^{(k)}$, and the number of concurrent active users, $N_j^{(usr)}$, sharing the base station resources in the j th epoch. The overall data throughput in the j th epoch, R_j, is:

$$R_j = \Delta B_j / \Delta t \tag{5.29}$$

Note that throughput R_j covers all the sectors of the base stations BS_0 to BS_6 for which data is collected. The average base station throughput (bit/s) over the duration, T, of the simulation, is as follows:

$$\langle R \rangle_{BS,T} = \frac{B_{Tot}}{7 N_{Tot} \Delta t} = \frac{\sum_{j=1}^{N_{Tot}} \sum_{k=1}^{N_{Tot}^{(usr)}} \Delta B_j^{(k)}}{7 N_{Tot} \Delta t} \tag{5.30}$$

The reason for including $N_{BS} = 36$ base stations around the central base station, BS_0, is to ensure the data collected for the inner 7 base stations (BS_0 to BS_6) includes an accurate representation of the interference power from their surrounding base stations in a "real" network situation. The presence of users serviced by base stations BS_7 to BS_{36} provides an accurate representation of the interference power they radiate into in the inner 7 cells as those base stations service their users. It is assumed that any base stations further out do not produce any significant interference power in the inner 7 cells; indeed, even the influence of the outermost base stations BS_{19} to BS_{36} is minimal.

Examples of the data the modeller can collect at each epoch of the simulation is detailed in Table 5.3.

Additional data that can be recorded when the simulation is completed is given in Table 5.4

TABLE 5.3
Data Collected in Each Epoch of the Simulation

Data Collected at End of Each Epoch, j

1. The identifying index number, k, for each new active user.

2. The number of new users that become active in the epoch, $N_j^{(usr)+}$. This number is generated by the Poisson process and epoch length constrains $N_j^{(usr)+}$ to $0 \le N_j^{(usr)+} \le 2$ *Comment – this is a temporary record that is discarded once users have been added*

3. Location, $z^{(k)} = (x^{(k)}, y^{(k)})$, of each user that becomes active in this epoch

4. The index number of current epoch, $j_{on}^{(k)}$, to identify the current epoch as when the k th user became active. This corresponds to the service start time for the k th user.

5. The identity of the base station, $m^{(k)}$, and sector, $p^{(k)}$, serving each new user, k, that comes on line during the current epoch

6. Update $N_{m,p}^{(usr)}$ the number of users being served by the sector p of base station m

7. Update $N_{j,m,p,PDSCH}^{(usr)}$ the number of PDSCH REs allocated to users in sector p of base station m

8. Calculate and record the values of $SINR^{(k)}$, and Spectral Efficiency, $SE^{(k)}$, for each new user that becomes active during the epoch. It is assumed these values do not change while user k is active. That is, the SINR and SE of the k th user is set to the value it has at the start of the $j_{on}^{(k)}$ th epoch, which is when the k th user goes on line:

$$SINR^{(k)} = SINR_{j_{on}^{(k)}}^{(k)}, \qquad SE^{(k)} = SE_{j_{on}^{(k)}}^{(k)} \qquad (5.31)$$

Where $SINR_{j_{on}^{(k)}}^{(k)}$ and $SE_{j_{on}^{(k)}}^{(k)}$ are the values of the SINR and SE at the location of the k th user in the $j_{on}^{(k)}$ th epoch which is the epoch in which they become active

9. Calculate the data downloaded for each user, k, during the current epoch, j

$$\Delta B_j^{(k)} = \delta t_{TTI} \delta f SE^{(k)} N_{j,m,p,PDSCH}^{(usr)} \qquad (5.32)$$

Where $N_{j,m,p,PDSCH}^{(usr)}$ is given by (5.14) and $\delta t_{TTI} \delta f SE^{(k)}$ is the number of bits per PDSCH RE for the spectral efficiency $SE^{(k)}$. If user k is off line, then $\Delta B_j^{(k)} = 0$

10. Update the cumulative downloaded data for each user, k, up to the current epoch, j

$$B_j^{(k)} = \sum_{i=j_{on}^{(k)}}^{j} \Delta B_i^{(k)} = SE^{(k)} \delta t_{TTI} \delta f \sum_{i=j_{on}^{(k)}}^{j} N_{i,m,p,PDSCH}^{(usr)} \qquad (5.33)$$

If $B_j^{(k)} \ge B^{(usr)}$, the download is completed and the user k goes off line.

11. Update the total amount of time each active user has been on line, up to the j th epoch

$$\Delta t_{on,j}^{(k)} = \left(j - j_{on}^{(k)} \right) \Delta t \qquad (5.34)$$

The download to users for which $\Delta t_{on,j}^{(k)} > \Delta t_{max,dl}$ seconds is terminated and they are sent off line, irrespective of whether or not their download has been completed. The download is recorded as terminated, but the user-accumulated data is included in the total traffic volume.

(Continued)

TABLE 5.3 (*Continued*)
Data Collected in Each Epoch of the Simulation

Data Collected at End of Each Epoch, *j*

12.   For all users, k, who go off line in this epoch, record the index of the epoch of completion (or termination) $j_{off}^{(k)}$. This corresponds to the service finish time for the k th user.

13.   Update the current active user number, $N_{j,m,p}^{(usr)}$, of the base stations and sectors This includes the number of users coming on line, $N_j^{(usr)+}$ and going off line, $N_j^{(usr)-}$.

14.   If this is the first epoch that satisfied the "steady-state condition" then set the starting time, t_{start}, for recording results.

15.   Update the value of $\mu_{BS,m,p,j}$ for the j th epoch for all base stations, m, and sectors, p.

$$M_{BS,m,p,j} = \sum_{i=1}^{j} \mu_{BS,m,p,i} \qquad (5.35)$$

This quantity gives the cumulative number of epochs during which the sector p of base station m has been downloading data.

16.   Update the total number of users who have come on line from epoch $j = 0$ up to this epoch, $N_{Tot,j}^{(usr)}$,

$$N_{Tot,j}^{(usr)} = \sum_{i=0}^{j-1} N_i^{(usr)+} + N_j^{(usr)+} \quad j = 1,2,\dots \qquad (5.36)$$

If the $N_{Tot,j}^{(usr)} > N^{(usr)}$ then terminate the simulation.

TABLE 5.4
Data Collected at the End of the Simulation

Data Collected at End of the Simulation

1.   Download service time, $\Delta t_{dl}^{(k)}$, for each user:

$$\Delta t_{dl}^{(k)} = \left(j_{off}^{(k)} - j_{on}^{(k)} \right) \Delta t = N_{dl}^{(k)} \Delta t \qquad (5.37)$$

2.   Throughput rate, $R^{(k)}$, achieved for each user:

$$R^{(k)} = \sum_{j=j_{on}^{(k)}}^{j_{off}^{(k)}} B_j^{(k)} / \Delta t_{dl}^{(k)} = B_{Tot}^{(k)} / \Delta t_{dl}^{(k)} \qquad (5.38)$$

3.   The average SINR and spectral efficiency over all users:

$$\langle SINR \rangle^{(usr)} = \frac{1}{N_{Tot}^{(usr)}} \sum_{k=1}^{N_{Tot}^{(usr)}} SINR^{(k)}, \qquad \langle SE \rangle^{(usr)} = \frac{1}{N_{Tot}^{(usr)}} \sum_{k=1}^{N_{Tot}^{(usr)}} SE^{(k)} \qquad (5.39)$$

Note these are unweighted user averages. The above spectral efficiency of the k th user, $SE^{(k)}$, is related to the system average spectral efficiency, $\langle SE \rangle_{RF}$, via (4.59). The system spectral efficiency $\langle SE \rangle_{RF}$, given in (4.28), is required for determining the base station 100% utilisation throughput, C_1, given in (4.31). The relationship between or the throughput, $R(t)$ and $SE^{(usr)}$ is given in (4.57) from which we see that the utilisation of the user, $\mu_{BS}^{(k)}$ is involved.

(Continued)

TABLE 5.4 (*Continued*)
Data Collected at the End of the Simulation

Data Collected at End of the Simulation

4. Total number of epochs during which each base station sector was actively transmitting data to users:

$$M_{BS,m,p} = \sum_{j=0}^{N_{Tot}} \mu_{BS,m,p,j} \tag{5.40}$$

Note: because $\mu_{BS,m,p,j}$ only takes on values of 0 or unity, this sum gives the total number of epochs each base station sector is actively transmitting data to users.

This also gives the number of epochs for which the base station is idle: $N_{Tot} - M_{BS,m,p}$.

5. The average utilisation of each base station sector and base station:

$$\langle \mu_{BS,m,p} \rangle_T = \sum_{j=1}^{N_{tot}} \mu_{BS,m,p,j} / N_{tot} \tag{5.41}$$

$$\langle \mu_{BS,m} \rangle_T = \sum_{p=1}^{N_{sec}} \langle \mu_{BS,m,p} \rangle_T / N_{sec} \tag{5.42}$$

where N_{sec} is the number of sectors in a base station.

6. The total volume of data downloaded by all users over the total duration of the simulation:

$$B_{Tot} = \sum_{k=1}^{N_{Tot}^{(usr)}} B_{Tot}^{(k)} \tag{5.43}$$

7. Average base station throughput, $\langle R \rangle_{BS}$ across the 7 base stations (BS_0 to BS_6) for which data is collected:

$$\langle R \rangle_{BS} = \frac{B_{Tot}}{7 N_{Tot} \Delta t} = \sum_{j=1}^{N_{Tot}} \sum_{k=1}^{N_{Tot}^{(usr)}} \Delta B_j^{(k)} \Big/ 7 N_{Tot} \Delta t \tag{5.44}$$

8. Average service rate per user:

$$\langle \lambda \rangle^{(usr)} = N_{Tot}^{(usr)} / N_{Tot} \Delta t \tag{5.45}$$

9. Total time to deliver the number of downloads requested. From (5.25) this is $N_{Tot} \Delta t$
10. The number of successful downloads, N_{sccs}.
11. The number of failed downloads, N_{fail}.

From the utilisation values in (5.40) the total energy consumed by the base station is:

$$Q_{BS,m} = 3 N_{ant} N_{Tot} \Delta t P_{idle} + \Delta t \Delta_{P,m} P_{max,out} \sum_{j=1}^{N_{Tot}} \sum_{p=1}^{3N_{ant}} \mu_{BS,m,p,j}$$

$$= \left(3 N_{Tot} N_{ant} P_{idle} + \Delta_{P,m} P_{max,out} \sum_{p=1}^{3N_{ant}} M_{BS,m,p} \right) \Delta t \tag{5.46}$$

Separating the total energy consumed into the "idle" and "incremental" components can provide further insight. From (5.46) the energy consumption due to the idle power component of the base station operation is $3 N_{Tot} \Delta t P_{idle}$. The energy consumption due to the incremental component of base station operation is:

$$\Delta_{P,m} P_{max,out} \Delta t \sum_{p=1}^{3N_{ant}} M_{BS,m,p} \tag{5.47}$$

Using this with (5.43), the energy per bit, H, for the base stations included in the simulation is given by:

$$H = \sum_{m=0}^{N_{BS}} Q_{BS,m} \Bigg/ B_{Tot} \tag{5.48}$$

where N_{BS} is the number of base stations for which data is collected in the simulation. The energy efficiency can also be expressed in terms of the average data throughput over the duration of the simulation, $\langle R \rangle_T$. Using this, we have:

$$H = \sum_{m=0}^{N_{BS}} Q_{BS,m} \Bigg/ \langle R_T \rangle T \tag{5.49}$$

where:

$$\langle R \rangle_T = \frac{B_{Tot}}{T} = \frac{1}{N_{Tot}\Delta t} \sum_{j=1}^{N_{Tot}} \Delta B_j = \frac{1}{N_{Tot}} \sum_{j=1}^{N_{Tot}} R_j \tag{5.50}$$

Form (5.49) can be used to calculate the energy efficiency over a diurnal cycle below.

5.7   CALCULATING DIURNAL CYCLE RESULTS

The results listed in Tables 5.3 and 5.4 are for a given constant user traffic demand $D^{(usr)}$. As shown in Figure 5.2, the value of $D^{(usr)}$ varies depending upon the stage of the diurnal cycle. Therefore, to calculate results over a diurnal cycle, an appropriate average of the results from the simulations for differing $D^{(usr)}$ levels needs to be constructed. Adopting the diurnal cycle shown in Figure 5.2, there are 5 levels of user demand given by $X_S D^{(usr)}$, in the diurnal cycle corresponding to the five time durations, T_S, over the 24 hour period.

5.7.1   DIURNAL CYCLE TOTALS OR AVERAGES

The results from the simulation typically have the form of either a total or an average. We use the generic notation $Y_{S,j}$ to represent the results for quantity Y_S in epoch j and the S subscript indicates this result is given by the simulation with user demand level, $X_S D^{(usr)}$. Some quantities also include a k-index that covers the user population and/or p and m indices that cover the base stations and/or base station sectors. Unless required, these subscripts will be suppressed to simplify the equations.

Setting the service request rate to correspond to user demand to $X_S D^{(usr)}$ and adopting the 5 stage representation of a diurnal cycle shown in Figure 5.2, the simulations will provide a collection of output data for a mobile network for each stage of the diurnal cycle. Examples of this data are listed in Tables 5.2 through Table 5.4. These data represent the steady-state response of network to user demand $X_S D^{(usr)}$.

The results of the simulations are for $N^{(usr)}_{Tot}$ service requests over a time duration of N_{Tot} epochs. However, the values $N^{(usr)}_{Tot}$ and N_{Tot} may not correspond to their appropriate values for the stages S of the diurnal cycle. The recorded results are typically either a total (such as total data downloaded) or an average (such as average user throughput). The simulation gives results for N_{Tot} epochs are for a steady state, hence we can use their average value to "re-scale" these results to approximate the corresponding result for the number of epochs, $N_{Tot,S} = T_S/\Delta t_S$, in the duration of stage S of the diurnal cycle. In this T_S and Δt_S are the stage and epoch durations corresponding to stage S of the diurnal cycle, respectively.

Consider a simulation output parameter $Y_{S,j}$ where S corresponds to the stage of the diurnal cycle and j is the epoch index. We have from the simulation of stage S, which ran for duration $T = N_{Tot}\Delta t$ with user demand, $X_S D^{(usr)}$:

$$\sum_{j=1}^{N_{Tot}} Y_{S,j} = N_{Tot} \langle Y_S \rangle_T \tag{5.51}$$

However, for stage S of the diurnal cycle, we require the j-sum to be over $N_{Tot,S}$ epochs to correspond to stage duration $T_S = N_{Tot,S}\Delta t_S$. To re-scale the sum from N_{Tot} to $N_{Tot,S}$, we use the fact that the simulation is in steady state and therefore:

$$\langle Y_S \rangle_T \approx \langle Y_S \rangle_{T_S} \tag{5.52}$$

That is, with the system in steady state the averages over durations T and T_S will be approximately equal. A detailed justification of this is provided in the Appendix.

With this, we can re-scale the results that are totals using:

$$\sum_{j=1}^{N_{Tot,S}} Y_{S,j} \approx N_{Tot,S} \langle Y_S \rangle_T = \frac{N_{Tot,S}}{N_{Tot}} \sum_{j=1}^{N_{Tot}} Y_{S,j} \tag{5.53}$$

To calculate the averages over the full diurnal cycle, we split the sum over the full cycle into the five separate stages. Using the average results for each and (5.53) to provide approximations for the summations in each stage, we have:

$$\langle Y \rangle_D = \frac{\sum_{j=1}^{N_{Diurnal}} Y_j}{N_D} \approx \frac{\sum_{S=1}^{5} N_{Tot,S} \langle Y_S \rangle_T}{\sum_{S=1}^{5} N_{Tot,S}} = \frac{\sum_{S=1}^{5} \frac{T_S}{\Delta t_s} \langle Y_S \rangle_T}{\sum_{S=1}^{5} \frac{T_S}{\Delta t_s}} \tag{5.54}$$

where N_D is the total number of epoch steps over the full diurnal cycle given by the sum of the $N_{Tot,S}$ of each stage and we have used $N_{Tot,S} = T_S/\Delta t_S$ for each stage. The Δt_S differ between the stages because we have $\Delta t_S \lambda_S = \Delta t_S X_S \lambda = C$ a constant across all the stages of the simulation and given by (5.13). This constant is set by the probability of requests occurring during the epoch Δt_S. This probability is constant across all stages. We express the durations of the stages of the diurnal cycle as a fraction of 24 hours; $T_S = \rho_S \times (24 \text{ hours})$. Therefore, $T_S/\Delta t_S = (24 \text{ hours})\rho_S X_S \lambda/C$ giving:

$$\langle Y \rangle_D = \sum_{S=1}^{5} \rho_S X_S \langle Y_S \rangle_T \bigg/ \sum_{S=1}^{5} \rho_S X_S \tag{5.55}$$

The example values of X_S across the diurnal cycle are given in Figure 5.2.

By its definition, $D^{(usr)}$ is the average user demand over diurnal cycle, T_D; therefore, we have:

$$\frac{1}{T_D} \sum_{S=1}^{5} X_S D^{(usr)} T_S = \frac{D^{(usr)}}{T_D} \sum_{S=1}^{5} X_S \rho_S T_D = D^{(usr)} \tag{5.56}$$

Therefore:

$$\sum_{S=1}^{5} \rho_S X_S = 1 \tag{5.57}$$

For the quotient of two quantities, Y and Z, following the same approach as immediately above, we have:

$$
\begin{aligned}
\langle Y \rangle_D / \langle Z \rangle_D &= \sum_{S=1}^{5} \rho_S X_S \langle Y_S \rangle_T \Big/ \sum_{S=1}^{5} \rho_S X_S \langle Z_S \rangle_T \\
&= \sum_{S=1}^{5} \rho_S X_S H_S \langle Z_S \rangle_T \Big/ \sum_{S=1}^{5} \rho_S X_S \langle Z_S \rangle_T
\end{aligned}
\tag{5.58}
$$

where we have introduced quotient $H_S = \langle Y_S \rangle / \langle Z_S \rangle$. Also:

$$\langle Y/Z \rangle_D = \sum_{S=1}^{5} \rho_S X_S \langle Y_S/Z_S \rangle_T \tag{5.59}$$

From (5.54) we see that the $\rho_S X_S$ multipliers reflect the fact that each stage of the diurnal cycle has a different duration, $T_S = \rho_S 24$ hours, and a different service request rate, $\lambda_S = X_S \lambda$. Both these factors need to be accounted for when averaging over the diurnal cycle.

Comparing (5.58) and (5.59) we see that dealing with a quotient of averages is quite different to dealing with the average of a quotient. Therefore, the modeller needs to decide which is the more appropriate quantity to use as an outcome of the model.

Several examples of network results presented in this chapter are as follows:

a. Average user SINR, $\langle SINR \rangle^{(usr)}$
b. Average throughput per user, $\langle R \rangle^{(usr)}$
c. Average base station utilisation, $\langle \mu_{BS} \rangle$
d. Proportion of failed downloads, F_{dl}
e. Average Base Station Throughput, $\langle R \rangle_{BS}$
f. Energy efficiency, H

The approach required to calculate each of these over the full diurnal cycle will be presented to provide examples of how such results are calculated.

5.7.2 Average User SINR, $\langle SINR \rangle^{(usr)}$

The SINR experienced by the k th user, $SINR^{(k)}$, is constant over every epoch during which they are downloading. Therefore over a full diurnal cycle, using (5.55) we have:

$$
\begin{aligned}
\langle SINR \rangle_D^{(usr)} &= \sum_{k=1}^{N_D^{(usr)}} SINR^{(k)} \Big/ N_D^{(usr)} = \sum_{S=1}^{5} \sum_{k=1}^{N_{Tot,S}^{(usr)}} SINR^{(k)} \Big/ \sum_{S=1}^{5} N_{Tot,S}^{(usr)} \\
&\approx \sum_{S=1}^{5} \rho_S X_S \langle SINR_S \rangle^{(usr)}
\end{aligned}
\tag{5.60}
$$

5.7.3 AVERAGE THROUGHPUT PER USER

The average throughput per user, $\langle R \rangle^{(usr)}$ has to be carefully defined. Because users are randomly located across the service area, we can expect some to have a high SINR and hence a high throughput, whereas others will have a low throughput. Further, the throughput for a given user may vary epoch to epoch as other users come off line and go on line. This means global averaging of total downloaded bits over total simulation duration will not encompass the full range of throughput values. The throughput of the k th user is defined in (5.38) where $B_{Tot}^{(k)}$ is given in (5.27).

The details of the calculation are in the Appendix. The end result is as follows:

$$\langle R \rangle_D^{(usr)} \approx \sum_{S=1}^{5} \rho_S X_S \langle R_S \rangle^{(usr)} \tag{5.61}$$

5.7.4 AVERAGE BASE STATION UTILISATION

Base station utilisation in the j th epoch is $\langle \mu_{BS,m,p,j} \rangle$. Using (5.55) we get for average base station utilisation:

$$\langle \mu \rangle_{BS,D} \approx \frac{1}{3 \times 7 N_{ant}} \sum_{S=1}^{5} \rho_S X_S \sum_{m=0}^{6} \sum_{p=1}^{3 N_{ant}} \langle \mu_{BS,S,m,p} \rangle_T \tag{5.62}$$

In which the data collection was for the inner cluster of base stations, BS_0 to BS_6.

5.7.5 PROPORTION OF FAILED DOWNLOADS

The proportion of failed downloads is the ratio of failed downloads to total downloads is given by the ratio:

$$F_{Prop} = \frac{N_{fail}}{N_{dl}} \tag{5.63}$$

where N_{fail} is the number of failed downloads and N_{dl} is the total number of downloads.

We can express N_{fail} as the sum of failed downloads over the epochs. That is:

$$N_{fail,S} = \sum_{j=1}^{N_{Tot,S}} N_{fail,S,j} \approx N_{Tot,S} \langle N_{fail,S} \rangle_T \tag{5.64}$$

The total number of downloads in state S of the diurnal cycle, $N_{dl,S}$, equals the total number of user download requests $N_{Tot,S}^{(usr)}$. Every download request is served, of which some fail and the rest are successful. Therefore, using (5.81) we have:

$$F_{Prop,S} \approx \langle N_{fail,S} \rangle_T \tag{5.65}$$

Using (5.58)

$$F_{Prop,D} = \sum_{S=1}^{5} N_{fail,S} \Bigg/ \sum_{S=1}^{5} N_{dl,S}$$

$$\approx \sum_{S=1}^{5} \rho_S X_S \langle N_{fail,S} \rangle_T = \sum_{S=1}^{5} \rho_S X_S F_{Prop,S} \tag{5.66}$$

5.7.6 Average Base Station Throughput

Using (5.30) and (5.55) we have for the average base station throughput over the diurnal cycle:

$$\langle R \rangle_{BS,D} = \sum_{S=1}^{5} \rho_S X_S \langle R \rangle_{BS,T} \tag{5.67}$$

5.7.7 Diurnal Cycle Energy Efficiency

Calculating the average energy efficiency, H, over a diurnal cycle is more complicated because it involves quotient, that being total energy consumed divided by total data downloaded. In this case the simulation provides an energy efficiency, H_S, for each stage of the diurnal cycle, given by (5.48). The overall energy efficiency of the network equals the total energy consumed across all base stations over a diurnal cycle divided by the total data downloaded in the network over a diurnal cycle. That is:

$$H_D = \sum_{S=1}^{5} \sum_{m=0}^{N_{BS}} Q_{BS,S,m} \left/ \sum_{S=1}^{5} B_{Tot,S} \right. \tag{5.68}$$

where $Q_{BS,S,m}$ is the energy consumption of the m th base station during stage S of the diurnal cycle and $B_{Tot,S}$ is the total data downloaded during stage S of the diurnal cycle given by (5.26).

There are several approaches to calculating the overall energy efficiency. Using (5.46) we can write for $Q_{BS,S,m}$ in (5.68):

$$\begin{aligned} Q_{BS,S,m} &\approx N_{Tot,S} \Delta t_S \langle P_{BS,S,m} \rangle_T = N_{Tot,S} \langle Q_{BS,S,m} \rangle_T \\ &= N_{Tot,S} \Delta t_S \left(P_{idle,m} + \Delta_{P,m} P_{max,out} \sum_{p-1}^{3} \langle \mu_{BS,S,m,p} \rangle_T \right) \end{aligned} \tag{5.69}$$

where $\langle \mu_{BS,S,m,p} \rangle_T$ is the mean base station utilisation result from the simulation of stage S and we have used (5.76) to rescale that result to give the sum over the duration T_S. Similarly for the downloaded data we have:

$$B_{Tot,S} = \sum_{j=1}^{N_{Tot,S}} \Delta B_{S,j} \approx N_{Tot,S} \langle \Delta B_S \rangle_T = N_{Tot,S} \Delta t_S \langle R_S \rangle_T = T_S \langle R_S \rangle_T \tag{5.70}$$

where $\langle \Delta B_S \rangle_T$ is the average data download per epoch and $\langle R_S \rangle_T = \langle \Delta B_S \rangle_T / \Delta t_S$, is the average download throughput given by the simulation corresponding to stage S of the diurnal cycle.

This gives for the energy efficiency during stage S of the diurnal cycle

$$H_S \approx \frac{\sum_{m=0}^{N_{BS}} \langle Q_{BS,S,m} \rangle_T}{\langle \Delta B_S \rangle_T} = \frac{\sum_{m=0}^{N_{BS}} \langle P_{BS,S,m} \rangle_T}{\langle R_S \rangle_T} \tag{5.71}$$

It is more convenient to represent the download data using the average download throughput, $\langle R_S \rangle_T$, than $\langle \Delta B_S \rangle_T$. Focussing on the form of H_S expressed in terms of $\langle R_S \rangle_T$ and using (5.58) the energy efficiency over the diurnal cycle can also be expressed as

$$
\begin{aligned}
H_D &\approx \sum_{S=1}^{5} \rho_S X_S \sum_{m=0}^{N_{BS}} \langle P_{BS,S,m} \rangle_T \Big/ \sum_{S=1}^{5} \rho_S X_S \langle R_S \rangle_T \\
&= \sum_{S=1}^{5} \rho_S X_S H_S \langle R_S \rangle_T \Big/ \sum_{S=1}^{5} \rho_S X_S \langle R_S \rangle_T
\end{aligned}
\tag{5.72}
$$

5.8   EXAMPLE SIMULATIONS

In this section, we present a representative series of simulations that span three of the geographical categories listed in Table 5.1.

1. Dense urban, with 10,000 potential service users per km^2, and a base station spacing of 0.5 km,
2. Urban, with 1000 users per km^2 and base station spacing of 1 km, and
3. Suburban, with 300 users per km^2, and base station separation of 1.73 km.

The "suburban" geography might describe an outer suburb or a rural town, while the "urban" geography might describe an inner-city suburb.

The simulation follows the flowchart in Table 5.8, with the additional parameters for the geographical categories as listed in Tables 5.5 and 5.6.

The propagation model used is the Stanford (SUI) model as described in Section 4.15.2 [14]. In particular, the Intermediate path loss/Terrain Type B in [14] is used. The SUI model differs from some other models in that it sets a non-zero mean for shadow fading 9dB fixed shadow component plus a variation around this mean. This brings the loss vs. distance curve for the SUI model into close alignment with the more commonly used values of 3GPP [4], GreenTouch [5] and other path loss models [16, 18]. The variable component is a random value with a log-normal distribution (i.e. Gaussian in dB) having zero mean and 8dB standard deviation. This broadly aligns the variation in shadow fading with other models which range anywhere between 3 and 10 dB.

As discussed in Section 4.6.1.2, the relationship between the SINR and resulting Spectral Efficiency is provided by an appropriately chosen "link curve". The choice of link curve depends upon multiple factors including the geography, receiver technology, multipath fading, range of modulation formats and system type (MIMO, SISO, SIMO).

Unfortunately, there are very few published link curves, and derivation of a link curve for particular scenarios involves complex link models and substantial computation resources. Figure 4.4 in Section 4.6.1.2 showed a number of curves relating Spectral Efficiency to SINR. For this simulation, we utilise two of the curves in that Figure to model this relationship. These curves and their underlying equations were presented in Morgensen et al. [20], and were discussed in Section 4.6.1.2.

We are principally interested modelling in a basic 2x2 MIMO network. When a user's signal conditions are good, LTE MIMO operates with two independent data streams between the base station transmitter and the user's receiver. If conditions are poor, the system resorts to a SIMO mode, with the two base station transmitters sending the same data stream to the user's receiver and thus achieving a higher signal strength. To emulate this process, we calculate the spectral efficiency that would be achieved in each of these two modes, and use the higher of these two values. Both of these modes use the same formula adapted from [20], albeit with different parameter values, and allow for modulation formats up to 64QAM:

$$SE\left(\text{bits/RE}\right) = \min\left(StreamsBW_{Eff} \log_2\left(1 + \frac{SINR}{SINR_{Eff}}\right), Cap \right) \qquad (5.73)$$

For the MIMO mode, $Streams = 2$, $BW_{eff} = 0.56$, $SINR_{eff} = 2$ and $Cap = 7.37$.

For the SIMO mode, $Streams = 1$, $BW_{eff} = 0.62$, $SINR_{eff} = 1.8$, and $Cap = 3.83$

The parameter Cap introduces a limit to the calculated spectral efficiency in each case, based on the available modulation and coding format.

Although considerably more complex than the Static Model of Chapter 4, this model is still just an intermediate step towards a full-blown industry simulation. We use the Morgensen *et al.* link curve model because it is the only easily accessible, publicly available model for the generation of the Spectral Efficiency for 2 × 2 MIMO that includes multipath fading for each stream.

Each simulation is run for 90,000 user entries to the wider service area, which results in approximately 15,000 users of the central core of base stations for whom data is recorded.

The results presented in this chapter have been plotted against average traffic demand per user, $D^{(usr)}$. The modeller may decide to use another parameter for the independent variable such as the total user demand per base station, given by the product of the average number of active users per base and $D^{(usr)}$. Another option is the demand per km², given by the total user demand per base station divided by the base station cell area. Using quantities such as these can assist the process of network dimensioning, which seeks to balance network download throughput with total user demand in an endeavour to attain an acceptable performance for users. These alternatives can be derived by appropriate rescaling of the $D^{(usr)}$ axis.

5.8.1 Results from Suburban Geography Simulation

With the range of parameters recorded as listed in Tables 5.5 and 5.6, there are a BS throughput number of options available for reporting performance. Some of these are specific to the user(s) whilst the others are related to the performance of the network as a whole. Figure 5.6 shows a sample of the results for a Suburban geography network simulation. The plots a) to f) in Figure 5.6 show examples of simulation results for a range of "average" user demand 1.5 kbit/s to 85 kbit/s. Naturally only "active" users are receiving traffic, but the service area includes many inactive users at any given time. This "average demand" includes both inactive as well as active users, and represents the long-term traffic demand of the user population as discussed in Section 5.4.1.

The results are for 2 × 2 MIMO unless the SINR is below a threshold value, in which case it will automatically drop from 2 layers to 1, being SIMO, to provide transmit and receive diversity. In this single-layer case, the same data is sent by both sector transmitters in two streams, adding to the received power and providing a level of channel diversity. The threshold value of the SINR for this transition is approximately 10.5 dB as seen in Figure 4.4 as the crossover value for BLAST2x2 and SIMO.

TABLE 5.5

Additional Parameters for Geographical Categories Listed in Table 5.1

Geographical Category	Dense Urban (DU)	Urban (U)	Suburban (SU)
Base Station antenna down-tilt, degrees	15	11	6
Operating frequency, MHz	2000	2000	2000
Service Area for simulation	3km square	6km square	10km square

TABLE 5.6

Common Parameters for Simulations of the Geographical Categories Listed in Table 5.5

Common Features

Base Station antenna height, metres	30
Base station antenna horizontal beam width, degrees	70
Base station antenna vertical beam width, degrees	10
Base station, antenna max. gain dBi	14
User antenna height metres	1.5
User antenna gain dBi	0
Building penetration loss	20 dB
Shadowing fading mean value/standard deviation (based on SUI model, log-normal distribution)	9 dB/8 dB
Base Station Max Power Consumption, W	684
Base Station Idle Power Consumption, W	156
Fixed Overhead % of total REs	10.92
Variable Overhead % of active REs	13.36

The performance of a Suburban network with the chosen model variables is summarised by the graphs in Figure 5.6.

Figure 5.6a shows a steady reduction in the typical SINR experienced by users as user demand across the network increases. With increasing demand, the probability of adjacent base stations being active and causing interference increases, thus lowering the SINR. With networks in which the cell areas and the inter base station distance are smaller, the average reduction in the strength of the received signal and of the interfering signals would closely track each other. In this case the service is "interference limited". However, in the Suburban geography shown here, the receiver noise has become more significant.

Figure 5.6b shows the average data throughput for downloads received by users across the network, as the network demand, corresponding to $D^{(usr)}$, increases. The achieved data rate decreases as a result of decreasing SINR (as shown in a) of the Figure) but is further amplified by the increasing number of concurrent active downloads which share the available PDSCH REs.

Figure 5.6c shows the utilisation of the base station increasing steadily with demand. For networks in which saturation begins to set in, the traces will flatten out at high user demand.

Figure 5.6d shows the proportions of such failed downloads for each of the 2×2 MIMO and the SIMO case. There will always be a spread in the SINR values across users and hence in the data throughput achieved across the cohort of active network users. Consequentially some users may find their download fails to be completed within a pre-set maximum time. A service quality measure is adopted which removes from the service area (also called "file dropping") users whose download of the file $B^{(usr)}$ (16 Mbit) has not finished within this pre-set maximum, $\Delta t_{max,dl}$ (4 seconds). This equivalent to a 4 Mbit/s data rate which in this model is considered to represent unsatisfactory network performance. The file drop might be initiated by either the network or a frustrated user. File dropping metrics are frequently used to represent the impact of increasing demand on network performance [4, 21–23]. The modeller may adopt a different file dropping criteria.

Figure 5.6e shows the total throughput per base station, as a function of user traffic demand. As the network saturates, there is a smaller number of unused PDSCH REs available to meet an increase in the user demand. This results in the base station throughput trace flattening at higher user demand. This plot shows an average taken over all user downloads and all base station sectors. In general, not all of the sectors nor user downloads have the same download performance; hence, the transition into saturation need not be abrupt.

Figure 5.6f shows the average energy per received bit across the user downloads in the simulation.

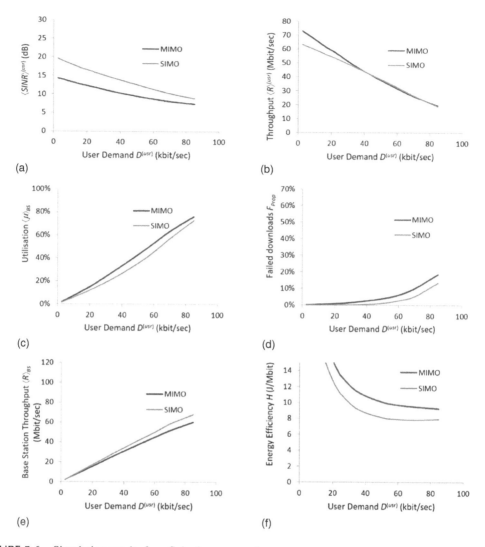

FIGURE 5.6   Simulation results for a Suburban geography.

Overall, relative to the geographies with smaller inter base stations distances, the Suburban geography exhibits lower SINR and user achieved data rates. Also, relative to the other geographies the base station utilisation and download termination rates are higher, and the combined effect of these has led to a degradation in energy efficiency, that is, a higher energy per bit.

5.8.2   RESULTS FROM URBAN GEOGRAPHY SIMULATION

The results for Urban geography shown in Figure 5.7 follow similar trends as with the Suburban geography. However, for the Urban geography the smaller inter-site distance and the smaller spread of distances between users and their serving base station lead to increased signal strengths. This means the achieved SINR and user achieved data rates are higher, the base station utilisation and download termination rates are lower, and the combined effect of these has led to better energy efficiency, that is lower energy per bit relative to the Suburban geography.

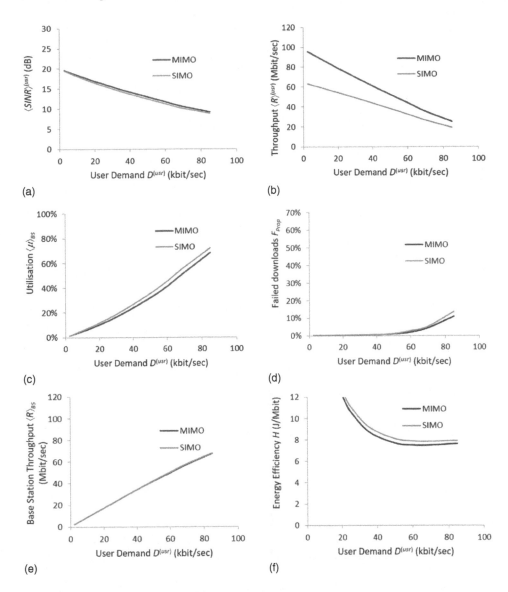

FIGURE 5.7   Simulation results for an Urban geography.

5.8.3   Results from Dense Urban Geography Simulation

The Simulations for a Dense Urban network present a similar pattern of results to those for the urban and suburban networks, but with increasing complexity. Whereas with the urban network simulation we assumed a population density of 1000 users per square km, for the dense urban network that rises to 10,000 users per square km. The same range of average user traffic demand, $D^{(usr)}$, is assumed, so that the demand on the base stations increases 10 times. To support that increased demand, more base stations are provided at a reduced inter-site distance of 500 m, but this is not sufficient of its own to meet the demand. It will be demonstrated in Chapter 6 how a network of lower power small cells can assist in meeting the higher demand without incurring the high energy cost.

The graphs in Figure 5.8 show the same sequence of performance metrics for a base station network in the dense urban environment in which no small cells are deployed. Figure 5.8a and b show SINR and user bit rate performance decreasing rapidly with network demand, and reaching a plateau

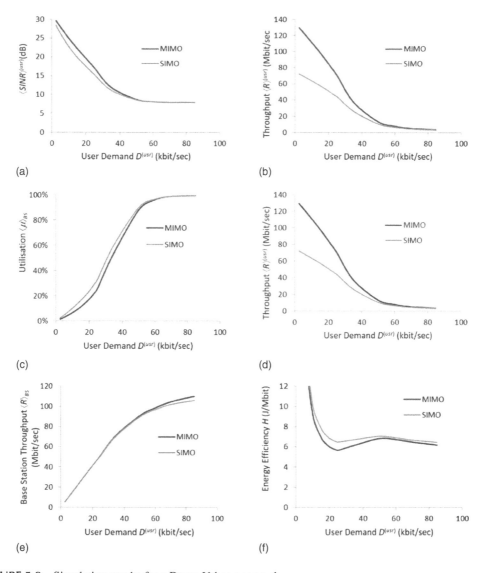

FIGURE 5.8 Simulation results for a Dense Urban geography.

at relatively low traffic levels. Figure 5.8c shows that base station utilisation quickly approaches congestion. Figure 5.8d in turn shows that the percentage of downloads terminated, because the achieved data rate falls below the "acceptable" requirement of 4 sec for a 16 Mbit file, is relatively high, reaching some 40% of downloads at the per user demand level estimated for the year 2020 [5, 7].

Figure 5.8d shows that base station throughput increases much quicker with per user demand than the other geographies. This is due to the significantly higher user density and results in the base stations reaching their maximum capacity (particularly for SIMO) at a lower per user demand.

The energy efficiency trace in Figure 5.8f requires some explanation. At low demand and traffic levels, the consumption is dominated by the base station idle power, which when amortised across the base station traffic results in the decrease in energy per bit with increasing user demand. Whenever the base station is transmitting user data, additional power is consumed, and at low traffic levels the incremental energy consumed per bit transmitted is approximately constant. As the traffic levels increase, the achieved bit rate decreases due to the combined effects of decreasing

SINR and of increasing resource sharing; each download takes longer, and the incremental energy consumption increases. At around the 40 kbit/s average demand level, the proportion of downloads terminated due to poor performance is increasing, the base station capacity is getting close to saturation, although its total data throughput continues to rise slowly with demand. The overall average incremental energy per bit decreases slowly in this phase, but the use of averages here conceals much of the detail.

The performance levels shown in Figure 5.8 are for a simulation of a network operating well beyond sensible traffic levels, without the benefit of complementary protocols or technologies. The use of small, energy-efficient short-range cells to assist in handling some the macro base station traffic in dense areas, will be discussed in Chapter 6. In that chapter, we will see how the use of small cells results in greatly improved performance of the dense urban macro base station network, when for example 2/3 of its traffic is offloaded to a complementary small cell network. This architecture becomes practical in situations where, although population and traffic densities may be high, they are not uniformly distributed but concentrated into clusters.

5.9 MULTIPLE INPUT MULTIPLE OUTPUT (MIMO)

This discussion is intended just to create an awareness of MIMO, as it will be encountered in any mobile system publications. More detailed discussion is beyond the objectives of this book; indeed there are many textbooks dealing just with MIMO and its many options [21–25].

The Static Model in Chapter 4 demonstrated several of the issues in predicting and characterising wireless network performance. The model relies on a knowledge of receiver location, and hence the path loss between transmitter and receiver and the gain of the transmitter's antenna in the direction of the receiver to determine received signal strength. It also undertakes a calculation of the signals from other transmitters that interfere with the desired signal to provide an estimation of the signal-to-interference and noise ratio at user locations.

However, the signal at the receiver in an urban environment is rarely the result of just a single or direct path but an aggregation of paths involving reflections from buildings, other structures, terrain, and even vegetation. Consequentially signals arrive at the receiver with different propagation delays as well as signal strengths. The multiple reflections combine constructively or destructively depending upon the phase differences. This effect, known as multipath fading, was discussed in Section 4.6.1.2 and 4.7.1. An allowance for multipath fading is included in link curves and is a substantial contributing factor to the difference between the predicted path attenuation models and the free-space propagation model given in (4.44). Moreover, the propagation delay between any pair of signal paths may, for example, cause constructive interference at one frequency within the signal frequency range and destructive interference at another frequency. This is described as frequency selective fading.

Wireless systems with multiple antennas offer several ways to achieve improvements in system performance, taking advantage of different paths between transmitting and receiving antennas. With two or more antennas at a transmitter site each sending the same signal, differently encoded, to a receiver with a single antenna, the multiple signals can be combined at the receiver to achieve a modestly improved performance. With a single transmitter but two receiving antennas, the signals at the receiving antennas can be selected, or in some instances combined, to also achieve a better performance. These techniques are, respectively, called Transmit Diversity and Receive Diversity, and are of most benefit in improving reception in difficult cell-edge areas. Any gain in network throughput is generally modest.

When systems involve at least two transmitting and two receiving antennas, MIMO systems can enable significant increases in system capacity. For a system with M antennas at the transmitter and N antennas at the receiver, that is $M \times N$ MIMO system, a total of K independent data streams can be conveyed between the transmitter and receiver, where K is the smaller of the antenna numbers M and N. In principle, the capacity of the base station can be increased by a factor of K. However this

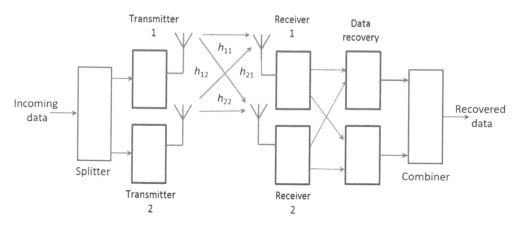

FIGURE 5.9 The four channels, h_{11}, h_{12}, h_{21}, h_{22}, between transmitter and receiver in the simplest representation of MIMO.

relies on the multiple paths between transmitter antennas and receiver antennas being separable at the receiver, and this is generally imperfect. That imperfection reduces the quality of the demodulated signal.

The multiple paths is illustrated for 2x2 MIMO in Figure 5.9, with the parameters h_{11}, h_{12}, h_{21} and h_{22} representing different characteristics of the four paths between two transmitters and two receivers

In its simplest form, a MIMO receiver relies on the four possible paths (also referred to as "channels") between the two transmitters and two receivers having different characteristics in terms of received signal strength and path delay. Using the received Reference Signal REs embedded in the LTE frame, as described in Section 4.6, the receiver is able to estimate the characteristics of the four channels (as denoted by h_{ab} in Figure 5.9) to partially cancel the cross channel interference. This is not effective when there is little difference between the channel characteristics.

For single-user MIMO as modelled, a single data stream is split into two component streams, which are pre-coded and transmitted towards the receiver. The pre-coding involves each transmitter output sending a blend of the two-component signals using one of a set range of codes from a "codebook" [22, 23, 26, 27]. The receiver in turn detects the signals form each transmitter, and uses its copy of the codebook codes to recover the two-component streams, and construct a replica of the original data stream. Feedback to the base station from the receiving equipment is used to select a pre-coding matrix from the codebook code set.

In general, recovery is not perfect because of the different characteristics of the four paths, (and imperfect channel) leading to some errors in the received stream.

MIMO as described may provide additional transmission capacity to one user (single-user MIMO), or to re-use the same transmission capacity to serve multiple users (multi-user MIMO). More elaborate MIMO options are available using up to eight transmitters. The service to any one user is limited by the numbers of transmitting and receiving antennas, i.e. by the smaller of the transmitting and receiving antennas. Even so, in the case of two receiving antennas a wider range of transmitters with consequently different transmission paths becomes available. The wider choice of transmitters in turn enables a higher-performing pairing of transmitter and receiver antennas to be achieved. Further possibilities include the use of collaboration between base stations, or the multiple transmitting antennas in a single base station to create and steer a narrower transmission beam. That could produce a higher signal strength in one direction, but more commonly to produce a signal null in a particular direction so as to reduce interference to users in difficult areas. Base stations may also collaborate using the improved capacity by each sending the same data to a user in a difficult location.

The simulations described in this chapter represent only the first steps in assessing the performance of wireless networks, sufficient to achieve an estimate of performance and hence energy consumption and efficiency. These simulations cover only signal strength and interference statistics. They do not include full matrix-based channel delay models necessary to fully model fading or MIMO performance; this involves much more complex computing models. Instead, this model uses the link curves, such as BLAST 2x2 and SIMO shown in Figure 4.4, to account for effects such as multipath fading, local geography and surrounds. A disadvantage of this is that the modeller has to find a link curve that adequately reflects the network circumstances the modeller is considering.

5.10   KEY DIFFERENCES BETWEEN THE SIMPLE AND ADVANCED MODELS

The Static Model in Chapter 4 uses Utilisation based traffic model which is a variant of the Full Buffer Traffic model for representing user demand [28] (see Section 4.5). In contrast, the Advanced Model in this chapter uses an FTP traffic model [28] (also see Section 4.5). In the Static Model a group of users (typically 10 per sector) are randomly located in the three sectors of BS_0 (see Figure 4.8) and continuously download data for the duration of the simulation. The users' locations are uniformly but randomly distributed so that the SINR distribution in the cell of BS_0 adequately "samples" the representative range of SINR values throughout the cell. The simulation is a time-independent model in that there are no changes in the modelled conditions for the duration of the simulation. In the usual Full Buffer Traffic model, the base station is operating at 100% utilisation whereas the Utilisation-based model can be run with base station operating at a lower utilisation.

The radiated power from adjacent cells is adjusted by their utilisation factor to emulate low traffic conditions while the central cell whose, performance is being evaluated, runs with $\mu_{BS,0}$ proportion of RBs available for carrying traffic. By setting base station utilisation at less than 100%, as seen from (4.17), the simulation represents transmitting only a proportion of available PDSCH REs from use. In turn, that will result in a quasi-linear relationship between user download data rate and, base station total throughput data rate and utilisation as seen in (4.56) and (4.57).

The Static Model involves multiple runs with each run relocating the cohort of users randomly across the cell. This is done to ensure an adequate sample of SINR values across locations in the cell. Alternatively, the multiple groups of users can be located around the cell of BS_0 in a single run with the data collected and analysed separately for each group. In either case, the base station utilisation is the sum of the user utilisations for each group.

In contrast, in the Advanced Model users do not download continuously. Rather, the model simulates users entering the service area, requesting and downloading a file, then exiting the simulation. The new user requests occur at a rate that commensurate with a pre-set per-user monthly data demand and stage of the diurnal cycle being modelled. In the Advanced Model, the locations of users are distributed using a uniform, random distribution throughout the service area. Upon completion of a download, a user "leaves" the service area. In the Advanced Model a base station will terminate a user download if the download time exceeds a pre-set maximum duration. Although the Advanced Model is run as a steady-state model, it does include temporal dynamics in that when users enter and leave the service area, the base stations re-allocate resources in accordance with those changes.

This is an important distinction between the two models. From the perspective of the Advanced Model, the continuous downloading used in the Static Model can be interpreted as users restarting their download of a file as soon as the previous file download has been completed. However, this does not provide a simplification that reduces the Advanced to a subset of the Static Model. Each new user request in the Advanced Model is located at a new randomly selected position. Therefore, the locations of the download sites are continuously changing as the simulation progresses. In the Static Model all downloads are to the same locations for the entirety of the simulation.

In both the network models (Simple and Advanced), active users are typically allocated an equal share of a base station transmitter's PDSCH REs to support their download. However, in the

Advanced Model the number of active users may vary between epochs. The Static Model has a constant number of users.

The differences in how downloads are simulated in the Simple and Advanced models give rise to different measures of network and user performance. For example, the Static Model assumes a cell utilisation level, but does not develop a metric for the average duration of a download, nor a relationship between average user demand ($D^{(usr)}$) and average user throughput ($\langle R \rangle^{(usr)}$), or include a method to identify poorly performing or failed downloads [29]. It should also be noted that Advanced Model begins with an assumption of user demand level, and can provide these plus a range of other outputs not available with the Static Model.

5.10.1   UTILISATION AND USER DEMAND IN THE SIMPLE AND ADVANCED MODELS

In its essence, the Static Model provides a relationship between base stations' utilisation, μ_{BS}, (which can be taken as the sum of base station user utilisations for a given number of users, each user contributing $\mu_{BS}^{(k)}$ to the overall utilisation) and the base station's download throughput and power consumption. The Model starts with setting the μ_{BS} values for all of BS_0–BS_{18}. The value of μ_{BS} for base stations BS_1 to BS_{18} sets their transmit power which contributes to the interference power experienced by signals transmitted from BS_0 to its users. This cumulative interference power sets the SINR experienced by BS_0 users; hence the Spectral Efficiency of their download and, in turn, their download throughput $R^{(k)}$. The utilisation $\mu_{BS,0}$ sets the proportion of PDSCH REs to be used to transmit data (after allowing for overheads) and hence influences $R^{(k)}$ as shown in (4.56). The sum of these throughputs determines the total base station throughput, R. The base station throughput has units of bit/s and with the Static Model corresponds to the given base station utilisation via (4.57). For a single layer per sector, the base station power consumption is related to utilisation via (4.18).

In contrast, the Advanced Model provides a relationship between average user demand across the network, $D^{(usr)}$, and the resulting base station energy or power consumption given by (5.46) and base station download throughput given by (5.44). Base station utilisation is given by (5.42). From Figures 5.6c, 5.7c and 5.8c it can be seen the relationship between $D^{(usr)}$ and average base station utilisation, $\langle \mu \rangle_{BS}$, in the Advanced Model is not linear, particularly when the user demand is high as in the case of a Dense Urban environment.

Because of the significant differences between the two models no simple, general relationship exists between the user utilisation, μ_{BS}, input into the Static Model and average user demand, $D^{(usr)}$, input into the Advanced Model. Such a relationship is likely to be determined only by running the a simulation of the Static Model to determine the $SINR^{(k)}$ distribution in the BS_0 coverage area. Once this distribution is known the appropriate link curve can be applied to determine a value for average base station throughput per user $\langle R \rangle^{(usr)}$. One may then endeavour identify this with the required average per user demand, $D^{(usr)}$, that will drive the network to that base station utilisation μ_{BS} for all base stations. This type of relationship cannot be implemented before the models are run. Rather, they will provide a relationship after the models' results have been generated.

Alternatively, the Static Model can be used to calculate the base station throughput, and other model outputs, for a range of base station utilisations. The Advanced Model can also be used to generate the corresponding outputs, including base station utilisation, for a range of download user demands. Using the relationship between $D^{(usr)}$ and $\langle \mu \rangle_{BS}$, provided by the Advanced Model (as in Figures 5.6c, 5.7c and 5.8c) one can look for correlations with the corresponding utilisation from the Static Model. That is, we set the μ_{BS} input into the Static Model to equal the $\langle \mu \rangle_{BS}$ output by the Advanced Model.

Because the Static Model has no mechanism to represent congestion (e.g. the download failure rate in the Advanced Model), it is likely any comparison will only work for low utilisations and for Advanced Model outputs that are averaged over the total simulation duration. The Static Model will never encounter congestion for input utilisation $\mu_{BS} \leq 1$ meaning it will never incur a demand greater than the base station's capacity. Another way of looking at this is that the required per user demand corresponding to $\mu_{BS} \leq 1$ will always be such C_1 of the base station will not be exceeded.

Overall, the stochastic and time dynamic features that are fundamental to the Advanced Model introduce processes that cannot be represented in the Static Model due to its time-independent nature.

5.10.2 Averaging Times and Throughput

With the Static Model, most outputs are averaged over a Radio Frame. In particular, user and base station download throughput ($R^{(k)}(t_u)$ and $R(t_u)$, respectively) as seen in (4.56) and (4.57), respectively. Both similarly depend on the product of utilisation and spectral efficiency and so display similar dependence on utilisation. This is because the premise of the Static Model is that the proportion of PDSCH REs available to users is set, and limited, by the model's input utilisation. Thus per user rate and per base station rate are set limited by the input utilisation.

With the Advanced Model, the download throughput of each user download, $R^{(k)}$, is an average for that user being averaged over a file download time of tens to hundreds of epochs for that individual user. The final result, $\langle R \rangle^{(usr)}$, is the average of several thousands of individual downloads as in (5.38) below. The nature of the simulation allows other download statistics to be compiled and reported.

The base station download throughput, $\langle R_{BS} \rangle$, is the average over the duration of the simulation as in (5.30). This gives the results shown in Figure 5.6b showing average per user throughput, $\langle R \rangle^{(usr)}$, reducing with increasing user demand, in contrast to Figure 5.6e which shows average base station throughput, $\langle R_{BS} \rangle$, increasing with increasing user demand. Similar results are seen in plots b) and e) of Figures 5.7 and 5.8. From these results, we see that while the achieved per user average throughput reduces with increasing utilisation, the total per base station average throughput increases.

The per user average download rate decreases because, as $D^{(usr)}$ increases so does the request rate, λ, meaning the PDSDCH resources have to be shared ever more user requests. Further, more users are active simultaneously across the network, leading to more interference and consequently lower spectral efficiency. This latter effect is the dominant contribution, especially at lower user demands. These in turn mean that the download of $B^{(usr)}$ bits to a user will take longer, hence reducing $\langle R \rangle^{(usr)}$. In contrast, over the total simulation time, more data will be downloaded to the cohort of users thereby increasing $\langle R \rangle_{BS}$.

The differences between the models' throughput results arise from fundamental aspects of the simulations. The Static Model defines utilisation as a proportion of PDSCH REs allocated within a Radio Frame (see (4.22)). The base station continuously downloads data at a fixed utilisation and averages the results over a Radio Frame to avoid issues that would otherwise arise from the variation in the number of PDSCH REs between symbol durations. The Advanced Model takes a rather different approach, operating at either maximising base utilisation (i.e. $\mu_{BS} = 1$) or zero utilisation. The base station operates at maximum utilisation during each epoch it is transmitting to users, in order to service users as quickly as possible, there-by making resources available for the next service request. In the Advanced Model utilisation is defined as the proportion of the simulation time the base station is operating at $\mu_{BS} = 1$ as in (5.41) and (5.42).

However the network or per-base station throughput results contrast to the per-user throughput results. When comparing the Advanced Model results in Figure 5.6c and e we see an approximately linear relationship between average base station utilisation, $\langle \mu \rangle_{BS}$ and average base station throughput, $\langle R \rangle_{BS}$. The Advanced Model averages both of these quantities over the duration of the simulation and across base stations. At low utilisation, we expect there will be a linear relation between base station utilisation and throughput for the Static Model as shown in (4.30). However, as the load on the network increases, from Figure 5.7c, e, Figure 5.8c and e, we see that the relationship between utilisation and base station throughput with the Advanced Model becomes increasingly nonlinear.

In general the Full Buffer Model is considered to provide more optimistic results than the FTP model [28, 30]. Ameigeiras *et al.*, in a paper that compares traffic models, state "full buffer model assumption may not yield the desired performance benefits in realistic scenarios where the finite buffer model [the Advanced Model] is more appropriate due to its birth and death process" Similarly, Navarro-Ortiz *et al.*, when surveying traffic models for 5G state, "the usage of the well-known full buffer traffic model may provide non-realistic performance evaluations".

This is not unexpected. Given that the standard Full Buffer Model assumes 100% utilisation, is likely to be downloading to more users than are simultaneously active with the Advanced Model and does not include dynamics that can lead to network congestion. As discussed above, the Static Model, allowing for less than 100% utilisation may be expected to produce some results that align with the Advanced Model at low utilisations.

5.10.3 USER COUNT

One drawback in the Static Model is that the number of active users downloading is assumed to be known a priori and is an input to the simulation. This sets per user throughput for the simulation, given by the total base station throughput divided by the (pre-set) number of active users. The number of users input by the modeller may bear little or no resemblance to the actual number of active users.

In contrast, the average user demand volume together with the chosen file size determine the service request, λ, rate in the Advanced Model, thus indirectly setting an average number of concurrent users. It is important to note that Advanced Model is based on finite-size file downloads whereas the Static Model assumes continuous downloads (at a pre-set base station utilisation). Results from the Advanced Model give time-varying numbers of concurrent users, averaging approximately 6 users per sector for a Dense Urban geography with average user download request rate, $D^{(usr)}$, of 50 kbit/s and the file size for the simulation is 16 Mbit. For smaller file sizes, the average number of concurrent users rises accordingly. Over a simulation time scale of seconds to minutes, a base station serving a Dense Urban area is likely to be serving a total of many tens to hundreds of users. Typically with many concurrent users the base station applies a "round-robin" approach, in which a few users are served simultaneously during each sub-frame duration. In each step of the round-robin, a different set of users is served, so that over the full round-robin cycle many users are sequentially served.

Because the Advanced Model has a form of congestion control (i.e. terminating slower downloads) and users leave the service area once their session is completed, this model will settle towards a steady-state number of active users in response to the service request rate. This means that the per user throughput predicted by simulation of the Static Model may not be easily compared to that obtained using the Advanced Model.

5.10.4 SINR AND CONGESTION

In both models, the SINR at the location of a user, $SINR^{(k)}$, is dependent upon the utilisation of the active base stations not serving that user. Once $SINR^{(k)}$ is calculated, both models rely on a link model to estimate the corresponding Spectral Efficiency for that user. However, in a given "run" of the Static Model, a user is continuously downloading at the same location where they were initially placed. There is no completion or termination of downloads, nor generation of new service requests as in the Advanced Model and therefore the Static Model can result in larger total download throughput.

The Advanced Model terminates a download when it has exceeded a maximum time, $t_{max,dl}$ (set to 4 s). This is a form of "congestion control" not included in the Static Model. Early termination of (slow) downloads has the effect of improving the simulation network performance, in particular throughput and energy efficiency. This results from slower downloads being culled hence removing poor-performing downloads. An outcome of this is lower download volumes at locations with poor SINRs, relative to locations with good SINRs, which typically complete their download. This feature helps mitigate congestion in the Advanced Model, allowing a base station to maintain its throughput as well as user throughput. It should be noted that the time-determined "termination" of downloads itself may not directly reflect "real network" behaviours, where slow downloads may be terminated due to user frustration, or by the network clearing over-sized data packet queues.

5.10.5 BASE STATION ENERGY EFFICIENCY

Base station energy efficiency performance is calculated either as the ratio of power consumed to total throughput (Power/Throughput) or the ratio of energy expended to total bits downloaded (energy expended/total bits). Because the Static Model is a time-invariant steady-state model, it effectively uses the Power/Throughput quotient. Although this can be converted to (energy expended)/(total bits) the conversion is trivial in that the denominator and numerator are multiplied by the same time duration.

As explained above with the Static Model the power consumption and throughput of BS_0 used to calculate the energy efficiency is averaged over multiple runs of the simulation with the users re-located around the service area in each run. This provides a fair representation of the energy efficiency.

In contrast, the Advanced Model includes time dynamics and the definition of energy efficiency contains some subtleties. The Advanced Model covers multiple base stations whose activity status at any time is generally random and which may not be active for equal times (i.e. different base stations have $\mu_{BS,m,p} = 1$ for differing numbers of epochs). The energy efficiency is calculated across a network that may consider just a single central base station, a cluster of seven base stations (i.e. one central station with one inner ring of stations) or include additional rings, of base stations as the modeller desires. Performance is generally assessed for a network rather than for a specific base station. If one confines their attention to one station such as BS_0 the modeller must decide whether to calculate its energy efficiency over the full duration of the simulation, or only over the epochs for which BS_0 has $\mu_{BS,0,p} \neq 0$. This issue is avoided by focusing on the energy efficiency of the entire network over the entire simulation duration. While the performance is commonly expressed as an energy-per-bit metric, the alternative metric of power for a given throughput rate is possible, but both would normally be averaged across a simulation time and a number of base stations.

Because the Static Model involves continuous downloading, rather than responding to user requests as with the Advanced Model, it is expected that the energy efficiency value provided by the Static Model may be higher than that calculated using a simulation of the Advanced Model [28].

Table 5.7 lists some of the key differences between the Simple and Advanced Models.

TABLE 5.7
Some of the Key Different Features of the Two Models

Model Feature	Static Model	Advanced Model	Comments
Traffic model input parameter	User utilisation	Average user demand	No simple relationship between these input parameters, but may be available for low utilisation.
Model dynamics	None	Random process for user service requests	Both are steady state but the Static Model has no time dynamics at all. Advanced Model includes variations due to stochastic process for service requests.
Averaging times	Radio Frame	Epoch duration & simulation duration	Advanced Model has two averaging times. Static Model has one.
Data collection area	BS_0	BS_0 to BS_6	Both include additional base stations to generate interference power

(Continued)

TABLE 5.7 (*Continued*)
Some of the Key Different Features of the Two Models

Model Feature	Static Model	Advanced Model	Comments
Congestion control	No	Terminates users with too slow download	Neither model provides an operator policy to deal with congestion.
Proportion of Failed Download	No	Provides a metric for amount of congestion	The Static Model does not have the concept of terminating downloads that are excessively slow, i.e. takes longer than 4s as assumed in the Advanced Model.
Randomised inputs	User locations	User locations and user request times	Both typically use uniform random distribution of users. Advanced Model uses Poisson process for service requests
User sessions	Continuous downloading	User download sessions have finite duration	Advanced Model has users enter and leave the service area. Static Model has a fixed number of users remain in the service area.
User locations	Fixed	New user requests can occur at different locations	Advanced model has each new user randomly located in the service area.
Varying network load	Provided by varying user utilisation, $\mu^{(k)}_{BS}$	Provided by varying user download request data rate, $D^{(usr)}$	Relating $\mu^{(k)}_{BS}$ to $D^{(usr)}$ is likely non-trivial except possibly at low utilisation
Number of active users	Fixed	Varying	Advanced Model models users entering the service area with a Poisson distribution and leaving the service area upon session completion. When Advanced Model reaches steady state, the number of users active in a given service area will stabilise to a steady-state value commensurate with the load demand being simulated

5.11   TECHNICAL SUMMARY

The simulation process is summarised in the flow chart in Table 5.8. In regard to collection of data from the simulation, Step 12 collects the data listed in Table 5.3 and Step 26 collects the data listed in Table 5.4.

TABLE 5.8
Flow Chart Summarising the Advanced Model for a Base Station within a Mobile Network

Step	Model Component
1.	Decide on the type of region in which the mobile network is to operate. In particular, user density and environment (e.g. flat open country, hilly country, urban area with low-rise buildings, etc.)
2.	Decide on a range of average download traffic volumes per user. For example, the user population might consume an average of say 10 kbit/s/user at low traffic times and 75 kbit/s/user at the busiest times of the day
3.	Decide on a mobile base station layout. For generic simulations, a hexagonal layout of base stations is most commonly used, with a fixed inter-base-station distance. Select base station locations and antenna beam orientations (see step 6) below.

(Continued)

TABLE 5.8 (*Continued*)
Flow Chart Summarising the Advanced Model for a Base Station within a Mobile Network

Step	Model Component
4.	Decide on the size of the area to be simulated.

A cluster of 7 base stations using a hexagonal cell array, one central and six surrounding that one, will generally give adequate results. A second ring of base stations surrounding the cluster, also supporting user traffic, is needed to ensure that all 7 of the inner cluster of base stations exhibit similar performance. A third ring as shown in Figure 5.3 could be included but may not be warranted.

E.g. a densely populated urban area might include 10,000 users per square km. The average traffic per user might run to 50 kbit/s/user, averaged across the population including active and non-active users.

Adopting base inter-site distance of 500 m, then each base station will serve approximately 0.217 km^2, or 1.52 km^2 for the 7 inner base stations. On average, each base station would need to handle 0.217 × 10,000 × 50 kbit/s = 108 Mbit/s.

| 5. | Decide on an area surrounding the base station cluster for the simulation. |

E.g., with hexagonal base station array and inter base station distance of 500 m a central cluster of 7 BS plus an outer ring of 12 fit neatly into a square "playground" 2.5 km on each side.

| 6. | In general, each base station includes three sets of transmitters, receivers, and antennas. For modelling purposes, three antennas with beams spaced at 120 degrees is generally used, with beams oriented to align with the gaps between beams of adjacent sites, as shown in Figure 5.3. 3GPP has recommended commonly used antenna patterns in both lateral and vertical directions [4]. See Table A.2.1.1-2 of 3GPP TR 36.814 V9.2.0 (2017-03) |

| 7. | Generate a grid of locations in the playground from which a user might request service. |

The spacing of these sites of approx. 1% of the Base station inter-site distance is suggested to ensure that potential gaps in beam coverage are not missed.

| 8. | Decide on the user traffic model to be used. Two such models will be encountered in the literature: |

a. "Full Buffer" model, [31–34] described in Chapter 4 and
b. "FTP Traffic" model [3, 5, 7–10] described in this chapter

a. Full Buffer model. This detail is included only for completeness. The "Static Model" described in Chapter 4 is based on a Full Buffer Model. Typically, in this model ten user sites are selected at random in each base station sector, and all download data continuously. The ten users are in locations with different distances from the base station, consequentially receive different signal power levels from the base station. The users share the capacity of the serving base station, and experience different levels of interference from other base stations. Interference power degrades the performance of the receiver, as does the sharing of base station capacity with other users.

This model is very fast in generating a result, but must be re-run many of times with different user layouts until the results converge to stable values. The model generates only a maximum-capacity metric which is likely to be quite optimistic.

b. FTP model. This is the model described in this chapter. A set of potential users "request" the download of a file, then cease activity once the download is completed. The requests are made by a randomly selected users in the playground. Each user is located at a different site on a grid of possible user locations. The requests are made at random intervals governed by a Poisson process whose mean value is related to the traffic level at which the network is being assessed.

E.g. If the playground is 2km square, with 10,000 users/km^2, the playground includes 40,000 users. At 50 kbit/s/user, that represents a total of 2 Gbit/s. If each user is to request a 2 MByte file, or 16 MBits, then this traffic level would be represented with a Poisson process with a mean request rate of 125 requests/s, or a mean interval between requests of 8 milliseconds.

| 9. | Select a "propagation model". This quantifies the degradation of strength of a base station signal as it propagates to the receiver [14–17, 19, 35–38]. |

There are many possible propagation models, some very simple but valid for just one or a limited range of signal frequencies and for one type of region. Others very complex involving statistical path vector modelling. The model used in this text is based on work by Stanford University (SUI model), used by IEEE for network and service modelling This model has the advantage of being usable over most of the frequencies in use today, and over a range of regional terrain types. Other models are more appropriate for simulations of dense urban areas

(Continued)

TABLE 5.8 (*Continued*)

Flow Chart Summarising the Advanced Model for a Base Station within a Mobile Network

Step	Model Component
10.	Select a "throughput model" to relate the SINR to the Spectral Efficiency.
	Examples of throughput models are given in Figure 4.4. Unlike propagation loss, no standard receiver throughput models have been widely adopted. This is because different receiver implementations may include better filtering of extraneous signals, have better noise performance, be able to better synchronise with the transmitting station, among many others. Therefore, the modeller needs to make a value judgement in how their model will relate the SINR at the user to Spectral Efficiency for data downloads. An example of such relationships is given in Tables 4.7 and 4.8.
11.	Set up base station transmit parameters (transmit power in active and idle states).
	Set up an initial condition with all base stations in their idle state
12.	Set up a "simulation clock" to provide a series of equal duration time "epochs". The epoch duration may be set to one or a few milliseconds.
	At the end of each time epoch generated by the software clock the following is done:

- For each user, determine how many other users are currently sharing the base station and antenna beam,
- Divide the available PDSCH REs transmitted in that beam by the number of sharing users
- For each user, increment the total bits received by the product of the allocated PDSCH REs and the user's Spectral Efficiency
- If the cumulative number of bits received by that user now exceeds the desired file size, record the results and user parameters, remove the user from the playground, reallocate the beam PDSCH REs among remaining users of that beam, or reset the beam to the idle state if it has no active users.
- In the case of a user failing to complete the file download within a designated time (e.g. 4 s for a 16 Mbit file), the performance achieved for that user is deemed inadequate, the download terminated and the user is removed from the playground
- maintain record of which base station beams are active and serving any user during the time epoch

Step	Model Component
13.	Set up a Poisson process that generates a sequence of starting times for user requests within the simulation duration.
	This Poisson process is set with a mean interval consistent with the average rate determined by the process described in 8b) above.
14.	Start the simulation, the Poisson process and simulation clock
15.	When Poisson process indicates the time for adding a user is reached, pick a random user location within the playground and place a new user at that location to commence a download. If it is not time to add a user, perform the steps outlined in step 12), then skip to step 24).
16.	Calculate the bearing of the user from each base station. Use this to calculate the horizontal radiation pattern value, $\Theta_{m,p}(z^{(k)})$ of the transmitting antenna in the direction of the user. Similarly, calculate the vertical angle of the user from the base station antenna to determine the vertical radiation pattern value, $\Phi_m(z^{(k)})$ for the user. Repeat this for every antenna in the playground.
17.	From the user and base station locations, calculate the distance of the user from each base station. Use that distance to calculate the propagation loss, $L(r_m^{(k)}, f, h_b, h_r, F, v)$ appearing in (5.20) from each base station to the user, including shadowing loss and building penetration loss.
18.	Determine the strength of the signals being received by the user from each beam of each sector, with base stations and beams active if already serving another user, or idle if not.
	Calculate the interference power, P_{interf}, received by the user using (5.21).
19.	Using (5.20) calculate the power of the signal, $P_{Rx,m,p}(z^{(k)})$ the user would receive from the beam of each sector if that base station and sector were running at transmit full power and serving the user.
20.	From 18) and 19), calculate the SINR, using (5.22), for the user provided by each base station and sector, for the user.
21.	Select the base station and sector giving the greatest SINR. Set this base station and antenna to serve that user.
22.	Using the throughput model chosen in 10) and the SINR chosen in 21) to determine a Spectral Efficiency to calculate the number of bits the user would receive for each of the PDSCH REs allocated to the user by the base station during the current epoch.

(Continued)

TABLE 5.8 (*Continued*)

Flow Chart Summarising the Advanced Model for a Base Station within a Mobile Network

Step	Model Component
23.	At the end of the current epoch, follow the procedure described in 12) to collect the results for the current epoch and remove users from the playground.
24.	Commence the next epoch and continue running the main program as from step 14), assessing the status of each user and each base station at every clock epoch, and adding users at intervals given by the Poisson process. As more users enter the simulation, the performance achieved by each user will be degraded. This is in part due to more users sharing the PDSCH RE capacity of the beam they are sharing. Further, because more active users equates to more active beams, hence more interference due to other services. This, in turn, causes a lower Spectral Efficiency being achieved, and hence longer download times
25.	Continue until the simulation has recorded results for the required number of user downloads
26.	Analyse the results for the required statistics. This would include for example;

- the download time for each user,
- the download rate achieved by each user,
- the average SINR and Spectral Efficiency achieved across users. For the network as a whole;
- the total volume of bits delivered,
- the average service rate,
- the total time to deliver the number of downloads requested,
- the total number of time epochs during which each of the network beams was active.

From the last two values, the base station utilisation is calculated, and hence the amount of energy consumed. Taking this in conjunction with the first of these values, we obtain the network Joule-per-Megabit efficiency result for the traffic level selected in steps 8).

5.12 EXECUTIVE SUMMARY

Two approaches to modelling of mobile networks are presented in this book. Although the Static Model of Chapter 4 can provide a "back of the envelope" estimation of mobile network energy efficiency, it is based on a significant number of simplifications to provide that estimate. Further it is relatively limited in the range of outputs that it can provide to describe the power consumption of a mobile network. On the other hand, the Advanced Model, although still a significant simplification, can provide much more information and flexibility in the outputs it provides. However, the Advanced Model requires a much greater investment in time and programming effort. This chapter provides a somewhat detailed guide on how to construct an Advanced Model; however, the modeller must still construct much of the detailed aspects of the model by themselves.

If one requires very approximate results with minimal delay, then the Static Model is likely a feasible option. However, if time is not urgent, then the Advanced Model can provide much more insight into the power consumption, operation and energy efficiency of a mobile network.

5.13 APPENDIX

5.13.1 STEADY STATE AND USING AVERAGES

Referring to Figure 5.4 and the corresponding discussion, in each simulation the data is collected while the respective networks are in steady state. A system is in steady state when the parameters that describe the state of the system do not change with over time [39]. Applying this to the network simulations, this corresponds to not changing over epoch index, j, covering the time steps in the simulation. In these simulations some results involve averaging over the user populations, $\langle Y_j \rangle^{(usr)}$, others involve averaging over base stations and/or base station sectors, $\langle Y_j \rangle_{BS}$. We represent all these averaged quantities with the generic form $\langle Y \rangle$. In all these cases, these parameters will vary from epoch to epoch even though the overall simulation is, on average over time, in steady state.

Therefore, we need to account for the fact, even in steady state, the value Y_j may not equal Y_{j+1}, although these fluctuations are small in relative terms. To allow for this, we specifically define "steady state" for the simulations in terms of the results of the simulation, Y_j. The simulation is in steady state over epoch index values j to k when localised averages of Y_j across index value ranges, N_1 and N_2 satisfies

$$\langle Y \rangle_{N_1} = \frac{1}{2N_1} \sum_{i=j-N_1}^{j+N_1} Y_i \approx \frac{1}{2N_2} \sum_{i=k-N_2}^{k+N_2} Y_i = \langle Y \rangle_{N_2} \tag{5.74}$$

for epoch index ranges $1 \ll N_1$, N_2 and $j+N_1$, $k+N_2 \leq N_{Tot}$. This equation states that "steady state" requires the average simulation result $\langle Y_j \rangle$ when also averaged from index value ranges $j - N_1$ to $j + N_1$ to be approximately the same as when averaged from index value $k - N_2$ to $k + N_2$. That is, the system is in "steady state" over the epoch index range j to k when averages of these quantities are approximately constant over the epoch index range j to k. The greater the difference between index values j and k the longer the duration of steady-state behaviour.

Therefore, provided that N_1 and N_2 satisfy the inequalities above, we can use the steady-state average over N_1 epochs to approximate the average results over N_2 epochs. From this we also have for steady state:

$$\sum_{j=1}^{N_2} Y_j \approx N_2 \langle Y \rangle_{N_1} \tag{5.75}$$

In particular, choosing $N_1 = N_{Tot}$ with $N_{Tot}\Delta t_S = T$ the duration of the simulation and $N_2 = N_{Tot,S}$ such that $N_{Tot,S}\Delta t_S = T_S$, the duration of stage S of the diurnal cycle, we have

$$\sum_{j=1}^{N_{Tot,S}} Y_j \approx N_{Tot,S} \langle Y \rangle_T \tag{5.76}$$

Thus provided the simulation has reached steady state, an averaged value for a parameter over an appropriate interval may be used to calculate the summed value of the parameter over the whole of its corresponding portion of the diurnal cycle.

5.13.2   Calculating Average per User Results

A per user result has the form $\langle Y \rangle^{(usr)}$ given by

$$\langle Y \rangle_D^{(usr)} = \sum_{k=1}^{N_D^{(usr)}} Y^{(k)} \bigg/ N_D^{(usr)} = \sum_{k=1}^{N_D^{(usr)}} \sum_{j=1}^{N_{Tot,D}} Y_j^{(k)} \bigg/ N_D^{(usr)} = \sum_{S=1}^{5} \sum_{k=1}^{N_{Tot,S}^{(usr)}} Y_S^{(k)} \bigg/ \sum_{S=1}^{5} N_{Tot,S}^{(usr)} \tag{5.77}$$

where $Y_j^{(k)}$ is the result for the k th user in the j th epoch, $N_D^{(usr)}$ is the total number of users served over the diurnal cycle and $N_{Tot,Dl}$ is the total number of epochs over the duration of the diurnal cycle. The second sum corresponds to $Y^{(k)}$ being the cumulative sum of $Y_j^{(k)}$ over $N_{Tot,D}$ epochs covering the full diurnal cycle.

Of particular interest is the user throughput given in (5.38). Using this definition, for stage S of the diurnal cycle, we have

$$\langle R_S \rangle^{(usr)} = \sum_{k=1}^{N_{Tot,S}^{(usr)}} R^{(k)} \Bigg/ N_{Tot,S}^{(usr)} = \frac{1}{N_{Tot,S}^{(usr)}} \sum_{k=1}^{N_{Tot,S}^{(usr)}} \sum_{j=j_{on}^{(k)}}^{j_{off}^{(k)}} \frac{\Delta B_j^{(k)}}{\Delta t_S N_{dl}^{(k)}} \tag{5.78}$$

where $N_{dl}^{(k)}$ is the number of epochs taken to download $B_{Tot}^{(k)}$ bits, that is $N_{dl}^{(k)} = j_{off}^{(k)} - j_{on}^{(k)}$ Therefore, $\Delta B_j^{(k)}/\Delta t_S N_{dl}^{(k)}$ is the data rate (bit/s) for the download to the k th user.

From (5.78) we can see that there will be potentially significant variations in $R^{(k)}$. If the k th user experiences a high SINR, their download will operate at a high spectral efficiency. Therefore, $\Delta B_j^{(k)}$ will be large and the number of epochs, $N_{dl}^{(k)}$, required to download file size $B^{(usr)}$ will be small. In contrast for a different user, k', located in an area of low SINR their download will operate at a low spectral efficiency. Consequentially, their $\Delta B_j^{(k')}$ will be smaller and hence the number of epochs required, $N_{dl}^{(k')}$, to download $B^{(usr)}$ data will be much greater. In addition, if the number of users concurrently on line with user k is much less than with user k', the difference in throughputs will be greater because k will be allocated more PDSCH REs than k'.

Note that that $\langle R \rangle^{(usr)}$ has the form $\langle \Delta B_j^{(k)}/N_{dl}^{(k)} \rangle$ not $\langle \Delta B_j^{(k)} \rangle / \langle N_{dl}^{(k)} \rangle$. In wireless literature the ratio $\langle \Delta B_j^{(k)}/N_{dl}^{(k)} \rangle$ is often referred to as the "call average throughput" and $\langle \Delta B_j^{(k)} \rangle / \langle N_{dl}^{(k)} \rangle$ is referred to as the "flow average throughput" or "flow throughput" [40]. This distinction is important because the average of the quotient properly accounts for the variation in, $R_j^{(k)}$, whereas the quotient of the averages does not. The issue of whether it is more appropriate to calculate $\langle Y \rangle / \langle Z \rangle$ or $\langle Y/Z \rangle$ is discussed in more detail below in this Appendix.

Determining the average per user result over the full diurnal cycle is complicated by the fact that there is no j-index in $Y^{(k)}$ to construct the sum over the stages of the diurnal cycle. However, the number of users, $N_{Tot,S}^{(usr)}$ for each stage of the simulation is dependent upon $N_{Tot,S}$ because new users are added per epoch according to the Poisson process in (5.13). To determine how to re-scale the $\langle Y_S \rangle^{(usr)}$ to align with the durations of the diurnal cycle stages, we need to determine the number of users that will be serviced over each time duration T_S. Unless the modeller carefully constructs the simulation for each user demand level, $X_S D^{(usr)}$, the number of users in the simulation run over N_{Tot} epochs is unlikely to align with the number of users corresponding to duration T_S.

Because of the Poisson random process for service requests and the details of the simulation, the actual number of users $N_{Tot,S}^{(usr)}$ whose data is collected is not precisely related to the total number of epochs, $N_{Tot,S}$, over which the simulation is run. With an average service request rate λ_S, we would expect the number of user service requests across the entire service area will be approximately given by:

$$N_S^{(usr)} \approx T_S \lambda_S = N_{Tot,S} \Delta t_S \lambda_S \tag{5.79}$$

However, when the simulation terminates it includes users beyond user index $k = N_S^{(usr)}$ therefore, (5.79) may not reflect the actual relationship between $N_S^{(usr)}$ and $N_{Tot,S}$. This is because the simulation termination process ensures the system remains steady-state operation, as described in Section 5.5. The download session for the last user for whom data is collected finishes at epoch $j_{off}^{\left(N_{Tot}^{(usr)} \right)}$. This means the simulation runs for total duration

$$T_S' = j_{off}^{\left(N_{Tot}^{(usr)} \right)} \Delta t_S = N_{Tot}' \Delta t_S \tag{5.80}$$

The only data recorded by the simulation is for users located in the inner cluster of 7 base stations. This needs to be accounted for when relating the number of users to the duration of the simulation. The effective service request rate for the population of users in the inner cluster is $\lambda'_S \approx \lambda 7 A_{BS}/A_{SA}$. Therefore, over duration of data collection T'_S:

$$N_{Tot,S}^{(usr)} \approx T'_S \lambda'_S = j_{off}^{\left(N_{Tot,S}^{(usr)}\right)} \Delta t_S \lambda'_S = N'_{Tot,S} \Delta t_S \lambda'_S \qquad (5.81)$$

There is an additional inaccuracy in this approximation because users close to the border of the inner cluster of 7 base stations may be served by base stations on either side of that border. Depending upon the SINR values for signals received from the various base stations in the vicinity of that user, a user close to that border may be served by a base station within the cluster of 7 or outside the cluster. If the user is served by one of the inner 7 cluster, their data is collected and they are recorded as one of the $N_{Tot,S}^{(usr)}$ users. However, if that user is served by a base station outside the inner 7 cluster, their data is not recorded and they are not in the $N_{Tot,S}^{(usr)}$ users. Although this adds a further variation to, multiple $N_{Tot,S}^{(usr)}$ simulations have shown that even at high shadowing loss (which induces large variations in SINR values), the deviation away from the value in (5.81) is well below 10%.

Using this in (5.81) gives:

$$\sum_{k=1}^{N_{Tot,S}^{(usr)}} Y_S^{(k)} \approx T'_S \lambda'_S \left\langle Y_S \right\rangle^{(usr)} \qquad (5.82)$$

Therefore, because all the data is only collected over the inner 7 base stations we have:

$$\begin{aligned} Y_D^{(usr)} &= \sum_{k=1}^{N_D^{(usr)}} Y^{(k)} \Bigg/ N_D^{(usr)} = \sum_{S=1}^{5} \sum_{k=1}^{N_{Tot,S}^{(usr)}} Y^{(k)} \Bigg/ \sum_{S=1}^{5} N_{Tot,S}^{(usr)} \\ &\approx \sum_{S=1}^{5} \rho_S X_S Y_S^{(usr)} \end{aligned} \qquad (5.83)$$

5.13.3 $\langle Y \rangle / \langle Z \rangle$ OR $\langle Y/Z \rangle$

The energy efficiency of the network is given by the quotient of averages, $H = \langle Q \rangle / \langle B \rangle$. In contrast, the average throughput per user is given by the average of a quotient, $\langle R \rangle^{(usr)} = \langle \Delta B / N_{dl} \Delta t \rangle$. The quantities $\langle Q/B \rangle$ and $\langle \Delta B \rangle / \langle N_{dl} \Delta t \rangle$ have the same dimensions as H and $\langle R \rangle^{(usr)}$, respectively. However, these latter quantities will generally give somewhat different results. Therefore, when dealing with quotient based results, the modeller needs to carefully consider which of $\langle Y \rangle / \langle Z \rangle$ or $\langle Y/Z \rangle$ to use.

All embracing advice cannot be provided; however, there are some basic principles that can help.

- For quotient based quantities that will later be combined to provide "overall" results typically are better defined using $\langle Y \rangle / \langle Z \rangle$. For example, energy efficiency over a diurnal cycle can be more easily constructed using $\langle Y \rangle / \langle Z \rangle$ for each of the diurnal cycle stages. Constructing an overall result using $\langle Y/Z \rangle$ can be somewhat complicated.
- Quantities that are expected to exhibit significant variation are better defined using $\langle Y/Z \rangle$. The quotient $\langle Y \rangle / \langle Z \rangle$ effectively averages out significant variations in the quotient Y/Z. If those variations are important for understanding the results, then $\langle Y/Z \rangle$ will exhibit those variations.
- The information that the quotient result expected to provide. In some cases, although one form may provide easier mathematical manipulation, it will not provide the information required. For example, the energy efficiency can be defined using $\langle Q/B \rangle$ and this definition does provide

information that is averaged out using $\langle Q \rangle / \langle B \rangle$. However, in many cases the understanding and information provided by $\langle Q \rangle / \langle B \rangle$ is sufficient. Therefore, it is likely more productive to use the form $\langle Q \rangle / \langle B \rangle$.

- In contrast, when considering user throughput, the form $\langle \Delta B \rangle / \langle N_{dl} \Delta t \rangle$ averages out too much important information compared to $\langle \Delta B / N_{dl} \Delta t \rangle$. In particular the impact of significant variation in $SINR^{(k)}$ across users. This variation is crucial to understanding the relationship between user location and the performance they experience. Therefore, when constructing an average across users, this variation should be included and hence the form $\langle \Delta B / N_{dl} \Delta t \rangle$ is more appropriate.

5.13.4 INTERNATIONAL STANDARDS

This chapter has continued the discussion of mobile base station energy consumption in Chapter 4, extending the scope to a network of base stations, and including a wider range of factors affecting their energy consumption. Techniques such as this help to develop an understanding of how consumption is affected by factors such as user location, network congestion, traffic load, and even load variation over a diurnal cycle.

Complementing this type of analysis, the International Telecommunications Union (ITU-T) and the European Telecommunications Standards Institute (ETSI) have jointly developed standards defining efficiency metrics for mobile networks, their measurement, and their reporting [41–43]. These are of interest, but principally aimed at network operators. Broadly, these involve recording key performance parameters for a number of base stations over periods of a week, a month, or a year. The performance parameters include total energy consumption and total data traffic, plus other metrics such as coverage and traffic delay. Part of the reporting includes the demographic and topographic characteristics of the service area.

Although these standards are intended for use by network operators, they may be of interest to more technology-inclined readers. While [41] covers early releases of LTE, [42, 43] cover newer variants of LTE including the use of centralised cloud-computing for traffic processing and network control.

REFERENCES

[1] Third Generation Participation Project (3GPP), "3rd Generation Participation Project Global Initiative," 2020. [Online]. Available: https://www.3gpp.org/. [Accessed 19 August 2020].

[2] ITU-R, "International Telecommunications Union Radiocommunication Sector," 2020. [Online]. Available: https://www.itu.int/en/ITU-R/Pages/default.aspx. [Accessed 15 August 2020].

[3] G. Auer *et al.*, "INFSO-ICT-247733 EARTH Deliverable D 2. 3 Energy efficiency analysis of the reference systems, areas of improvements and target breakdown," 2020.

[4] Third Generation Participation Project (3GPP), "TR 36.814 V9.2.0 (2017-03) Technical Specification Group Radio Access Network; Evolved Universal Terrestrial Radio Access (E-UTRA); Further advancements for E-UTRA physical layer aspects (Release 9)," Third Generation Participation Project, 2017.

[5] GreenTouch, "GreenTouch Final Results from Green Meter Research Study: Reducing the Net Energy Consumption in Communications Networks by up to 98% by 2020," GreenTouch, 2015.

[6] GreenTouch, "GreenTouch," Bell Laboratories, 2015. [Online]. Available: http://www.bell-labs.com/greentouch/. [Accessed 23 February 2021].

[7] O. Blume *et al.*, "*Energy Efficiency of LTE networks under traffic loads of 2020*," in *The Tenth International Symposium on Wireless Communication Systems*, 2013.

[8] M. Ismail *et al.*, "A Survey on Green Mobile Networking: From The Perspectives of Network Operators and Mobile Users," *EEE Communications Surveys & Tutorials*, vol. 17, no. 3, p. 1535, 2015.

[9] F. Salem *et al.*, "*Energy consumption optimization in 5G networks using multilevel beamforming and large scale antenna systems*," in *IEEE Wireless Communications and Networking Conference (WCNC)*, 2016.

[10] NTT DOCOMO, Alcatel-Lucent, Alcatel-Lucent Shanghai Bell, "R1-11 3495 "Base Station Power Model"," 2011.

[11] S. Ben Fredj et al., "Statistical Bandwidth Sharing: A Study of Congestion at Flow Level," *ACM SIGCOMM Communication Review*, vol. 31, no. 4, p. 111, 2001.

[12] C. Chan et al., "Assessing Network Energy Consumption of Mobile Applications," *EEE Communications Magazine*, vol. 53, no. 11, p. 182, 2015.

[13] Third Generation Participation Project (3GPP), "TR 38.901 V16.1.0 Technical Specification Group Radio Access Network; Study on channel model for frequencies from 0.5 to 100 GHz (Release 16)," Third Generation Participation Project, 2019.

[14] V. Erceg et al., "Channel Models for Fixed Wireless Applications," IEEE 802.16 Broadband Wireless Access Working Group, 2001.

[15] R. Jain, "Channel Models: A Tutorial, V1," 21 February 2007. [Online]. Available: https://www.cse.wustl.edu/~jain/cse574-08/ftp/channel_model_tutorial.pdf. [Accessed 15 August 2020].

[16] V. Abhayawardhana et al., "*Comparison of Empirical Propagation Path Loss Models for Fixed Wireless Access Systems*," in *IEEE 61st Vehicular Technology Conference*, 2005.

[17] A. Zreikat and M. Djordjevic, "Performance Analysis of Path loss Prediction Models in Wireless Mobile Networks in Different Propagation Environments," in *Proceedings of the 3rd World Congress on Electrical Engineering and Computer Systems and Science (EECSS'17)*, 2017.

[18] P. Sharma and R. Singh, "Comparative Analysis of Propagation Path loss Models with Field Measured Data," *International Journal of Engineering Science and Technology*, vol. 2, no. 6, p. 2008, 2010.

[19] I. Khan and S. Kamboh, "Performance Analysis of Various Path Loss Models for Wireless Network in Different Environments," *International Journal of Engineering and Advanced Technology (IJEAT)*, vol. 2, no. 1, p. 161, 2012.

[20] P. Morgensen et al., "*LTE Capacity Compared to the Shannon Bound*," in *IEEE 65th Vehicular Technology Conference (VTC2007 Spring)*, 2007.

[21] T. Ali-Yahiya, *Understanding LTE and its Performance*, Springer, 2011.

[22] H. Holma and A. Toskala, Eds., *LTE for UMTS – OFDMA and SC-FDMA Based Radio Access, John Wiley & Sons*, 2009.

[23] S. Sesia et al., Eds., *LTE – The UMTS Long Term Evolution: From Theory to Practice*, John Wiley & Sons, 2009.

[24] C. Cox, *An Introduction to LTE, LTE-Advanced, SAE, VoLTE and 4G Mobile Communications* (2nd Edition), John Wiley & Sons, 2014.

[25] A. Mohammadi and F. Ghannouchi, *RF Transceiver Design for MIMO Wireless Communications*, Lecture Notes in Electrical Engineering. Vol. 145, Springer-Verlag, 2012.

[26] E. Dahlman et al., *$G LTE/LTE-Advanced for Mobile Broadband*, Academic Press, 2011.

[27] Third Generation Participation Project (3GPP), "TS 38.214 V15.7.0 (2019-09) Technical Specification Group Radio Access Network; NR: Physical Layer procedures for data (Release 15)," Third Generation Participation Project, 2019.

[28] P. Ameigeiras et al., "Traffic Models Impact on OFDMA Scheduling Design," *EURASIP Journal on Wireless Communications and Networking*, vol. 2012, p. 61, 2012.

[29] C. Mueller, "On the Importance of Realistic Traffic Models for Wireless Network Evaluations: COST 2100 TD(10)12039," Institute of Communication Networks and Computer Engineering, Universitat Stuttgart, 2010.

[30] J. Navarro-Ortiz et al., "A Survey on 5G Usage Scenarios and Traffic Models," *IEEE Communications Surveys & Tutorials*, vol. 22, no. 2, p. 905, 2020.

[31] D. Lopez-Perez et al., "*Optimization Method for the Joint Allocation of Modulation Schemes, Coding Rates, Resource Blocks and Power in Self-Organizing LTE Networks*," in *30th IEEE International Conference on Computer Communications (IEEE INFOCOM 2011)*, 2011.

[32] J. Liu et al., "*Design and Analysis of LTE Physical Downlink Control Channel*," in *IEEE 69th Vehicular Technology Conference - VTC Spring 2009*, 2009.

[33] D. Lee et al., "Inter-Cell Interference Coordination for LTE Systems," in *IEEE Global Communications Conference (GLOBECOM)*, 2012.

[34] O. Ramos-Cantor et al., "Centralized coordinated scheduling in LTE-Advanced networks," *EURASIP Journal on Wireless Communications and Networking*, 2017, p. 122, 2017.

[35] M. Hata, "Empirical Formula for Propagation Loss in Land Mobile Radio Services," *IEEE Transactions on Vehicular Technology*, vol. 29, no. 3, p. 317, 1980.

[36] Z. Naseem *et al.*, "Propagation Models for Wireless Communication System," *International Research Journal of Engineering and Technology (IRJET)*, vol. 5, no. 1, p. 237, 2018.

[37] T. Chrysikos and S. Kotsopoulos, *"Site-specific Validation of Path Loss Models and Large-scale Fading Characterization for a Complex Urban Propagation Topology at 2.4 GHz,"* in *Proceedings of the International Multi-Conference of Engineers and Computer Scientists 2013 (IMECS 2013)*, 2013.

[38] F. Khan, *LTE for 4G Mobile Broadband: Air Interface Technologies and Performance*, Cambridge University Press, 2009.

[39] P. Gagniuc, *Markov Chains: From Theory to Implementation and Experimentation*, John Wiley & Sons, 2017.

[40] P. Fitzpatrick and M. Ivanovich, *"On Approximating Throughput in Wireless Systems with Complex Rate Variability QoS,"* in *Australasian Telecommunication Networks and Applications Conference (ATNAC 2008)*, 2008.

[41] ITU-T Recommendation L.1330, "Energy efficiency measurement and metrics for Telecommunication networks, " 2015 [Online]. Available at: https://www.itu.int/rec/T-REC-L.1330-201503-I/en Accessed 15 June 2021.

[42] ITU-T Recommendation L.1331, "Assessment of mobile network energy efficiency," 2020 [Online]. Available at: https://www.itu.int/rec/T-REC-L.1331-202009-I/en. Accessed 16 June 2021.

[43] ETSI Standard ES 203 228 V1.3.1 (2020-10), "Environmental Engineering (EE); Assessment of mobile network energy efficiency," 2020 [Online]. Available at: https://www.etsi.org/deliver/etsi_es/203200_20 3299/203228/01.03.01_60/es_203228v010301p.pdf. Accessed 16 June 2021.

6 Heterogeneous Mobile Network

In localised regions with high population densities such as shopping precincts, air terminals or railway stations, local "small cell" mobile base stations may be required. In a shopping centre, a small cell may be deployed in a food hall where significant numbers of customers frequently gather. The small cells are designed to cover high traffic corridors, or areas where a high concentration of traffic may occur frequently or time-to-time. Examples of "time-to-time" situations include sporting or entertainment events or annual cultural or community events.

Very often, these small cells are located within the coverage area of a larger (macro-) cell and the small cell may be energised only when the traffic demand in the localised area around the small cell base station reaches levels above the norm for the macro base station. This combination of small cells within macro cells is referred to as a "heterogeneous" mobile network because it includes base stations of different types: macro and small.

Heterogeneous mobile networks are often deployed to improve coverage or to increase capacity to meet localised traffic demands. Especially in periods of high traffic demand, some of the traffic in the area can then be "off loaded" onto the small cells. The remainder of the traffic in the wider service area is carried by the macro base stations. Fifth Generation (5G) mobile networks designed to meet today's increasing traffic demand levels will include extensive heterogeneous networks to provide both better coverage and exploit the lower power consumption of small cells [1–5].

To estimate the power consumption of a heterogeneous network, the model described here will leverage off the Advanced Model described in Chapter 5. The heterogeneous network simulation is essentially an application of the Advanced Model, but in which some users are served by a small cell network and the remaining are served by a macro cell network.

Heterogeneous networks can incorporate a small cell component in several different deployment models. In each case, small cells are deployed within an area served by a macro base station network. The small cells may form a separate sub-network using a different frequency band, in which case there will be no interference between signals in the macro and small cell domains. Alternatively, the small cells may share the same space and the same frequency band in a single joint network, in which case there will be mutual interference between the small cell and macro cell signals. There is also a heterogeneous network option, placing some small cells in a separate sub-network and others in the joint network.

In modelling the separate small cell network case, simulations for each of the macro and small cell sub-networks can be run separately, albeit with coordination of traffic demand levels. In modelling the single joint network and heterogeneous networks, a more complex model which includes interference between small cells and macro cells is needed, but many of the design factors and equations are common to each deployment type. In this chapter, we for the most part will discuss the design factors for the heterogeneous network solution.

The modeller can stipulate which small cells are to operate in the "small cells only" sub-network and which are to operate in the joint "small and macro cell" sub-network. With this separation each simulation follows the methodology of the Advanced Model in Chapter 5.

DOI: 10.1201/9780429287817-6

The traffic demands of the two components differ, reflecting the fact that the small cells deal with a higher demand density (i.e. users/km^2) than the macro cell network. The results of these two simulations can then be combined to provide the overall simulation results. This approach is similar to that of Blume *et al.* [6].

To clarify the types of mobile networks discussed in this chapter we have the following:

- Small cell network is a network of small cell base stations operating on its own separate frequency band
- Macro cell network is a network of macro cell base stations operating on its own separate frequency band
- Joint network is a network consisting of both small and macro cells all of which operate on its own given frequency band
- Heterogeneous network consists of two networks with overlapping coverage, one being a small cell network on one frequency band and the second being a joint or macro cell network operating on a separate frequency band.

6.1   MODELLING APPROACH AND DETAILS

Chapter 5 discussed the simulation of macro base station networks in three of four geographic area categories. The graphs for the macro network performance in a dense urban geographic area, in Figure 5.8 of that chapter, demonstrated that this network was capable of supporting only modest network demands. For example, at the estimated year 2020 average user demand of 50 kBit/s, Figure 5.8c indicates the base station utilisation running at 90% of its capability.

The Dense Urban model presumed a user population density of 10,000 per km^2, and base stations in a hexagonal array with 0.5 km inter-site distance. The notional coverage area of a base station is thus 0.217 km^2, or 0.0723 km^2 per sector, and one sector would serve 723 user locations. This means a notional traffic demand of 36.2 Mbit/s per sector.

Assuming a small cell coverage range of 50 m, the cell coverage area would be 0.00785 km^2. At 10,000 users/km^2, the cell coverage would be 78.5 users and the notional demand level 3.93 Mbit/s. Thus three small cells would serve only 11.8 Mbit/s of the notational traffic demand i.e. 32.5% of the macro station demand. Referring again to Figure 5.8, this would improve overall performance, but not greatly.

As discussed earlier, deploying small cells is a technique to provide additional support to the macro base station network in areas of high population density or high traffic demand. For example, if a sector of a base station included three hotspots with a population density of 20,000 users per km^2, which could be covered by three small cells, some 23.6 Mbit/s of demand or approximately 65% of the sector traffic could be offloaded, enabling a substantial performance improvement.

Thus the introduction of small cells to a macro base station network can improve network performance, but significant improvements can be achieved only in zones with high population density (or high per-user demand)

The introduction of small cells into the Advanced Model introduces extra complications. Firstly, the type of model for deploying small cells within the macro base station network is to be decided. If the modelling is for a specific area, the population or traffic hotspots will be known and both numbers and locations for small cells selected accordingly. Otherwise, a generic assessment of a small cell or heterogeneous network might be made on the basis of randomised small cell location selection. This would also entail running a simulation several times and averaging results, to avoid accidental location selection bias. Some of these small cells may form a separate small cell network, where sufficient spectrum is available and where small cell performance is paramount. Other small cells might be part of a joint network. Alternatively all small cells might form a single joint network

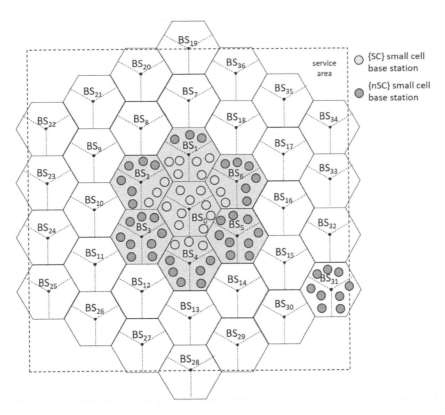

FIGURE 6.1 An example of the small cell layout of a heterogeneous network. In this network, there are three small cells in each macro cell sector. The small cells indicated with light shaded circles are in the "small cell only sub-network", {*SC*}. The small cells indicated with dark shading are in the "not small cell only sub-network", {*nSC*}. The {*nSC*} joint sub-network includes all the macro cell bases stations BS_0 to BS_{36}. These base stations operate in the same frequency band as the small cells in {*nSC*}. In this model, every macro cell base station sector has three small cells; however, they are only shown for the inner cluster of cells and BS_{31}.

with the macro base stations. Figure 6.1 shows a heterogeneous network, with a separate small cell network serving the highest traffic region in the vicinity of base station BS_0, and a joint network in the balance of the service area. In this figure and throughout this chapter, we will where necessary distinguish the separate small cell network by the symbol {*SC*} and the joint small-cell-plus-macro network by the symbol {*nSC*} (standing for not the small cell only network) because it may either a joint network or macro cell only network.

Further complications arising from including small cell networks include the need to: cluster users and traffic in the proximity of the small cells, use different equipment parameters including signal power levels for small cells and apply different propagation models for small cells due to different antenna heights. That said, if the reader is familiar with the material in Chapter 5 and they wish to acquire an overview of the modelling process, they are referred the Technical Summary in Section 6.9 to get a step-by-step overview of the model's structure before reading the details of how the small cell component is constructed.

In the sections given subsequently, we first describe how small cells and users are located within the service area. With that process described, we then discuss how the model deals with the serving of users in the heterogeneous network using two independent simulations.

6.2   USER AND CELL LOCATIONS ACROSS THE HETEROGENEOUS NETWORK

As an overview of this subsection, a process to develop a set of locations for small cells within the base station network and a set of locations for users within both the small cell and macro networks are described. In the case of a generic simulation for a network including small cells, the cells need to be located randomly within each base station sector, for each simulation run, but subject to a few conditions. Users' potential site locations are constructed to surround each small cell base station, and potential user locations across the entire service area set as described in Chapter 5. To simplify the overall simulation, a scheme to a priori split the user population into two sub-populations is described.

These two sub-populations are: Group 1 consisting of users who are served only by a pre-determined set of small cells, which form the separate small cell only sub-network and which we denote with $\{SC\}$. Group 2 consisting of users are of those served by the rest of the network, which includes the small cells not in $\{SC\}$ and all the macro cells. We denote this sub-network with $\{nSC\}$ standing for "not the Small Cell only network" which is a joint network consisting of macro and small cells.

An example end result of these preliminary tasks is displayed in Figure 6.2, which illustrates a central base station with three sectors, three small cells within each sector, an array of user sites within each small cell coverage area, and an array of user sites to be served by the macro base station. In each case, only a few of the potential sites are shown.

6.2.1   CONSTRUCTING A HETEROGENEOUS NETWORK

This section describes the process for constructing a heterogeneous network layout and then introduces the potential user locations and the user service requests for the simulation. The small cell layout and a pattern of potential user locations is constructed in a manner that is independent of the

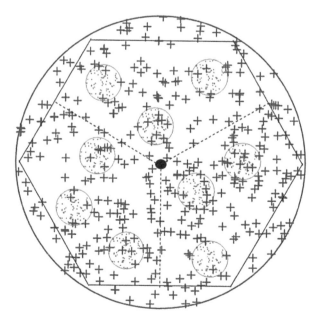

FIGURE 6.2   An example of Group 1 and Group 2 users in a macro cell that contains nine small cells. Group 1 users, indicated by small dots, are located within the small cells. Group 2 users, indicated by "+" symbol are distributed evenly across the entire cell. Although there are some Group 2 users located in a small cell, they will be served by a macro cell.

selection of small cells for the small cell network. This allows the modeller to construct a simulation of a small cell only sub-network and/or a small and macro cell "joint" network consisting of both small and macro cells. Once the small cell layout and potential user locations have been determined, then the modeller can decide which small cells to assign to a small cell only sub-network and which are assigned to the joint network.

6.2.1.1   Step 1: Developing a Small Cell Location Grid

A small cell layout needs to meet a number of criteria. Small cells are to be located within the coverage area of each sector of a base station, separated from the base station by a certain minimum distance, and likewise separated from each other by a minimum spacing. To this end, a grid of possible cell locations is developed using a rectilinear (x,y) grid centred on a particular base station BS_0, thus ensuring that the potential locations are evenly spaced. The origin of this coordinate system, $(0,0)$, is set as the location of the base station BS_0.

The grid spans the range along the x-axis from $x = -ISD/2$ to $x = ISD/2$, and along the y axis from $y = 0$ to $y = ISD/2$, where ISD is the inter-site distance between the macro base stations. The separation of the grid points, Δx in the x-direction and Δy in the y-direction, is set be a few percent of ISD. The exact spacing is not critical provided is small relative to ISD.

Each of the grid points must satisfy the following requirements:

(a) The distance of the grid point is greater than a minimum distance MD, (say 50 m) from the base station location at the origin,
(b) The distance of the grid point is less than $ISD/2$ from the base station location,
(c) The angle between a line from BS_0 to the grid point is less than 58 degrees from the y-axis in either direction.

If the grid point fails to satisfy any of these requirements, it is culled from the set of grid points. These tests ensure that the potential small cell location grid points fall within the first sector of BS_0. Test (a) ensures the small cell base station is not too close to the macro cell base station located at the origin of the (x,y) grid. Test (b) ensures the small cell base station is located at a radial distance within the coverage range of BS_0. Test (c) ensures that the small cell base station is located within the angular coverage area of sector 1 of BS_0. Sector 1 of BS_0 has angular coverage ranging up to 60 degrees away from the y-axis in clockwise and anti-clockwise directions as shown in Figure 6.3.

These locations are given an identification index number $\{w\}$ where $w = 1, 2, 3, ..., N_{Sctr,grid}$ where $N_{Sctr,grid}$ is the number of small cell location grid points in the sector. The coordinates (x_w,y_w) for each grid point relative to the origin at BS_0 are also recorded.

This set of grid points covers sector $p = 1$ of BS_0. To construct the grid points to cover the other two sectors, $p = 2$ and 3, of BS_0 we rotate this first pattern by +120 degrees, to cover sector $p = 2$, and by −120 degrees, to cover sector $p = 3$. These grid locations will be allocated corresponding identifying duplets $\{p,w\}$ and coordinates $(x_{p,w},y_{p,w})$ covering all three sectors in the cell of BS_0. Index p corresponds to the sector and w to the location of the grid point in that sector. This results in a three sector grid pattern of potential small cell base station locations as shown in Figure 6.3.

Location grid patterns for cells associated with other macro cell base stations are constructed by translating the pattern developed for all sectors of BS_0 by the x and y distances required to centre this pattern on each of the surrounding macro base stations at which small cells are to be located. In the example shown in Figure 6.4, small cells have been located in macro cell base stations BS_0 to BS_6. With a grid centred on each base station, the location of each grid point is allocated an identifying triplet $\{m,p,w\}$ where m identifies the base station (BS_0 to BS_6), p identifies the sector and w identifies the location of the grid point in the sector. Each triplet identifies a corresponding location $(x_{m,p,w},y_{m,p,w})$ at which a small cell base station may be located.

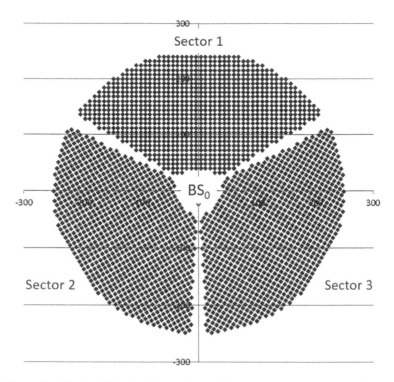

FIGURE 6.3 Example of (x,y) grid points that can be used to locate small cell base stations in the sectors of a macro cell base station. The grid points are evenly distributed over the area of each cell sector. This enables a uniform probability density function to be used to locate the small cell base stations. The grid also includes a zone close to the macro base station in which no small cell base station is to be located.

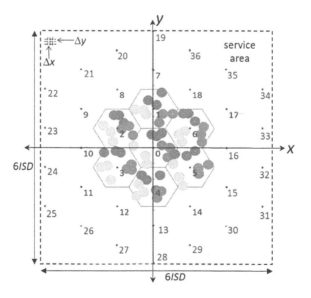

FIGURE 6.4 Example of small cell deployment in a service area covering 36 macro cell base stations. Small cells are randomly located in every sector of every macro cell base station (BS_0 to BS_{36}) across the entire service area. In the figure only the small cells located in the coverage areas of the central macro cell base station and the ring of six surrounding macro base stations are shown. Small cell diameter SCD = 50 m and inter-small cell spacing of ISCD = 50 m minimum. The small cells are colour coded with three small cells randomly located in each sector of the macro cell in which they are placed. The service area is a 6ISD by 6ISD square.

6.2.1.2   Step 2: Placing Small Cell Base Stations onto the Grid

The next step is to actually locate small cells in the sectors of the macro base stations in the service area. If the small cell locations have been pre-determined, because the geographical distribution of users is well understood, then the locations can be inserted accordingly. Examples of this include the locations of small cells in entertainment arenas, transport hubs or shopping complexes. However, often the modeller will be interested in the energy efficiency advantages of small cells in more generic situations [6, 7].

To construct a generic small cell network model, a number $N_{SC} \geq 1$ of small cell base stations are to be randomly located in each sector, so that the model provides generally representative results. In this situation, the process for locating the small cell base stations is more complicated than having predetermined small cell locations. Further, to generate a comprehensive set of results for a given number, N_{SC}, of small cells, it is best to repeat the small cell simulation multiple times, each with a different set of locations for the N_{SC} small cells.

The task is to randomly position a number, N_{SC}, of small cell base stations at sites inside each sector of each macro cell in the service area, and this is achieved using the potential cell location grid points developed in Section 6.2.1.1. An example of small cell locations is shown in Figure 6.4, in which small cells have been located in sectors of the central base station and inner-most ring of base stations in the service area. The modeller may introduce small cell base stations into more macro cell sectors using the same procedure as described here.

The locations of the small cells are to be random but satisfy the following requirements:

(a) All of the N_{SC} small cells in each sector have their base station located within the coverage area of that sector of the macro base station.
(b) Every small cell base station must be located beyond a minimum distance, *MD*, from every macro cell base station.
(c) Every small cell base station must be located beyond a minimum inter-small cell distance, *ISCD*, from every other small cell base station.

Any small cell location that fails any of these requirements is discarded. Although it is relatively easy to satisfy (a), (b) and (c) for simulations with small cell diameter, *SCD*, and *ISCD* much smaller than the inter-site distance, *ISD*, between the macro cell base stations, it can be difficult to do so for *SCD* or *ISCD* values closer to *ISD* and for simulations involving many small cells in each sector.

The small cells will be located on one of the $N_{Sctr,grid}$ locations constructed in each sector as shown in Figure 6.4. In a given sector, each point has an index value, $w = 1, 2, ..., N_{Sctr,grid}$. The probability of locating a small cell location on the w th grid point in a sector is set to be $1/N_{Sctr,grid}$. Thus, the modeller can use a uniform random distribution number generator to generate a number between 1 and $N_{Sctr,grid}$ to select the grid point for the location of the first small cell base station. Let this randomly generated number between 1 and $N_{Sctr,grid}$ be denoted as $w(1)$ meaning it is the first of a sequence of random numbers generated in this manner. This is then repeated for every sector of the required number of base stations resulting in one small cell base station per sector for each of these base stations. Allocating a small cell base station in the macro cell base station sectors with small cells will give $3N_{MC,SC}$ triplets $\{m,p,w(1)\}$ defining (x,y) grid locations $(x_{m,p,w(1)}, y_{m,p,w(1)})$ of the first small cell location in each sector of macro base stations (where $N_{MC,SC}$ = number of macro cells with small cells in each of the three sectors in their coverage area). In these triplets, m denotes the base station ($m = 0, 1, ...,6$) and p denotes the sector, $p = 1, 2, 3$.

For each of these small cell base stations, conditions (b), and (c) must be checked. Condition (b) in particular is checked using the inequality:

$$\sqrt{\left(x_{m,p,w(1)} - x_m\right)^2 + \left(y_{m,p,w(1)} - y_m\right)^2} \geq MD \qquad (6.1)$$

For each value $m = 0, 1, \ldots, 6$ and $p = 1, 2, 3$. Each small cell location is thus checked against the (x_m, y_m) location of the m th macro base station. If the location $(x_{m,p,w(1)}, y_{m,p,w(1)})$ does not satisfy this inequality, it is discarded and another random number is generated to provide a new location for the small cell base station. This is repeated until there is one small cell base station located in every macro cell sector. In defining the grid of possible locations in $N_{Sctr,grid}$, condition (a) has already been met.

If the number of small cell base stations per sector is $N_{SC} = 1$, then the location process is finished. However, typically $N_{SC} > 1$ and can be $N_{SC} \gg 1$. Therefore, the location of the first small cell base station in a given sector is $(x_{m,p,w(1)}, y_{m,p,w(1)})$. With the first small cell base station in each sector located, the locations of subsequent $N_{SC} - 1$ small cell base stations need to be determined. Using the same probability density of $1/N_{Sctr,grid}$ a random number generator is used to give a value $w(2)$ between 1 and $N_{Sctr,grid}$. This gives location $(x_{m,p,w(2)}, y_{m,p,w(2)})$ of the second small cell base station in each sector of each macro cell.

Constraint (b) is checked applying (6.1) to $(x_{m,p,w(2)}, y_{m,p,w(2)})$. If the location of the second small cell base station fails this constraint, it is discarded and another location is generated using the probability density of $1/N_{Sctr,grid}$. If constraint (b) is satisfied, then constraint (c) is checked.

Constraint (c) is more complicated to deal with. The location of the second small cell base station at $(x_{m,p,w(2)}, y_{m,p,w(2)})$ must be further than $ISCD$ away from the first located at $(x_{m,p,w(1)}, y_{m,p,w(1)})$ in that sector. In addition, it must also be further than $ISCD$ away from every other already located small cell location; $(x_{m',p',w(1)}, y_{m',p',w(1)})$ and $(x_{m',p',w(2)}, y_{2,m',p',w(2)})$ for every other sector, p', and cell, m', for the macro cells. That is:

$$\sqrt{\left(x_{m,p,w(2)} - x_{m',p',w'}\right)^2 + \left(y_{m,p,w(2)} - y_{m',p',w'}\right)^2} > ISCD \qquad (6.2)$$

for all values of m', p' and w' that correspond to already located macro cell base stations.

If the second small cell base station fails constraint (c), then that location is discarded and another generated using the $1/N_{Sctr,grid}$ probability density. If the location satisfies all three constraints, it is kept and recorded as the location of the second small cell base station in that sector.

Constraints (a) to (c) are applied to every randomly generated small cell base station location until N_{SC} small cells have been added to each macro cell sector. The requirements for each subsequently added small cell base station is that it must be in a macro base station sector, at least MD away from every macro cell base station and at least $ISCD$ away from every other small cell base station.

As the number of small cell base stations increases, and if $ISCD$ is not much smaller than ISD, it can occur that this process will fail to run to completion to produce N_{SC} small cell base station locations. If need be, the process may have to be re-started from scratch. However, experience has shown that this is not a common occurrence and, given its simplicity, it provides an acceptably quick random allocation of small cell base stations. The small cell base station locations shown in Figure 6.4 were generated in this manner in a matter of seconds.

Assuming that every sector of the macro cell base stations with small cells has N_{SC} small cells, at the end of this process, there will be $3N_{MC,SC}N_{SC}$ triplets $\{m,p,w\}$ identifying each small cell base station, m and p identifying the base station and sector in which the small cell is located and w identifying the location on the grid point (as in Figure 6.3) on which the w th small cell base station in that macro base station sector is located. The global position of the small cell, relative to the origin at BS_0, is given by $(x_{m,p,w}, y_{m,p,w})$.

At this point, we have generated a pattern of small cells located within each sector of the macro base station network. The next stages are to develop a series of potential user locations surrounding each small cell site and assign small cells among the various network types.

6.2.1.3 Step 3: Constructing the Small Cell User Location Grid

With the small cells positioned, the next step is to set up a location grid for the potential user locations within each small cell. This is effectively a repeat of the steps given earlier for setting up the small cell locations.

In this model, all of the small cells have the same small cell diameter, *SCD*. The grid of possible user locations is generated around the notional centre of a small cell. Again a rectilinear coordinate system is adopted such that the grid covers the *x*-axis range $x = -SCD/2$ to $x = SCD/2$, and *y*-axis range $y = -SCD/2$ to $y = SCD/2$, with a grid point separations, Δx and Δy set to be a fraction δ (typically around 3%) of *SCD*. Each grid point is tested and culled if either:

(a) Its distance from the small cell centre exceeds *SCD*/2
(b) Its distance from the small cell centre is less than a pre-set minimum distance $MD_{SC} = \varepsilon SCD/2$. In the simulation results to be reported here, a value of 100m for SCD has been used and the multiplier is set to $\varepsilon = 0.2$, in keeping with the 3GPP nomination of a 10m separation for modelling of outdoor small cell deployments. [13] However as with other simulation parameters, these values are chosen at the modeller's discretion.

With the small cell diameter *SCD* set, the small cell coverage area is $\pi SCD^2/4$ and an (x,y) rectilinear position grid of potential user sites within the small cell coverage area is set up. The grid of points excludes a circular region immediately around the small cell base station of radius $MD_{SC} = \varepsilon SCD/2$. This reflects the fact that user cannot be arbitrarily close to the small cell bases station. Thus the grid covers an annulus around the small cell base station. The number of grid points, $N_{SC,grid}$, within a single small cell is given by the integer part of the small cell area multiplied by the density of points per unit area:

$$N_{SC,grid} \approx \left\lfloor \left(\left(\frac{SCD}{2} \right)^2 - \left(\frac{MD_{SC}}{2} \right)^2 \right) \frac{\pi}{(\delta SCD)^2} \right\rfloor = \left\lfloor \left(\left(\frac{SCD}{2} \right)^2 - \left(\frac{\varepsilon SCD}{2} \right)^2 \right) \frac{\pi}{(\delta SCD)^2} \right\rfloor$$
$$= \left\lfloor \frac{(1-\varepsilon^2)\pi}{4\delta^2} \right\rfloor \tag{6.3}$$

where $\lfloor x \rfloor$ is the largest integer less than *x*. The precise value of $N_{SC,grid}$ depends upon the details of implementing the model. An example of the resulting user locations centred on the small cell base station is shown in Figure 6.5. Each grid point is given allocated an index number, $v = 1, 2, \ldots N_{SC,grid}$.

With the location of the small cell base stations previously determined in Step (2), the grid for user locations constructed in this step is copied to surround each small cell base station. A small

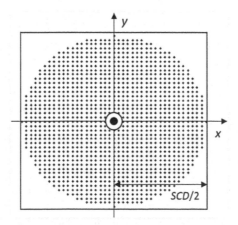

FIGURE 6.5 Grid points of potential (x,y) locations of users in a small cell. The small cell base station is in the centre and is surrounded by a circular exclusion area where no users can be located.

cell user location grid is thus centred on the location of each small cell base stations, $(x_{m,p,w},y_{m,p,w})$. Using the location of the small cell base station, $(x_{m,p,w},y_{m,p,w})$, relative to the global origin at BS_0, the location of each user grid point relative to BS_0 can be determined. Therefore each potential small cell user location can be identified with the quadruplet, $\{m,p,w,v\}$ giving its location relative to BS_0 as $(x_{m,p,w,v},y_{m,p,w,v})$.

Across the network, if $N_{MC,SC}$ macro base stations have N_{SC} small cells in each of their three sectors, the total number of small cell grid points, available for user locations, will be $N_{SC,grid}^{(usr)}$ given by

$$N_{SC,grid}^{(usr)} = 3N_{MC,SC}N_{SC}N_{SC,grid} \tag{6.4}$$

Each of these $N_{SC,grid}^{(usr)}$ potential user sites has a unique index number $\{m,p,w,v\}$.

For example, for a small cell with $SCD = 110$ m, $MD_{SC} = 8$ m and an (x,y) grid spacing of 4 m, from (6.3), we have $N_{SC,grid} \approx 590$ potential user location sites to each cell. The precise number of sites will depend upon the details of the implementation of the grid setup. For example, if small cells are placed only on the seven central base stations, there are 21 base station-sectors each with three small cells giving a total of $N_{SC,grid}^{(usr)} \approx 3 \times 7 \times 3 \times 590 = 37{,}170$ possible user location sites in the simulation.

6.2.1.4   Step 4: Constructing the Macro Cell User Location Grid

The construction of the grid for the location of macro cell users is the same as in the Advanced Model of Chapter 5. Applying (5.1) to the macro cell network, the number of macro cell grid user locations, $N_{MC,grid}^{(usr)}$ is given by:

$$N_{MC,grid}^{(usr)} = \left\lfloor \frac{\left(N_{ISD}^2 - \pi\left(N_{MC}+1\right)\left(MD_{MC}/ISD\right)^2\right)}{\delta_{MC}^2} \right\rfloor \tag{6.5}$$

where N_{ISD} is the number of inter-macro cell site distances used for the length and width of the service area, N_{MC} is the number of macro cells in the simulation, MD_{MC} is the minimum distance a user can be located from macro cell base station and δ_{MC} is the grid spacing ($\Delta x = \Delta y = \delta_{MC}ISD$) for the macro cell location grid.

This grid is used to locate users across the entire service area. Consequently, this grid provides locations for some of the joint network users. Other users of the joint will be located on the small cell grid, as described in the following steps.

6.2.1.5   Step 5: Determining the Number of Small Cell and Macro Cell Users

With all the grid locations for the small cell and macro cell networks set up, we now allocate users to these cells and the two sub-networks. We want to include the range of sub-network configurations mentioned earlier. That is, configurations range from one extreme of a single joint network that includes all the small and macro cells to the other extreme of two totally independent networks; the separate small cell network consisting of only small cells and the other consisting of only macro cells. To do this, we separate the total number of users in the service area into two groups: Group 1 users are those users served only by a pre-determined set of small cells which form the separate small cell only sub-network. We denote this sub-network with $\{SC\}$. Group 2 users are of those served by the rest of the network, which includes the small cells not in $\{SC\}$ and all the macro cells. We denote this sub-network with $\{nSC\}$ standing for "not the small cell only network" which is a joint network consisting of macro and small cells.

In this network simulation, the $\{SC\}$ small cells of the small-cell-only sub-network are deployed to cluster in and around BS_0 where the user density may be high enough to require an independent small cell network. These small cells are identified with light shading in an example network shown in Figure 6.1. These small cells operate on a separate frequency independently of the $\{nSC\}$ joint sub-network.

The remaining small cells (with dark shading) are part of the $\{nSC\}$ joint sub-network which also includes the macro cell base stations operating on the same frequency band as the small cells in $\{nSC\}$.

To simplify this model, all base stations sectors have the same number of small cells, including BS_7 to BS_{36}. A set of small cells in this part of the network is only shown for BS_{31}. The modeller may decide to change this detail. However, to do so will require additional details to an already complicated model. One significant advantage of not having small cells in every macro cell sector across the service area is the reduction in run-time of the simulation.

The extreme configuration of a single $\{nSC\}$ joint network means there are no Group 1 users (because $\{SC\}$ has no members) and all users are in Group 2. The other extreme of two independent networks has Group 1 users being served only by the small cells (that is $\{SC\}$ includes all the small cells) and Group 2 users being served only by the macro cells [8].

We assume that users assigned to Group 1 will be served only by small cells in $\{SC\}$, even if the SINR at the user's location achieved by transmissions from a cell not in $\{SC\}$ is greater than that for transmissions from a small cell in $\{SC\}$. Similarly, users allocated to Group 2 will be served only by cells in $\{nSC\}$ even if a cell in $\{SC\}$ could provide a better SINR. This allocation of users into Group 1 and Group 2 is done a priori, meaning it is an input to the model and not determined within the model. Thus users in Group 1 will be served only by small cells in $\{SC\}$, and users allocated to Group 2 will only be served by cells in $\{nSC\}$, regardless of whether the alternate network might provide a better SINR. This is done to control the numbers of users and hence traffic loading in each network.

Some heterogeneous network simulations take a different approach by implementing a bias on the SINR to force the allocation of users to the small cells. We do not implement a bias; however, some simulations do. For example, GreenTouch implemented an SINR bias of 5 dB towards small cells for their technology roadmap [7] and Blume *et al* implement a bias of up to 15 dB [6]. That is, a user was allocated to a small cell provided that the SINR when serviced by a small cell was no poorer than 15 dB less than the SINR that would be provided by the macro cell. The reasoning for the bias in these simulations has been to "encourage" the migration of traffic to the more energy-efficient small cells.

The total number of Group 1 and Group 2 users in the service area, $N_{SA}^{(usr)}$, is given by the product of the total service area, A_{SA}, and average user density across the service area, $U^{(usr)}$. The average download data rate per user, $D^{(usr)}$, and download file size, $B^{(usr)}$, are then introduced to determine the service request rate averaged across the service area as given by (5.11):

$$\lambda = A_{SA}U^{(usr)}D^{(usr)}\big/B^{(usr)} = N_{SA}^{(usr)}D^{(usr)}\big/B^{(usr)} \qquad (6.6)$$

For example, adopting the Dense Urban geography (see Table 5.1), the user density is $U^{(usr)} = 10{,}000$ users/km² and the macro cell inter-site distance is $ISD = 500$ m. Adopting a square service area that is $6ISD$ each side, $A_{SA} = 9$ km². With a download traffic demand of $D^{(usr)} = 49.5$ kbit/s per user and a file size of $B^{(usr)} = 16$ Mbits using (6.6), the average service request rate across the service area is $\lambda = 278.4$ requests/s.

Small cell user requests arise from a higher user population density in the areas located in the small cells. In the simulation, this higher density is generated by taking a proportion of the user population from a macro cell coverage area and effectively relocating those users, and their traffic, to locations clustered around the small cells. Although this will change the density of users in the small cells, the total number of users and hence average user density over the total service area will remain the same. The proportion of users located into a small cell is set by introducing a user "allocation factor", φ, giving the number of users placed into the areas covered by small cells, $N_{SC}^{(usr)}$:

$$N_{SC}^{(usr)} = \varphi N_{SA}^{(usr)} = \varphi U^{(usr)} A_{SA} \tag{6.7}$$

The number of remaining users, who become macro cell users, is:

$$N_{MC}^{(usr)} = (1-\varphi) N_{SA}^{(usr)} = (1-\varphi) U^{(usr)} A_{SA} \tag{6.8}$$

The allocation factor, φ, thus partly determines the number of active users in the vicinity of a small cell. (Note that this is not the same as the number of potential user sites.) The macro cell users, $N_{MC}^{(usr)}$, are evenly spread across the entire service area hence their density is:

$$U_{MC}^{(usr)} = \frac{N_{MC}^{(usr)}}{A_{SA}} = (1-\varphi) U^{(usr)} \tag{6.9}$$

Note that because macro cell users are evenly spread across the entire service area, there will be some macro cell users located in small cell service areas.

Recalling that Group 1 users are served by the small cell only network $\{SC\}$ and Group 2 users by the remaining joint network, $\{nSC\}$, this means the population of users in the area covered by small cells may consist of both Group 1 and Group 2. For a small cell in $\{SC\}$, the users in the area of that small cell will consist of the Group 1 users allocated to $\{SC\}$ plus a few Group 2 users who, because the $N_{MC}^{(usr)}$ users are evenly spread over the entire service area, A_{SA}, happen to be located in that small cell. We see an example of this in Figure 6.2 in which the small cells include some "+" symbols representing Group 2 users located within a small cell.

Therefore, the total number of users located within the total area covered by small cells, $N_{SC,Area}^{(usr)}$, is the sum of all Group 1 users and Group 2 users who are located in a small cell area. Therefore, this total number of users, $N_{SC,Area}^{(usr)}$, is given by (6.7) plus the number of Group 2 users located in a small cell area. The number of Group 2 users located in a small cell area is given by their density (6.9) multiplied by the area of a small cell, A_{SC}, multiplied by N_{SC} small cells in each of the three sectors multiplied by the $N_{MC,SC}$ macro base stations which have small cells located in them. Summing these Group 1 and Group 2 users thus gives:

$$\begin{aligned} N_{SCArea}^{(usr)} &= N_{SC}^{(usr)} + U_{MC}^{(usr)} 3 N_{MC,SC} N_{SC} A_{SC} \\ &= U^{(usr)} \left(\varphi A_{SA} + (1-\varphi) 3 N_{SC} N_{MC,SC} A_{SC} \right) \end{aligned} \tag{6.10}$$

Note that we have $N_{MC,SC} \leq N_{MC}$, that is, the number of macro cells that have small cells, $N_{MC,SC}$, is less than or equal to the number of macro cells, N_{MC}. This gives an overall user density in the small cell service area, $U^{(usr)}{}_{SC}$ of:

$$U_{SC}^{(usr)} = \frac{N_{SCArea}^{(usr)}}{3 N_{SC} N_{MC,SC} A_{SC}} = U^{(usr)} \left(\varphi \left(\frac{A_{SA}}{3 N_{SC} N_{MC,SC} A_{SC}} - 1 \right) + 1 \right) \tag{6.11}$$

Of the users in the small cell areas, some are served by the small cell and others by the macro cells. Because $A_{MC}/3 N_{SC} N_{MC,SC} A_{SC} > 1$, we have $U_{SC}^{(usr)} > U^{(usr)} > U_{MC}^{(usr)}$. That is, the small cells have a higher user density than the average across the service area, which in turn is higher than the density in the macro cell areas surrounding the small cells. The total number of users across the entire service area remains $A_{SA} U^{(usr)}$. This process of a priori allocating users into small cells increases the user density in the small cell areas. This allocation process results in the user density for the small cells including those in $\{SC\}$ simulation given in (6.11) being greater than the average user density in the macro cells in the $\{nSC\}$ joint network simulation given in (6.9). This reflects the real world situation we wish to model with the user density in the small cells representing "high density" locations relative to the rest of the network.

An example of this for a single macro cell with three sectors and containing nine small cells is shown in Figure 6.2. In this example, all of the small cells are in $\{SC\}$, Group 1 users are shown as dots, and are located only inside one of the nine small cells. Group 2 users are depicted with "+" symbols and are evenly distributed across the entire cell including within some small cells. Consequentially, there are some Group 2 users located inside a small cell area. Although these users are located in a small cell, they will be served by the macro cell.

With the user densities set, we need to determine expressions for the request rates for the two components, $\{SC\}$ and $\{nSC\}$. As described earlier, the model is set up with Group 1 users served by $\{SC\}$ and Group 2 users served by $\{nSC\}$. The total number of small cell base stations across the service area is $N_{SC,Tot} = 3N_{MC,SC}N_{SC}$. Not all of these small cells are in $\{SC\}$. Let $N_{\{SC\}}$ be the number of small cells in $\{SC\}$. Then, the number of small cells in $\{nSC\}$ is $N_{\{nSC\},SC} = N_{SC,Tot} - N_{\{SC\}}$. The average user service request rate, $\lambda_{\{SC\}}$, for the $\{SC\}$ network is set by the number of users in $\{SC\}$, $N_{\{SC\}}^{(usr)}$ giving:

$$\lambda_{\{SC\}} = N_{\{SC\}}^{(usr)} \frac{D^{(usr)}}{B^{(usr)}} = N_{SC}^{(usr)} \frac{N_{\{SC\}}}{N_{SC,Tot}} \frac{D^{(usr)}}{B^{(usr)}} = \varphi \frac{N_{\{SC\}}}{N_{SC,Tot}} N_{SA}^{(usr)} \frac{D^{(usr)}}{B^{(usr)}} \tag{6.12}$$

In this, we have assumed that small cell users are equally distributed among all the small cells in the service area. This means the average service request rate for the remaining users who are in $\{nSC\}$ is:

$$\lambda_{\{nSC\}} = \left(N_{SA}^{(usr)} - N_{\{SC\}}^{(usr)}\right) \frac{D^{(usr)}}{B^{(usr)}} = \left(1 - \varphi \frac{N_{\{SC\}}}{N_{SC,Tot}}\right) N_{SA}^{(usr)} \frac{D^{(usr)}}{B^{(usr)}} \tag{6.13}$$

The effect of (6.12) and (6.13) is to "off load" user requests from the macro cells into the $\{SC\}$ network. This has been achieved by allocating the proportion φ of total users to be located in small cell. This allocation is done in (6.7) and (6.8). With the users allocated, a number, $N_{\{SC\}}$, of small cells were then chosen to constitute the $\{SC\}$ sub-network with the remaining small cells being in $\{nSC\}$. The service request rate for users to be served by $\{SC\}$, $\lambda_{\{SC\}}$ is given by (6.12) and the service request rate for users in $\{nSC\}$ is given by (6.13). Bringing these factors together, we can define an "off load ratio" $\varphi_{\{SC\}}$ for the $\{SC\}$ sub-network by:

$$\varphi_{\{SC\}} = \varphi \frac{N_{\{SC\}}}{N_{SC,Tot}} \tag{6.14}$$

With this definition, the number of users "offloaded" to be served by the $\{SC\}$ network, $N_{\{SC\}}^{(usr)}$, and the number remaining in the $\{nSC\}$ joint network, $N_{\{nSC\}}^{(usr)}$, respectively are given by

$$N_{\{SC\}}^{(usr)} = \varphi_{\{SC\}}N_{SA}^{(usr)}$$
$$N_{\{nSC\}}^{(usr)} = \left(1-\varphi_{\{SC\}}\right)N_{SA}^{(usr)} \tag{6.15}$$

We see that service request rate $\lambda_{\{SC\}}$ corresponds to the service requests offloaded to the small cell only network $\{SC\}$, whereas the service requests corresponding to $\lambda_{\{nSC\}}$ will be served by either a macro cell or a small cell in $\{nSC\}$. For $\lambda_{\{nSC\}}$, we need to account for the fact that a portion of these request will be by users not located in a small cell. This issue is addressed in Step (8).

6.2.1.6   Step 6: Allocating Grid Locations to {*SC*} and {*nSC*} Sub-Networks

Step (1) to Step (4) of the simulation process covered the preliminary tasks of setting up the locations of the small cell within the macro cell sectors, constructing the user location grids within the small cells and macro cells in the service area. Step (5) separated the total number of users in the service area into those to be served by the small cell only network, {*SC*}, and those to be served by the joint network, {*nSC*}. The potential user locations produced by Step (1) to Step (4) have now to be allocated to the two sub-network components, {*SC*} and {*nSC*} for the corresponding Group 1 and Group 2 users to occupy.

To do this, the modeller chooses which of the small cells are to be placed into the small cell sub-network {*SC*}. This selection depends upon the type of heterogeneous network layout the modeller wishes to simulate. For example, typically the small cells in the {*SC*} network will be clustered around high use density areas, not widely scattered across the entire service area. An example of this is shown in Figure 6.4 in which the {*SC*} network consists of the small cell base stations clustered in the inner group of BS_0 to BS_6. There may be small cells located in the remaining base station cell areas (BS_7 to BS_{36}) but these small cells will be allocated to the {*nSC*} sub-network.

Let the number of small cells allocated to the {*SC*} sub-network be $N_{\{SC\}}$. The total number of small cells $N_{SC,Tot}$ is the sum of small cell base stations in {*SC*} and small cell base stations in {*nSC*}. That is:

$$N_{SC,Tot} = N_{\{SC\}} + N_{\{nSC\},SC} \tag{6.16}$$

where $N_{\{nSC\},SC}$ is the number of small cells in the {*nSC*} joint network.

All of the macro cells are in {*nSC*}; therefore, the total number of base stations in {*nSC*} is:

$$N_{\{nSC\}} = N_{MC} + N_{\{nSC\},SC} \tag{6.17}$$

With $N_{\{SC\}}$ small cells allocated to the {*SC*} sub-network, the number of potential user location grid points available in the {*SC*} network is:

$$N^{(usr)}_{\{SC\},grid} = N_{\{SC\}} N_{SC,grid} \tag{6.18}$$

All of the macro cell user location grid points are in {*nSC*} and to that we add the number of grid points in the $N_{\{nSC\},SC}$ small cells that are allocated to the {*nSC*} joint network. This gives a total of:

$$N^{(usr)}_{\{nSC\},grid} = N^{(usr)}_{MC,grid} + N_{\{nSC\},SC} N_{SC,grid} \tag{6.19}$$

available user grid locations in the {*nSC*} joint network.

6.2.1.7   Step 7: Group 1 User Request Rate and Location in the {*SC*} Network

The previous steps have set locations and user numbers for the two sub-networks. Focussing on the {*SC*} sub-network, note that the number of user sites, $N^{(usr)}_{\{SC\},grid}$ is not the same as the number of small cell users $N^{(usr)}_{\{SC\}}$. The number of users is set by the modeller based upon the amount of data they wish to collect for analysis. The random selection of a site for the location of a user download request means that in any simulation run, many sites may not be used, and other sites used several times.

As with the Advanced Model, the small cell simulation epoch time step, $\Delta t_{\{SC\}}$, is set by the modeller so as to ensure a low probability that multiple service requests are generated by the Poisson process within a time step for a given service request rate, $\lambda_{\{SC\}}$. The software created by the modeller needs to cope with the possibility of multiple such requests; a short time step allows simpler software. As a compromise, we have allowed for up to two new requests, $\Delta N_j^{(usr)+}$, to occur in the duration of the epoch. The corresponding maximum allowable epoch time step, $\Delta t_{\{SC\}}$, is determined by requiring it to satisfy:

$$Prob\left(\Delta N_j^{(usr)+} \leq 2\right) > 0.95$$

$$= e^{-\lambda_{\{SC\}}\Delta t_{\{SC\}}} \sum_{n=0}^{2} \frac{\left(\lambda_{\{SC\}}\Delta t_{\{SC\}}\right)^n}{n!} = e^{-\lambda_{\{SC\}}\Delta t_{\{SC\}}} \left(1 + \lambda_{\{SC\}}\Delta t_{\{SC\}} + \frac{\left(\lambda_{\{SC\}}\Delta t_{\{SC\}}\right)^2}{2}\right) \qquad (6.20)$$

Each new Group 1 user is randomly located at one of the $N_{\{SC\},grid}^{(usr)}$ small cell sites. The locations sites are identified by an $(x_{m,p,w,v}, y_{m,p,w,v})$ location in which the base station index, m, and sector index, p, values are set by the determination of which small cells are allocated to the $\{SC\}$ network. These locations correspond to an identifying index, $\{m,p,w,v\}$ as described in Step 3 above.

To generate the uniform random (x,y) locations of users in $\{SC\}$, we ascribe an integer from 1 to $N_{\{SC\}.grid}^{(usr)}$ to each index value $\{m,p,w,v\}$ for the $\{SC\}$ network, accounting for the allocation of specific m and p values to correspond to the $\{SC\}$ network. We then use a uniform random number generator to provide a random number between 1 and $N_{\{SC\},grid}^{(usr)}$ to choose the location of each user.

For example, if there are 36 macro base stations in the simulation, each with three sectors and it is decided the inner seven base stations (BS_0 to BS_6) will have all their small cells in $\{SC\}$, and if there are four small cells per macro cell sector, then m will have seven values ranging from 0 to 6, p will range have three values (1, 2 and 3) and w will have four values (1, 2, 3 and 4). With $N_{SC,grid}$ locations in each small cell, v ranges from 1 to $N_{SC,grid}$. This gives the total number of possible user locations in $\{SC\}$ of $N_{\{SC\},grid}^{(usr)} = 7 \times 3 \times 4 \times N_{SC,grid}$. Each number in this range will be identified with a unique location on the grid corresponding to the pre-set values of m and p to provide the possible locations of users in $\{SC\}$. Note: apart from the use of a uniform random distribution for the location of users, the initial user site location made no specific attempt to evenly "share" users across $\{SC\}$ cells.

Upon entry of a user with a service request into the service area, the signal level received from every base station in $\{SC\}$ at the new user's location $(x^{(k)}, y^{(k)})$ is recorded.

6.2.1.8   Step 8: Group 2 User Request Rate and Location in the {nSC} Joint Network

For Group 2 users, the $\{nSC\}$ simulation epoch time step, $\Delta t_{\{nSC\}}$, is set by the modeller using the same approach as for Group 1 user. Adapting (6.20) for Group 2, (6.21) sets the condition used to determine the maximum allowable value for the epoch duration $\Delta t_{\{nSC\}}$:

$$Prob\left(\Delta N_j^{(usr)+} \leq 2\right) > 0.95$$

$$= e^{-\lambda_{\{nSC\}}\Delta t_{\{nSC\}}} \sum_{n=0}^{2} \frac{\left(\lambda_{\{nSC\}}\Delta t_{\{nSC\}}\right)^n}{n!} = e^{-\lambda_{\{nSC\}}\Delta t_{\{nSC\}}} \left(1 + \lambda_{\{nSC\}}\Delta t_{\{nSC\}} + \frac{\left(\lambda_{\{nSC\}}\Delta t_{\{nSC\}}\right)^2}{2}\right) \qquad (6.21)$$

Group 2 users are distributed across the macro cells and the small cells in $\{nSC\}$. This distribution is not uniform because the user density within the small cells in $\{nSC\}$ is higher than that in the areas outside the small cells. To reflect this, the process for locating Group 2 users across the network involves extra steps compared to the location process for $\{SC\}$ users. The first step is to separate out the user requests generated by (6.21) into two sub-groups. The users in one sub-group will be located in the macro cell location grid. The second sub-group will be located in the small cell location grid.

To determine which sub-group a new request is allocated into, a uniformly random number, ν, between 0 and 1 is generated. If ν satisfies the condition:

$$\nu \leq \frac{N_{\{nSC\},SC}^{(usr)}}{N_{\{nSC\}}^{(usr)}} = \frac{\varphi N_{SA}^{(usr)}\left(\dfrac{N_{\{nSC\},SC}}{N_{SC,Tot}}\right)}{\left(1 - \varphi_{\{SC\}}\right)N_{SA}^{(usr)}} = \frac{\varphi N_{\{nSC\},SC}}{\left(N_{SC,Tot} - \varphi N_{\{SC\},SC}\right)}, \tag{6.22}$$

then that user is located in a small cell in $\{nSC\}$. That is, if the random number ν is less than or equal to the ratio of number of users in small cells in $\{nSC\}$, $N_{\{nSC\},SC}^{(usr)}$, to the total number of users in $\{nSC\}$, $N_{\{nSC\}}^{(usr)}$, then that user is located in a small cell in $\{nSC\}$. The number of users in the small cells in $\{nSC\}$, $N_{\{nSC\},SC}^{(usr)}$, is given by the number of users allocated to small cells multiplied by the proportion of small cells in $\{nSC\}$ which gives the numerator in (6.22). For the denominator, the number of users in $\{nSC\}$ is given by (6.15). We use (6.14) to simplify the expression giving the right-hand side of (6.22).

For Group 2 users that are located in a small cell in $\{nSC\}$, their site is identified by an $(x_{m,p,w,v}, y_{m,p,w,v})$ location in a small cell in $\{nSC\}$. The process for determining the user's location is the same as with the Group 1 users above;, however, the range of values of m and p will now correspond to base station sectors with small cells in $\{nSC\}$.

Continuing with the example above, we have four small cells in each of the three sectors of base stations BS_7 to BS_{36}. Therefore, m has 30 values (7 to 36), p has three values, v has four values giving $N_{\{nSC\},grid}^{(usr)} = 30 \times 3 \times 4 \times N_{SC,grid}$.

If the value of ν does not satisfy the condition in (6.22), then that user will be located in the macro cell grid. We then use a uniform random number generator to choose number between 1 and $N_{MC,grid}^{(usr)}$ to select a location for the user in the macro cell grid.

Upon entry in the service area, the signal level received from every base station in $\{nSC\}$ at the new user's location $(x^{(k)}, y^{(k)})$ is recorded and used to determine the serving station SINR.

With the user service request rates and locations determined, the heterogeneous network model is composed of the $\{SC\}$ and $\{nSC\}$ networks each of which are simulated separately. Then the results are combined to calculate the heterogeneous network performance. The performance of both $\{SC\}$ and $\{nSC\}$ networks are modelled with a range of user demand levels $D^{(usr)}$ representing different stages of the diurnal cycle. The simulations can be repeated multiple, N_{sim}, times to ensure the results are indicative for the given number, N_{SC}, of small cells in each macro cell sector. For each of the N_{sim} runs the small cells are relocated within the macro cell sectors.

A Group 2 user may be located in a small cell area; however, a Group 1 user cannot be located outside a small cell area. The coverage area of the $\{nSC\}$ joint network may overlap with that of the $\{SC\}$ network, and both networks may share the local demand. However, if a particular service request is generated by a Group 1 user, it must be handled by a cell in the $\{SC\}$ network. Likewise, a Group 2 user request must be served by an $\{nSC\}$ base station, which could be either a small cell or macro cell station. This is a common approach when determining small cell locations [8–11]. The purpose of the small cells is to reduce the load on the macro cell network in areas where the demand density (requests/s/km^2) is too great for the macro cell alone to accommodate.

6.2.2 MODEL SEPARATION

In summary, the Steps described earlier construct a heterogeneous network model in which the small cells in sub-network $\{SC\}$ are assumed to operate in a different frequency band to the cells in sub-network $\{nSC\}$. This allows the heterogeneous network model to be separated into two independent simulation components because there will be no signal interference between the two sub-networks $\{SC\}$ and $\{nSC\}$. In other words, only the macro and small base stations in $\{nSC\}$ will generate interference signals to transmissions of other macro and small base stations in $\{nSC\}$. Separately, only small cells in $\{SC\}$ will generate interference signals to other small cells in $\{SC\}$. (There are no macro cells in $\{SC\}$). Thus the SINR values for the users serviced by the $\{nSC\}$ cells will be influenced only by transmissions from other $\{nSC\}$ cells. Similarly, the SINR for users serviced by a $\{SC\}$ cells will only be influenced by transmissions from other $\{SC\}$ cells. This assumption enables the heterogeneous network simulation to be effectively split into two independent component network simulations: a $\{nSC\}$ joint sub-network component and a $\{SC\}$ sub-network component. Thus, two independent components can be modelled separately and their results brought together at the end of the simulations.

This heterogeneous network simulation significantly leverages off the Advanced Model. That is, the download traffic demand is modelled using a Poisson process to generate the occurrence of new user service requests across the entire service area. However, in this heterogeneous network simulation some of the generated requests will be by users allocated to be served by base stations in $\{SC\}$ and the remaining users will be served by base stations in $\{nSC\}$. The portion of user service requests allocated the $\{SC\}$ network is set by an "offload ratio", $\varphi_{\{SC\}}$. The remaining portion of service requests, $(1 - \varphi_{\{SC\}})$, will be served by the $\{nSC\}$ joint network. This allocation of users relates to the network that serves them not their specific location in the service area. In particular, although all users allocated to the $\{SC\}$ network are located in a small cell, users allocated to the $\{nSC\}$ joint network can be located anywhere across the entire service area including within small cells. Because of this, the location of service requests in the $\{nSC\}$ simulation need to be weighted using (6.22) so as to reflect the user densities around the small cells relative to the remaining areas in the overall service area.

6.2.3 Small Cell Base Station Power

The power model for the small cell base stations is different to that of the macro cell base stations because they do not require the environmental control and do not include the feed loss from the power amplifier to the antenna. Also, small cells typically have sub-system parameters different to macro cells and commonly use omnidirectional antennas so effectively have only one sector. Using the form given in (4.11) and (4.13), this gives:

$$P_{SC}\left(\mu_{SC}\right) = \frac{1}{\left(1 - \sigma_{DC,SC}\right)\left(1 - \sigma_{MS,SC}\right)}\left(P_{RF,SC} + P_{BB,SC} + \frac{P_{idle,out,SC} + \mu_{SC}\Delta P_{out,SC}}{\eta_{PA,SC}}\right) \quad (6.23)$$

where $\Delta P_{out,SC} = P_{max,out,SC} - P_{idle,out,SC}$. The transmit power from the base station is:

$$P_{out,SC}\left(\mu_{SC}\right) = \mu_{SC}\Delta P_{out,SC} \quad (6.24)$$

where μ_{SC} is the utilisation of the small cell base station. In this model, μ_{SC} is set to unity in every epoch in which the small station is transmitting to users, as with the Advanced Model of Chapter 5.

In the model presented here, the output transmit power levels are set to $P_{max,out,SC} = 1$ W (30 dBm) when transmitting data, $\mu_{SC} = 1$ and idle transmit power of $P_{idle,SC} = 0.25$ W (24 dBm), due to the signalling overhead. The idle power does not contribute to interference as none of this power is transmitted concurrent with PDSCH REs. The corresponding small cell electrical power consumption levels are $P_{SC}(\mu_{SC} = 0) = 1.8$ W and $P_{SC}(\mu_{SC} = 1) = 6.5$ W. These values are based on values quoted for a pico cell in [14], but adjusted to model a single channel SISO system, with a single power amplifier delivering 1W output power instead of two amplifiers delivering 0.5W each.

6.3   RUNNING THE SIMULATION COMPONENTS

With the user locations set for the $\{SC\}$ and $\{nSC\}$ components, the two simulations can be run independently. Both simulations are run in the same manner as the Advanced Model in Chapter 5. The values of $\lambda_{\{SC\}}$ and $\lambda_{\{nSC\}}$ and the Poisson process for introducing new user requests, and the simulations set epoch time steps, $\Delta t_{\{SC\}}$ and $\Delta t_{\{nSC\}}$ are given by (6.20) and (6.21) respectively.

The target number of users over which simulation data is collected can be set by the modeller to $N^{(usr)}_{\{SC\},data}$ and $N^{(usr)}_{\{nSC\},data}$ for the $\{SC\}$ and (nSC) networks respectively, and the simulations are run for durations as shown in Figure 5.4. The sample size $N^{(usr)}_{\{SC\},data}$ and $N^{(usr)}_{\{nSC\},data}$ can be chosen independently of each other. The only requirement is that they are sufficiently large to ensure the results of the simulations are representative of the service area performance.

It is important to note that $N^{(usr)}_{\{SC\},data}$ and $N^{(usr)}_{\{nSC\},data}$ are not to be confused with $N^{(usr)}_{\{SC\}}$ and $N^{(usr)}_{\{nSC\}}$. The former are the numbers of user requests over which the simulation is run. The latter are the number of user sites in the $\{SC\}$ and $\{nSC\}$ sub-networks. Because the simulation runs over an extended duration, we have $N^{(usr)}_{\{SC\},data} \gg N^{(usr)}_{\{SC\}}$ and $N^{(usr)}_{\{nSC\},data} \gg N^{(usr)}_{\{nSC\}}$.

As with the Advanced Model in Chapter 5, at the time of termination of the simulation, there may be some users who have commenced their download but their session is terminated with the simulation termination. Their data and duration of service are included in the data collection and therefore the total number of users in the small cell and macro cell simulations are $N^{(usr)}_{Tot,\{SC\},data} > N^{(usr)}_{\{SC\},data}$ and $N^{(usr)}_{Tot,\{nSC\},data} > N^{(usr)}_{\{nSC\},data}$ respectively. Also, as in Chapter 5, larger service request rates $\lambda_{\{SC\}}$ and $\lambda_{\{nSC\}}$ result in a larger difference between the target number of users served in the simulation, $N^{(usr)}_{\{SC\},data}$, $N^{(usr)}_{\{nSC\},data}$, and the actual number served, $N^{(usr)}_{Tot,\{SC\},data}$, $N^{(usr)}_{Tot,\{nSC\},data}$, respectively. However, we still always have the difference between the target and actual as small compared to the target number of service requests.

The SINR that each new user in Group 1 would receive from each of the $\{SC\}$ based on current cell activity status is calculated upon entry, to determine which base station serves that new user. Similarly, the SINR that each new user in Group 2 would receive from each of the $\{nSC\}$ based on current cell activity status is calculated to determine which base station serves that new user. The SINR calculation involves calculating the received signal power from each base station, dependent on the signal propagation loss, the beam pattern of each base station transmitting antenna in vertical and horizontal planes, a building penetration loss, plus a random additional "shadowing" factor appropriate to the $\{SC\}$ and $\{nSC\}$ networks. For each possible serving base station, its received signal power and an interference power from every other base station in the appropriate network, depending on its current activity status, is calculated. From these results, the SINR at the k th user's location $(x^{(k)}, y^{(k)})$ for the signal from every base station for the appropriate network is evaluated and the most appropriate serving base station selected. The interference noise in the j th epoch of the simulation for the k th user in the $\{SC\}$ network being served by small cell with index $w^{(k)}$, is given by:

$$P^{(k)}_{Interf\{SC\},j} = \left(\sum_{w=1}^{N_{\{SC\}}} \mu_{SC,w,j} W_{SC}\left(w, z^{(k)}\right) - W_{SC}\left(w^{(k)}, z^{(k)}\right) \right) P_{max,out,SC} \qquad (6.25)$$

In this $\mu_{SC,w,j}$ is the utilisation of the w th small cell base station in $\{SC\}$ for the j th epoch; and $N_{\{SC\}}$ is the number of small cell base stations in the $\{SC\}$ network. Note: the utilisation of the small cell w is unity if it is serving a user in epoch j, i.e. $\mu_{SC,w,j} = 1$. We also note that the $w^{(k)}$ th small cell is operating because it is serving user k; therefore, the utilisation of the small cell $w^{(k)}$ is unity.

The channel function $W_{SC}(w, z^{(k)})$ represents the relationship between transmit power, $P_{max,out,SC}$, from the w th small cell base station in $\{SC\}$ and received power at $z^{(k)}$, $P_{Rx,w}(z^{(k)})$:

$$P_{Rx,w}\left(z^{(k)}\right) = P_{max,out,SC} W_{SC}\left(w, z^{(k)}\right) = P_{max,out,SC} G_{Tx} G_{Rx} \Phi_w\left(z^{(k)}\right) L_{SC}\left(r_w^{(k)}, f_i, h, d, F, \upsilon\right) \qquad (6.26)$$

The terms in this equation are defined in Section 5.5. The term $W_{SC}(w^{(k)}, z^{(k)})$ represents the channel function and $w^{(k)}$ the w index for the small cell base station serving the k th user. L_{SC} is the corresponding small cell path loss. Note that L_{SC} will have a form and parameters corresponding to a small cell model.

The k th user's serving base station is then selected, either on the basis of giving the best SINR to the new user, or simply giving the highest signal strength. The base station in $\{SC\}$ delivering the highest SINR is selected to service each new Group 1 user. This gives SINR for the k th user served by the $\{SC\}$ network:

$$SINR_{\{SC\}}^{(k)} = \frac{W_{\{SC\}}\left(w^{(k)}, z^{(k)}\right) P_{max,out,SC}}{P_{Interf\{SC\}}^{(k)} + P_{RxN}} \tag{6.27}$$

Following the same approach as in Chapter 5, in this equation we assume that the SINR for the k th user remains constant during the download time, at its value in the epoch when the k th user commences being served by the $\{SC\}$ network.

For users of the $\{nSC\}$ joint network, the base station delivering the highest SINR is selected to service each new Group 2 user. The serving base station may be a small cell or a macro cell in $\{nSC\}$. Because of this, the interference power for user k served by a base station in the $\{nSC\}$ joint network needs to account for this:

$$P_{Interf\{nSC\},j}^{(k)} = \sum_{m=0}^{N_{MC}} \sum_{p=1}^{3} \mu_{MC,m,p,j} W_{MC}\left(m, p, z^{(k)}\right) P_{max,out,MC} +$$
$$+ \left(\sum_{w=1}^{N_{\{nSC\},SC}} \mu_{SC,w,j} W_{SC}\left(w, z^{(k)}\right) - W_{SC}\left(w^{(k)}, z^{(k)}\right) \right) P_{max,out,SC} \qquad a)$$
$$P_{Interf\{nSC\},j}^{(k)} = \sum_{m=0}^{N_{MC}} \sum_{p=1}^{3} \left(\mu_{MC,m,p,j} W_{MC}\left(m, p, z^{(k)}\right) - W_{MC}\left(m^{(k)}, p^{(k)}, z^{(k)}\right) \right) P_{max,out,MC} +$$
$$+ \sum_{w=1}^{N_{\{nSC\},SC}} \mu_{SC,w,j} W_{SC}\left(w, z^{(k)}\right) P_{max,out,SC} \qquad b) \tag{6.28}$$

In this equation, the form a) is used when the k th user is served by a the $w^{(k)}$ th small cell and the form b) is used if user k is served by the $p^{(k)}$ th sector of the $m^{(k)}$ th macro base station. Also, N_{MC} is the number of macro cell base stations in the $\{nSC\}$, $N_{\{nSC\},SC}$ is the number of small cells in the $\{nSC\}$ joint network, $\mu_{MC,m,p,j}$ is the utilisation of sector p of macro base station m during epoch j, $\mu_{SC,w,j}$ is the utilisation of the w th small cell in $\{nSC\}$ during epoch j. The channel function $W_{MC}(m,p,z^{(k)})$ is for the p th sector of the m th macro base station in $\{nSC\}$ at the location $z^{(k)}$; similarly $W_{SC}(w,z^{(k)})$ is the channel function for the w th small cell base station in $\{nSC\}$ at location $z^{(k)}$. If user k is being served by a macro base station in $\{nSC\}$, then we use the indices $m^{(k)}, p^{(k)}$. If user k is being served by a small cell in $\{nSC\}$, we use index $w^{(k)}$. Also, we have for the users served by the $\{nSC\}$ joint network:

$$SINR_{\{nSC\}}^{(k)} = \frac{W_{SC}\left(w^{(k)}, z^{(k)}\right) P_{max,out,SC}}{P_{Interf\{nSC\}}^{(k)} + P_{RxN}^{(k)}} \qquad a)$$
$$SINR_{\{nSC\}}^{(k)} = \frac{W_{MC}\left(m^{(k)}, p^{(k)}, z^{(k)}\right) P_{max,out,MC}}{P_{Interf\{nSC\}}^{(k)} + P_{RxN}^{(k)}} \qquad b) \tag{6.29}$$

As before, the form a) is for the k th user being served by a small cell in $\{nSC\}$ and form b) is for the k th user being served by a macro cell in $\{nSC\}$.

The value of $SINR^{(k)}$ is mapped to $SE^{(k)}$ using an appropriate link curve as discussed in Section 4.6.1.2. Depending upon the technology (e.g. SISO, SIMO, MIMO) and the user's local environment under consideration, the modeller may be able to select one of the link curve examples presented in Figure 4.4 or seek a more appropriate alternative from the references listed in that section.

A user "leaves" the service area upon the completion of their download of the $B^{(usr)}$ sized file or when their download time has exceeded the $\Delta t_{max,dl}$ time limit. Statistics are compiled for the ensemble of small cells within the service area, for each of the N_{sim} runs of the small cell model. The file download rates could vary from small cell to small cell because of the small cell location, proximity to other small cells and the randomness in the active user numbers in each small cell. These factors affect the user's SINR, and hence the resulting Spectral Efficiency.

To develop generic results, the simulation components are re-run multiple, N_{sim}, times each with a different set of small cell locations, to provide a reasonably generic set of results. This entails implementing Step (2) for each re-run, to relocate the small cell base stations for both the $\{SC\}$ and $\{nSC\}$ sub-networks.

6.4   COLLECTION AND ANALYSIS OF SIMULATION RESULTS

With the separation of the heterogeneous network simulation into two independent components, the collection of results can be likewise undertaken separately. However, the final analysis is done for the total heterogeneous network. Therefore, the two sets of results need to be combined in a manner that reflects the overall user demand in the heterogeneous network.

The simulation components can be run N_{sim} times with the same number of small cells, N_{SC}, randomly re-located within macro cell sectors as described in Section 6.2.1. The locations of the macro base stations do not change. The results are collected over all the runs as tabulated as in Tables 5.2 and 5.3. Because each run has a different set of locations for the N_{SC} small cells, there will be a spread in the results.

Setting a higher number of runs N_{sim} will reduce this spread. However the spread needs to be taken into account when presenting and interpreting the results.

6.4.1   Simulation Component Results

There are multiple parameters over which the sets of the simulations' output data can be averaged. These include averages over users, k, across base stations in the $\{SC\}$ and $\{nSC\}$ networks, simulation runs N_{sim} and simulation epochs in the u th simulation run, $N_{Tot,u}$. To assist with reading the equations, we provide a brief review of the notation used for the averaging. With each averaging, the notation will indicate the parameter over which the average has been calculated.

Averages over the users are indicated with the form $\langle Y \rangle^{(usr)}$ where Y is the quantity averaged and the superscript $^{(usr)}$ outside the averaging angular brackets, $\langle . \rangle$, indicates the average is over the users, $k = 1, 2, ..., N^{(usr)}$.

An average of quantity Y over the number the base stations in $\{SC\}$ will be denoted as $\langle Y \rangle_{\{SC\}}$ with the subscript $_{\{SC\}}$ indicating the average is over $q = 1, 2, ..., N_{\{SC\}}$.

Similarly, averaging over the base stations in the $\{nSC\}$ joint network is denoted with $\langle Y \rangle_{\{nSC\}}$ where the average is over $q = 1, 2, ..., N_{\{nSC\}}$, where $N_{\{nSC\}}$ is the number of base stations in the $\{nSC\}$ joint network.

An average of quantity Y over the N_{sim} simulation runs is denoted $\langle Y \rangle_{sim}$ with the subscript $_{sim}$ outside the averaging angular brackets, indicating an average over $u = 1, 2, ..., N_{sim}$.

An average over a simulation run (i.e. over epoch index, $j = 1, 2, ..., N_{Tot}$) will be denoted $\langle Y \rangle_T$ where T represents the duration of the simulation. In cases where multiple averaging has been done, each will be indicated.

Any parameter that is not averaged over remains inside the angular brackets.

In addition to the averaging, we also need to indicate over which component the calculation is being undertaken. If the result, Y, is from the $\{SC\}$ network it will include a subscript $_{\{SC\}}$ inside the averaging brackets. Likewise, a result from the $\{nSC\}$ joint network will have a subscript $_{\{nSC\}}$ inside the averaging angle brackets. Finally, for the diurnal cycle averaged results, both $\{SC\}$ and $\{nSC\}$ components include simulations for the five stages of the diurnal cycle. When the stage number is relevant, the result will included a subscript S to denote the stage of the diurnal cycle. The stage subscript will be suppressed unless it is required for clarity.

For example, the average over users and base stations in $\{SC\}$ in the u th run of the small cell component is $\left\langle Y_{\{SC\},u} \right\rangle^{(usr)}_{\{SC\}}$. The $\{SC\}$ inside the averaging brackets indicating this applies to the $\{SC\}$ network, the u inside the averaging brackets meaning this average applies to a specific run of the $\{SC\}$ network simulation. The $^{(usr)}$ outside means this result has been averaged over the users and the $_{\{SC\}}$ outside the averaging brackets means this parameter has been averaged over all the base stations in the $\{SC\}$ network.

If that output were also averaged over all the N_{sim} simulation runs of the $\{SC\}$ component, it will then be denoted as $\left\langle Y_{\{SC\}} \right\rangle^{(usr)}_{\{SC\},sim}$, where the $_{sim}$ outside the averaging brackets indicates the average over the simulation runs.

Because the small cells are randomly located in the macro cell sectors, it is instructive to compare results averaged across the total number of Group 1 users, $N^{(usr)}_{Tot,\{SC\},u}$, for each $u = 1, 2, \ldots, N_{sim}$, of the $\{SC\}$ simulations. This gives the average:

$$\left\langle Y_{\{SC\},u} \right\rangle^{(usr)} = \frac{1}{N^{(usr)}_{Tot,\{SC\},u}} \sum_{k=1}^{N^{(usr)}_{Tot,\{SC\},u}} Y^{(k)}_{\{SC\},u} \tag{6.30}$$

Because the simulation involves a Poison random process, the total number of users in the simulation may differ between runs; hence, the user number $N^{(usr)}_{Tot,\{SC\},u}$ requires a u-index. Examples of this type of result are the average user SINR, $\langle SINR_{\{SC\},u} \rangle^{(usr)}$, and throughput, $\langle R_{\{SC\},u} \rangle^{(usr)}$, for each of eight runs of the $\{SC\}$ component as shown in Figure 6.6a and b.

These results are from a simulation of a Dense Urban geographical area for offload ratio $\varphi_{SC} = 0.67$. For the plots in Figure 6.6, simulations were run for four small cells deployed per macro cell sector, each small cell with a small cell diameter $SCD = 50$ m. The small cell locations in each sector were randomly placed with an inter-cell distance $(ISCD)$ greater than or equal to 50 m. In this simulation series, the network component $\{SC\}$ consists of all and only the small cells and the component $\{nSC\}$ consists of all and only the macro cells in the heterogeneous network.

Averaging over the N_{sim} runs give the overall averaged results, $\left\langle Y_{\{SC\}} \right\rangle^{(usr)}_{sim}$, for the $\{SC\}$ network component of the network simulation, we have:

$$\left\langle Y_{\{SC\}} \right\rangle^{(usr)}_{sim} = \frac{1}{N_{sim}} \sum_{u=1}^{N_{sim}} \left\langle Y_{\{SC\},u} \right\rangle^{(usr)} = \frac{1}{N_{sim}} \sum_{u=1}^{N_{sim}} \frac{1}{N^{(usr)}_{Tot,\{SC\},u}} \sum_{k=1}^{N^{(k)}_{Tot,\{SC\},u}} Y^{(k)}_{\{SC\},u} \tag{6.31}$$

The macro cell network component $\{nSC\}$ does not include any small cells and therefore that component is only run once and we have:

$$\left\langle Y_{\{nSC\}} \right\rangle^{(usr)} = \frac{1}{N^{(usr)}_{Tot,\{nSC\}}} \sum_{k=1}^{N^{(usr)}_{Tot,\{nSC\}}} Y^{(k)}_{\{nSC\}} \tag{6.32}$$

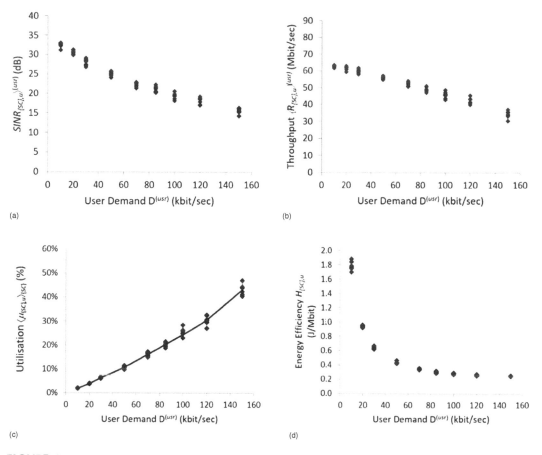

FIGURE 6.6   Example of spread of results of $N_{sim} = 8$ simulation runs with different small cell locations within the macro cell network with fixed inter-small cell distance, ISCD, and number of small cells, N_{SC}.

Some results are not derived on a per-user basis. For example, the base station utilisations, $\langle \mu_{\{SC\},u} \rangle_{\{SC\}}$, in Figure 6.6c is an average over the total number of small cells. For example, $\{SC\}$ consists of the three sectors of the inner seven base stations, then the number of small cells in $\{SC\}$ is $N_{\{SC\}} = 7 \times 3 \times N_{SC}$. That is:

$$\left\langle \mu_{\{SC\},u} \right\rangle_{\{SC\},T} = \frac{1}{N_{\{SC\}}} \sum_{q=1}^{N_{\{SC\}}} \left\langle \mu_{\{SC\},q,u} \right\rangle_T \tag{6.33}$$

where $\langle \mu_{\{SC\},q,u} \rangle_T$

$$\left\langle \mu_{\{SC\},q,u} \right\rangle_T = \frac{1}{N_{Tot,\{SC\},u}} \sum_{j=1}^{N_{Tot,\{SC\},u}} \mu_{\{SC\},q,u,j} \tag{6.34}$$

is the average utilisation of the q th small cell in the u th run over the $\{SC\}$ simulation duration defined in (5.41). The duration of the u th simulation run is $T_{\{SC\},u} = N_{Tot,\{SC\},u} \Delta t_{\{SC\}}$.

Also plotted in Figure 6.6c is the trend line for the average utilisation, $\langle \mu_{\{SC\}} \rangle_{\{SC\},sim}$, over both small cells, $q = 1, 2, \ldots, N_{\{SC\}} = N_{SC}$ (in this case) and simulation runs, $u = 1, 2, \ldots, N_{sim}$, given by:

$$\left\langle \mu_{\{SC\}} \right\rangle_{\{SC\},sim,T} = \frac{1}{N_{sim}} \sum_{u=1}^{N_{sim}} \left\langle \mu_{\{SC\},u} \right\rangle_{\{SC\},T} \tag{6.35}$$

The energy efficiencies of the $\{SC\}$ and $\{nSC\}$ networks for each u th simulation run are given by:

$$H_{\{SC\},u} = \sum_{q=1}^{N_{\{SC\},SC}} Q_{\{SC\},q,u} \left/ \sum_{q=1}^{N_{\{SC\},SC}} B_{\{SC\},q,u} = \sum_{q=1}^{N_{\{SC\},SC}} P_{\{SC\},q,u} \right/ R_{\{SC\},u}$$

$$H_{\{nSC\},u} = \left(\sum_{m=1}^{N_{MC}} Q_{MC,m,u} + \sum_{q=1}^{N_{\{nSC\},SC}} Q_{\{nSC\},q,u} \right) \left/ \left(\sum_{m=1}^{N_{MC}} B_{MC,m,u} + \sum_{q=1}^{N_{\{nSC\},SC}} B_{\{nSC\},q,u} \right) \right. \qquad (6.36)$$

$$= \left(\sum_{m=1}^{N_{MC}} P_{MC,m,u} + \sum_{q=1}^{N_{\{nSC\},SC}} P_{\{nSC\},q,u} \right) \left/ R_{\{nSC\},u} \right.$$

In this, $Q_{\{SC\},q,u}$ and $P_{\{SC\},q,u}$ are the energy and power consumption of the q th small cell in the $\{SC\}$ network for the u th simulation run, respectively. $B_{\{SC\},q,u}$ is the total amount of data downloaded by the q th base station in $\{SC\}$ for the u th simulation run. $R_{\{SC\},u}$ is the total download throughput of the $\{SC\}$ network in the u th simulation run. We have $P_{\{SC\},u} = Q_{\{SC\},u}/\Delta t_{\{SC\}}$, and $R_{\{SC\},u} = \Sigma_q B_{\{SC\},q,u}/\Delta t_{\{SC\}}$ the total of all the downloaded data across $\{SC\}$ divided by the epoch interval $\Delta t_{\{SC\}}$.

Similarly, $Q_{MC,m,u}$ and $P_{MC,m,u}$ are the energy consumption of the m th macro base station in $\{nSC\}$. Also $Q_{\{nSC\},q,u}$ and $P_{\{nSC\},q,u}$ are the energy consumption of the q th small cell base station in $\{nSC\}$. In these summations, N_{MC} is the number of macro cells and $N_{\{nSC\},SC}$ the number of small cell base stations in $\{nSC\}$. We have $N_{MC} + N_{\{nSC\},SC} = N_{\{nSC\}}$. Next, $B_{\{nSC\},m,u}$ and $B_{MC,q,u}$ are the amount of data downloaded by the m th macro cell and q th small cell in run u of the simulation, respectively. Finally $R_{\{nSC\},u}$ is the total download throughput of the $\{nSC\}$ joint network in the u th simulation run. $P_{\{nSC\},u} = Q_{\{nSC\},u}/\Delta t_{\{nSC\}}$ and $R_{\{nSC\},u} = \Sigma_q B_{\{nSC\},q,u}/\Delta t_{\{nSC\}}$. Figure 6.6d shows the energy efficiency $H_{\{SC\},u}$ for $N_{sim} = 8$ simulation runs for the Dense Urban simulation described earlier.

Figure 6.6a and b shows the results $\langle SINR_{\{SC\},u} \rangle^{(usr)}$ and the user-achieved service rate $\langle R_{\{SC\},u} \rangle^{(usr)}$ respectively, for each of the N_{sim} simulation runs each with a different random cell placement. Figure 6.6c shows the total small cell network utilisation per macro cell and Figure 6.6d shows the small cell energy efficiency $H_{\{SC\},u}$ for each of the N_{sim} runs. These results have been generated with 4 small cells and 1,000 service requests. The reduced number of service requests was chosen to demonstrate the potential spread in the results across the eight simulation runs if an insufficient service request number was selected. For the usual service request count of around 16,000, as used in generating results shown later in this chapter, a much smaller spread would be seen. Viewing the results in Figure 6.6a–c, it can be seen that for each value of $D^{(usr)}$ along the x-axis, there are several values for results plotted on the y-axis. As explained in Section 6.2.1, the small cell locations are randomly redistributed for each of the N_{sim} runs of the small cell network simulation. The spread arises because, although the number of small cells in each sector, N_{SC}, is constant, the users are randomly located in small cells in each run, and the locations these small cells also randomly change for each of the N_{sim} runs. Consequentially each of the N_{sim} runs will provide a different averaged value. Figure 6.6 shows that the results of the 1,000 service requests are reasonably closely clustered, for a given network configuration and the results for 16,000 are even more closely clustered. From these results, we can conclude the small cell location does not significantly change the result $Y^{(usr)}$. Based on these results, we can have confidence that the small cell runs provide generally indicative results.

6.5 CALCULATING DIURNAL CYCLE RESULTS

To cover the diurnal cycle, the small cell network simulations are repeated for each stage, S, of the diurnal cycle. For these stages the request rate, $\lambda_{\{SC\}}$, is rescaled by a value X_S to replicate the stages of the diurnal cycle as shown in Figure 6.7. Therefore, for stage S of duration T_S we have $\lambda_{\{SC\},S} = X_S \lambda_{\{SC\}}$

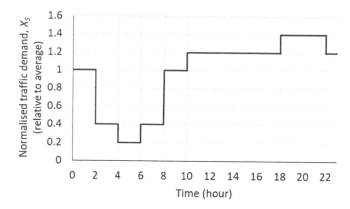

FIGURE 6.7 Example diurnal cycles for mobile network power modelling. This cycle is based upon a Dense Urban area with the traffic scaled relative to the mean traffic over the diurnal cycle. This is a step-wise approximation of the measured cycle reported in [6].

and $\Delta t_{\{SC\},S} = C/\lambda_{\{SC\}}$ with C a constant set by the condition (6.20). Similarly, for the stages of the $\{nSC\}$ component, $\lambda_{\{nSC\},S} = X_S\lambda_{\{nSC\}}$ and $\Delta t_{\{nSC\},S} = C/\lambda_{\{nSC\}}$

6.5.1 DIURNAL CYCLE RESULTS

The components involve data from N_{sim} simulations for each of the diurnal cycle stages. The data from these multiple runs is required in calculating the diurnal cycle results. We recall that for all the simulations, the data is collected during the simulation steady state. Therefore, following the approach in Section 5.7.1, we can use average results $Y_{\{SC\},S,j}$ and $Y_{\{nSC\},S,j}$ from the simulations over various N_{Tot} epochs to approximate the summations for each stage of the diurnal cycle. That is using the fact that for a steady state $\langle Y \rangle_T \approx \langle Y \rangle_{TS}$ we have:

$$\sum_{j=1}^{N_{Tot}} Y_j = N_{Tot} \langle Y \rangle_{Ts} \approx N_{Tot} \langle Y \rangle_T \tag{6.37}$$

In this Y_j represents the recorded data at each epoch (j index) for the stage S of the diurnal cycle for the small cell component. The average $\langle Y \rangle_T$ comes from the simulation over N_{Tot} epochs and is used to approximate the average over duration T_S of stage S of the diurnal cycle. This is used to approximate the j-sum of the $Y_{\{SC\},S,j}$ and $Y_{\{nSC\},S,j}$ that we use to combine the results of the multiple simulations, the diurnal cycle stages and, ultimately, the heterogeneous networks.

This approximation can be applied to the combined data from all the N_{sim} simulation results of each network for stage S of the diurnal cycle. In this case $\langle Y_{\{SC\},S} \rangle_T$ and $\langle Y_{\{nSC\},S} \rangle_T$ are given by:

$$\langle Y_{\{A\},S} \rangle_T = \langle Y_{\{A\},S} \rangle_{T,sim} \approx \frac{1}{N_{sim}} \sum_{u=1}^{N_{sim}} \frac{1}{N_{Tot,\{A\},S,u}} \sum_{j=1}^{N_{Tot,\{A\},S,u}} Y_{\{A\},S,u,j} \tag{6.38}$$

where $\{A\}$ stands for $\{SC\}$ or $\{nSC\}$ for each of the two network components. In this $Y_{\{A\},S,u,j}$ is the result for epoch j in run u for stage S of the diurnal cycle for the A network component. $N_{Tot,\{A\},S,u}$ is the number of epochs for which data was collected in the u th run of the A component simulation corresponding to diurnal stage S. All N_{sim} simulation runs for stage S of the diurnal cycle use the same epoch duration, $\Delta t_{\{A\},S}$; hence they have the same value of $\Delta t_{\{A\},S} = C/X_S\lambda_{\{A\}} = C/\varphi_{\{A\}}X_S\lambda$

(where $\lambda = A_{SA} U^{(usr)} D^{(usr)} / B^{(usr)}$) and the duration of stage S is given by $\rho_S \times 24$ hours. Therefore, using (5.55) we have for the average over the full diurnal cycle, $\langle Y_{\{A\}} \rangle_D$:

$$\left\langle Y_{\{A\}} \right\rangle_D \approx \sum_{S=1}^{5} \varphi_{\{A\}} \rho_S X_S \left\langle Y_{\{A\},S} \right\rangle_T \Bigg/ \sum_{S=1}^{5} \varphi_{\{A\}} \rho_S X_S = \sum_{S=1}^{5} \rho_S X_S \left\langle Y_{\{A\},S} \right\rangle_T \tag{6.39}$$

With this we can use the corresponding forms: (5.61) for the average user throughput, (5.60) for SINR, (5.62) for base station utilisation, (5.63) for proportion of download failures and (5.68) or (5.72) for energy efficiency of each of the small cell component of the simulation. Also we have used:

$$\sum_{S=1}^{5} \rho_S X_S = 1 \tag{6.40}$$

6.6 HETEROGENEOUS NETWORK RESULTS

Two sets of results for the heterogeneous network can be calculated. The first set presents the relationship between the simulation outputs and the overall user demand, $D^{(usr)}$. This set provides plots of the form shown in Figure 6.6 in which the overall heterogeneous network performance parameters are displayed as functions of $D^{(usr)}$, with $N_{\{SC\}}^{(usr)}$ users allocated to $\{SC\}$ network and $N_{\{nSC\}}^{(usr)}$ users allocated to the $\{nSC\}$ joint network as given by (6.15). This is equivalent to implementing a service request rate $\lambda_{\{SC\}}$, given by (6.12) for the $\{SC\}$ network and service request rate $\lambda_{\{nSC\}}$, given by (6.13) for the $\{nSC\}$ joint network. In this case, the results are typically averaged over the users or averaged over time (equivalently, epochs).

The second set gives the average values over a diurnal cycle accounting for the different stage durations of the diurnal cycle. In this case, all the results are averaged over all the diurnal cycle.

Following from Chapter 5, the network results presented in this chapter are

(a) Average throughput per user, $\langle R \rangle^{(usr)}$
(b) Average user SINR, $\langle SINR \rangle^{(usr)}$
(c) Average base station utilisation, $\langle \mu_{BS} \rangle$
(d) Proportion of failed downloads, F_{dl}
(e) Energy efficiency, H

6.6.1 USER DEMAND, $D^{(USR)}$ RESULTS

The details of combining the small cell and macro cell results to show the heterogeneous network results dependence on the user demand are shown in the Appendix. We have:

$$\left\langle Y_{Het} \left(D^{(usr)} \right) \right\rangle_T \approx \varphi_{\{SC\}} \left\langle Y_{\{SC\}} \left(\varphi_{\{SC\}} D^{(usr)} \right) \right\rangle_T + \left(1 - \varphi_{\{SC\}} \right) \left\langle Y_{\{nSC\}} \left(\left(1 - \varphi_{\{SC\}} \right) D^{(usr)} \right) \right\rangle_T \tag{6.41}$$

In this we have shown the explicit dependence on $D^{(usr)}$. The value of $Y_{\{SC\}}(\varphi_{\{SC\}} D^{(usr)})$ is given by running the $\{SC\}$ simulation component for actual $\{SC\}$ user demand $\varphi_{\{SC\}} D^{(usr)}$ not the overall service user demand $D^{(usr)}$. Similarly, $Y_{\{nSC\}}((1-\varphi_{\{SC\}}) D^{(usr)})$ is given by running the $\{nSC\}$ simulation component for the actual $\{nSC\}$ user demand $(1-\varphi_{\{SC\}}) D^{(usr)}$. By combining these resulting $Y_{\{SC\}}$ and $Y_{\{nSC\}}$ according to (6.41), we get the result for the heterogeneous network for overall network user demand $D^{(usr)}$. Likewise:

$$\left\langle Y_{Het} \left(D^{(usr)} \right) \right\rangle_T \Bigg/ \left\langle Z_{Het} \left(D^{(usr)} \right) \right\rangle_T \approx$$

$$\frac{\varphi_{\{SC\}} \left\langle Y_{\{SC\}} \left(\varphi_{\{SC\}} D^{(usr)} \right) \right\rangle_T + \left(1 - \varphi_{\{SC\}} \right) \left\langle Y_{\{nSC\}} \left(\left(1 - \varphi_{\{SC\}} \right) D^{(usr)} \right) \right\rangle_T}{\varphi_{\{SC\}} \left\langle Z_{\{SC\}} \left(\varphi_{\{SC\}} D^{(usr)} \right) \right\rangle_T + \left(1 - \varphi_{\{SC\}} \right) \left\langle Z_{\{nSC\}} \left(\left(1 - \varphi_{\{SC\}} \right) D^{(usr)} \right) \right\rangle_T} \tag{6.42}$$

$$\left\langle \left(Y\left(D^{(usr)}\right) \Big/ Z\left(D^{(usr)}\right)\right) \right\rangle_{Het} \Big/_T \approx \varphi_{\{SC\}} \left\langle \left(Y\left(\varphi_{\{SC\}}D^{(usr)}\right) \Big/ Z\left(\varphi_{\{SC\}}D^{(usr)}\right)\right) \right\rangle_{\{SC\}} \Big/_T$$
$$+\left(1-\varphi_{\{SC\}}\right)\left\langle \left(Y\left(\left(1-\varphi_{\{SC\}}\right)D^{(usr)}\right) \Big/ Z\left(\left(1-\varphi_{\{SC\}}\right)D^{(usr)}\right)\right) \right\rangle_{\{nSC\}} \Big/_T \tag{6.43}$$

Using these forms, we can construct plots for the heterogeneous simulation results showing the dependence on overall network demand, $D^{(usr)}$.

6.6.2 Diurnal Cycle Results

We now bring together the small cell and macro cell network components to derive the overall heterogeneous network results averaged over the full diurnal cycle. The details of this calculation are in the Appendix. Denoting the average over a diurnal cycle with $\langle Y \rangle_D$ we have:

$$\langle Y_{Het} \rangle_D \approx \varphi_{\{SC\}}\langle Y_{\{SC\}} \rangle_D + \left(1-\varphi_{\{SC\}}\right)\langle Y_{\{nSC\}} \rangle_D \tag{6.44}$$

$$\langle Y_{Het} \rangle_D \Big/ \langle Z_{Het} \rangle_D \approx \frac{\varphi_{\{SC\}}\langle Y_{\{SC\}} \rangle_D + \left(1-\varphi_{\{SC\}}\right)\langle Y_{\{nSC\}} \rangle_D}{\varphi_{\{SC\}}\langle Z_{\{SC\}} \rangle_D + \left(1-\varphi_{\{SC\}}\right)\langle Z_{\{nSC\}} \rangle_D} \tag{6.45}$$

$$\langle \left(Y/Z\right)_{Het} \rangle_D \approx \varphi_{\{SC\}}\langle \left(Y/Z\right)_{\{SC\}} \rangle_D + \left(1-\varphi_{\{SC\}}\right)\langle \left(Y/Z\right)_{\{nSC\}} \rangle_D \tag{6.46}$$

In Chapter 5, this list included the "Average Base Station Throughput". For the heterogeneous network, the average overall base stations may not be appropriate because of the significant difference between the small cell and macro cell base station maximum capacities and the implementation of offloading. The forms for calculating each of the earlier mentioned five results a) to e) will be described.

6.6.3 Average Throughput per User

As shown in the Appendix, we have:

$$\left\langle R_{Het}\left(D^{(usr)}\right) \right\rangle^{(usr)} \approx$$
$$\varphi_{\{SC\}}\left\langle R_{\{SC\}}\left(\varphi_{\{SC\}}D^{(usr)}\right) \right\rangle^{(usr)} + \left(1-\varphi_{\{SC\}}\right)\left\langle R_{\{nSC\}}\left(\left(1-\varphi_{\{SC\}}\right)D^{(usr)}\right) \right\rangle^{(usr)} \tag{6.47}$$

$$\left\langle R_{Het} \right\rangle_D^{(usr)} \approx \varphi_{\{SC\}}\left\langle R_{\{SC\}} \right\rangle_D^{(usr)} + \left(1-\varphi_{\{SC\}}\right)\left\langle R_{\{nSC\}} \right\rangle_D^{(usr)} \tag{6.48}$$

In this we have used (6.31) with $\langle R_{\{A\}} \rangle^{(usr)} = \langle R_{\{A\}} \rangle_{sim}^{(usr)}$.

6.6.4 Average User SINR

Similarly for the SINR:

$$\left\langle SINR_{Het}\left(D^{(usr)}\right) \right\rangle^{(usr)}$$
$$\approx \varphi_{\{SC\}}\left\langle SINR_{\{SC\}}\left(\varphi_{\{SC\}}D^{(usr)}\right) \right\rangle^{(usr)} + \left(1-\varphi_{\{SC\}}\right)\left\langle SINR_{\{nSC\}}\left(\left(1-\varphi_{\{SC\}}\right)D^{(usr)}\right) \right\rangle^{(usr)} \tag{6.49}$$

$$\left\langle SINR_{Het} \right\rangle_D^{(usr)} \approx \varphi_{\{SC\}}\left\langle SINR_{\{SC\}} \right\rangle_D^{(usr)} + \left(1-\varphi_{\{SC\}}\right)\left\langle SINR_{\{nSC\}} \right\rangle_D^{(usr)} \tag{6.50}$$

In this we have used (6.31) with $\langle SINR_{\{A\}} \rangle^{(usr)} = \langle SINR_{\{A\}} \rangle_{sim}^{(usr)}$.

6.6.5 AVERAGE BASE STATION UTILISATION

Using (6.44):

$$\left\langle \mu_{Het}\left(D^{(usr)}\right)\right\rangle_T \approx \varphi_{\{SC\}}\left\langle \mu_{\{SC\}}\left(\varphi_{\{SC\}}D^{(usr)}\right)\right\rangle_T + \left(1-\varphi_{\{SC\}}\right)\left\langle \mu_{\{nSC\}}\left(\left(1-\varphi_{\{SC\}}\right)D^{(usr)}\right)\right\rangle_T \qquad (6.51)$$

$$\left\langle \mu_{Het}\right\rangle_D \approx \varphi_{\{SC\}}\left\langle \mu_{\{SC\}}\right\rangle_D + \left(1-\varphi_{\{SC\}}\right)\left\langle \mu_{\{nSC\}}\right\rangle_D \qquad (6.52)$$

In this, we have set $\langle\mu_{\{A\}}\rangle = \langle\mu_{\{A\}}\rangle_{\{A\},sim,T}$ from (6.35).

6.6.6 PROPORTION OF FAILED DOWNLOADS

Using the same approach as in Section 5.7.5, we have:

$$F_{Het,Prop}\left(D^{(usr)}\right) = \frac{N_{\{SC\},fail}\left(\varphi_{\{SC\}}D^{(usr)}\right) + N_{\{nSC\},fail}\left(\left(1-\varphi_{\{SC\}}\right)D^{(usr)}\right)}{N_{\{SC\}}^{(usr)} + N_{\{nSC\}}^{(usr)}}$$

$$\approx \varphi_{SC}\left\langle N_{\{SC\},fail}\left(\varphi_{\{SC\}}D^{(usr)}\right)\right\rangle_T + \left(1-\varphi_{\{SC\}}\right)\left\langle N_{\{nSC\},fail}\left(\left(1-\varphi_{\{SC\}}\right)D^{(usr)}\right)\right\rangle_T \qquad (6.53)$$

Similarly for the diurnal cycle:

$$F_{Het,Prop,D} = \frac{\displaystyle\sum_{S=1}^{5} N_{\{SC\},fail,S} + N_{\{nSC\},fail,S}}{\displaystyle\sum_{S=1}^{5} N_{\{SC\},S}^{(usr)} + N_{\{nSC\},S}^{(usr)}}$$

$$\approx \frac{\displaystyle\sum_{S=1}^{5}\left(\varphi_{\{SC\}}\rho_S X_S \left\langle N_{\{SC\},fail}\right\rangle_T + \left(1-\varphi_{\{SC\}}\right)\rho_S X_S \left\langle N_{\{nSC\},fail}\right\rangle_T\right)}{\displaystyle\sum_{S=1}^{5}\rho_S X_S} \qquad (6.54)$$

$$= \varphi_{\{SC\}}F_{\{SC\},Prop,D} + \left(1-\varphi_{\{SC\}}\right)F_{\{nSC\},Prop,D}$$

6.6.7 ENERGY EFFICIENCY

Using (6.45) for both the forms $H = \langle Q\rangle_T/\langle B\rangle_T = \langle P\rangle_T/\langle R\rangle_T$ where $\langle Y\rangle_T$ is the average of quantity Y per epoch averaged over all the epochs in time duration T:

$$H_{Het}\left(D^{(usr)}\right) = \left\langle Q_{Het}\left(D^{(usr)}\right)\right\rangle_T \Big/ \left\langle B_{Het}\left(D^{(usr)}\right)\right\rangle_T$$

$$\approx \frac{\varphi_{\{SC\}}\left\langle Q_{\{SC\}}\left(\varphi_{\{SC\}}D^{(usr)}\right)\right\rangle_T + \left(1-\varphi_{\{SC\}}\right)\left\langle Q_{\{nSC\}}\left(\left(1-\varphi_{\{SC\}}\right)D^{(usr)}\right)\right\rangle_T}{\varphi_{\{SC\}}\left\langle B_{\{SC\}}\left(\varphi_{\{SC\}}D^{(usr)}\right)\right\rangle_T + \left(1-\varphi_{\{SC\}}\right)\left\langle B_{\{nSC\}}\left(\left(1-\varphi_{\{SC\}}\right)D^{(usr)}\right)\right\rangle_T}$$

$$H_{Het}\left(D^{(usr)}\right) = \left\langle P_{Het}\left(D^{(usr)}\right)\right\rangle_T \Big/ \left\langle R_{Het}\left(D^{(usr)}\right)\right\rangle_T \qquad (6.55)$$

$$\approx \frac{\varphi_{\{SC\}}\left\langle P_{\{SC\}}\left(\varphi_{\{SC\}}D^{(usr)}\right)\right\rangle_T + \left(1-\varphi_{\{SC\}}\right)\left\langle P_{\{nSC\}}\left(\left(1-\varphi_{\{SC\}}\right)D^{(usr)}\right)\right\rangle_T}{\varphi_{\{SC\}}\left\langle R_{\{SC\}}\left(\varphi_{\{SC\}}D^{(usr)}\right)\right\rangle_T + \left(1-\varphi_{\{SC\}}\right)\left\langle R_{\{nSC\}}\left(\left(1-\varphi_{\{SC\}}\right)D^{(usr)}\right)\right\rangle_T}$$

$$H_{Het,D} = \langle Q_{Het} \rangle_D / \langle B_{Het} \rangle_D \approx \frac{\varphi_{\{SC\}} \langle Q_{\{SC\}} \rangle_D + \left(1 - \varphi_{\{SC\}}\right) \langle Q_{\{nSC\}} \rangle_D}{\varphi_{\{SC\}} \langle B_{\{SC\}} \rangle_D + \left(1 - \varphi_{\{SC\}}\right) \langle B_{\{nSC\}} \rangle_D}$$

$$H_{Het,D} = \langle P_{Het} \rangle_D / \langle R_{Het} \rangle_D \approx \frac{\varphi_{\{SC\}} \langle P_{\{SC\}} \rangle_D + \left(1 - \varphi_{\{SC\}}\right) \langle P_{\{nSC\}} \rangle_D}{\varphi_{\{SC\}} \langle R_{\{SC\}} \rangle_D + \left(1 - \varphi_{\{SC\}}\right) \langle R_{\{nSC\}} \rangle_D} \tag{6.56}$$

In some cases, it is preferable to express the energy efficiency in terms of the energy efficiencies of the small and macro cell networks and corresponding download throughputs $\langle R_{\{SC\}} \rangle$ and $\langle R_{\{nSC\}} \rangle$. For this approach, we have:

$$
\begin{aligned}
H_{Het,D} &= \frac{\langle P_{Het} \rangle_D}{\langle R_{Het} \rangle_D} \approx \frac{\varphi_{\{SC\}} \langle P_{\{SC\}} \rangle_D + \left(1 - \varphi_{\{SC\}}\right) \langle P_{\{nSC\}} \rangle_D}{\langle R_{Het} \rangle_D} \\
&= \frac{\varphi_{SC} \sum_{S=1}^{5} H_{\{SC\},S} \rho_S X_S \langle R_{\{SC\},S} \rangle_T + \left(1 - \varphi_{\{SC\}}\right) \sum_{S=1}^{5} H_{\{nSC\},S} \rho_S X_S \langle R_{\{nSC\},S} \rangle_T}{\langle R_{Het} \rangle_D}
\end{aligned}
\tag{6.57}
$$

In this $\langle R_{Het} \rangle_D$ is given by (6.44):

$$\langle R_{Het} \rangle_D \approx \varphi_{\{SC\}} \langle R_{\{SC\}} \rangle_D + \left(1 - \varphi_{\{SC\}}\right) \langle R_{\{nSC\}} \rangle_D \tag{6.58}$$

Viewing the terms in the summations over S, we see that the term $\rho_S X_S \langle R_{\{SC\},S} \rangle$ represents the amount of data downloaded to users in the $\{SC\}$ component of the simulation during stage S of the diurnal cycle. The term $\rho_S X_S \langle R_{\{nSC\},S} \rangle$ is the corresponding quantity for the $\{nSC\}$ component. Therefore, the quotient $\rho_S X_S \langle R_{\{A\},S} \rangle / \langle R_{Het} \rangle_D$ is the proportion of total data downloaded to users in $\{A\}$ network component during stage S of the diurnal cycle. We can re-write (6.57) in the form:

$$H_{Het,D} \approx \varphi_{\{SC\}} \sum_{S=1}^{5} H_{\{SC\},S} \frac{\rho_S X_S \langle R_{\{SC\},S} \rangle_T}{\langle R_{Het} \rangle_D} + \left(1 - \varphi_{\{SC\}}\right) \sum_{S=1}^{5} H_{\{nSC\},S} \frac{\rho_S X_S \langle R_{\{nSC\},S} \rangle_T}{\langle R_{Het} \rangle_D} \tag{6.59}$$

From this we see that the overall energy efficiency of the heterogeneous network is, as expected, somewhat dependent upon the relative portions total network traffic downloaded by the small cell and macro cell networks over the diurnal cycle.

6.7   HETEROGENEOUS NETWORK SIMULATION

Using the approach described earlier, a heterogeneous network simulation was constructed. In this simulation, all the small cells are in the $\{SC\}$ sub-network (hence $\varphi_{\{SC\}} = \varphi$) and all the macro cells in the $\{nSC\}$ sub-network. That is, the overall network consists of a small cell sub-network that operates on frequencies that do not overlap with the frequencies used in the macro cell sub-network.

TABLE 6.1

Parameter Values for Advanced Model of a Dense Urban Mobile Simulation Using Only Macro Cells

Area Size and Traffic Demand (Dense Urban)

Inter-site distance of macro cell base stations, ISD (km)	0.5
Total Service Area, A_{SA} (km^2)	6.50
User download file size, $B^{(usr)}$ (Mbyte)	2
Area of macro base station hexagonal cell, A_{MC} (km^2)	0.22
Dense Urban population density (users/km^2) ($U_{BS}^{(usr)}$)	10,000
Average download per inhabitant per month, $GPM^{(usr)}$ (GB/month)	16.3
Average download demand per user, $D^{(usr)}$ (kbit/s)	50.31
Average service request rate over total service area, λ (requests/s))	204.38

TABLE 6.2

Example Parameter Values for Heterogeneous Network Based on Dense Urban Geography

Heterogeneous Network Overall Data

Small cell diameter, SCD (m)	100
Small cell offload ratio, $\varphi = \varphi_{\{SC\}}$	0.667
Total users located in a macro cell sector, $U^{(usr)}A_{MC}/3$	722
Users served by a macro base station sector, $(1-\varphi)U^{(usr)}A_{MC}/3$	240.4
Users served by small cells in a sector, $\varphi U^{(usr)}A_{MC}/3$	481.6
Average user demand in a macro cell sector, $U^{(usr)}A_{MC}D^{(usr)}/3$ (Mb/s)	36.31
Average user demand served by macro cell sector, $(1-\varphi)U^{(usr)}A_{MC}D^{(usr)}/3$ (Mb/s)	12.09
Average user demand served by small cells in a sector, $\varphi U^{(usr)}A_{MC}D^{(usr)}/3$ (Mb/s)	24.22
Average service request rate in a macro base station sector, $\lambda A_{MC}/3A_{SA}$ (requests/s)	2.27
Average service request rate served by macro base station sector, $(1-\varphi)\lambda A_{MC}/3A_{SA}$ (requests/s)	0.76
Average service request rate served by small cells in a sector, $(1-\varphi)\lambda A_{MC}/3A_{SA}$ (requests/s)	1.51

These values are calculated on a per macro cell basis and for offload ratios $\varphi = 0.67$. These are given to provide the reader with example values. All average values are averaged over a diurnal cycle.

The parameter values shown in Table 6.1 for the macro cell sub-network and Table 6.2 for the small cell sub-network for scenarios of $N_{SC} = 2$, 3 and 4 and off load ratio $\varphi_{SC} = 0.667$. The macro cell sub-network uses 2 × 2 MIMO and the small cell sub-network uses SISO.

For each of these configurations, the small cells use a single omnidirectional antenna and an output transmit power level of 1 W (30 dBm) when transmitting data, $\mu_{SC} = 1$, the idle transmit power level 0.25 W (24 dBm), due to the signalling overhead. This power does not contribute to interference as none of this power is transmitted concurrently with PDSCH REs occurs when $\mu_{SC} = 0$. The corresponding small cell electrical power consumption levels are $P_{SC}(\mu_{SC} = 1) = 6.5$ W and $P_{SC}(\mu_{SC} = 0) = 1.8$ W. These values are based on values quoted for a pico cell in [14], but adjusted to model a single channel SISO system, with a single power amplifier delivering 1W output power instead of two amplifiers delivering 0.5W each.

The available radio frequency spectrum is 20 MHz across the small cells and the entire spectrum is available to each small cell. It is assumed there is frame synchronism between cells, as for Advanced Model in Chapter 5. The frequency spectrum used by the macro cells does not overlap with that used by the small cells enabling the use of two independent simulation components. The link curves used to relate the SINR to the Spectral Efficiency are based on Morgensen *et al.* [12].

With the small cells randomly located within the macro cell sectors, collecting a representative set of data has the added complication that it must be run for a number, N_{sim}, times, each with a different set of randomly generated small cell locations. A typical number of repeat simulations is $N_{sim} = 8$. An example set of small cell locations within a set of macro cells is shown in Figure 6.4 in which $N_{SC} = 3$. The small cells are placed only in sectors of the central macro base station, BS_0, and surrounding ring of macro cell base stations, BS_1 to BS_6. Figure 6.4 shows that the coverage area of a small cell may overlap with its neighbours' and include regions beyond the inner cluster of macro cells. Overlapping occurs more frequently when number of small cells deployed, N_{SC}, or cell diameter, SCD, is increased or when the minimum inter-small cell distance, $ISDC$, is decreased. Because the small cells use the same radio frequency bands, this overlapping is likely to degrade average small cell user SINR.

To enable comparisons between various heterogeneous network configurations, several simulations were undertaken. These included small cell counts of $N_{SC} = 2$, 3 and 4 in each macro cell sector. The radius of coverage of each small cell is set to $SCD = 50$ m. The simulations also covered several values of minimum inter-small cell base station distance, $ISCD = 50$ m, 75 m or 100 m. With these variations, the total number of small cell simulation runs for each chosen traffic demand level was:

(number of N_{SC} values) × (number of I_{SCD} values) × $N_{sim} = 3 \times 3 \times 8 = 72$. The data recorded from these runs is collectively analysed.

6.7.1 SMALL CELL NETWORK COMPONENT RESULTS

This section will provide a more extensive analysis of small cell network performance characteristics to provide a better appreciation of how key parameters, including the traffic demand, number cells deployed within a sector and minimum inter-cell spacing, affect energy efficiency. These results focus on the small cell sub-network component of a heterogeneous network and hence assume the small cell network is dealing with the off loaded traffic only. The x-axis shows the average demand per user and this is the same regardless whether these users are in the {SC} or {nSC} networks. The reason for this approach to plotting the small cell results is to readily compare the performance of the small cell and macro cell component contributions of the overall heterogeneous network. Figure 6.12 shows the performance of the macro base station network after off-loading 66.7% of the normal Dense Urban traffic as complementary data series.

In the following discussions, the graphs were derived from simulations run for a minimum of $N_{sim} = 8$ different small-cell deployments, and for a minimum of $N_{Tot}^{(usr)} = 15,000$ user downloads in each case. The randomness in the cell location, user location and user activity can cause some variation in the results.

The plots show the following small cell results for each number of small cells per macro cell sector:

(a) Average throughput per user, $\langle R_{\{SC\}} \rangle_{sim}^{(usr)}$,

(b) Average user SINR, $\langle SINR_{\{SC\}} \rangle_{sim}^{(usr)}$,

(c) Average small cell utilisation, $\langle \mu_{\{SC\}} \rangle_{\{SC\},sim}$,

(d) Proportion of failed downloads, $F_{Prop, \{SC\}}$,

(e) Total Small Cell Base Station Throughput for each macro cell, $3N_{SC}\langle R \rangle_{\{SC\}}$,

(f) Average energy efficiency for each of the small cell spacing values, $\langle H_{\{SC\}} \rangle_{\{SC\},sim}$,

Item (e) in this list is the total throughput (bit/s) of all the small cells located in each macro cell area. We adopt this metric because it more closely reflects the offloading of users to the small cells rather than per small cell. Because small cells are introduced to reduce the load on the macro cell, the total throughput of the small cells provides a better measure of how the deployment of small cells assists with dealing with high user demand.

The a) plots in Figures 6.8 through 6.10 all show a significant reduction in average throughput per user, $\langle R_{\{SC\}} \rangle^{(usr)}_{\{SC\},sim}$ as user demand, $D^{(usr)}$, increases. The quantity $\langle R_{\{SC\}} \rangle^{(usr)}_{\{SC\},sim}$ is the average data rate achieved when servicing user requests. In contrast, as explained in Section 5.4.1, $D^{(usr)}$ is the average demand data rate representing user requests. The reduction in throughput as user demand increases arises because at low demand the network is lightly loaded. This results in a higher SINR (as shown in the b) plots), hence higher spectral efficiency. Therefore, the amount of

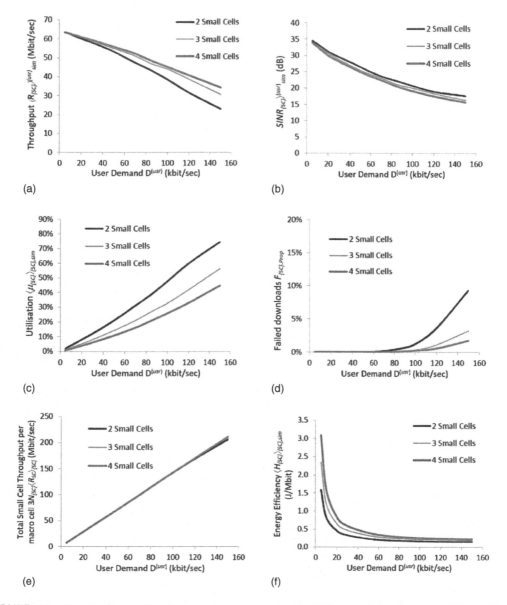

FIGURE 6.8 Results for small cell network component with SCD = 100 m, minimum ISCD = 50 m, $\varphi = 0.67$ with 2, 3 and 4 small cells per macro cell sector.

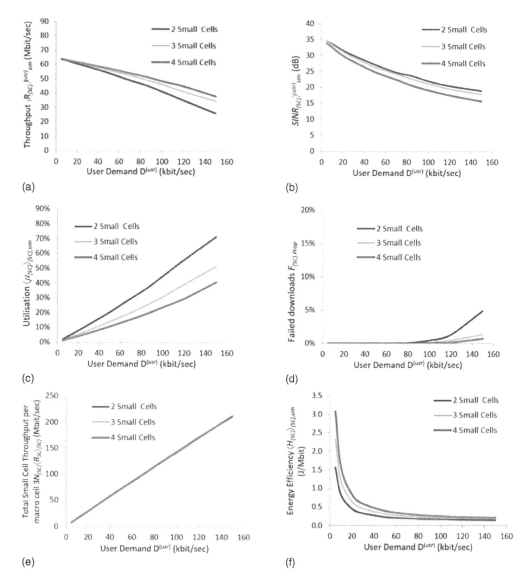

FIGURE 6.9 Results for small cell network component with SCD = 100 m, minimum ISCD = 75 m, $\varphi = 0.67$ with 2, 3 and 4 small cells per macro cell sector.

data per PDSCH RE is larger and the amount of data downloaded per TTI is greater. This in turn results in a higher average throughput in bits per second. At higher demand levels, the network is more heavily loaded resulting in a reduction in SINR, hence in spectral efficiency and therefore less data per PDSCH RE and a lower throughput.

The sharing of PDSCH REs across multiple users at higher traffic levels further reduces per-user throughput. With fewer small cells, the probability of cell capacity sharing is greater, hence simulations with two small cells show faster throughput degradation at increasing demand levels than four small cells. This reduction in capacity (two cell vs four cell) offsets the advantage in SINR that two-cell cases have over the four cell cases at high traffic levels. Further, when traffic demand is shared across a larger number of cells, individual cell utilisation is lower.

The b) plots in Figures 6.8 through 6.10 show that across all the small cell configurations, the SINR, $\left\langle SINR_{\{SC\}} \right\rangle_{sim}^{(usr)}$ decreases with increasing traffic demand. This is because of increased

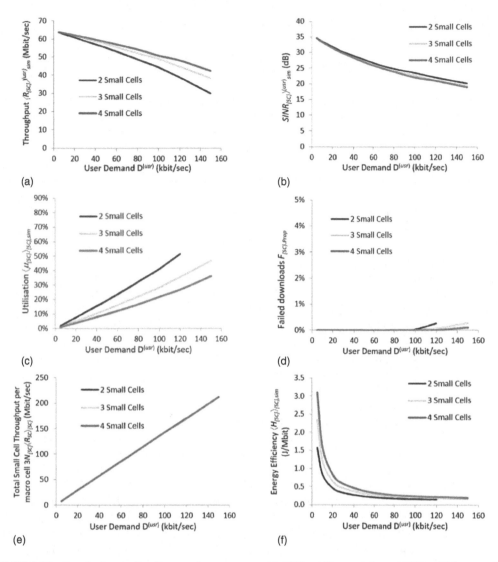

FIGURE 6.10   Results for small cell network component with SCD = 100 m, minimum ISCD = 100 m, $\varphi = 0.67$ with 2, 3 and 4 small cells per macro cell sector.

interference as adjacent cells serve their users, thereby transmitting for longer periods. As demand levels increase from a low level, cells are also increasingly serving a greater number of user requests simultaneously. The result is that cells are active for longer periods as well as more frequently, leading to a more pronounced interference, hence SINR degradation with increasing traffic level. The reduction in SINR is slightly greater for larger N_{SC}. This is because the N_{SC} small cells are all located within the area of a macro cell sector. Consequentially the larger N_{SC}, the closer the small cells will be located on average, resulting in a higher interference power from other small cells.

The decrease in SINR with demand level leads to the selection of a more conservative modulation and coding format (lower MCS level), and hence lower per-user throughput. Consequentially more PDSCH REs are required to download a file, resulting in further increased cell utilisation. This means that the incremental energy consumption increases. Counterbalancing this increase, at higher demand levels there are more bits travelling through the network for the same idle power consumption, and the overall effect is a lower energy-per-bit.

Increasing the minimum inter-cell spacing, *ISCD*, will have effect of improving SINR perfor-mance since interfering signals from adjacent cells are being attenuated by the increased distance they have to travel to a user. This leads to an improvement in user throughput as well as reductions in download termination rate, cell utilisation, power consumption and energy per bit performance of a cell.

Viewing the results for SINR values, we see they are highest when the number of cells in the region is lowest, due to the greater separation of cells and thus a smaller level of interference from adjacent cells. Similarly, the SINR values are greater with increased *ISCD*.

The (c) plots in Figures 6.8 through 6.10 show that small cell utilisation increases with user demand, as one would expect. They also show that per-cell utilisation is lower with greater numbers of small cells, because when traffic demand is shared across a larger number of cells, individual cell utilisation is lower.

The proportion of failed downloads, $F_{\{SC\},Prop}$, shown in the (d) plots, is the ratio of downloads that are terminated before completion to the total download requests. As can be seen, these plots have a common characteristic in that the proportion of download failures remains close to zero until a certain value of demand $D^{(usr)}$ and then increases relatively quickly beyond that value. This trend indicates a transition from congestion free operation to the onset of congestion. We can also see that the proportion of failures is greater for smaller *ISCD* and smaller N_{SC}. These features occur because smaller *ISCD* causes higher interference power from nearby small cells, hence lower spectral effi-ciency which results in longer download times for user requests. In turn, this leads to a more failed downloads.

The higher failure proportion with fewer small cells arises because there are less small cells to share the load. Consequently, the utilisation of small cells will be higher (as shown in the (c) plots), leading to less PDSCH REs is available to each user. This leads to longer download times and hence the higher failure proportion of downloads.

The d) plots show the proportion of download service requests that fail to reach their target data volume in time. Recall that a cell is recorded as "utilised" if it is serving least one user. However even at moderate average utilisation levels, the random assignment of service download requests across the user population leads to an irregular distribution of user activity across the set of small cells at certain times. Thus in addition to the higher probability of cell congestion when there are fewer small cells, some of those cells will encounter even higher traffic demand arising from that irregular distribution. This leads to a sufficient number of download failures, due to exceeding the maximum service download time, to affect the failure statistics even at moderate overall utilisation levels. Nonetheless, across all of Figures 6.8d, 6.9d, and 6.10d, the failure rate is significantly lower than was seen for a dense urban macro-base-station deployment shown in Figure 5.8d.

The e) figures show the base station total throughput as a function of average user demand. These show the expected trend of throughput increasing with demand. These are in contrast with the same macro base station characteristic in Figure 5.8e, which shows the effect of network congestion on base station throughput.

The relationship between base station energy efficiency, $\langle H_{\{SC\}}\rangle_{\{SC\},sim}$, and user demand, $D^{(usr)}$, shown in the sequence of f) plots, is little more subtle. At low traffic demand per user, $D^{(usr)}$, the base station utilisation is close to zero. Therefore, the base station idle power term in (6.23) is dominant and $\langle H_{\{SC\}}\rangle_{\{SC\},sim}$ is approximately inversely proportional downloaded volume of data, $\langle B_{\{SC\},Tot}\rangle_{sim}$, as seen in low demand regions of Figures 6.8f, 6.9f and 6.10f. For higher traf-fic demand levels, the idle component of energy consumption is being amortised over a greater number of bits. Moreover, each base station is active for a greater proportion of time. At high demand levels, the diminishing amortised idle power and the increasing demand-driven incre-mental power consumption in (6.23) approach a balance. The result is an efficiency curve that asymptotes towards a lower limit.

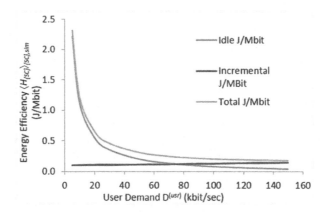

FIGURE 6.11 Energy per bit for 3 small cells with inter-small cell distance 75 m. The energy per bit is separated out into the contribution from idle power which is independent of utilisation, $\langle \mu_{SC} \rangle$ and the contribution from incremental power which is dependent upon utilisation.

Figure 6.11 again shows the energy efficiency characteristic but separates out the idle and incremental power contributions, in this case for a $N_{SC} = 3$ cell per sector deployment with minimum inter-small cell spacing of $ISCD = 75$ m. In this figure, we see that the idle power contribution to the energy per bit reduces with increased user traffic demand as described earlier.

The incremental contribution to energy efficiency is more difficult to analyse because it is dependent on a range of factors including traffic demand level, $D^{(usr)}$, number of cells, N_{SC}, minimum inter-cell spacing, $ISCD$, the spectral efficiency experienced by users, $\langle SINR_{\{SC\}} \rangle^{(usr)}$, proportion of download failures and the number of concurrent users (in the j th epoch) being served by each small cell.

As the user demand increases, the average SINR reduces thereby reducing the average spectral efficiency of the downloads. This increases the number of PDSCH REs required to download $B^{(usr)}$ size file which increases the energy expended to download the file. Hence increasing Joules/bit and so making the energy efficiency metric worse.

The impact of failed downloads on energy efficiency is subtle. A failed download corresponds to a session downloading fewer than the required $B^{(usr)}$ bits within the maximum allowed time duration $\Delta t_{max,dl}$. The higher failure rate means a lower total volume of data is downloaded for the number of download requests and total time duration of those downloads. The energy expended is dependent upon the total duration of downloads and so reducing the volume of data downloaded will increase the energy per bit, $H_{\{SC\}}$. Mitigating this is the fact that terminated downloads correspond to those with a lower spectral efficiency. Those downloads are energy inefficient for the reasons given earlier. Therefore, terminating them can assist in removing energy inefficient downloads from the service area rather than have them continue.

The overall impact of this is an increase in the incremental contribution to the energy efficiency as seen in Figure 6.11.

6.7.2 MACRO CELL NETWORK COMPONENT RESULTS

The macro cell component of the simulation is effectively the Advanced Model described in Chapter 5, but with the portion of traffic $\varphi_{\{SC\}}$ off loaded to the small cells leaving portion $(1 - \varphi_{\{SC\}})$ for the macro cell to handle. In the simulation reported on here, the offload was 67% leaving the macro cell network dealing with 33% of the total traffic. The reason for this approach to plotting the macro cell results is to reflect the performance of the macro cell component of the heterogeneous network.

The results for this component are shown in Figure 6.12.

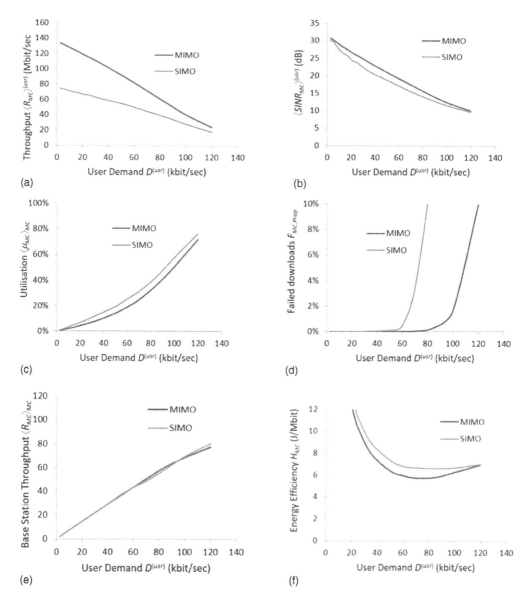

FIGURE 6.12 Macro cell network component results for Dense Urban geography with an off load ratio $\varphi = 0.667$.

6.7.3 HETEROGENEOUS NETWORK RESULTS

6.7.3.1 User Demand, $D^{(usr)}$ Results

The user demand dependent results for a heterogeneous mobile network with 50 m, 75 m and 100 m inter small cell spacing ($ISCD$) are shown in Figures 6.13, 6.14 and 6.15, respectively, each for 2, 3 and 4 small cells.

   In these figures, the results for a macro cell network covering the same service area are also shown to provide a comparison of the performance of the two network types; heterogeneous and macro cell networks. We can see how the deployment of small cells deals with much greater levels of user demand more proficiently than the macro cell only network.

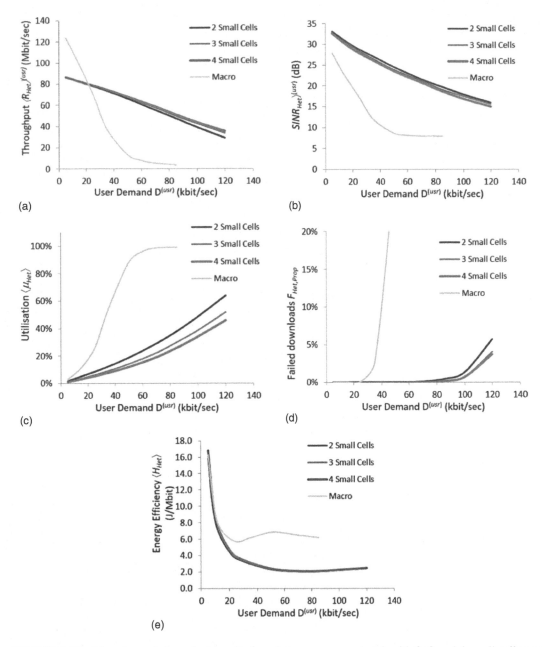

FIGURE 6.13   User demand dependent results for a heterogeneous network with 2, 3 and 4 small cells per macro cell sector with the inter-small cell distance (ISCD) of 50 m. Also shown are the corresponding results for a macro cell network covering the service area.

6.7.3.2   Diurnal Cycle Results

We now use the equations in Section 6.6 to combine the results of the two components above to determine numerical results for the heterogeneous network over the course of a day. To do this we require the values of normalised per user demand for each stage, X_S, the stage duration, T_S, and the proportion of the diurnal cycle, ρ_S for each stage, S, of the diurnal cycle as shown in Table 6.3. We adopt a value of $D^{(usr)} = 50$ kbit/s.

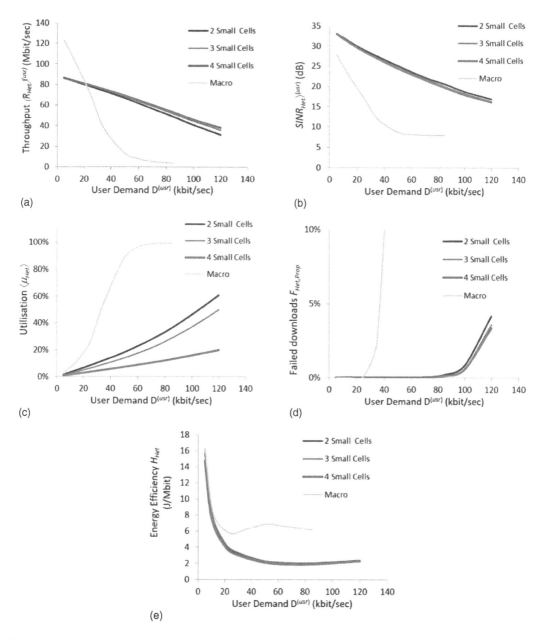

FIGURE 6.14 User demand dependent results for a heterogeneous network for 2, 3 and 4 small cells per macro cell sector with the inter-small cell distance (ISCD) of 75 m. Also shown are the corresponding results for a macro cell network covering the service area.

The diurnal cycle results for the heterogeneous network are shown in Table 6.4a–e. Also included are the corresponding values for a macro cell only network covering the same service area.

The diurnal cycle results also show the advantages of deploying small cells in areas of high demand. As can be seen, without the introduction of small cells, the macro cell only network struggles to accommodate the demand. The introduction of small cells addresses this situation.

The results in Table 6.4 also show that the energy efficiency of a heterogeneous network is relatively insensitive to the number of small cells deployed or the setting of their inter-site distances.

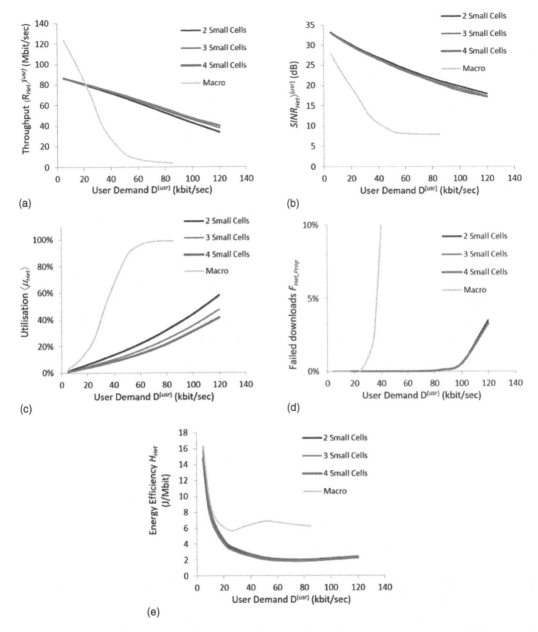

FIGURE 6.15 User demand dependent results for a heterogeneous network for 2, 3 and 4 small cells per macro cell sector with the inter-small cell distance (ISCD) of 100 m. Also shown are the corresponding results for a macro cell network covering the service area.

This is because the energy efficiency of the macro cells is the dominant contributor to the overall efficiency. This can be deduced from the energy efficiency of a macro-cell-only network shown Table 6.4. When its traffic is partially offloaded and hence the station is operating at a lower utilisation rate, the energy efficiency of a macro cell is expected to be lower, i.e. higher than the 6.64 J/Mbit shown. Since macro cells only carry 1/3 of the traffic volume after offload, its traffic weighted contribution to the overall energy efficiency would be approximately 2.21 J/Mbit.

TABLE 6.3

Stages of the Diurnal Cycle

S (stage index)	X_S (normalised user demand)	Duration T_S (hours)	ρ_S (portion of diurnal cycle)
1	0.2	2	1/12
2	0.4	4	1/6
3	1	4	1/6
4	1.2	8	1/3
5	1.4	6	¼

Source: Adopted by Blume *et al.* [6] and GreenTouch [7].

TABLE 6.4(a)

User Throughput for a Heterogeneous Network for Given Number of Small Cells per Macro Cell Sector (N_{SC}) and Inter-Small Cell Distance (ISCD)

$\left\langle R_{Het}\right\rangle_D^{(usr)}$ (Mb/s)		ISCD (m)		
		50	75	100
	2	61.5	62.4	63.3
Cells	3	63.2	64.2	64.9
N_{SC}	4	63.9	65.0	65.9
	Macro	14.6		

TABLE 6.4(b)

User SINR for a Heterogeneous Network for Given Number of Small Cells per Macro Cell Sector (N_{SC}) and Inter-Small Cell Distance (ISCD)

$\left\langle SINR_{Het}\right\rangle_D^{(usr)}$ (dB)		ISCD (m)		
		50	75	100
	2	23.4	24.0	24.5
Cells	3	22.9	23.6	24.0
N_{SC}	4	22.5	23.4	24.0
	Macro	9.2		

TABLE 6.4(c)

Average Base Station Utilisation for a Heterogeneous Network for Given Number of Small Cells per Macro Cell Sector (N_{SC}) and Inter-Small Cell Distance (ISCD)

$\langle \mu_{Het} \rangle_D$		ISCD (m)		
		50	75	100
	2	23.1%	22.3%	21.6%
Cells N_{SC}	3	17.7%	17.1%	16.8%
	4	15.0%	14.5%	14.1%
	Macro	88.9%		

TABLE 6.4(d)

Proportion of Failed Downloads for a Heterogeneous Network for a Given Number of Small Cells per Macro Cell Sector (N_{SC}) and Inter-Small Cell Distance (ISCD)

$F_{Het,Prop,D}$		ISCD (m)		
		50	75	100
	2	0.07%	0.03%	0.01%
Cells N_{SC}	3	0.03%	0.01%	0.01%
	4	0.02%	0.01%	0.01%
	Macro	39.85%		

TABLE 6.4(e)

Energy Efficiency for a Heterogeneous Network for Given Number of Small Cells per Macro Cell Sector (N_{SC}) and Inter-Small Cell Distance (ISCD)

$H_{Het,D}$ (J/Mbit)		ISCD (m)		
		50	75	100
	2	2.37	2.37	2.37
Cells N_{SC}	3	2.42	2.42	2.42
	4	2.47	2.47	2.47
	Macro	6.64		

6.8 SCALING FOR DIFFERENT OFFLOAD RATIOS AND DOWNLOAD VOLUME

Whilst simulation results have been presented in detail for a single demand offload ratio, it is a straightforward task to use these results to synthesise the results for other offload ratios and gain a view as to the relative benefit of different offload ratios.

From (6.68) in the Appendix, the results from a heterogeneous network simulation have the form:

$$\left\langle Y_{Het}\left(D^{(usr)}\right)\right\rangle \approx \varphi_{\{SC\}}\left\langle Y_{\{SC\}}\left(\varphi_{\{SC\}}D^{(usr)}\right)\right\rangle + \left(1-\varphi_{\{SC\}}\right)\left\langle Y_{\{nSC\}}\left(\left(1-\varphi_{\{SC\}}\right)D^{(usr)}\right)\right\rangle \tag{6.60}$$

The service request rate for the small cell and macro cell components of the heterogeneous network models are simplified variants of (6.12) and (6.13), reproduced here:

$$\lambda_{\{SC\}} = \varphi_{\{SC\}}D^{(usr)}\frac{A_{SA}U^{(usr)}}{B^{(usr)}} \tag{6.61}$$

$$\lambda_{\{nSC\}} = \left(1-\varphi_{\{SC\}}\right)D^{(usr)}\frac{A_{SA}U^{(usr)}}{B^{(usr)}} \tag{6.62}$$

In these equations the user density $U^{(usr)}$ is set by the geographical category which, in combination with the service area A_{SA}, determines the total user population, $N_{SA}^{(usr)}$. The geographical category also sets the inter-site distance, *ISD*, which strongly influences the SINR of macro cell users and hence the network performance of the macro cells. Similarly, for the small cell network, the number and size of the small cells strongly influence the SINR seen by the small cell users and hence the small cell network performance. The other key factor that impacts the SINR is the utilisation of surrounding base stations which, in turn, is determined by the user demand, $D^{(usr)}$.

Four factors, the service area, geographic category (thus user density), modelled file size, and user demand come together in the service request rate, λ.

With the A_{SA}, $U^{(usr)}$ and other geographical parameters set, and download file size, $B^{(usr)}$, fixed, we see from (6.61) and (6.62) that the small cell service request rate, $\lambda_{\{SC\}}$, is determined by the product $\varphi_{\{SC\}}D^{(usr)}$ and, likewise, the service request rate for the macro cell network, $\lambda_{\{nSC\}}$, is determined by the product $(1 - \varphi_{\{SC\}})D^{(usr)}$.

With all the simulation geographical parameters set, the heterogeneous network performance is determined by the values of $\lambda_{\{SC\}}$ and $\lambda_{\{nSC\}}$. This provides an opportunity to "rescale" the user demand, $D^{(usr)}$, to provide results for different off load ratios, $\varphi_{\{SC\}}$, provided that the overall service request rates $\lambda_{\{SC\}}$ and $\lambda_{\{nSC\}}$ are kept constant. For example, it is instructive to compare the simulation result $\langle Y_{Het}\rangle$ with an off load ratio $\varphi_{\{SC\}}$ to the result with a different off load ratio $\varphi'_{\{SC\}}$. To make this comparison, we introduce adjusted user demands $D_{\{SC\}}'^{(usr)}$ and $D_{\{nSC\}}'^{(usr)}$ for fixed request rates, $\lambda_{\{SC\}}$ and $\lambda_{\{nSC\}}$, respectively, such that:

$$\lambda_{\{SC\}} = \varphi_{\{SC\}}D^{(usr)}\frac{A_{SA}U^{(usr)}}{B^{(usr)}} = \varphi'_{\{SC\}}D_{SC}'^{(usr)}\frac{A_{SA}U^{(usr)}}{B^{(usr)}}$$

$$\lambda_{\{nSC\}} = \left(1-\varphi_{\{SC\}}\right)D^{(usr)}\frac{A_{SA}U^{(usr)}}{B^{(usr)}} = \left(1-\varphi'_{\{SC\}}\right)D_{\{nSC\}}'^{(usr)}\frac{A_{SA}U^{(usr)}}{B^{(usr)}} \tag{6.63}$$

That is:

$$\varphi_{\{SC\}}D^{(usr)} = \varphi'_{\{SC\}}D_{\{SC\}}'^{(usr)}$$

$$\left(1-\varphi_{\{SC\}}\right)D^{(usr)} = \left(1-\varphi'_{\{SC\}}\right)D_{\{nSC\}}'^{(usr)} \tag{6.64}$$

To ascertain the impact of changing the off load ratio from $\varphi_{\{SC\}}$ to $\varphi'_{\{SC\}}$, we compare the result in (6.60) with the result given by:

$$\left\langle Y_{Het}\left(D'^{(usr)}\right)\right\rangle \approx \varphi'_{\{SC\}}\left\langle Y_{\{SC\}}\left(\varphi'_{\{SC\}}D'^{(usr)}\right)\right\rangle + \left(1-\varphi'_{\{SC\}}\right)\left\langle Y_{\{nSC\}}\left(\left(1-\varphi'_{\{SC\}}\right)D'^{(usr)}\right)\right\rangle \quad (6.65)$$

The outputs of the simulations including $\langle SINR_{\{SC\}}\rangle^{(usr)}$, $\langle R_{\{SC\}}\rangle^{(usr)}$, $\langle \mu_{\{SC\}}\rangle$, etc., are plotted as functions of user demand, $D^{(usr)}$, for a given off load ratio, $\varphi_{\{SC\}}$ as shown in Figures 6.8, 6.12 and 6.13. Using (6.64) to establish the results for an off load ratio $\varphi'_{\{SC\}}$, we re-scale the x-axis from $D^{(usr)}$ to $D'^{(usr)}_{\{SC\}}$ in the small cell result plots and $D'^{(usr)}_{\{nSC\}}$ in the macro cell result plots with these values given by:

$$D'^{(usr)}_{\{SC\}} = \frac{\varphi_{\{SC\}}}{\varphi'_{\{SC\}}}D^{(usr)}$$

$$D'^{(usr)}_{\{nSC\}} = \frac{\left(1-\varphi_{\{SC\}}\right)}{\left(1-\varphi'_{\{SC\}}\right)}D^{(usr)} \quad (6.66)$$

These results can be combined to provide the corresponding heterogeneous network results.

This allows the modeller to take the results of simulations with off load $\varphi_{\{SC\}}$ and use rescaling of the per user demand from $D^{(usr)}$ to $D'^{(usr)}$ to approximate simulation results for different offload ratio $\varphi'_{\{SC\}}$. The validity of this approach is indicated in Figure 6.16. The solid traces on these plots show the dependence of the throughput per user, SINR and energy efficiency on $D^{(usr)}$ developed directly by simulation for off load values $\varphi_{\{SC\}} = 0.667$ and $\varphi'_{\{SC\}} = 0.50$ from a heterogeneous network simulation. The dashed line shows the dependence for $\varphi'_{\{SC\}} = 0.50$ developed by using (6.66) to rescale the results of the simulation for $\varphi_{\{SC\}} = 0.667$ to $\varphi'_{\{SC\}} = 0.50$. As can be seen, the rescaled results are very close to the simulation results for $\varphi'_{\{SC\}} = 0.50$.

This offload rescaling technique also can be applied to the calculation of the overall diurnal cycle results for different off load ratios. The off load ratio used to produce Figures 6.8 through 6.10 and Figure 6.12 was $\varphi_{\{SC\}} = 0.67$. The original user demand values over the diurnal cycle are given in the column headed $X_S D^{(usr)}$ in Table 6.5.

If we wish to consider the diurnal cycle network performance with a new offload ratio of $\varphi'_{SC} = 0.5$ we rescale the small cell and macro cell user demand to:

$$D'^{(usr)}_{\{SC\}} = \frac{\varphi_{\{SC\}}}{\varphi'_{\{SC\}}}D^{(usr)} = \frac{0.67}{0.5}D^{(usr)} = 1.34D^{(usr)}$$

$$D'^{(usr)}_{\{nSC\}} = \frac{\left(1-\varphi_{\{SC\}}\right)}{\left(1-\varphi'_{\{SC\}}\right)}D^{(usr)} = \frac{0.33}{0.5}D^{(usr)} = 0.66D^{(usr)} \quad (6.67)$$

This gives the values shown in Table 6.6. Using the rescaled values, $D'^{(usr)}_{\{SC\}}$ along the x-axis in Figure 6.8 for the $N_{SC} = 3$ small cells per macro cell sector with $ISDC = 50$ m and rescaled values $D'^{(usr)}_{\{nSC\}}$ along the x-axis in the macro cell results shown in Figure 6.12, we get the values in Table 6.7 for each stage of the diurnal cycle. In this table the small cells utilise SISO and macro cells utilise MIMO protocols.

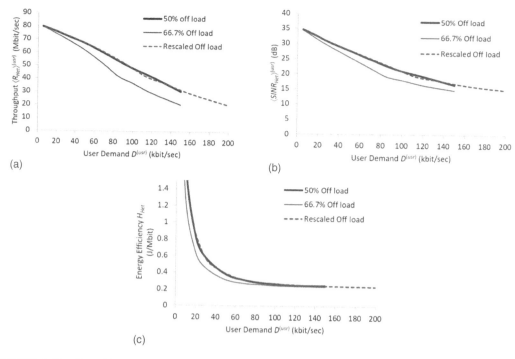

(a) (b)

(c)

FIGURE 6.16 Rescaling of simulation results to provide results for different off load ratios. This example shows the results for $N_{SC} = 3$ small cells per macro cell sector, ISCD = 50 m with rescaling from $\varphi_{\{SC\}} = 0.667$ to $\varphi_{\{SC\}} = 0.50$. The plots show that rescaling $D^{(usr)}$ using (6.67) provides accurate results for the value of $\varphi_{\{SC\}} = 0.50$.

TABLE 6.5

Diurnal Cycle Use Demand Values and Stage Durations for Off Load Ratio $\varphi_{\{SC\}} = 0.667$ Used to Give Simulation Outputs Shown in Figures 6.8 through 6.10 and 6.12

S (stage index)	X_S (normalised user demand)	$X_S D^{(usr)}$ (kbit/s)	Duration T_S (hours)	ρ_S (portion of diurnal cycle)
1	0.2	10	2	1/12
2	0.4	20	4	1/6
3	1	50	4	1/6
4	1.2	60	8	1/3
5	1.4	70	6	1/4

Using (6.39) for $\{SC\}$ and $\{nSC\}$ and (6.44), we bring these quantities together to give the diurnal cycle results for the heterogeneous network shown in Table 6.8. Also shown are the corresponding results for off load ratio $\varphi_{\{SC\}} = 0.67$.

From Table 6.8, we can see the advantage of deploying small cells to accommodate areas of high traffic in that the use of off load to small cells can significantly improve overall network performance. Further, off-loading a majority of the traffic to the small cells can make a significant improvement.

TABLE 6.6

Rescaled Diurnal Cycle User Demands for Small Cell and Macro Cell Component Simulations and Stage Durations for Off Load Ratio $\varphi_{\{SC\}}$ = 0.5 for Simulation Outputs

S (stage index)	X_S (normalised user demand)	$X_S D'^{(usr)}_{\{SC\}}$ kbit/s	$X_S D'^{(usr)}_{\{nSC\}}$ (kbit/s)	Duration T_S (hours)	ρ_S (portion of diurnal cycle)
1	0.2	13.34	6.66	2	1/12
2	0.4	26.68	13.32	4	1/6
3	1	66.7	33.3	4	1/6
4	1.2	80.04	39.96	8	1/3
5	1.4	93.38	46.62	6	1/4

TABLE 6.7

Rescaled Diurnal Cycle Results for Small Cell and Macro Cell Component Simulations and Stage Durations for Off Load Ratio $\varphi_{\{SC\}}$ = 0.5 for Simulation Outputs

Stage (S)	$\langle R_{\{SC\}} \rangle^{(usr)}$ (Mbit/s)	$\langle R_{\{nSC\}} \rangle^{(usr)}$ (Mbit/s)	$\langle SINR_{\{SC\}} \rangle^{(usr)}$ (dB)	$\langle SINR_{\{nSC\}} \rangle^{(usr)}$ (dB)	$H_{\{SC\}}$ (J/Mbit)	$H_{\{nSC\}}$ (J/Mbit)
1	79.14	149.45	33.33	29.36	1.21	35.01
2	77.65	143.61	31.11	28.10	0.64	17.24
3	73.97	123.18	25.49	23.82	0.31	7.66
4	62.49	115.80	23.65	22.37	0.29	6.76
5	51.02	107.57	21.80	20.97	0.27	5.96

TABLE 6.8

Heterogeneous Network Results for N_{SC} = 3 Small Cells per Macro Cell Sector, ISCD = 50 m for Off Load Ratios $\varphi_{\{SC\}}$ = 0.50 and 0.667 (The Results for $\varphi_{\{SC\}}$ = 0.5 Were Calculated Using the Re-Scaling of the User Demand. Also Listed Are the Results for a Macro Cell Only Network)

φ_{SC}	$R_{HetD}^{(usr)}$ (Mbit/s)	$SINR_{HetD}^{(usr)}$ (dB)	$H_{Het,D}$ (J/Mbiit)
0.667	63.2	22.9	2.42
0.50	89.12	23.29	4.06
Macro cells only	$R_{MCD}^{(usr)}$ (Mbit/s)	$SINR_{MCD}^{(usr)}$ (dB)	$H_{MC,D}$ (Joule/bit)
-	14.6	9.17	6.64

6.9 TECHNICAL SUMMARY

A heterogeneous network that includes the deployment of small cells in areas of high demand can provide significant improvements in network performance, including energy efficiency. The Advanced Model described in Chapter 5 can be used in the modelling of a heterogeneous network by splitting the network into two separate components: a small cell network component and the remaining network component which may be combination of macro and small cells. These two components are modelled independently using two separate implementations of the Advanced Model. This approach provides flexibility for the modeller to study a small cell network that consists of a sub-set of all the small cells in the service area. This type of network is becoming increasingly common practice, and will be expanded with the rollout of 5G networks. The small cell network component is denoted $\{SC\}$ and the remaining network component is denoted $\{nSC\}$.

The separation into two networks is done by allocating a proportion, $\varphi_{\{SC\}}$, of overall users have their traffic "off loaded" from the macro cell network to the small cell network $\{SC\}$. This portion of users set the "service request rate" according to (6.12). The details of how the locations of small cell users are distributed randomly across the small cells in the service area are described in Section 6.2.1.

The approach taken in this work for simulating a heterogeneous mobile network can be summarised as follows:

1. Small cell sub-network $\{SC\}$ operates in frequency bands outside the frequency bands of the remaining network $\{nSC\}$ covering the same area. Hence from a signal interference perspective, the $\{SC\}$ and $\{nSC\}$ sub-networks can be treated as independent systems and analysed separately. Full reuse of the frequency spectrum allocated to each of the two sub-networks is also assumed.
2. Because the $\{SC\}$ component frequency bands are outside those of the $\{nSC\}$ component, the heterogeneous network simulation is split into two independent component simulations. One for the $\{SC\}$ network and a separate simulation for the $\{nSC\}$ joint network.
3. Population densities and traffic demand levels in the real world are not uniform. The heterogeneous network model includes small cells deployed to assist meeting the traffic demand in at locations with very high demand density. In simulations, deployments of 2, 3 or 4 small cells per macro base station sector are often used, but up to 10 cells per sector can be included [13]. For the given geographical category, the density of users served by the macro cell network and small cell network are given by (6.9) and (6.11), respectively. These equations reflect that the user density in the small cells is greater than that in the macro cells. The number of users served by the $\{SC\}$ and $\{nSC\}$ networks is given by (6.15).
4. Where the simulation is designed to represent a service area in which specific geographic regions having elevated traffic areas are known, these and a matching small-cell layout can be designed. However, for generic simulations, a randomised location of small cells with a proportion of the service requests forced onto the coverage area of the small cells can be used to model the service quality and energy efficiency gains that small cells can achieve.
5. Simulations are run for a pre-determined set of user average traffic demand levels. The service area is set to a "Dense Urban" geographical category defined in Table 5.1 of because it is the high population density of Dense Urban geographies that requires small cells to assist with accommodating high user traffic. The macro cell base stations in generic simulations are deployed in a series of hexagonal "rings", with a central macro base station (BS_0) and two rings $(BS_1$ to $BS_{18})$ or even three rings $(BS_1$ to $BS_{36})$ being common [13]. (See Figure 6.4)
6. Each small cell is served by an omnidirectional antenna. That is, the small cell antenna will transmit to and receive from the full 360 degrees of directions around the base station antenna. Hence, small cells have one sector in the simulations reported here. MIMO-equipped small cell deployments are also possible.

7. For simulations using the FTP traffic models, 3GPP has adopted a file size ($B^{(usr)}$) of 2 MByte (16 Mbit) to be downloaded to the user [13], but any value more relevant to the application at hand could be used. Having established the service area dimensions, population density, a per-user traffic demand, and file size, all of the parameters are set for a macro network model.

8. Based on the chosen inter-site distance of the macro cell base stations, ISD, the overall size of the service area of the model is chosen to be large enough to include the coverage areas of all the macro cells included in the simulation. Thus for example, a service area or "playground" of 3 km × 3 km would conveniently cover the three ring network of macro base stations with $ISD = 0.5$ km as shown in Figure 6.4.

9. A fixed number, N_{SC}, of small-cells (for example 2, 3 or 4) are deployed within each sector of macro base stations in the central service area. The locations of the small cells are randomly selected, subject to some minimum spacing rule (e.g. small-cells must be at least 50 m, 75 m or 100 m apart). In general, small cells will cover areas of higher user-density or higher traffic. The number deployed depends on the traffic level. Details of small cell placement are in Section 6.2.1.

10. In the $\{nSC\}$ joint network simulation component, locations from which users might request the service are distributed across the total service area in both small and macro cells. Details are in Step (8) of Section 6.2.1.8. For the $\{SC\}$ network simulation, user locations are clustered in the vicinity of each small cell base station. In terms of the traffic demand model, user traffic demand in the $\{nSC\}$ joint network component is modelled as a series of file download requests at rate $\lambda_{\{nSC\}}$, given by (6.13), from user locations across the entire service area. The traffic demand in the $\{SC\}$ network component is modelled as a series of download requests, with request rate $\lambda_{\{SC\}}$ given by (6.12) from users at random locations that are clustered around the small cell base stations. The $\{SC\}$ traffic level is represented as a proportion of the entire service area traffic that is "a priori" set to be serviced by the small cells. That proportion is referred to as the "offload ratio", $\varphi_{\{SC\}}$.

11. As with the Advanced Model, the heterogeneous network user grid system is implemented to locate user requests. However, this grid system is more complicated than the simple square rectilinear (x,y) grid constructed for the Advanced Model. The heterogeneous network consists of small cells that are randomly located within the sectors of the network macro cells and users are randomly distributed across the small and macros cells. A pre-set off load portion of users is a priori allocated to be served by the $\{SC\}$ component. This requires the construction of three rectilinear grids. The first grid covers each sector of each macro cell to provide possible locations of the small cell base stations within each macro cell sector. The second grid is constructed to provide possible locations of users who are a priori allocated to be served by the small cell network. The third grid covers the entire service area. This grid provides potential locations for users served by the macro cell network and is constructed as described in the Advanced Model of Chapter 5. The method for construction of these grids is given in Section 6.2.1.

12. The first (x,y) grid provides potential locations for the required number of small cells positioned within each macro base station coverage area. This grid provides positions within each sector of a macro cell with each sector having its own grid. Small cell base stations are positioned at randomly selected locations within each macro base station sector. They are positioned beyond a minimum separation from the central macro cell base station, MD, and preserve a minimum inter-small cell distance, $ISCD$, between the small cell base stations. Small cells may overlap and some small cells may include areas outside the base station sector in which their base station is located. Examples of this are seen in Figure 6.4 in which the some small cells in the sector of BS_1 overlap and some small cells extend beyond the border of the BS_1 macro cell.

13. The second (x,y) grid provides potential user locations and is implemented to cover each small cell. This is a collection of multiple rectilinear (x,y) grids each covering circular area given by

the small cell diameter, *SCD*. There is a grid for each small cell and the locations in these grids are the basis for the small cell network component of the model. This grid will provide locations for all users in {*SC*} and a proportion of users in {*nSC*}.

14. The third rectilinear (x,y) grid covers the entire heterogeneous network service area to provide potential locations of new service requests that are not off loaded to the {*SC*} network.

15. Just as with the Advanced Model, each potential user site of this grid is allocated a unique identifying index used to locate that site within the overall service area. When new users who are not off loaded to the {*SC*} sub-network are introduced into the service area, they are placed on one of the (x,y) locations on this grid and that location is recorded for that user. Because users who are not off loaded to a small cell are served by a {*nSC*} base station, the grid index includes the identification of which cell and sector is serving that user with the appropriate *n*-tuple and index number.

16. When a traffic request is a priori allocated to the {*SC*} network, the request is assigned to one of the $N^{(usr)}_{\{SC\},grid}$ potential user sites within the area covered by the {*SC*} network, where $N^{(usr)}_{\{SC\},grid}$ is given by (6.18). The service request rate for the {*SC*} network is given by (6.20). The {*SC*} base station that serves that user is set to be the station that provides the greatest SINR at that location.

17. The traffic requests allocated to the {*nSC*} joint network are positioned at random on one of the $N^{(usr)}_{\{nSC\},grid}$ potential user sites where $N^{(usr)}_{\{nSC\},grid}$ is given by (6.19). The service request rate for the {*nSC*} joint network is given by (6.21). The {*nSC*} base station that serves that user is set to be the station that provides the greatest SINR at that location which could be a macro or small cell base station.

18. For the {nSC} network, a proportion of user traffic requests are located in a small cell in {*nSC*} as given by (6.22).

19. The utilisation behaviour of a small cell at different traffic demand levels is dependent on the cell location (within a macro cell sector), distance from neighbouring small cells and the configuration of that small cell including its bandwidth, signalling overheads and antenna settings. For the results presented in this text, each small cell is assumed to employ a 20 MHz frequency spectrum and 19% of its resource elements are reserved for signalling overheads. (That is, $\sigma_{PDSCH} = 0.81$ corresponding to a single antenna port with CFI = 2.)

20. To generalise the heterogeneous network simulation results for any specific inter-small cell distance, *ISCD*, the results are averaged over the small cells in the heterogeneous network and over multiple, N_{sim}, simulation runs, with random re-location of the small cells for each run in each of the {*SC*} and {*nSC*} simulations. This averaging involves:

    a. Averaging the utilisation of the N_{SC} small cells in each sector of the inner base stations (BS_0 to BS_6) in the service area for each simulation run.

    b. Averaging over the N_{sim} small cell network simulation runs, each with a different set of small cell locations within the service area.

21. The transmitted power, $P_{max,out}$, of a small cell is modelled in the same manner as that for a macro cell in Chapter 5, which is linearly varying with cell utilisation given in (6.24), where $P_{max,out}$ is the maximum transmitted power and μ_{SC} is the utilisation of the small cell base stations. In this model μ_{SC} is set to unity in every epoch in which the small station is transmitting to users, as with the Advanced Model of Chapter 5.

22. The conversion from $P_{out}(\mu_{SC})$ to the power supplied to a small cell $P_{SC}(\mu_{SC})$ follows the same approach as in Chapters 4 and 5, but noting that there are no σ_{feed} or σ_{cool} terms as shown in (6.23).

23. The determination of $P_{idle,out}$ follows the same procedure as outlined in Chapters 4 and 5. Essentially it includes the power used in transmission of the load-independent signalling

overheads for the given small cell configuration as well as the usual ancillary equipment power consumption.

24. Using the same approach as for the macro base station models in Chapters 4 and 5, this model computes the SINR, $SINR^{(k)}$, of each user, k, served by a $\{SC\}$ and $\{nSC\}$ base station. The SINR is then converted into a Spectral Efficiency, $SE^{(k)}$, which together with the number of PDSCH REs that user has been allocated (as described in Sections 5.4.3 and 5.5) determine the user's throughput.

25. The results for each of five stages of the diurnal cycle are collected for the small cell and macro cell network, where the user demand is given by $X_S D^{(usr)}$ shown in Figure 6.7 and Table 6.3.

26. The simulation outputs from the two components of the simulation: the $\{SC\}$ component and $\{nSC\}$ component are combined to provide the corresponding overall performance for the heterogeneous network.

27. The heterogeneous network simulation provides two types of results for each output parameter. The first is the dependence of the output parameter on user demand, $D^{(usr)}$. The second is the overall diurnal cycle average of the output parameter.

28. The dependence of the heterogeneous simulation outputs on user demand are calculated using the formulae (6.41), (6.42) and (6.43).

29. The heterogeneous simulation overall diurnal cycle results are calculated using the formulae (6.44), (6.45) and (6.46).

30. Specific results for the following are described:
    a. Average throughput per user, $\langle R \rangle^{(usr)}$, in Section 6.6.3
    b. Average user SINR, $\langle SINR \rangle^{(usr)}$, in Section 6.6.4
    c. Average base station utilisation, $\langle \mu_{BS} \rangle$, in Section 6.6.5
    d. Proportion of failed downloads, F_{dl}, in Section 6.6.6
    e. Energy efficiency, H, in Section 6.6.7

31. The results for one offload ratio $\varphi_{\{SC\}}$ can be used to approximate the results for a different off load ratio, $\varphi'_{\{SC\}}$ by re-scaling the user demand, $D^{(usr)}$ to values $D'^{(usr)}_{\{SC\}}$ (for small cell component) and $D'^{(usr)}_{\{nSC\}}$ (for macro cell component) respectively given in (6.66)

6.10 EXECUTIVE SUMMARY

Heterogeneous mobile networks consist of a combination of macro cells and small cells. The small cell component of the network is deployed to accommodate locations within the macro cell network that may exhibit higher levels of user demand. Examples include sport stadia, retail shopping precincts, entertainment complexes and inner city areas.

To model heterogeneous networks, the techniques of the Advanced Model described in Chapter 5 can be utilised. However, to simplify the model, the heterogeneous network can be split into two independent component networks: a small cell sub-network component and the remaining (macro and small) cell sub-network component. On the assumption that the frequency bands used in these two component sub-networks do not overlap, they can be separately modelled and their results can be combined to model the overall heterogeneous network.

The overall heterogeneous network model as described involved two separate applications of the Advanced Model. However, the introduction of small cells requires the construction of a more complex set of user locations because the heterogeneous network consists of multiple small cell service areas scattered across the macro cells.

The process for combining the results of the two component networks is relatively straightforward and can provide results for each stage of the diurnal cycle or for the overall diurnal cycle.

6.11 APPENDIX

6.11.1 Constructing $D^{(usr)}$ Results

To construct heterogeneous network results that show the dependence on $D^{(usr)}$, we combine the corresponding $D^{(usr)}$ results for the macro and small cell network components. Examples of these results for a macro cell network are presented in Chapter 5 and for the small cell and macro network components of a heterogeneous network are shown in Figures 6.8 through 6.10 and 6.12 of this chapter. The role of per-user traffic demand $D^{(usr)}$ in the simulation is in determining the service request rate, λ. This rate is dependent upon $D^{(usr)}$ via (6.6) in which the user density, $U^{(usr)}$, is set by the geographical category. The service area, A_{SA}, and the download file size $B^{(usr)}$ are set by the modeller. The number and spacing of macro base stations to be included in the simulation are also set by the modeller, either to align with a known demography or for compatibility with other parties' simulation scenarios. In a heterogeneous network simulation we also have a relationship between the $D^{(usr)}$ and service request rates of the $\{SC\}$ network, $\lambda_{\{SC\}}$, and $\{nSC\}$ joint network, $\lambda_{\{nSC\}}$, via (6.12) and (6.13) respectively. Because $D^{(usr)}$ characterises the average demand across the entire heterogeneous network simulation, we express the simulation results as functions of $D^{(usr)}$ as shown in the figures.

For per-user results such as $\langle SINR \rangle^{usr}$ and $\langle R \rangle^{(usr)}$, we have to weight the small cell and macro cell results for the number of users using $N_{Tot,\{SC\}}^{(usr)} = \varphi_{\{SC\}} N_{Tot}^{(usr)}$ and $N_{Tot,\{nSC\}}^{(usr)} = (1 - \varphi_{\{SC\}}) N_{Tot}^{(usr)}$. This gives:

$$
\begin{aligned}
\left\langle Y_{Het}\left(D^{(usr)}\right)\right\rangle^{(usr)} &= \frac{\displaystyle\sum_{k=1}^{N_{Tot,\{SC\}}^{(usr)}} Y_{\{SC\}}^{(k)}\left(\varphi_{\{SC\}} D^{(usr)}\right) + \sum_{k=1}^{N_{Tot,\{nSC\}}^{(usr)}} Y_{\{nSC\}}^{(k)}\left(\left(1 - \varphi_{\{SC\}}\right)D^{(usr)}\right)}{N_{Tot,\{SC\}}^{(usr)} + N_{Tot,\{nSC\}}^{(usr)}} \\[2mm]
&\approx \frac{N_{Tot,\{SC\}}^{(usr)}\left\langle Y_{\{SC\}}\left(\varphi_{\{SC\}} D^{(usr)}\right)\right\rangle^{(usr)} + N_{Tot,\{nSC\}}^{(usr)}\left\langle Y_{\{nSC\}}\left(\left(1 - \varphi_{\{SC\}}\right)D^{(usr)}\right)\right\rangle^{(usr)}}{N_{Tot,\{SC\}}^{(usr)} + N_{Tot,\{nSC\}}^{(usr)}} \\[2mm]
&= \varphi_{\{SC\}}\left\langle Y_{\{SC\}}\left(\varphi_{\{SC\}} D^{(usr)}\right)\right\rangle^{(usr)} + \left(1 - \varphi_{\{SC\}}\right)\left\langle Y_{\{nSC\}}\left(1 - \varphi_{\{SC\}}\right)\left(D^{(usr)}\right)\right\rangle^{(usr)}
\end{aligned}
\tag{6.68}
$$

In this, we have explicitly shown the dependence on user demand, $D^{(usr)}$, and the demand values used in the simulation components.

For the quantities that are averaged over time for each stage of the diurnal cycle, we use (5.81) from Section 5.12.2 we have:

$$
\begin{aligned}
N_{Tot,\{SC\},S}^{(usr)} &\approx N'_{Tot,\{SC\},S}\Delta t_{\{SC\},S}\lambda'_{\{SC\},S} = N'_{Tot,\{SC\},S}C \\
N_{Tot,\{nSC\},S}^{(usr)} &\approx N'_{Tot,\{nSC\},S}\Delta t_{\{nSC\},S}\lambda'_{\{nSC\},S} = N'_{Tot,\{nSC\},S}C
\end{aligned}
\tag{6.69}
$$

where C is a constant. This gives:

$$
\frac{N_{Tot,\{SC\},S}^{(usr)}}{N'_{Tot,\{SC\},S}} \approx \frac{N_{Tot,\{nSC\},S}^{(usr)}}{N'_{Tot,\{nSC\},S}} \approx C
\tag{6.70}
$$

Using this we have:

$$\left\langle Y_{Het}\left(D^{(usr)}\right)\right\rangle_T = \frac{\sum\limits_{j=1}^{N'_{Tot,\{SC\}}} Y_{\{SC\},j}\left(\varphi_{\{SC\}}D^{(usr)}\right) + \sum\limits_{j=1}^{N'_{Tot,\{nSC\}}} Y_{\{nSC\},j}\left(\left(1-\varphi_{\{SC\}}\right)D^{(usr)}\right)}{N'_{Tot,\{SC\}} + N'_{Tot,\{nSC\}}}$$

$$\approx \frac{N^{(usr)}_{Tot,\{SC\}}\left\langle Y_{\{SC\}}\left(\varphi_{\{SC\}}D^{(usr)}\right)\right\rangle_T + N^{(usr)}_{Tot,\{nSC\}}\left\langle Y_{\{nSC\}}\left(\left(1-\varphi_{\{SC\}}\right)D^{(usr)}\right)\right\rangle_T}{N^{(usr)}_{Tot,\{SC\}} + N^{(usr)}_{Tot,\{nSC\}}} \tag{6.71}$$

$$= \varphi_{\{SC\}}\left\langle Y_{\{SC\}}\left(\varphi_{\{SC\}}D^{(usr)}\right)\right\rangle_T + \left(1-\varphi_{\{SC\}}\right)\left\langle Y_{\{nSC\}}\left(\left(1-\varphi_{\{SC\}}\right)D^{(usr)}\right)\right\rangle_T$$

where we have used $N^{(usr)}_{Tot,\{SC\}} = \varphi_{\{SC\}}N^{(usr)}_{Tot}$ and $N^{(usr)}_{Tot,\{nSC\}} = \left(1-\varphi_{\{SC\}}\right)N^{(usr)}_{Tot}$

6.11.2 Constructing Heterogeneous Network Results $\langle Y_{Het}\rangle_D$ from $\langle Y_{MC}\rangle_D$ and $\langle Y_{SC}\rangle_D$

The small cell and macro cell network simulation components provide averaged results for each of the stages, S, of the diurnal cycle. That is, results of the form: $\langle Y_{X,S}\rangle$, $\langle Y_{X,S}\rangle/\langle Z_{X,S}\rangle$ and $\langle(Y/Z)_{X,S}\rangle$ where $X = nSC$ for $\{nSC\}$ component results and $X = SC$ for $\{SC\}$ component results. The averaging is generally over the number of epochs, j, and/or number of users, k, or both. In this section we focus on the averaging over epoch index, j. The following section focuses on averaging over user index, k.

To construct the diurnal cycle results, we start with five sets of results for each of the macro cell network and small cell network. We have to combine these 10 results to construct a heterogeneous result over a diurnal cycle.

For the separate small cell and macro cell simulation components, combining results $\langle Y_{\{A\},S}\rangle$ for a given $X = SC, MC$ and $S = 1$ to 5 to give the diurnal result, $\langle Y_{\{A\}}\rangle$, is described in Section 5.7.1. In that case, the process combined quantities over different, non-overlapping time durations, T_S, for S = 1 to 5. Combining small cell and macro cell components differs because this involves combining averages over the same time durations of the stages of the diurnal cycle. That is, to calculate the heterogeneous results $\langle Y_{Het,S}\rangle$ we need to combine $\langle Y_{\{SC\},S}\rangle$ and $\langle Y_{\{nSC\},S}\rangle$ both with $S = 1, 2, ..., 5$. We then combine the $\langle Y_{Het,S}\rangle$ to get the overall result $\langle Y_{Het}\rangle_D$ for the diurnal cycle.

To show how to construct this we start with $\langle Y_{Het}\rangle$ and focus on averaging over time (or equivalently, over epochs) of a diurnal cycle. We have:

$$\langle Y_{Het}\rangle_D = \frac{\sum\limits_{j=1}^{N_D} Y_{Het,j}}{N_D} = \frac{\sum\limits_{S=1}^{5}\left(\sum\limits_{j=1}^{N_{Tot,\{SC\},S}} Y_{\{SC\},S,j} + \sum\limits_{j=1}^{N_{Tot,\{nSC\},S}} Y_{\{nSC\},S,j}\right)}{\sum\limits_{S=1}^{5}\left(N_{Tot,\{SC\},S} + N_{Tot,\{nSC\},S}\right)} \tag{6.72}$$

In this, N_D is the number data values over the diurnal cycle for the heterogeneous network which is given by the total number of data values for the small cell and macro cell components over the diurnal cycle. That is:

$$N_D = \sum\limits_{S=1}^{5}\left(N_{Tot,\{SC\},S} + N_{Tot,\{nSC\},S}\right) \tag{6.73}$$

To ensure the averaging is over the appropriate time durations, T_S, the number of data points for each S satisfy:

$$N_{Tot,\{SC\},S}\Delta t_{\{SC\},S} = N_{Tot,\{nSC\},S}\Delta t_{\{nSC\},S} = T_S = \rho_S \times (24 \text{ hours}) \tag{6.74}$$

In this equation we know the epoch durations of for the small cell and macro cell simulations, and require epoch duration for the heterogeneous network to all satisfy requirement:

$$\Delta t_{\{SC\},S}\lambda_{\{SC\},S} = \Delta t_{\{SC\},S}\varphi_{\{SC\}}X_S\lambda = C = \Delta t_{\{nSC\},S}\lambda_{\{nSC\},S} = \Delta t_{\{nSC\},S}\left(1-\varphi_{\{SC\}}\right)X_S\lambda \tag{6.75}$$

where C is a constant set by the Poisson process requirement (6.20) and (6.21). Using (6.37), we have:

$$
\begin{aligned}
\langle Y_{Het}\rangle_D &\approx \frac{\sum_{S=1}^{5}\left(N_{Tot,\{SC\},S}\left\langle Y_{\{SC\},S}\right\rangle + N_{Tot,\{nSC\},S}\left\langle Y_{\{nSC\},S}\right\rangle\right)}{\sum_{S=1}^{5}\left(N_{Tot,\{SC\},S} + N_{Tot,\{nSC\},S}\right)} \\
&= \frac{\sum_{S=1}^{5}\left(\varphi_{\{SC\}}\rho_S X_S\left\langle Y_{\{SC\},S}\right\rangle + \left(1-\varphi_{\{SC\}}\right)\rho_S X_S\left\langle Y_{\{nSC\},S}\right\rangle\right)}{\sum_{S=1}^{5}\rho_S X_S} \\
&= \varphi_{\{SC\}}\left\langle Y_{\{SC\}}\right\rangle_D + \left(1-\varphi_{\{SC\}}\right)\left\langle Y_{\{nSC\}}\right\rangle_D
\end{aligned}
\tag{6.76}
$$

where we have used (6.39) and (6.40).

Therefore, we also have:

$$
\begin{aligned}
\langle Y_{Het}\rangle_D / \langle Z_{Het}\rangle_D &\approx \frac{\sum_{S=1}^{5}\left(\varphi_{\{SC\}}\rho_S X_S\left\langle Y_{\{SC\},S}\right\rangle + \left(1-\varphi_{\{SC\}}\right)\rho_S X_S\left\langle Y_{\{nSC\},S}\right\rangle\right)}{\sum_{S=1}^{5}\left(\varphi_{\{SC\}}\rho_S X_S\left\langle Z_{\{SC\},S}\right\rangle + \left(1-\varphi_{\{SC\}}\right)\rho_S X_S\left\langle Z_{\{nSC\},S}\right\rangle\right)} \\
&= \frac{\varphi_{\{SC\}}\left\langle Y_{\{SC\}}\right\rangle_D + \left(1-\varphi_{\{SC\}}\right)\left\langle Y_{\{nSC\}}\right\rangle_D}{\varphi_{\{SC\}}\left\langle Z_{\{SC\}}\right\rangle_D + \left(1-\varphi_{\{SC\}}\right)\left\langle Z_{\{nSC\}}\right\rangle_D}
\end{aligned}
\tag{6.77}
$$

$$
\begin{aligned}
\left\langle (Y/Z)_{Het}\right\rangle_D &\approx \frac{\sum_{S=1}^{5}\left(\varphi_{\{SC\}}\rho_S X_S\left\langle (Y/Z)_{\{SC\},S}\right\rangle + \left(1-\varphi_{\{SC\}}\right)\rho_S X_S\left\langle (Y/Z)_{\{nSC\},S}\right\rangle\right)}{\sum_{S=1}^{5}\rho_S X_S} \\
&= \varphi_{\{SC\}}\left\langle (Y/Z)_{\{SC\}}\right\rangle_D + \left(1-\varphi_{\{SC\}}\right)\left\langle (Y/Z)_{\{nSC\}}\right\rangle_D
\end{aligned}
\tag{6.78}
$$

6.11.3 Constructing Heterogeneous Network per User Results

To undertake this calculation we use the results from Section 5.13.2. We have for the heterogeneous network:

$$
\langle Y_{Het} \rangle^{(usr)} = \frac{\sum_{k=0}^{N_D^{(usr)}} Y^{(k)}}{N_D^{(usr)}} = \frac{\sum_{S=1}^{5} \left(\sum_{k=1}^{N_{Tot,SC,S}^{(usr)}} Y_{\{SC\},S}^{(k)} + \sum_{k=1}^{N_{Tot,nSC,S}^{(usr)}} Y_{\{nSC\},S}^{(k)} \right)}{\sum_{S=1}^{5} \left(N_{Tot,\{SC\},S}^{(usr)} + N_{Tot,\{nSC\},S}^{(usr)} \right)}
\tag{6.79}
$$

Using the approach applied in Section 5.13.2:

$$
\sum_{k=1}^{N_{Tot,\{SC\}S}^{(usr)}} Y_{\{SC\}}^{(k)} \approx T_S' \lambda_S' \left\langle Y_{\{SC\},S} \right\rangle^{(usr)} = \varphi_{\{SC\}} \rho_S X_S \left\langle Y_{\{SC\},S} \right\rangle^{(usr)} \frac{\lambda \left(24\,hours \right)}{C}
$$
$$
\sum_{k=1}^{N_{Tot,\{nSC\}S}^{(usr)}} Y_{\{nSC\}}^{(k)} \approx T_S' \lambda_S' \left\langle Y_{\{nSC\},S} \right\rangle^{(usr)} = \left(1 - \varphi_{\{SC\}}\right) \rho_S X_S \left\langle Y_{\{nSC\},S} \right\rangle^{(usr)} \frac{\lambda \left(24\,hours \right)}{C}
\tag{6.80}
$$

and we have $N_{Tot,\{SC\},S}^{(usr)} + N_{Tot,\{nSC\},S}^{(usr)} = N_{Tot,S}^{(usr)}$. Using these in (6.79) gives:

$$
\langle Y_{Het} \rangle_D^{(usr)} \approx \frac{\sum_{S=1}^{5} \left(\varphi_{\{SC\}} \rho_S X_S \left\langle Y_{\{SC\},S} \right\rangle^{(usr)} + \left(1 - \varphi_{\{SC\}}\right) \rho_S X_S \left\langle Y_{\{nSC\},S} \right\rangle^{(usr)} \right)}{\sum_{S=1}^{5} \rho_S X_S}
\tag{6.81}
$$
$$
= \varphi_{\{SC\}} \left\langle Y_{\{SC\}} \right\rangle^{(usr)} + \left(1 - \varphi_{\{SC\}}\right) \left\langle Y_{\{nSC\}} \right\rangle^{(usr)}
$$

References

[1] M. Usama and M. Erol-Kantarci, "A Survey on Recent Trends and Open Issues in Energy Efficiency of 5G," *Sensors*, vol. 19, p. 3126, 2019.
[2] GSMA, "Road to 5G: Introduction and Migration," GSMA, 2020.
[3] P. Gammel *et al.*, "5G in Perspective A Pragmatic Guide to What's Next, White Paper," Skyworks, 2017.
[4] D. Tripathi and J. Reed, "5G Evolution – On the Path to 6G, White Paper," Rohde & Schwarz GmbH & Co., 2020.
[5] AT&T, "Will 5G replace WiFi?, White Paper," AT&T, 2019.
[6] O. Blume *et al.*, "*Energy Efficiency of LTE networks under traffic loads of 2020*," in *The Tenth International Symposium on Wireless Communication Systems*, 2013.
[7] GreenTouch, "GreenTouch Final Results from Green Meter Research Study: Reducing the Net Energy Consumption in Communications Networks by up to 98% by 2020," GreenTouch, 2015.
[8] Q. Mu *et al.*, "*Small Cell Enhancement for LTE-Advanced Release 12 and Application of Higher Order Modulation*," in *Mobile and Ubiquitous Systems: Computing, Networking, and Services. MobiQuitous 2013. Lecture Notes of the Institute for Computer TableSciences, Social Informatics and Telecommunications Engineering*, 2013.

[9] P. Chand *et al.*, *"Energy Efficient Performance Analysis of Downlink CoMP in Heterogeneous Wireless Network,"* in *9th International Conference on Industrial and Information Systems (ICIIS)*, 2014.

[10] M. Alsharif *et al.*, "Small Cells Integration with the Macro-Cell Under LTE Cellular Networks and Potential Extension for 5G," *Journal of Electrical Engineering & Technology*, vol. 14, p. 2455, 2019.

[11] W. Ahmad *et al.*, "Study Paper on HetNet: The LTE Perspective," Telecom Engineering Centre Khurshid Lal Bhavan, Janpath, New Delhi, 2016.

[12] P. Morgensen *et al.*, *"LTE Capacity Compared to the Shannon Bound,"* in *IEEE 65th Vehicular Technology Conference (VTC2007 Spring)*, 2007.

[13] Third Generation Participation Project (3GPP), "TR 36.814 V9.2.0 (2017-03) Technical Specification Group Radio Access Network; Evolved Universal Terrestrial Radio Access (E-UTRA); Further advancements for E-UTRA physical layer aspects (Release 9)," Third Generation Participation Project, 2017.

[14] B. Delaillie *et al.*, *"A Flexible and Future-Proof Power Model for Cellular Base Stations,"* in *IEEE 81st Vehicular Technology Conference (VTC Spring)*, 2015.

7 Traffic Models for Networks and Services

We saw in the previous chapter that many network elements' power consumption is dependent upon their traffic throughput. Therefore, to construct a model for network and service power consumption, we require knowledge of the network and service traffic to be used in the power models. Multiple publications have used traffic growth over time to estimate changes in network power consumption and energy efficiency (also referred to as "energy intensity") [1, 2] or to undertake life-cycle assessment [3]. In some publications, traffic modelling is undertaken to provide estimates for future power consumption [4–6]. To construct these estimates, we require traffic forecasts into the future. Although there is a useful amount of publicly available global and regional forecast traffic data, they typically only provide very general statistics and trends, rather than detailed time-of-day traffic forecasts [7–10].

To model energy efficiency of services, we require time-of-day traffic data over the full daily (diurnal) cycle of traffic. This is because, as we will see in later chapters, the energy efficiency of a service can be highly dependent upon whether or not that service's diurnal-cycle traffic is aligned in time with the other traffic in the network: that is, whether or not that service's diurnal-cycle traffic has its maxima and minima at the same time of day as the other traffic in the network.

Detailed diurnal-cycle traffic data is available; however, they are either recent or historical [11, 12]. Forecasts of future detailed (hour by hour) diurnal-cycle traffic data is rarely available. This means diurnal-cycle data required for detailed network and service power and energy efficiency forecasting is very unlikely to be easily sourced. Therefore, in this chapter, we describe a technique for constructing forecasts for future diurnal-cycle traffic that requires access to current or recent diurnal-cycle data and forecast values only for future peak and mean traffic. Using this approach, the monthly or yearly peak and mean traffic estimates can be calculated from their respective growth rates. These rates are fairly readily available from a range of organisations and websites [13–15]. The resulting, hour-by-hour traffic forecasts assume there is no dramatic change in the network or user behaviour (e.g. dramatic change in user behaviour caused by the advent of an unexpectedly popular new service or decline in demand for an existing service, or innovation).

The content of the following sections of this chapter involves a significant amount of mathematical analysis and derivation. For many readers, this level of detail may not be of interest. Therefore, for those readers who do not wish to engage in that level of detail, a summary of results is presented in Section 7.6. The summary is drafted to enable a user to apply the forecasting methods presented below with an understanding of their range of application and limitations.

7.1 DIURNAL-CYCLE TRAFFIC AND POWER CONSUMPTION

An important application of modelling is to provide forecasts for possible future growth in network power and energy consumption. As mentioned previously, this is important for network design and estimation of the future environmental footprint of a network or service.

Recall from Chapter 3 that the generic form for the power consumption of most network elements has the form:

$$P(t) = P_{idle} + ER(t) \tag{7.1}$$

For a network with N_E network elements, the total network power will have the form:

$$P_{Ntwk}(t) = \sum_{j=1}^{N_E} \left[P_{idle,j} + E_j R_j(t) \right] \tag{7.2}$$

where the subscript j indicates the respective values of the corresponding quantities for the j th element. For a network, we are interested in the peak power consumption because of the need to ensure enough electrical power is available to accommodate times of peak traffic. We are also interested in the total energy consumption over a time duration (such as a year) as a contributor to the network's environmental footprint. From (7.2) we see that peak power will occur at the time of peak traffic, t_{peak}:

$$R_{peak} = R(t_{peak}) \tag{7.3}$$

To calculate the energy consumption, $Q_{Ntwk}(t,T)$, over time duration T commencing at time t, we integrate (7.2)

$$Q_{Ntwk}(t,T) = \sum_{j=1}^{N_E} \left(P_{idle,j} T + E_j \int_s^{s+T} R_j(s) ds \right) = T \sum_j^{N_E} \left(P_{idle,j} + E_j \langle R_j;t \rangle_T \right) \tag{7.4}$$

where we have denoted the mean traffic through the j th network element over time interval t to $t + T$ by $\langle R_j;t \rangle_T$,

$$\langle R_j;t \rangle_T = \frac{1}{T} \int_t^{t+T} R_j(s) ds \tag{7.5}$$

Because both the peak and mean traffic are of interest, we will initially focus on modelling the evolution of these two quantities.

7.2 AVERAGING DIURNAL-CYCLE TRAFFIC DATA

Telecommunications network traffic can be studied over several timescales:

1. Sub-second to minute timescale in which the instantaneous data rate may vary significantly as various users transfer data in bursts with the data rate significantly reduced between bursts. Aggregate traffic (such as in the metro/edge and core networks) displays rapid and significant variation: for example, Netnod traffic [12].
2. Multi-minute to hourly timescale during which the traffic, when averaged over the timescale of between minutes and hours, displays an approximately cyclic behaviour on a daily timescale (diurnal cycle), as shown in Figure 7.1 [16, 17].
     On this scale there may be atypical variations in traffic level due to atypical events: for example, a major sporting event or a newsworthy incident.
3. Monthly-to-yearly timescale in which the traffic, when averaged on a timescale of day to week, often displays a (exponential) growth on a month to yearly timescale [7, 18].

The traffic behaviour on each of these timescales impacts the network design and deployment and hence power consumption. The monthly-to-yearly timescale influences the strategy for equipment deployment over the years of network operation. The hourly timescale provides the mean and peak traffic levels over the diurnal cycle. The diurnal peak traffic level is used to ascertain the amount of equipment required to be deployed during a given network upgrade. The sub-second timescale

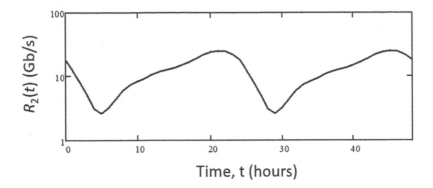

FIGURE 7.1   Typical diurnal cycles plotted over a 48-hour period.

variation plays a role in determining the maximum utilisation factor (ρ, defined below) that is used to determine the amount of capacity deployed relative to the diurnal-cycle peak traffic [19]. This parameter is introduced to accommodate random variations in the traffic at the short (order of hours or less) timescale (e.g. flash crowds, special events) [20].

To construct a relationship between these timescales, let $R_1(t)$ be the instantaneous data transfer rate at time t, $R_2(t)$ be the minute to hourly timescale rate, and $R_3(t)$ the day to weekly timescale rate. All three have units of bit/s. We have:

$$R_2(t) = \langle R_1; t \rangle_{T_1} = \frac{1}{T_1} \int_t^{t+T_1} R_1(s)\, ds \quad T_1 \sim minute\ to\ hour$$

$$R_3(t) = \langle R_1; t \rangle_{T_2} = \frac{1}{T_2} \int_t^{t+T_2} R_1(s)\, ds = \frac{1}{T_2} \sum_{n=0}^{(T_2/T_1)-1} \int_{t+nT_1}^{t+(n+1)T_1} R_1(s)\, ds$$

$$= \frac{1}{T_2} \sum_{n=0}^{(T_2/T_1)-1} T_1 \langle R_1; t+nT_1 \rangle_{T_1} = \frac{T_1}{T_2} \sum_{n=0}^{(T_2/T_1)-1} R_2(t+nT_1) \quad T_2 \sim day\ to\ week$$

(7.6)

where we have assumed T_2/T_1 is an integer. For example, we could set T_1 to be 1 hour and T_2 to be the number of hours in 7 days (168 hours).

7.3   MODELLING PEAK AND AVERAGE TRAFFIC GROWTH

We will focus on the timescales R_2 and R_3 over minutes to years. R_2 reflects the diurnal traffic, which is approximately cyclic, whereas R_3 reflects the monthly to annual growth in (mean) traffic. From data published by Cisco [7], we know that annual traffic has been exhibiting exponential growth. Therefore, we can write the following approximations for the cyclic behaviour of R_2 and the exponentially increasing profile of R_3:

$$\begin{aligned} R_2(t+T) &\approx R_2(t) & T &= 24\ hours \\ R_3(t+\tau) &\approx R_3(t)\beta^{\tau/T} & T &= 1\ year \end{aligned}$$

(7.7)

where β represents the annual mean traffic increase on the yearly scale.

The Cisco data also shows that the diurnal-cycle peak traffic is exponentially growing at a rate greater than the mean. Therefore, we also have:

$$R_2(t_{peak}+\tau) \approx R_2(t_{peak})\sigma^{\tau/T} \qquad T = 1\ year$$

(7.8)

where t_{peak} is the peak traffic time over the 24-hour diurnal cycle and σ is the annual rate of peak traffic increase on the yearly scale. The Cisco data [7] indicates that $\sigma > \beta$.

The relationships in (7.7) and (7.8) indicate that although the peak and mean traffic are increasing exponentially on the monthly and longer timescales, the traffic retains an approximately cyclic profile on the daily timescale. Therefore, for our purposes, the traffic profile R_2 will be represented using two timescales in a single equation: (1) a minute to hourly timescale denoted by t, which covers the traffic cycle over a day; and (2) a monthly and longer timescale denoted by τ, wherein the traffic is exponentially increasing over a year or more. Using this, R_2 can be expressed in terms of the diurnal-cycle peak traffic, $R_2(t_{peak},\tau)$, and a normalised diurnal-cycle variation, $f(t,\tau)$, relative to that peak value. That is:

$$R_2\left(t,\tau\right) = R_2\left(t_{peak},\tau\right)f\left(t,\tau\right) = R_{peak}\left(\tau\right)f\left(t,\tau\right) \tag{7.9}$$

where $R_{peak}(\tau) = R_2(t_{peak},\tau)$ is the peak diurnal-cycle traffic on timescale t. The normalised form, $f(t,\tau)$, describes the shape of the diurnal-cycle traffic and satisfies:

$$0 \leq f\left(t,\tau\right) \leq 1, \qquad f\left(t_{peak},\tau\right) = 1 \tag{7.10}$$

Away from the peak and trough of the diurnal cycle, these two timescales are distinguished by the inequality:

$$\left|\frac{\partial R_2\left(t,\tau\right)}{\partial t}\right| >> \left|\frac{\partial R_2\left(t,\tau\right)}{\partial \tau}\right| \tag{7.11}$$

That is, the rate of change of R_2 on the hourly timescale is typically much greater than the rate of peak traffic growth on the monthly timescale. On the τ timescale, we know the peak traffic is increasing exponentially, as described in (7.8), and therefore we have:

$$R_{peak}\left(\tau\right) \approx R_{peak}\left(\tau_0\right)\sigma^{(\tau-\tau_0)/T} \qquad T = 1 \; year \tag{7.12}$$

where τ_0 is the start year for the traffic growth modelling.

Because R_3 is the average of R_2 over duration T_2, we have:

$$\begin{aligned} R_3\left(\tau\right) &= \left\langle R_2\left(\tau\right)\right\rangle_{T_2} = R_{peak}\left(\tau\right)\left\langle f\left(\tau\right)\right\rangle_{T_2} \qquad T_2 = 24 \; hours \\ &= R_{peak}\left(\tau_0\right)\left\langle f\left(\tau_0\right)\right\rangle_{T_2} \beta^{(\tau-\tau_0)/T} = R_3\left(\tau_0\right)\beta^{(\tau-\tau_0)/T} \quad T = 1 year \end{aligned} \tag{7.13}$$

where

$$\left\langle f\left(\tau\right)\right\rangle_T = \frac{1}{T}\int_{\tau}^{\tau+T} f\left(t,\tau\right)dt \tag{7.14}$$

In summary, the following relationships apply amongst the various parameters above:

$$R_{peak}\left(\tau\right) = R_{peak}\left(\tau_0\right)\sigma^{(\tau-\tau_0)/T} \qquad T = 1 \; year \tag{7.15}$$

$$\begin{aligned} R_3\left(\tau\right) &= \left\langle R_2\left(\tau\right)\right\rangle_{T_2} = \beta^{(\tau-\tau_0)/T}\left\langle R_2\left(\tau_0\right)\right\rangle_{T_2} = \beta^{(\tau-\tau_0)/T}R_3\left(\tau_0\right) \\ &= \sigma^{(\tau-\tau_0)/T}R_{peak}\left(\tau_0\right)\left\langle f\left(\tau\right)\right\rangle_{T_2} = \beta^{(\tau-\tau_0)/T}R_{peak}\left(\tau_0\right)\left\langle f\left(\tau_0\right)\right\rangle_{T_2} \quad T_2 = 24 \, hours \end{aligned} \tag{7.16}$$

$$\left\langle f(\tau) \right\rangle_{T_2} = \left(\frac{\beta}{\sigma} \right)^{(\tau - \tau_0)/T} \left\langle f(\tau_0) \right\rangle_{T_2} \quad T_2 = 24 \ hours \tag{7.17}$$

In some cases, we will focus on the annual changes in traffic, in the year n after a start year 0. In this case, the terms $(\tau - \tau_0)/T$ will be replaced with integer $n = \lceil (\tau - \tau_0)/T \rceil$ corresponding to the number of years between time τ for the n th year and time τ_0 for the start time (where $\lceil x \rceil$ is the largest integer less than or equal to x). Let $R_{peak}^{[n]}$ denote the peak traffic at the start of year n and $f^{[n]}(t)$ denote the normalised diurnal cycle at the start of year n. Therefore, using the equations above, we can write for the peak traffic at the start of year n:

$$R_{Peak}^{[n]} = R_{Peak}^{[0]} \sigma^n \tag{7.18}$$

$$\left\langle R^{[n]} \right\rangle_{T_2} = \beta^n \left\langle R^{[0]} \right\rangle_{T_2} = \sigma^n R_{Peak}^{[0]} \left\langle f^{[n]} \right\rangle_{T_2} = \beta^n R_{Peak}^{[0]} \left\langle f^{[0]} \right\rangle_{T_2} \tag{7.19}$$

The value of $\langle f^{[n]} \rangle_{T2}$ is given by:

$$\left\langle f^{[n]} \right\rangle_{T_2} = \left(\frac{\beta}{\sigma} \right)^n \left\langle f^{[0]} \right\rangle_{T_2} \tag{7.20}$$

Below we use the discrete $[n]$ forms to model several scenarios for deploying upgrades and equipment additions to the network on an annual basis.

7.4   MODELLING TIME-OF-DAY TRAFFIC GROWTH

As discussed above, peak and average network traffic values provide important information on network power and energy consumption because they parametrise traffic throughput (*peak* providing maximum power provisioning requirements and *average* leading to a measure of the environmental footprint). However, there are cases where we need the details of the traffic load over a full diurnal cycle: for example, quantifying the benefits of using low-power states in network equipment when traffic levels are low [21, 22]. An important reason for modelling the full diurnal cycle is assessing the power consumption of a service, which can be highly dependent on its traffic as well as other network traffic. Estimating the power and energy consumption of a service can be more complicated because the diurnal traffic cycle of a specific service may not time align with the overall network traffic. As will be discussed below, this can affect the allocation of network power to that service.

Therefore, it is very advantageous to have a method for constructing a credible forecast for the time-of-day traffic over a full diurnal cycle. Unfortunately, this can be a very complicated matter with many influences that impact on time-of-day traffic values [23, 24]. It is unlikely that a highly accurate model can be constructed without the collection of a very large amount of data and the use of very sophisticated analysis. This makes constructing a rigorously accurate approach to forecasting future diurnal cycles very implausible. Therefore, the approach discussed here is a very simplified (although not simple) heuristic model that constructs a diurnal profile using just the peak and mean traffic growth rates. These two parameters are often found to be publicly available or can be determined from publicly available data [12–14, 25].

Our model uses traffic growth trend data such as that published by Cisco [7, 9]. That trend data indicates that the rate of annual growth in daily peak traffic is faster than the annual growth rate in daily average traffic [7]. This trend applies to the national and global traffic. Therefore, to apply this construction at a more local level would require confirmation that these relative growth trends for peak and average data apply at that level.

Before describing the details of this model, it is important to appreciate the assumptions used to construct the forecast. Publicly available data (such as the Cisco forecasts) indicates that peak-time traffic is growing at an annual rate that is greater than mean traffic. Assuming this is correct, the traffic growth rate at a specific time of day can be considered to be dependent upon the traffic at that time of day. Further, the growth rate has to be greater for times of higher traffic. This outcome can be secured by assuming the annual growth rate of diurnal-cycle traffic at time t, $0 \leq t \leq 24$ hours, is proportional to the traffic at that time. Mathematically, this can be expressed as follows:

$$\frac{\partial}{\partial \tau} R_2(t,\tau) = \varepsilon(t,\tau) R_2(t,\tau) \qquad (7.21)$$

where $\varepsilon(t_{peak},\tau) > \varepsilon(t,\tau)$ for all times t away from t_{peak}.

Another assumption is that future traffic growth is basically "business as usual" in that no new, disruptive influences come into play. (For example, one may have a significant economic downturn that will reduce traffic or the introduction of a new "fad" service that dramatically increases traffic.) This is a major simplification of the real situation, but there is effectively no way of modelling future events that are currently unknown.

Solving (7.21) for R_2 gives:

$$R_2(t,\tau) = \exp\left(\int_0^\tau \varepsilon(t,s)\,ds \right) R_2(t,0) \qquad (7.22)$$

Following the Cisco data, we denote the annual traffic growth rate at peak time, t_{peak}, as σ. Therefore, we need $\varepsilon(t_{peak},\tau) = \sigma^{t/T} > \varepsilon(t,\tau)$ for all $t \neq t_{peak}$ with $T = 1$ year. Using the Cisco trend results, we also have:

$$R_2(t_{peak},\tau) = \sigma^{\tau/T} R_2(t_{peak},0) = \exp\left(\frac{\tau}{T} \ln \sigma \right) R_2(t_{peak},0)$$

$$\langle R_2(\tau) \rangle_{T_2} = \beta^{\tau/T} \langle R_2(0) \rangle_{T_2} = \exp\left(\frac{\tau}{T} \ln \beta \right) \langle R_2(0) \rangle_{T_2} \qquad (7.23)$$

where $T = 1$ year and $T_2 = 24$ hours.

The conditions in (7.23) do not yield a unique solution to (7.22) that provides a traffic forecast over the full diurnal cycle. Therefore, we have to make some further assumptions. We refine our assumption that the growth rate for traffic at a specific time of day is proportional to the traffic at that time by giving $\varepsilon(t,\tau)$ a specific form:

$$\varepsilon(t,\tau) = \frac{\ln \sigma}{T} - \frac{\partial \kappa(\tau)}{\partial \tau}\big(f(t_{peak},0) - f(t,0) \big) = \frac{\ln \sigma}{T} - \frac{\partial \kappa(\tau)}{\partial \tau}\big(1 - f(t,0) \big) \qquad (7.24)$$

where we have used (7.10). By adopting this form, we are assuming the growth in traffic at a time t in the diurnal cycle is determined from the traffic at that time in year 0, $f(t,0)$, relative to the peak traffic in year 0, $f(t_{peak},0)$. The growth rate of traffic at time t is suppressed relative to the peak-time growth rate, σ. The degree of suppression is given by κ and will be set so that the average traffic growth rate is β. This accords with our assumption that the traffic growth is "business as usual", because the diurnal-cycle profile in year 0 effectively determines the growth at that time over the coming years.

Placing (7.24) into (7.22) gives:

$$R_2(t,\tau) = \exp\left(\frac{\ln\sigma}{T}\tau - \kappa(\tau)\big(1 - f(t,0)\big)\right)R_2(t,0)$$

$$= \sigma^{\tau/T}\exp\left(-\kappa(\tau)\big(1 - f(t,0)\big)\right)R_{peak}(0)f(t,0) \tag{7.25}$$

Looking at (7.25), we see that setting $t = t_{peak}$ gives us the required annual growth of traffic at peak time, in accord with (7.23). To determine the form for $\kappa(\tau)$, we use the fact that the average traffic has an annual growth rate given by (7.23). Therefore, we require:

$$\big\langle R_2(\tau)\big\rangle_{T_2} = \frac{R_{peak}(\tau)}{T_2}\int_0^{T_2}\exp\left(-\kappa(\tau)\big(1 - f(t,0)\big)\right)f(t,0)\,dt$$

$$= \beta^{\tau/T}\frac{R_{peak}(0)}{T_2}\int_0^{T_2}\big(1 - f(t,0)\big)\,dt \tag{7.26}$$

In this equation, only the quantity $\kappa(\tau)$ is unknown. Therefore, we numerically solve for $\kappa(n)$ for a given year n. To solve for $\kappa(n)$, we approximate the integral with a sum over a diurnal cycle using hourly traffic. The resulting equation to be solved for $\kappa(n)$ is:

$$\sum_{j=1}^{24}\exp\left(-\kappa(n)\big(1 - f(t_j,0)\big)\right)f(t_j,0) - \left(\frac{\beta}{\sigma}\right)^n\big(1 - f(t_j,0)\big) = 0 \tag{7.27}$$

The diurnal traffic at hour t_j for year n is then given by

$$R_2(t_j,n) = \sigma^n R_{peak}(0)f(t_j,0)\exp\left(-\kappa(n)\big(1 - f(t_j,0)\big)\right) \tag{7.28}$$

To demonstrate the results provided by this construction, in Figures 7.2 and 7.3 we have plotted four years of projected two-day diurnal cycles based on diurnal-cycle data in "Year 0". The Year 0 data is very typical of a generic diurnal cycle, with peak traffic around 10 pm, a significant reduction in traffic from then until the early morning, and then gradual growth from mid-morning back up to the peak. Adopting the 2019 Cisco data [7] and forecast for 2017 to 2022, the yearly growth parameter for peak traffic is $\sigma = 1.37$ and, for mean traffic, $\beta = 1.30$. Figure 7.2 uses a log vertical scale whereas Figure 7.3 uses a linear vertical scale for the same diurnal cycle.

In some cases, annual traffic growth is more realistically represented by linear growth. This may also apply if the annual traffic growth is exponential but at a very slow rate. This approximation applies because $(1 + x)^n \approx 1 + nx$ for $x \ll 1$. In this case, keeping with the requirement that peak-hour traffic is increasing more rapidly than average traffic, we replace (7.21) with:

$$\frac{\partial}{\partial\tau}R_2(t,\tau) = \varepsilon(t) \tag{7.29}$$

which gives:

$$R_2(t,\tau) = R_2(t,0) + \varepsilon(t)\tau \tag{7.30}$$

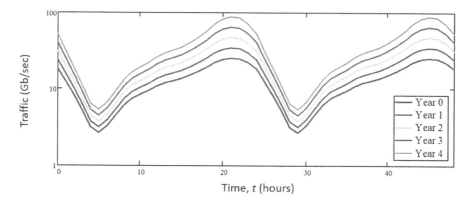

FIGURE 7.2 Diurnal cycles constructed using (7.28) plotted with a log vertical axis scale. The input data required to construct these cycles are the diurnal-cycle data for Year 0, the annual peak traffic growth rate ($\sigma = 1.37$ in this case [7]) and annual average traffic growth rate ($\beta = 1.30$ in this case [7]). It is also assumed no new services that have a dramatic impact on future traffic (years 1, 2, 3, 4) come into play.

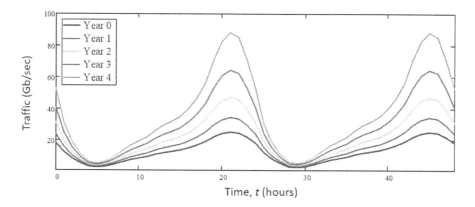

FIGURE 7.3 Diurnal cycles constructed using (7.28) for the same data as in Figure 7.2, however plotted with a linear vertical axis scale. With a linear vertical scale one can see the significant growth in peak traffic relative to the average traffic.

with the requirements:

$$\varepsilon\left(t_{peak}\right) = \sigma$$

$$\frac{1}{T_2} \int_0^{T_2} \varepsilon\left(t\right) dt = \beta \tag{7.31}$$

Following the approach given above, we give $\varepsilon(t)$ the form:

$$\varepsilon\left(t\right) = \sigma - \kappa\left(1 - f\left(t,0\right)\right) \tag{7.32}$$

Using (7.32) and solving for κ we get:

$$\kappa\left(t\right) = \frac{\sigma - \beta}{1 - \left\langle f\left(t,0\right)\right\rangle_{T_2}} \tag{7.33}$$

This gives for the diurnal cycle at time τ, with linearly increasing traffic:

$$R_2(t,\tau) = R_2(t,0) + \tau \left(\sigma - \frac{(\sigma - \beta)}{\left(1 - \langle f(t,0) \rangle_{T_2}\right)} (1 - f(t,0)) \right) \tag{7.34}$$

Based on the examples given above, we can see that the construction method given above can provide approximate shapes for future diurnal cycles. However, it must be used with due care. It should only be used if there is no access to detailed diurnal-cycle forecast data over several years.

7.5   DIURNAL TRAFFIC AND USER TYPES

The wired access network is often a significant contributor to telecommunications network energy consumption [4, 26]. A widely proposed strategy for reducing access network power consumption has been to put customer premises equipment into a low-power sleep state at times of inactivity [27–30]. To gain an insight as to the effectiveness of this strategy and its potential cost-effectiveness (given it may require retrofitting or upgrading of equipment in customer premises), we need an estimate of potential energy savings when idle or sleep modes are implemented.

To ascertain these potential energy savings, a customer traffic model over the diurnal cycle is needed to provide estimates of the amount of customer equipment that can be placed into a low-power state and the duration of this state. In this section, we will describe a model that provides such an estimate.

Again, we face the problem of the lack of availability of sufficiently detailed user traffic data, in this case for access networks. As before, this model is built using a limited amount of data on network traffic and counts of active users. To do this, we again use network traffic trends and several frequently reported observations.

As noted above, from Cisco's white papers, we know peak traffic growth is faster than mean traffic growth. We endeavour to create a simple model that can replicate this behaviour with the added detail of identifying user types. In this context, a "user type" is a group of users, all of whose traffic pattern over a diurnal cycle is sufficiently similar to allow their behaviour to be represented by a characteristic diurnal-cycle profile. We are primarily interested in two types of end users: (a) those whose diurnal cycle enables their customer premises equipment to be put into sleep mode during some hours of the day, without degrading their quality of service; and (b) those whose customer premises equipment must be left active all day. We want to split the total forecast access network traffic into these two end-user-type contributions. Doing this will provide traffic forecasts that enable calculation of potential energy savings using low-power states in the access network.

The forecasting of total traffic was discussed above. In this section, we will show that a diurnal traffic growth trend that has peak-hour traffic growing faster than average traffic justifies splitting the users into these two types.

Let $N(t,n)$ be the number of active users of an access network at time t during a diurnal cycle in year n from a start year that we set to be $n = 0$. Denoting by $A^{(k)}(t,n)$ the access speed (bit/s) of the k th user ($k = 1,2,\dots N(t,n)$), then we have, for the total access network traffic, $R'(t,n)$, in year n at time t:

$$R'(t,n) = \sum_{k=1}^{N(t,n)} A^{(k)}(t,n) + R_{oh} \tag{7.35}$$

The precise definition of the user "access speed" can be non-trivial. For example, it can be defined as the outcome of access speed measurements. In this case, the details of the measurement methodology will be important [31–34]. One can adopt the measurement provided by available packet analysers [35–37]. Another approach is to use download time for given file sizes [38]. In (7.35) R_{oh} is the network overhead traffic that is independent of the number of active users. For example, R_{oh} will include network monitoring and management traffic. We assume this traffic is relatively independent of the year, n, because being network management system traffic is unlikely to increase significantly over the years of operation of the network.

We absorb R_{oh} into $R'(t,n)$ by defining user-dependent network traffic $R(t,n) = R'(t,n) - R_{oh}$. We focus on $R(t,n)$ from now on.

We start by assuming there is only one type of user and their behaviour over the diurnal cycle does not change except for increasing traffic per user over the coming years (i.e. "business as usual"). With this assumption, we can define $A(t,n)$ be the mean access speed per active user (bit/s) at time t during the day in year n. That is:

$$A(t,n) = \frac{R(t,n)}{N(t,n)} \tag{7.36}$$

Using the mean access speed, the total traffic at time t in year n can be written as:

$$R(t,n) = N(t,n)A(t,n) \tag{7.37}$$

From this, the mean traffic (over a diurnal cycle) in year n can be expressed as:

$$\langle R(n) \rangle_{T_2} = \langle N(n)A(n) \rangle_{T_2} \tag{7.38}$$

where $T_2 = 1$ day. In this section, all averages will be taken over a day; therefore, we will drop the T_2 subscript for the rest of this section. The peak traffic at peak-traffic time, t_{peak}, can be expressed as follows:

$$R(t_{peak},n) = N(t_{peak},n)A(t_{peak},n) \tag{7.39}$$

Again, we refer to the Cisco traffic data, which shows that the annual growth of the mean and peak traffic is exponential, and so they can be expressed in the form:

$$R(t_{peak},n) \approx R(t_{peak},0)\sigma^n$$
$$\langle R(n) \rangle \approx \langle R(0) \rangle \beta^n \tag{7.40}$$

where $\sigma > \beta$.

From broadband growth statistics, we know that the number of users (N) and mean access speed per user (A) are also growing exponentially [10]. If the number of users is growing exponentially, we can write:

$$N(t,n) \approx N(t,0)\gamma^n \tag{7.41}$$

If we assume there is only one type of user and their behaviour over the diurnal cycle does not change except for exponentially increasing traffic per user over the coming years (i.e. "business as usual"), we can write:

$$A(t,n) \approx A(t,0)\delta^n \tag{7.42}$$

Based on these two assumptions, we have:

$$\langle N(n) \rangle \approx \langle N(0) \rangle \gamma^n, \quad N(t_{peak}, n) \approx N(t_{peak}, 0) \gamma^n$$
$$\langle A(n) \rangle \approx \langle A(0) \rangle \delta^n, \quad A(t_{peak}, n) \approx A(t_{peak}, 0) \delta^n \tag{7.43}$$

If this is an accurate representation of the user population (i.e. all users have a roughly similar diurnal-cycle profile), then we would have:

$$\langle R(n) \rangle = \langle N(n) A(n) \rangle \approx \langle N(0) \gamma^n A(0) \delta^n \rangle = \langle R(0) \rangle (\gamma\delta)^n$$
$$R(t_{peak}, n) = N(t_{peak}, n) A(t_{peak}, n) \approx N(t_{peak}, 0) \gamma^n A(t_{peak}, 0) \delta^n = R(t_{peak}, 0)(\gamma\delta)^n \tag{7.44}$$

That is, assuming all users are similar results in the peak and mean data rates increasing annually at the same rate, which is contrary to the Cisco network measurements. This means it is inappropriate to assume a homogeneous population in which all users are well represented by a single average quantity $A(t,n)$. The simplest extension of this assumption is to assume there are two groups of users, the users in each group being approximately homogeneous but the typical behaviour of the members of one group being distinctly different from the typical behaviour of the members of the other.

We now show that the observed traffic trend of mean traffic annual growth rate being less than peak traffic annual growth rate can be used to identify two distinct groups of users. Although multiple user types have been identified in published literature [39–41], in our case we are only interested in categorising the total population into two sub-population types. Further, these two types align with our interest in estimating the power savings that can be secured by placing inactive user equipment into a sleep mode.

Splitting the user population into two user "Types" enables us to match the mean and peak annual traffic growth rates to the overall traffic profile. We express the diurnal-cycle traffic $R(t,n)$ as the sum of the traffic generated by two user sub-populations:

$$R(t,n) = R_a(t,n) + R_b(t,n) = N_a(t,n) A_b(t,n) + N_b(t,n) A_b(t,n) \tag{7.45}$$

where N_a and A_a are the number of active users and mean access speed per user of Type (a) users, respectively. Similarly, N_b and A_b are the number of users and mean access speed per user of Type (b), respectively. Also, we write:

$$R_a(t,n) = N_a(t,n) A_a(t,n), \qquad R_b(t,n) = N_b(t,n) A_b(t,n) \tag{7.46}$$

We continue with the assumptions of exponential growth for both groups and that their behaviour over the diurnal cycle does not change except for increasing traffic per user over the coming years. Hence:

$$R_a(t,n) \approx R_a(t,0) \eta_a^n, \qquad R_b(t,n) \approx R_b(t,0) \eta_b^n \tag{7.47}$$

Therefore, from (7.40) and (7.47) as well as the observed growth rates for average and peak traffic, we have:

$$\frac{\langle R(n) \rangle}{\langle R(0) \rangle} = \frac{\langle R_a(n) \rangle + \langle R_b(n) \rangle}{\langle R(0) \rangle} \approx \frac{R_a(0)\eta_a^n}{\langle R(0) \rangle} + \frac{\langle R_b(0) \rangle \eta_b^n}{\langle R(0) \rangle} \approx \beta^n \tag{7.48}$$

and

$$\frac{R\left(t_{peak},n\right)}{R\left(t_{peak},0\right)} = \frac{R_a\left(t_{peak},n\right) + R_b\left(t_{peak},n\right)}{R\left(t_{peak},0\right)} \approx \frac{R_a\left(t_{peak},0\right)\eta_a^n}{R\left(t_{peak},0\right)} + \frac{R_b\left(t_{peak},0\right)\eta_b^n}{R\left(t_{peak},0\right)} \approx \sigma^n \quad (7.49)$$

For both (7.48) and (7.49) to both be true, we need Type (a) users to dominate one of these two equations and Type (b) users to dominate the other. If one type of users dominates both equations, the mean and peak growth rates will be equal, which we know is not the case. Therefore, without loss of generality, we can assume Type (a) users dominate the growth in average traffic:

$$\frac{\langle R(n) \rangle}{\langle R(0) \rangle} \approx \frac{\langle R_a(n) \rangle}{\langle R(0) \rangle} \approx \frac{\langle R_a(0) \rangle \eta_a^n}{\langle R(0) \rangle} \approx \beta^n \quad (7.50)$$

With that assumption, Type (b) users will dominate the growth in peak-hour traffic:

$$\frac{R\left(t_{peak},n\right)}{R\left(t_{peak},0\right)} \approx \frac{R_b\left(t_{peak},n\right)}{R\left(t_{peak},0\right)} \approx \frac{R_b\left(t_{peak},0\right)\eta_b^n}{R\left(t_{peak},0\right)} \approx \sigma^n \quad (7.51)$$

With this, we have:

$$\langle R_a(n) \rangle \approx \langle R_a(0) \rangle \eta_a^n >> \langle R_b(n) \rangle \approx \langle R_b(0) \rangle \eta_b^n$$
$$R_b\left(t_{peak},n\right) \approx R_b\left(t_{peak},0\right)\eta_b^n >> R_a\left(t_{peak},n\right) \approx R_a\left(t_{peak},0\right)\eta_a^n \quad (7.52)$$

Setting $n = 0$, we have:

$$\langle R(0) \rangle \approx \langle R_a(0) \rangle, \quad \eta_a \approx \beta$$
$$R\left(t_{peak},0\right) \approx R_b\left(t_{peak},0\right), \quad \eta_b \approx \sigma \quad (7.53)$$

This result is telling us the following:

(a) In total over the diurnal cycle, T, Type (a) users consume much more total traffic (i.e. total bits over time T) than Type (b) users:

$$\langle R_a(n) \rangle T >> \langle R_b(n) \rangle T \quad (7.54)$$

(b) At peak traffic time, Type (b) users have a much higher total data rate than Type (a) users. That is:

$$R_b\left(t_{peak},n\right) >> R_a\left(t_{peak},n\right) \quad (7.55)$$

Because $\sigma > \beta$, this model will only be applicable up to a certain number of years, n, because, even if $\langle R_a(0) \rangle \beta^n >> \langle R_b(0) \rangle \sigma^n$ for small integer values of n, this inequality will ultimately cease to hold unless the values for the base year ($n = 0$) are adjusted. We will limit our discussion to those years for which this inequality holds.

Both $R_a(t,n)$ and $R_b(t,n)$ will display approximately cyclic behaviour over the day time scale. That is, $R_a(t+T_2,n) \approx R_a(t,n)$ and $R_b(t+T_2,n) \approx R_b(t,n)$. In addition, the requirements in (7.52) will place constraints on the relative shapes of these diurnal cycles. In particular, using (7.54) and (7.55) we have:

$$\frac{R_b\left(t_{peak},n\right)}{\left\langle R_b\left(n\right)\right\rangle} >> \frac{R_a\left(t_{peak},n\right)}{\left\langle R_b\left(n\right)\right\rangle} >> \frac{R_a\left(t_{peak},n\right)}{\left\langle R_a\left(n\right)\right\rangle} > 1 \qquad (7.56)$$

This means the diurnal cycle of R_b is much deeper than that of R_a. That is, $R_b(t,n)$ has a much wider range of relative values than $R_a(t,n)$.

Although we now have some guidance regarding the relative traffic contributions of the two types of users to total traffic, to construct an estimate of potential power savings using low-power sleep states in user equipment, we need the numbers, $N_a(t,n)$ and $N_b(t,n)$, of the two types. To determine values for $N_a(t,n)$ and $N_b(t,n)$ we need additional information. To obtain this additional information we refer to Figure 7.6 in the Appendix. These plots are measured throughput, $R(N)$ (bit/s) as a function of the number of active users, N, for several different access networks. As shown in the Appendix, the plots in Figure 7.6 are very well described by a linear profile, as shown in Figure 7.4, with:

$$R\left(N\right) = H\left(N - N_{min}\right) + R_{min} + R_{oh} \qquad (7.57)$$

where $R(N)$ is the total network throughput for N active users, $H = (R_{peak} - R_{min})/(N_{peak} - N_{min})$ is the slope of the profile (bit/s/user), R_{min} is the minimum throughput (excluding R_{oh}) ($R_{min} + R_{oh} = B$ for the linear profile case in the Appendix) and N_{min} is the minimum number of active users over the diurnal cycle. This profile shape applies when the access network is not subject to congestion [38]. That is, when the traffic load on the network is well below the maximum load the network can accommodate. In this case, the average incremental access speed per user, H, is independent of the number of users, N, [38].

Measurements indicate that it is only when an access network load increases to a level too close to the network's maximum load that users find their access speed, is reduced as the network operator endeavours to distribute the remaining available bandwidth across the increasing number of active users. When a network is subject to congestion, users' access speeds typically reduce as the other users come online [38]. Therefore, by-and-large, because H is only modified when the network becomes congested, for a network that is not congested H is independent of time over the diurnal cycle.

On the basis that the population of users can be separated into Type (a) and Type (b), with no congestion we can set the average access data rates A_a and A_b across these two user types to

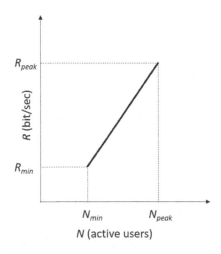

FIGURE 7.4   Linear approximation for the plot of traffic throughput, R, vs number of active users, N.

be approximately constant over the duration of their activity [39, 40]. That is, on average the members of each user Type desire similar access speeds and can secure their desired access speed, approximately A_a or A_b, at any time over the diurnal cycle. With no capping of access speeds when there is no congestion we have:

$$A_a\left(t,n\right) \approx A_a\left(n\right), \qquad A_b\left(t,n\right) \approx A_b\left(n\right) \tag{7.58}$$

With A_a and A_b constant over the diurnal cycle, result (7.58) becomes:

$$\frac{N_b\left(t_{peak},n\right)}{\left\langle N_b\left(n\right)\right\rangle} \gg \frac{N_a\left(t_{peak},n\right)}{\left\langle N_a\left(n\right)\right\rangle} > 1 \tag{7.59}$$

Therefore, the diurnal cycle of the number of Type (b) users, $N_b(t,n)$, is much more peaked around the peak traffic time than that of Type (a) users, $N_a(t,n)$. This, in combination with (7.54) means Type (a) users are typically active for a larger proportion of the diurnal cycle than Type (b) users, whereas most Type (b) users are active only during the hours around the peak traffic time. Further, (7.55) indicates that, around the time of peak traffic, the total traffic generated by Type (b) users is much greater than that generated by Type (a) users. This can occur because either, when Type (b) users are active they use much greater access data rates than Type (a) users (i.e. $A_b(n) \gg A_a(n)$) and/or there are many more active Type (b) users around t_{peak} than Type (a) users ($N_b(t_{peak},n) \gg N_a(t_{peak},n)$).

We also have from the inequalities above:

$$\frac{\left\langle N_a\left(n\right)\right\rangle}{\left\langle N_b\left(n\right)\right\rangle} \gg \frac{A_b\left(n\right)}{A_a\left(n\right)} \gg \frac{N_a\left(t_{peak},n\right)}{N_b\left(t_{peak},n\right)} \tag{7.60}$$

With our focus on energy, we further develop our understanding of these behaviours of these two user types to see what energy savings that may be attained by implementing low-power sleep states in the customer premises equipment of one or other of these two user types. To do this we need to focus on the numbers of each user type, $N_a(t,n)$ and $N_b(t,n)$ as given in (7.46). At this point we don't have any specific information on the average access speeds of these two groups, A_a and A_b. Therefore, we consider the two cases of: $A_a \gg A_b$ and $A_a \ll A_b$.

Case $A_a \gg A_b$: In this case from (7.55) we get:

$$\frac{N_b\left(t_{peak},n\right)}{N_a\left(t_{peak},n\right)} \gg \frac{A_a\left(n\right)}{A_b\left(n\right)} \gg 1 \tag{7.61}$$

In this situation although the access speed of Type (b) users is less than that of Type (a), enough Type (b) users come online around the peak time so that, as required by (7.55), their traffic dominates the network at that time.

Case $A_b \gg A_a$: In this case from (7.54),

$$\frac{\left\langle N_a\left(n\right)\right\rangle}{\left\langle N_b\left(n\right)\right\rangle} \gg \frac{A_b\left(n\right)}{A_a\left(n\right)} \gg 1 \tag{7.62}$$

In this situation, although the average number of active Type (a) users may be much greater over the diurnal cycle than Type (b) users, the access speed of Type (b) users is still large enough so that they dominate the traffic around peak times.

In all cases, we see that it is the Type (b) users that are mainly active around the time of peak traffic. This means that although there may be energy savings by placing Type (a) users' equipment into a sleep state, because Type (b) users are primarily active only around the peak traffic time, it is more likely the major savings will be attained by the network operator focussing on using sleep states in Type (b) user equipment. Starting with (7.2), we assume each user has one unit of customer premises equipment, that being the broadband gateway that connects the home to the access network. Therefore, the total number of users, $N_{tot} = N_E$ the number of customer premises equipment units. (In the following we drop the year label n to simplify the notation, with the understanding that the equations apply to a given year.) We separate the number of network elements, N_E, into two groups aligned with the two customer types. Let $N_{a,tot}$ be the total number of Type (a) users and $N_{b,tot}$ be the total number of Type (b) user. Let $N_{a,tot} = \chi N_{tot}$, hence $N_{b,tot} = (1 - \chi)N_{tot}$. The energy savings are attained by placing the customer premises equipment of inactive Type (b) users into a sleep state such that $P_{sleep} \ll P_{idle}$. Therefore, if there are $N_b(t)$ active Type (b) users at the time t, the power savings compared to no sleep state (i.e. all idle equipment consuming power P_{idle}) is:

$$P_{save}(t) = (P_{idle} - P_{sleep})(N_{b,tot} - N_b(t)) \qquad (7.63)$$

Therefore the energy savings over a diurnal cycle will be:

$$Q_{sav}(T) = (P_{idle} - P_{sleep})\left(N_{b,tot} - \int_0^T N_b(t)\,dt\right)$$
$$= (P_{idle} - P_{sleep})(N_{b,tot} - T\langle N_b\rangle) \qquad (7.64)$$

Therefore, we need an estimate of $\langle N_b\rangle$. If the access network is not congested the $R(N)$ diurnal-cycle profile is approximately linear as in (7.57) and we know: R_0, N_0, $R(t_{peak})$, $N(t_{peak})$ and the slope, H. To secure and estimate of $\langle N_b\rangle$ will require additional network measurements to identify the Type (b) users and their diurnal profile. If this information is not available, an approximation for $\langle N_b\rangle$ may be obtainable by using the diurnal-cycle data $R(N)$, plus several assumptions.

In summary, the user model described above provides a simple description that aligns with the currently available data on network traffic growth [7, 10] and user behaviour [35, 36, 39–42]. The model shows that if peak-hour traffic annual growth rate is consistently faster than average traffic annual growth rate then the population of users can be split into (at least) two types of users. A significant proportion of Type (a) users are active across most of, if not the full, diurnal cycle. These users are unlikely to be good candidates for saving energy by placing their equipment into a low-power sleep state for long durations. The vast majority of Type (b) users are active only around the time of peak traffic. Members of this group are likely to be good candidates for applying a low-power sleep state to their customer premises equipment to provide potentially reasonable energy savings.

The Cisco Visual Networking Index: Global Mobile Data Traffic Forecast Update, 2017–2022 [7] has identified typical Type (b) users as those customers who use their Internet connection to view streaming video content. The report states;

> One consequence of the growth of video in both fixed and mobile contexts is the resulting acceleration of busy-hour traffic in relation to average traffic growth. Video usage tends to occur during evening hours and has a "prime time," unlike general web usage that occurs throughout the day. As a result, more video usage means more traffic during the peak hours of the day.

This group has been identified in multiple surveys of Internet user behaviour in which it has been found that these users do not change the time of viewing video content (it remains around the hours around 8 pm to 11 pm), rather they change the source of the content from broadcast TV to video streaming [43].

Therefore, to construct this estimate we need to determine or estimate value of $\langle N_b \rangle$. We want to do this without having to undertake the collection of detailed information on the hourly behaviour of users. Of course, without this information the estimate of $\langle N_b \rangle$ will be an approximation the accuracy of which will depend on the amount of information we can extract from general knowledge of the network's diurnal cycle and Internet user types.

We start by adopting an approach used in several publications [29, 40, 42, 44]. These report that in many cases user traffic roughly follows widely reported phenomenon called the "Pareto Principle" also called "80/20 rule" [45] in that 80% of the total monthly traffic is due to 20% of the users. In communications networks, the 20% of the user population that generates 80% of the traffic are sometimes referred to as Elephants [39–41]. If we identify this group as Type (a) users then Type (b) users will constitute 80% of total users which are responsible for 20% of the total daily traffic (i.e. total downloaded bits). Further, that 20% of the total daily traffic is concentrated around the peak traffic hour. This aligns well with the Cisco report quoted above.

In some cases the Pareto Principle, is not expressly "80/20", for example, Yang *et al.* [42] give an example of a "95/5" rule in that 5% of applications contribute 95% of traffic. Kramer *et al.* [44] report that some network providers have observed "90/10" in that 10% of users generate 90% of traffic. Let N_{tot} be the total number of users, therefore $N_{tot} \geq N(t_{peak})$. Based on the discussion leading to (7.63) above, we generalise the "80/20 rule" to a "$(1 - \chi)/\chi$ rule" by introducing parameter χ such that χN_{tot} are Type (a) users (i.e. $N_{a,tot} = \chi N_{tot}$) and are responsible for proportion $(1 - \chi)$ of total user bits over a diurnal cycle, $\langle R \rangle T$. That is:

$$A_a \chi N_{tot} T \approx \left(1 - \chi\right)\langle R \rangle T \tag{7.65}$$

This requirement introduces a relationship between mean user Type numbers $\langle N_a \rangle$ and $\langle N_b \rangle$ and their access speeds A_a and A_b. We can see this using (7.45) from which we have $\langle R \rangle = A_a \langle N_a \rangle + A_b \langle N_b \rangle$ giving:

$$A_b \left(1 - \chi\right)\langle N_b \rangle \approx A_a \left(\chi N_{tot} - \left(1 - \chi\right)\langle N_a \rangle \right) \tag{7.66}$$

Applying the linear profile to (7.57) and separating into the two user types, we can write:

$$\Delta R\left(t\right) = R\left(t\right) - R_{min} = A_a \Delta N_a(t) + A_b \Delta N_b\left(t\right) \approx H\left(\Delta N_a\left(t\right) + \Delta N_b\left(t\right)\right) \tag{7.67}$$

In this we have defined $\Delta N_a(t) = N_a(t) - N_{a,min}$ and $\Delta N_b(t) = N_b(t) - N_{b,min}$. That is,

$$R_{min} = N_{a,min}A_a + N_{b,min}A_b + R_{oh} \tag{7.68}$$

where R_{oh} is overhead traffic required for monitoring and management of the network. If $A_a \approx A_b$, then the form in (7.67) immediately provides the appropriate linear profile $R(N) = HN + R_{min}$. However, published measurements indicate that in most cases $A_a \neq A_b$ [35, 39, 40, 42]. In this case, the linear profile indicates that, on average, the numbers of users of each type typically become active in proportion. That is, $\Delta N_b(t) \approx \zeta \Delta N_a(t)$ where $\zeta =$ constant. If this was not the case, then if one access rate is significantly greater than the other, e.g. $A_b \gg A_a$, then as Type (b) users come online around the peak traffic time (recalling $R_b(t_{peak}) \gg R_a(t_{peak})$), then we would see a significant increase in slope of the N vs $R(N)$ profile as the number of active users, N, increased. The value of ζ is given by:

$$\zeta = \frac{\Delta N_b \left(t_{peak}\right)}{\Delta N_a \left(t_{peak}\right)} = \frac{N_b \left(t_{peak}\right) - N_{b,min}}{N_a \left(t_{peak}\right) - N_{a,min}} \tag{7.69}$$

With constant ζ we can recast (7.67):

$$\Delta R(t) = \frac{(A_a + \gamma A_b)}{(1+\gamma)} \Delta N(t) = H \Delta N(t) \tag{7.70}$$

with $H = (A_a + \gamma A_b)/(1 + \zeta)$ and $\Delta N(t) = \Delta N_a(t) + \Delta N_b(t) \approx \Delta N_a(t)(1 + \zeta) = \Delta N_b(t)(1 + \zeta)/\zeta$.

From (7.64), we are seeking a means to provide an estimate for $\langle N_b \rangle$. From the definition of γ and with $\Delta N(t) = \Delta N_a(t) + \Delta N_b(t)$ we have:

$$\langle \Delta N_b \rangle = \zeta \langle \Delta N \rangle / (1 + \zeta) \tag{7.71}$$

Using (7.70) and the definition of ΔR and ΔN_b, we get:

$$\langle N_b \rangle = \frac{\zeta (\langle R \rangle - R_{min})}{(1+\zeta)H} + N_{b,min} \tag{7.72}$$

In this equation, although from the diurnal-cycle data we may have direct access to H, $\langle R \rangle$ and R_{min}, it is unlikely we will directly know the values for $N_{b,min}$ and ζ.

To determine values for these two parameters, we note the linear approximation $\Delta N_b(t)/\Delta N_a(t) \approx \zeta = $ constant is based upon the values $\Delta N_a(t_{peak})$ and $\Delta N_b(t_{peak})$. We assume $N_a(t_{peak}) \approx N_{a,\,tot}$ and $N_b(t_{peak}) \approx N_{b,\,tot}$ and apply the ratio γ to the total number of Type (a) and Type (b) users giving:

$$\Delta N_{b,tot} = N_{b,tot} - N_{b,min} \approx \zeta \Delta N_{a,tot} = \zeta (N_{a,tot} - N_{a,min}). \tag{7.73}$$

Also if we apply the "$(1 - \chi)/\chi$ rule" then $\Delta N_{a,\,tot} = \chi N_{tot} - N_{a,min}$. Bringing these together and noting $N_{b,\,tot} \approx N_{tot} - N_{a,\,tot} = (1 - \chi)N_{tot}$, we have:

$$N_{b,tot} = \zeta (N_{a,tot} - N_{a,min}) + N_{b,min} = \zeta (\chi N_{tot} - N_{a,min}) + N_{b,min} = (1 - \chi) N_{tot} \tag{7.74}$$

This gives an approximation for ζ:

$$\zeta \approx \frac{(1 - \chi) N_{tot} - N_{b,min}}{\chi N_{tot} - N_{a,min}} \tag{7.75}$$

Referring back to (7.72), we see that we still need values for $N_{a,min}$, $N_{b,min}$ and χ. To address this we need to make yet another approximation. In this case, we consider the inequality (7.59) from which we ascertained the diurnal cycle of Type (b) users is much deeper than that of Type (a) users. Based on this it is reasonable to assume the number of active Type (b) user at the time of minimum traffic, N_{min}, is much smaller than the number of active Type (a) users. That is, $N_{b,min} \ll N_{a,min}$. Hence $N_{min} \approx N_{a,min}$ and we have:

$$A_a \approx R_{min}/N_{a,min} \approx R_{min}/N_{min} \tag{7.76}$$

With (7.75) this approximation also gives:

$$\zeta \approx \frac{(1 - \chi) N_{tot}}{\chi N_{tot} - N_{min}} \tag{7.77}$$

And dropping $N_{b,min}$ from (7.72) we have:

$$\langle N_b \rangle \approx \frac{(1-\chi) N_{tot} (\langle R \rangle - R_{min})}{(N_{tot} - N_{min}) H} \tag{7.78}$$

In this form, we can determine all the terms from the R vs N diurnal cycle except χ. Therefore, to construct an estimate for the power savings that may be attained by placing Type (b) users' equipment into a sleep state will require information either about the value of χ. If this value can be determined, then we can use (7.78) along with the diurnal-cycle information will provide an estimation of power savings.

It is important to note the assumptions/approximations made to get to (7.78). These are:

1. The assumption the peak and mean traffic annual growth rates justify splitting the population of users into two distinct types of users whose behaviours align with the "$(1 - \chi)/\chi$ rule" given in (7.65)
2. The assumptions $N_a(t_{peak}) \approx \chi N_{tot}$ and $N_b(t_{peak}) \approx (1-\chi) N_{tot}$ to give (7.75)
3. The assumption $N_{b,min} \ll N_{a,min}$ which then gives (7.76) and (7.77)
4. We have access to a relatively accurate value for χ or A_b.

To assess the accuracy of the approximations applied above, a simulation of a diurnal cycle was constructed. The cycle assumed Type (a) and Type (b) users with set values of $N_a(t)$, $N_b(t)$ over the diurnal cycle, A_a, A_b and χ. The diurnal cycles and user data rates were set to conform with inequality (7.59). Several differently shaped diurnal cycles were used, based on the cycles referenced in the Appendix to this chapter. With these cycles set, the quantities $R(t)$, $R(t_{peak})$, R_{min}, H, ζ, N_{min} and $N(t_{peak})$, $\langle N_a \rangle$ and $\langle N_b \rangle$ were calculated from the diurnal-cycle data. Assuming knowledge of χ, the approximation (7.78) for $\langle N_b \rangle$ was then calculated and compared with the actual value for each of the diurnal cycles. A sensitivity analysis showed that even if the peak-time number of active users is only 70% of total users and $N_{b,min} \approx 0.3 N_b(t_{peak})$, the relative error using (7.78) was less than 20% for $\chi \leq 0.3$.

It is important to stress that this estimation of $\langle N_b \rangle$ is only very approximate. It is presented because often it is very difficult, if not impossible, to get sufficient detailed data to determine accurate values of parameters such as $\langle N_b \rangle$ that may be required to describe important aspects of the power and energy consumption of a network. In such cases, these expressions provide approximations only requiring several quantities, most of which can be determined from the diurnal-cycle data in many cases. We have also assumed the network is not congested and the Pareto Principle applies in that $\chi \leq 0.3$ of users are responsible for $(1 - \chi) \geq 0.7$ of the total diurnal-cycle traffic with the remaining fraction $(1 - \chi)$ of users being active primarily around peak traffic times. In addition to this, we need to ascertain the amount of overhead traffic, R_{oh}. This quantity should be determinable from information about the network protocol (e.g. ADSL, Ethernet, etc.) being used to manage the network.

7.6 TECHNICAL SUMMARY

This section contains a summary of the technical results derived in the chapter.

7.6.1 PEAK AND AVERAGE TRAFFIC GROWTH

The growth rates of both peak and average traffic are important. The monotonically increasing dependence of network power consumption on traffic means peak traffic will correspond to peak network power consumption. Therefore, when dimensioning the power supply for a network element, a facility or an overall network, we need an estimate of peak power consumption to ensure the power supply is adequate and not overloaded.

With regard to network elements and facilities, the local power supply will need to be designed with an adequate "safety margin" to ensure the power supply equipment is not overloaded at times of peak power demand. For a network element, peak power occurs at the time of peak traffic, t_{peak}, and is given by using this value in (7.1). For an overall network, it is important that the telecommunications network does not introduce an unacceptable load on the local or national power supply grid.

The average traffic growth rate is important because, with the approximately linear relationship between equipment power consumption and traffic throughput for network equipment, the average traffic provides an estimate of the energy consumption of the equipment over the averaging time. For a network consisting of N_E network elements, the total energy consumption, $Q_{Ntwk}(t,T)$, over duration t to $t + T$ is given by (7.4). The estimate for total energy, $Q_{Ntwk}(t,T)$, can then be used to estimate the total carbon emissions of that equipment or network over duration T. If we can forecast the average traffic for a future time, then (7.4) provides a procedure for estimating the network carbon footprint into the future.

Although network traffic, which we represent by $R_1(t)$ (bit/s), typically varies on a sub-second timescale, for the purposes of estimating future network power and energy consumption, traffic variations on longer timescales are the primary focus: averaged over minutes to hours, denoted by $R_2(t)$ and defined in (7.6); and days to weeks, denoted by $R_3(t)$ and also defined in (7.6). Traffic profile $R_2(t)$, averaged over an hourly timescale, is typically approximately cyclic over several consecutive diurnal cycles, as shown in Figure 7.1.

Diurnal-cycle traffic data can be accessed from a range of network operator websites [11, 12, 14, 16]. In some cases, the data will include the peak and average traffic over diurnal cycles [13]. This data can then be used to estimate diurnal-cycle peak power and energy consumption of the network.

Averaging traffic over multiple days or weeks, $R_3(t)$, gives an indication of traffic growth trends over months and years. Again, some websites provide this data for both peak and average traffic [13, 25]. Assuming historical network growth trends will continue (i.e. business as usual), these growth trends can be used to forecast expected growth in peak network power demand and network energy consumption.

The traffic growth trend is often represented as an annual multiplier as shown in (7.18) and (7.19). In some reports, this increase is expressed as a percentage increase, $x\%$, on the previous year's traffic. If x_{peak} and x_{ave} are the percentage annual growth rates for peak and average traffic, respectively, then $\sigma = (1 + 100x_{peak})$ and $\beta = (1 + 100x_{ave})$ [7]. Knowing the peak and average traffic growth rates, we can use to forecast peak and average traffic in coming years, assuming "business as usual" growth over the forecast years.

7.6.2   FORECASTING TIME-OF-DAY TRAFFIC GROWTH

In some cases, we may need to forecast the actual hour-by-hour diurnal-cycle traffic. For example, if equipment is to be placed into a sleep state during times of low traffic in the diurnal cycle, then hour-by-hour forecasts will be required to estimate potential future energy savings [46]. Hour-by-hour traffic forecasts can also be used to estimate energy efficiency (Joules/bit) of an Internet service. In this case, the hour-by-hour relative magnitudes of service traffic flows can have a significant impact on service energy efficiency [47].

The forecasting method presented here is based on extrapolating historical traffic data and trends and assumes traffic growth is "business as usual" in that there is no dramatic, unexpected change in the traffic demand or services that drive traffic growth.

To apply this method, the current (or "year 0") hour-by-hour (corresponding to t_j, $j = 0, 1, 2, ...,23$) diurnal cycle, $R_2(t_j,0)$, is required, along with the peak and average traffic growth rates, σ and β, respectively. Using this method, described in Section 7.4, the hour-by-hour (t_j) traffic in year n is then given by (7.28). In this equation, the parameter $\kappa(n)$ is the solution to the nonlinear equation (7.27).

This equation is based on the assumption that the traffic at time t_j in the diurnal cycle increases exponentially at a rate dependent upon the traffic at that time. That is $R_2(t_j,n) = R_2(t_j,0)\sigma_j^n$, where σ_j is dependent on the time of day and at peak time, t_{peak}, we have $\sigma_{peak} = \sigma$, the peak traffic growth rate. The equation is set up so that the solution, $\kappa(n)$, in year n will give the corresponding peak and average traffic flow volumes according to their corresponding growth rates, σ and β, respectively.

If the historical peak and average growth rates are linear (rather than exponential), then the appropriate forecast equation for diurnal-cycle traffic is (7.34).

7.6.3 User Types

The use of placing equipment into a sleep state to mitigate power consumption in access networks has been widely proposed [27–30]. Not all users have a time usage behaviour that is amenable to placing their customer equipment into a sleep state. Section 7.5 provides a method that uses the collection of traffic growth data, minimal diurnal-cycle data, and the application of the so-called "80/20" rule to construct a rough approximation for the number of users who are inactive for a significant proportion of the diurnal cycle (and thus their equipment could be placed into a sleep state during inactive times).

This approach separates the population of end users into two types:

Type (a) users: these are online and active for almost all of the day and therefore their customer equipment is not easily amenable to being placed into a sleep state without a potential reduction in Quality of Service.

Type (b) users: these users are inactive for a significant proportion of the day and therefore well suited for placing their customer equipment into a sleep state without a reduction in Quality of Service.

The estimation methodology provides the average number of Type (b) users who can have their equipment placed into a sleep state for a significant amount of time. There may well be some Type (a) users whose equipment can also be placed into a sleep state, but for a shorter overall duration.

In Section 7.5, it is shown that the user population of an access network for which the peak traffic is increasing at an annual rate that is distinctly greater than for average traffic (a relatively common situation) can be separated into Type (a) and Type (b) users.

Having access to the data required to closely identify and calculate the number of Type (b) users and their exact online activity is extremely unlikely. However, with several reasonable assumptions, we can construct an approximate forecast of the average number of each user type over future years.

We assume the following:

1. Peak-hour traffic $R_{peak}(n) \approx R_{peak}(0)\sigma^n$ and average traffic $\langle R(n) \rangle \approx \langle R(0) \rangle \beta^n$ are growing exponentially at annual rates σ and β, with $\sigma > \beta$.
2. The access network is not congested at any time of the day in the forecast year (denoted by year n). Consequentially, the relationship between the number of active users, $N(t,n)$, at time t over the diurnal cycle and the total traffic, $R(t,n)$, in year n is approximately linear. That is, $R(t,n) \approx H(n)N(t,n)$, where $H(n)$ is independent of t.
3. The access speed of each user type, when active, is approximately constant over the diurnal cycle. The access network being uncongested allows this to occur.

4. The 80/20 rule applies, in that the total number of Type (a) users (active and inactive) is $\chi N_{tot}(n)$, where $N_{tot}(n)$ is the total number of users (both active and inactive) served by the access network ($\chi \approx 0.2$). Hence, the total number Type (b) users is $(1 - \chi)N_{tot}(n)$. Further, in accordance with the 80/20 rule, the total traffic due to Type (a) users is $(1 - \chi)\langle R(n)\rangle T$ (over diurnal-cycle duration T) and the total traffic due to Type (b) users is $\chi\langle R(n)\rangle T$.

If these assumptions are appropriate, then the average number of Type (b) users, $\langle N_b(n)\rangle$, whose equipment can be placed into a low-power sleep mode for a significant amount of time, without an impact on Quality of Service, can be approximated by (7.78).

7.7   EXECUTIVE SUMMARY

An important application of modelling is to provide forecasts for possible future growth in network power and energy consumption. This is important for network design and estimation of the future environmental footprint of a network or service.

In broad terms, traffic behaviour on three timescales impacts the network design and deployment, and hence power consumption. The monthly-to-yearly timescale influences the strategy for equipment deployment over the years of network operation. The hourly timescale provides the mean and peak traffic levels over the diurnal cycle. The diurnal peak traffic level is used to ascertain the amount of equipment required to be deployed during a given network upgrade. The sub-second timescale variation plays a role in determining the maximum utilisation factor that is used to determine the amount of capacity deployed relative to the diurnal-cycle peak traffic. A maximum utilisation parameter is introduced to accommodate random variations in the traffic at the short (order hours or less) timescale (e.g. flash crowds, special events).

To model energy efficiency of services, we require time-of-day traffic data over the full daily (diurnal) cycle of traffic. This is because the energy efficiency of a service can be highly dependent upon whether or not the diurnal cycle of that service's traffic is time aligned with the other traffic in the network: that is, whether or not that service's diurnal-cycle traffic has its maxima and minima at the same time of day as the other traffic in the network.

Detailed diurnal-cycle traffic data that is available is typically recent or historical. Forecasts of future detailed (hour by hour) diurnal-cycle traffic data are rarely available. This means diurnal-cycle data required for detailed network and service power and energy efficiency forecasting is very unlikely to be easily sourced. Therefore, a technique for constructing forecasts for future diurnal-cycle traffic, which requires access to current or recent diurnal-cycle data and forecast values only for future peak and mean traffic, is described in this chapter. Using this approach, the monthly or yearly peak and mean traffic estimates can be calculated from their respective growth rates. These rates are fairly readily available from a range of organisations and websites. The resulting, hour-by-hour traffic forecasts provided by the method used in this chapter assume there is no dramatic change in the network or user behaviour (e.g. no advent of an unexpectedly popular new service or innovation, no dramatic change in user behaviour).

The access network is often a significant contributor to telecommunications network energy consumption. The use of placing equipment into a sleep state to mitigate power consumption in access networks has been widely proposed. Not all users have a time usage behaviour that is amenable to placing their customer equipment into a sleep state. To estimate potential energy savings that may be attained by implementing sleep states in access networks, we need to ascertain the number of users whose equipment can be placed into such a state. In Section 7.5, we constructed an estimate of the average minimum number of users whose behaviour enables their equipment to be placed into a sleep state. This value can then be used to derive a lower-bound estimate of potential energy savings through the use of sleep states.

A major difficulty with deriving such an estimate is the lack of easily accessible detailed user behaviour data. Therefore, the method that is described in Section 7.5 uses the collection of traffic growth data, minimal diurnal-cycle data and the application of the so-called "80/20" rule to construct a rough approximation for the number of users who are inactive for a significant proportion of the diurnal cycle. The equipment of these users is then amenable for placing into a sleep state during inactive times.

This approach separates the population of end users into two types. Type (a) users: these are online and active for almost all of the day and therefore their customer equipment is not easily amenable to being placed into a sleep state without a potential reduction in Quality of Service. Type (b) users: these users are inactive for a significant proportion of the day and therefore well suited for placing their customer equipment into a sleep state without a reduction in Quality of Service.

The estimation methodology provides the average number of Type (b) users who can have their equipment placed into a sleep state for a significant amount of time. There may well be some Type (a) users whose equipment can also be placed into a sleep state, but for a shorter overall duration.

It may occur that an access network does not have a sufficient number of Type (b) users to make the implementation of a sleep state cost effective. Therefore, before an estimate of the number of Type (b) users is undertaken, an analysis of diurnal-cycle traffic is needed to determine if there is a reasonable number of Type (b) users. In Section 7.5, it is shown that the user population of an access network for which the peak traffic is increasing at an annual rate that is distinctly greater than for average traffic (a relatively common situation) can be separated into Type (a) and Type (b) users. The task is then to construct an estimate for the number of Type (b) users.

7.8 APPENDIX: LINEAR RELATIONSHIP $R(T,N) = H(N)N(T,N)$

In this appendix, we show that diurnal-cycle data for the traffic, $R(t)$, and number of active users, $N(t)$, sourced from several publications all support the assumption used above that the relationship between $N(t)$ and $R(t)$ is approximately linear.

The plots in Figure 7.5, taken from published papers, contain examples of diurnal-cycle data from real networks [35, 36, 42].

a. *Diurnal-cycle data from Maier et al. [36]*
b. *Diurnal-cycle data from Kihl et al. [35]*
c. *Diurnal-cycle data from Yang et al. [42]*

From these plots we construct a profile of $R(t)$ vs $N(t)$ for each of the data sets above, as shown in Figure 7.6.

a. *R(t) vs N(t) for diurnal-cycle data in Figure 7.5a*
b. *R(t) vs N(t) for diurnal-cycle data in Figure 7.5b*
c. *R(t) vs N(t) for diurnal-cycle data in Figure 7.5c*

Figure 7.6 shows the $R(t)$ vs $N(t)$ plots for the three data sets in Figure 7.5 with the appropriate trend line equation that provides the best fit, as measured by the R^2 value. Although the best fits are not linear forms, Table 7.1 lists the R^2 value for a range of curve fit profiles. We see that in each case a linear fit is effectively as good as the best fit, in that the linear fit R^2 value is only marginally less than the best.

The corresponding coefficients for the curve fits are given in Table 7.2.

Therefore, we can see that all three data sets are very well represented by a linear curve. None of the data sets significantly deviate from a linear fit.

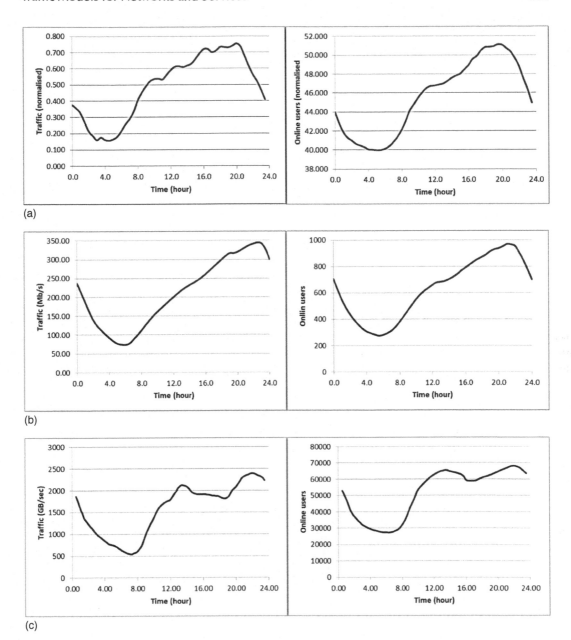

FIGURE 7.5 Diurnal-cycle data for traffic and active users from three separate publications.

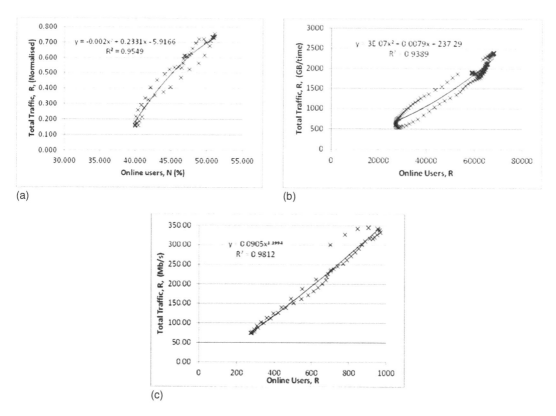

(a)

(b)

(c)

FIGURE 7.6 Plots of N vs R for the three data sets presented in Figure 7.5. Included in the plots are curve fits for the curve profile with the highest R^2 factor. See text for more details.

TABLE 7.1
R^2 Coefficients for Different Curve Fits to Diurnal-Cycle Data

R^2 Curve Fit Values	Plot (a)	Plot (b)	Plot (c)
Linear ($R = AN+B$)	0.942	0.963	0.934
Power ($R = AN^B + C$)	0.948	0.964	0.939
Exponential ($R = Ae^{BN}$)	0.871	0.955	0.916
Log ($R = A\ln(N) + B$)	0.948	0.932	0.918
Quadratic ($R = AN^2 + BN + C$)	0.955	0.964	0.939

TABLE 7.2
Curve-Fit Parameters for the Data Sets for Different Curve Fits

Fitted Curve Form	Plot (a)	Plot (b)	Plot (c)
Linear ($R = AN+B$)	$A = 0.0503, B = -1.802$	$A = 0.388, B = -34.4$	$A = 0.038, B = -405.24$
Power ($R = AN^B + C$)	$A = 1.19 \times 10^3, B = 2.478 \times 10^{-4}, C = -9.187 \times 10^3$	$A = 0.141, B = 1.138, C = -8.91$	$A = 2.036 \times 10^{-6}, B = 0.186, C = 345.1$
Exponential ($R = Ae^{BN}$)	$A = 0.0015, B = 0.124$	$A = 49.38, B = 0.0021$	$A = 308.3, B = 3 \times 10^{-5}$
Log ($R = A\ln(N) + B$)	$A = 2.277, B = -8.19$	$A = 212.5, B = -1144$	$A = 1701.2, B = -16808$
Quadratic ($R = AN^2+BN+C$)	$A = -0.002, B = 0.233, C = -5.92$	$A = 4 \times 10^{-5}, B = 0.338, C = -21.283$	$A = 3 \times 10^{-7}, B = 0.0079, C = 237.3$

REFERENCES

[1] J. Malmodin and D. Lunden, "The Energy and Carbon Footprint of the Global ICT and E&M Sectors 2010–2015," *Sustainability*, vol. 10, p. 3027, 2018.

[2] J. Aslan *et al.*, "Electricity Intensity of Internet Data Transmission: Untangling the Estimates," *Journal of Industrial Ecology*, vol. 22, no. 4, p. 785, 2017.

[3] J. Malmodin *et al.*, "LCA of Data Transmission and IP Core Networks," in *2012 Electronics Goes Green*, 2012.

[4] J. Baliga *et al.*, "Energy consumption of Optical IP Networks," *Journal of Lightwave Technology*, vol. 27, no. 13, p. 2391, 2009.

[5] D. Kilper *et al.*, "Power Trends in Communication Networks," *IEEE Journal Of Selected Topics In Quantum Electronics*, vol. 17, no. 2, p. 275, 2011.

[6] S. Anders *et al.*, "On Global Electricity Usage of Communication Technology: Trends to 2030," *Challenges*, vol. 6, p. 117, 2015.

[7] Cisco, "Cisco Visual Networking Index: Forecast and Trends, 2017–2022," Cisco White Paper, 2018.

[8] Cisco, "Cisco Visual Networking Index: Global Mobile Data Traffic Forecast Update, 2017–2022," Cisco White Paper, 2019.

[9] Cisco, "The Zettabyte Era: Trends and Analysis," Cisco White Paper, June 2017.

[10] Sandvine, "The Global Internet Phenomena Report," Sandvine, October 2018.

[11] Google, "Google IPV6 Statistics," Google, 10 July 2020. [Online]. Available: https://www.google.com/intl/en/ipv6/statistics.html. [Accessed 12 July 2020].

[12] Netnod, "Aggregate traffic graphs for Netnod," 12 July 2020. [Online]. Available: https://www.netnod.se/ix-stats/sums/. [Accessed 20 July 2020].

[13] DE-CIX, "DE-CIX Frankfurt statistics," DE-CIX, 12 July 2020. [Online]. Available: https://www.de-cix.net/en/locations/germany/frankfurt/statistics. [Accessed 12 July 2020].

[14] Hong Kong Internet Exhange, "HKIX Switching Statistics," Hong Kong Internet Exhange, 12 July 2020. [Online]. Available: https://www.hkix.net/hkix/stat/aggt/hkix-aggregate.html. [Accessed 12 July 2020]

[15] Netnod, "IX Statistics," Netnod, 2020. [Online]. Available: https://www.netnod.se/ix/statistics. [Accessed 12 July 2020].

[16] London Internet Exchange, "Linx," London Internet Exchange, 2020. [Online]. Available: https://www.linx.net/. [Accessed 12 July 2020].

[17] Cenic, "CENIC Network Statistics," Cenic, [Online]. Available: http://cricket.cenic.org/. [Accessed 12 July 2020].

[18] S. Korotky, "Semi-empirical Description and Projections of Internet Traffic Trends Using a Hyperbolic Compound Annual Growth Rate," *Bell Laboratories Technical Journal*, vol. 18, no. 3, p. 5, 2013.

[19] International Telecommunications Union, "Telecom Network Planning for evolving: Network Architectures Reference Manual; Document NPM/4.1," International Telecommunications Union.

[20] Cisco, "*Best Practices in Core Network Capacity Planning*," Cisco, 2013.

[21] F. Idzikowski *et al.*, "A Survey on Energy-Aware Design and Operation of Core Networks," *IEEE Communications Surveys & Tutorials*, vol. 18, no. 2, p. 1453, 2016.

[22] M. Mohamed *et al.*, "Bounds on GreenTouch GreenMeter Network Energy Efficiency," *Journal of Lightwave Technology*, vol. 36, no. 23, p. 5395, 2018.

[23] P. Laskot *et al.*, "Long-Term Drivers of Broadband Traffic in Next-Generation Networks," *Annals of Telecommunications*, vol. 70, p. 1, 2015.

[24] J. Marcus and D. Elixmann, "*Build It!... but What If They Don't Come?*," in *EuroCPR 2013*, 2013.

[25] London Internet Exchange, "Total LINX Traffic Flow," London Internet Exchange, 2020. [Online]. Available: https://portal.linx.net/. [Accessed 12 July 2020].

[26] V. Coroama *et al.* "The Energy Intensity of the Internet: Home and Access Networks," in *ICT Innovations for Sustainability, Advances in Intelligent Systems and Computing*, vol. 310, L. Hilty and B. Aebischer Eds. Springer, 2015, p. 137.

[27] R. Butta *et al.*, "An Energy Efficient Cyclic Sleep Control Framework for ITU PONs," *Optical Switching and Networking*, vol. 27, p. 7, 2018.

[28] R. Hirafuji *et al.*, "The Watchful Sleep Mode: A New Standard for Energy Efficiency in Future Access Networks," *IEEE Communications Magazine*, vol. 53, no. 8, p. 150, 2015.

[29] B. Lannoo *et al.*, "How Sleep Modes and Traffic Demands Affect the Energy Efficiency in Optical Access Networks," *Photonic Network Communications*, vol. 30, p. 85, 2015.

[30] S.-W. Wong *et al.*, "*Sleep Mode for Energy Saving PONs: Advantages and Drawbacks*," in *IEEE Globecom Workshops*, 2009.

[31] S. Sundaresan *et al.*, "Measuring Home Broadband Performance," *Communications of the ACM*, vol. 55, no. 11, p. 100, 2012.

[32] O. Goga and R. Teixeira, "Speed Measurements of Residential Internet Access, PAM 2012, Lecture Notes in Computer Science," in *International Conference on Passive and Active Network Measurement*, 2012.

[33] S. Bauer *et al.*, "*Understanding Broadband Speed Measurements*," in *Telecommunications Policy Research Conference*, 2020.

[34] OECD, "Access Network Speed Tests, OECD Digital Economy Papers, No.237," OECD, 2014.

[35] M. Kihl and *et al.*, "*Traffic Analysis and Characterization of Internet User Behaviour*," in *International Conference on Ultra-modern Telecommunications and Control Systems*, 2010.

[36] G. Maier *et al.*, "*On Dominant Characteristics of Residential Broadband Internet Traffic*," in *Proceedings of 9th ACM SIGCOMM Conference on Internet Measurement*, 2009.

[37] A. Vishwanath *et al.*, "Energy Consumption Comparison of Interactive Cloud-Based and Local Applications," *IEEE Journal on Selected Areas in Communications*, vol. 33, no. 4, p. 616, 2015.

[38] Ofcom, ""UK Home Broadband Performance: The performance of fixed-line broadband delivered to UK residential consumers," Ofcom, 2018.

[39] K.-C. Lan and J. Heidemann, "A Measurement Study of Correlations of Internet Flow Characteristics," *Computer Networks*, vol. 50, no. 1, p. 46, 2006.

[40] C. Pellicer-Lostao *et al.*, "Statistical User Model for the Internet Access," arXiv.org, 2007.

[41] P. Megyesi and S. Molnar, "*Analysis of Elephant Users in Broadband Network Traffic*," in *Advances in Communication Networking. EUNICE 2013. Lecture Notes in Computer Science*, 2013.

[42] J. Yang *et al.*, "Characterizing Internet Backbone Traffic from Macro to Micro," in *IEEE International Conference on Network Infrastructure and Digital Content*, 2009.

[43] J. Morley *et al.*, "Digitalisation, Energy and Data Demand: The Impact of Internet Traffic on Overall and Peak Electricity Consumption," *Energy Research & Social Science*, vol. 38, p. 128, 2018.

[44] G. Kramer *et al.*, "Evolution of Optical Access Networks: Architectures and Capacity Upgrades," *Proceedings of the IEEE*, vol. 100, no. 5, p. 118, 2012.

[45] J. Epstein and R. Axtell, *Growing Artificial Societies: Social Science from the Bottom-Up*, MIT Press, 1996.

[46] J. Elmirghani *et al.*, "GreenTouch GreenMeter Core Network Energy efficiency Improvement Measures and Optimization," *Journal of Optical Communications and Networking*, vol. 10, no. 2, p. A250, 2018.

[47] K. Hinton and F. Jalali, "*A Survey of Energy Efficiency Metrics*," in *5th International Conference on Smart Cities and Green ICT Systems (SMARTGREENS)*, 2015.

8 Network Power Consumption Modelling

In this chapter, we bring together the results derived in Chapters 3 and 7 to construct forecasts for network power consumption. Using the equipment power model developed in Chapter 3 along with the peak and average traffic growth models from Chapter 7, in this chapter we show how to construct network power models that can be used to forecast network power consumption. Such forecasts, while approximate, are of value by quantifying network power consumption in coming years. Such forecasts are of value because they will enable network operators to gauge the potential peak power requirements (to ensure enough power is available) and overall energy consumption (useful for carbon footprint estimates) as their network grows to accommodate increasing traffic demands.

The power models presented in this chapter are not precise. Rather they provide approximate values for peak and average power. The main advantage of the models presented here is that they do not require detailed information on all equipment in the network. The models use averaged equipment and network parameters and several general approximations from statistical theory. The veracity of these approximations is demonstrated with several numerical models described towards the end of the chapter.

To use this approach, several general equations relating to averaged equipment and network parameters for edge and core networks are derived in Section 8.4. These results are then used in Section 8.5 to forecast edge and core network average and peak power consumption years into the future. These forecasts enable the investigation of network power consumption for different network upgrade deployment strategies including the replacement of depreciated equipment. Thus they can be used for scenario planning of future network upgrades.

8.1 NETWORK AND SERVICE POWER MODELS

The power consumption modelling approach described in this chapter can be applied to both overall networks and specific services carried across the network. This chapter focuses on modelling total network power consumption. The application of this model to services will be described in detail in Chapter 10.

Put simply, estimating the power consumption of a network is effectively summing up the power consumption of all the network equipment. In contrast, estimating the power consumption of a service is not as simple, because the traffic flows generated by a service share the equipment through which it travels with many other services. Therefore, we need to adopt an approach for allocating a share of equipment power consumption to each service.

One important use of power models is to provide metrics for the energy efficiency (Watts/bit/s) of a network or service. Energy efficiency metrics are widely used because they provide a method for comparing network and service architectures. For example, consider two networks that have significantly different total power consumption values, P_1 and P_2 with $P_1 \gg P_2$. However, if the first network carries significantly more traffic, R_1, than the second, R_2, to the extent that $P_1/R_1 < P_2/R_2$, then the first network is generally considered to be more "energy efficient" than the second. Taking this a step further, the network operator then also needs to consider how the capacities of the two

DOI: 10.1201/9780429287817-8

networks align with traffic demand trends, to reach an optimised decision. This type of comparison is often used when planning future network or service architectures and technologies. The network model developed in this chapter is well suited for developing network and service energy efficiency metrics of this form.

8.2 A SURVEY OF NETWORK POWER MODELS

Exact evaluation of telecommunications network and service power consumption is extremely complicated, to the extent that direct measurement is impractical. Most telecommunications networks (and most certainly the Internet) and services involve too many network elements spread over too diverse geographical (or corporate) regions for a direct measurement to be feasible. Consequentially, estimates of network and service power consumption require a combination of collected data (typically equipment power profiles and some network architecture and traffic details) and a significant amount of mathematical modelling.

In this chapter, we will describe the construction of a "bottom-up" network power model that can provide forecasts for future network power consumption as the network grows to meet customer traffic demands.

Bottom-up models are one of three types of network power models that have been published over recent years. Many authors have categorised models as either "bottom-up" or "top-down" [1–4]. Coroama and Hilty have distinguished a third type of model, which they refer to as "model-based" [5]. These approaches are used to estimate both total network power or energy consumption (Watts or Joules) and network energy efficiency, sometimes called "energy intensity" (Watts/bit/s or Joules/bit). There are subtle differences between applying these approaches to quantifying total network power as compared to network energy efficiency.

In this chapter, we will adopt the bottom-up and top-down categorisations. Ishii *et al.* [1] provide succinct descriptions of bottom-up and top-down network power models. For bottom-up models [1]:

> The 'bottom-up' method assumes a network model based on network design principles, collects relevant data such as the power consumption of each individual network component and measured traffic volume, and deduces the overall energy consumption by combining the model and the data. Since the relationships among the network parameters and the network energy consumption are derived, it is possible to provide predictions according to various plausible network scenarios.

For top-down models [1]:

> On the other hand, the 'top-down' method is based on statistical information regarding network equipment shipments, sales revenue, or carriers' purchase volumes of electricity. The top-down method references more direct information on actual networks than the bottom-up method, and therefore incorporates some or all of the energy overheads of these networks.

Examples of bottom-up models include [6–12]. Examples of top-down models include [4, 13–18].

These different approaches have relative advantages and disadvantages. Both models have been used to forecast future power consumption; however, bottom-up models provide greater flexibility to incorporate different deployment strategies. Further, bottom-up models provide more information on the relative impacts of different network elements and architecture parameters: for example (as will be seen below), the influence of reducing idle power of network equipment and the application of different protection strategies.

Another advantage of bottom-up models is their ability to provide power consumption estimates for specific network services, because they can resolve down to the level of service traffic flows. Top-down models are typically based upon national or global network data, and hence do not allow for consideration of specific traffic flows. Rather, they typically estimate the power consumption of a service by providing an estimate of overall network energy efficiency (Watts/bit/s), which is multiplied by the bit-rate of the given service [19]. A bottom-up model can provide power estimates for specific paths or path types along which a service's traffic travels. This enables direct comparisons of different service architectures and network modifications.

On the other hand, bottom-up models typically assume network architectures that are significantly simpler than those of real networks. For example, they may not include network overlays deployed for specific services and protocols. Bottom-up models also typically require additional parameter estimates to include the impact of factors such as power provisioning, facility environmental control, future growth and over-subscription. In contrast, because top-down models are based on actual equipment inventories and more direct network data, they intrinsically include these types of overall network detail.

A challenge with top-down models is the collection of this data and whether or not it does adequately represent the operating network. For example, if the top-down approach is based on equipment sales, then it is assumed all the purchased equipment has been deployed and is in service. Assuming that the data collected for a top-down model is both up-to-date and representative of the operating network, top-down models are usually considered to provide more accurate power consumption estimates for established networks [1].

A comprehensive description and comparison of top-down and bottom-up models has been given by Schien and Preist [3]. By-and-large, all top-down metrics are similar, in that each top-down model:

> estimates the total energy use of an entire sub-system, such as 'all data centers' or 'the Internet,' measures or estimates the total quantity of a given service type provided (e.g. data transmitted), and divides the former by the latter to give the energy consumption per unit of service. [3]

In their analysis of top-down models, Schien and Preist explain:

> Top-down models are conducive to this whole systems perspective. Energy consumption in top-down models is usually estimated from market sales per device class and corresponding average power consumption to give a total over all considered device classes. By comparing this total with other macro-scale energy statistics, it is easy to sanity check them. And by treating the modelled system partly as a black box, they also do not require detailed knowledge on the network architecture. On the other hand, they cannot be used to evaluate changes to part of the network but only on trends of changing total network traffic or total energy consumption. [3]

Referring to bottom-up models, Schien and Preist's description is:

> A bottom-up model, in contrast, calculates the overall energy intensity from the sum of the energy intensity of the sub-system components – usually the physical devices in the network. These models have also been referred to as transactional models as they allocate energy consumption to the transaction of data from end to end. [3]

Their analysis of bottom-up models states:

> As they represent energy intensity on the level of the system components, they are more flexible than top-down models to evaluate change: they can be used to evaluate the impact of modifications to the system architecture and its components.
>
> Such models thus require detailed knowledge of the operation and design of networks. Although this knowledge is already held by network operators, and thus, in principle, it is possible to represent

each individual device in a bottom-up model, in practice network operators do not publicly disclose this information for business reasons. Instead, bottom-up models have been built based on implicit assumptions around the typical architecture of networks and are thus more difficult to validate. [3]

And they continue:

Bottom-up models facilitate the evaluation of alternative design choices if they represent the network components that are to be altered. The scope of an investigation thus affects the level of detail at which the network is modelled. For example, if the goal of a study is to evaluate savings from optical switching, these devices must be explicitly modelled; if not, fibre optic components might be modelled in less detail with average values for energy consumption and capacity. At the same time the level of detail of modelling is naturally constrained by the simultaneously increasing complexity of the model, which is particularly relevant for end-to-end models. [3]

In this work, we focus on developing models to provide forecasts for telecommunications network and service power consumption over future years. Consequentially, we develop a bottom-up model for network and service power consumption (Watts) and energy efficiency (Joules/bit).

8.3   ACCURACY VERSUS PRACTICALITY OF MODELLING

A challenge with any network power or energy efficiency model (both top-down or bottom-up) is how accurately the calculated values represent the real network power values. There is no easily applied process to check the accuracy, because accurately assessing the power consumption of telecommunications networks and services typically requires large amounts of data that are extremely difficult to access. Therefore, the approach adopted by most researchers is to compare different models to see how widely spread their values are. Multiple authors have applied this method [1–5].

Across these publications, several key determinant factors have been found to strongly influence the resulting estimate. These include:

- System boundary [4, 5]: This is widely accepted as the most important factor. It addresses the "boundary" that delineates what is included in the model. For example, does the model include home equipment? If so, what types of equipment: only PCs and laptops; or does it also include internet TVs, PDAs, smart appliances, etc.? A model for which the system boundary includes home networks will give a very different power consumption figure than one that stops at the household broadband gateway.
- Year of the estimate [4, 5]: As equipment improves over the years, its energy efficiency will improve. Therefore, estimates of power consumption undertaken in the early 2000s will typically give values greater than those undertaken in the 2010s. This issue also applies to the age of the equipment still in service.
- Representative model [4]: If the model has been designed to provide power and efficiency estimates for a specific network, then the model must be representative of the actual network.
- Parameter values [3, 4]: It is relatively obvious that the parameter values used in a power model will influence its accuracy. However, particular care must be taken when using a model to forecast power consumption. This is because most forecasts are based on exponential changes. Therefore, inaccuracies in these parameters can have a dramatic impact on medium- and longer-term forecasting.
- Inclusion of embedded energy [20]: Some estimates of network energy include the "life-cycle energy" of equipment, so-called embedded energy. Embedded energy includes the energy of fabrication, deployment, disposal and recycling, separate to operational or "use phase" energy.

When constructing network power models, it is important to consider each of these factors, especially when attempting to model a specific network. When modelling very large networks, such as the Internet, resolving these issues can be quite complicated [3–5].

8.4   EDGE AND CORE NETWORK POWER CONSUMPTION

To construct a model of network power consumption we need to describe its architecture. That is, how the various network elements are interconnected. We assume that the edge and core network consists of a collection of inter-connected nodes which house core and edge routers, as represented in Figure 8.1. We assume that all traffic enters and leaves the core network via one or more edge routers. Thus core routers are connected to one or more edge routers and vice versa. Any node through which traffic enters or leaves the core network will house both edge and core routers. The edge routers constitute a boundary around the core network, a bit like the surface of a bubble forms a boundary around its interior. All traffic that enters or leaves the core goes through an edge router.

The power model described here includes both edge and core network elements. However, one may constrain the model to just edge or core, if desired.

Using (3.10), the total network power consumption, at time t, will be given by

$$P_{Ntwk}(t) = M_{OH} \sum_{j=1}^{N_E} \left(P_{idle,j} + E_j R_j(t) \right) \tag{8.1}$$

where N_E is the number of network elements and $P_{idle,j}$, E_j and $R_j(t)$ are the element idle power, incremental energy per bit, and traffic throughput of the j th network element, respectively. The term M_{OH} is a multiplier used to represent the proportionate increase in power consumption due to overheads such as equipment redundancy, cooling, power supply and distribution losses in the facility that houses the equipment [6, 8, 10]. A single "typical" M_{OH} value is assumed, in preference to making M_{OH} specific to network sites, partly to avoid the added complexity, but also because the overhead consumption is a modest fraction of the whole equipment consumption tally. The summation in (8.1) is over of all the equipment that has been deployed in the nodes across the network with the knowledge of the traffic through each node.

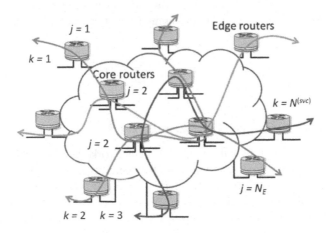

FIGURE 8.1   Edge and core network consisting of inter-connected routers spread across a geographical area. Paths through the network are provided by inter-connections between edge and core routers as well as between core routers.

Looking at a network from a broader perspective, its purpose is to provide data services to end users. These data services are manifest as traffic flows that are routed and transported through the network (between user end-points) by the routers and other equipment in the network. Therefore, the total traffic through the network, $R_{Ntwk}(t)$, can be expressed as the sum of service data traffic flows. In this, we include protection traffic, network management traffic and control traffic also as service flows. Hence, protection and management overheads are included in total network traffic. With this, we can express the total network traffic, $R_{Ntwk}(t)$, as the sum of service traffic flow volumes:

$$R_{Ntwk}(t) = \sum_{k=1}^{N^{(svc)}} R^{(k)}(t) \tag{8.2}$$

where $R^{(k)}(t)$ is the traffic due to service k (including protection and management traffic incorporated with that service) and $N^{(svc)}$ is the number of services transported through the network. We adopt the notation that a superscript index in parentheses, such as the k in $X^{(k)}$, applies to a service; and a subscript index, such as the j in X_j, applies to a network element. In later sections, we introduce a superscript in square brackets, such as n in $X^{[n]}$, which will be a year index for network growth over time.

For a precise evaluation of $P_{Ntwk}(t)$, we need a total inventory of the equipment in every node and their corresponding traffic throughputs. This will require details of the interconnections between routers within the nodes and across the network which, in turn, requires details of the interconnection architecture within and between nodes. This amount of information is unlikely to be easily (let alone publicly) available at any particular time and even less likely to be available over the years of network operation and upgrades. An alternative approach is to accept a loss in accuracy to allow a significant simplification of the network model that requires much less detailed information about network equipment and traffic flows. That is, we can derive an equation for the approximate total network power by averaging across the network elements and service traffic flows, in order to significantly reduce the amount of detailed information required the power model. This approach will be described below, after some preliminaries.

As we see in (8.1), the power consumption of a network is dependent upon the traffic flows, $R_j(t)$, through the network elements. However, from (8.2), the total network traffic, $R_{Ntwk}(t)$, is dependent upon the service flows, $R^{(k)}(t)$. In general service traffic flows through multiple network elements and by the same token, a network element might carry a very large number of service traffic flows Therefore total network traffic is not equal to the summation of network element traffic rates. In fact:

$$R_{Ntwk}(t) = \sum_{k=1}^{N^{(svc)}} R^{(k)}(t) \leq \sum_{j=1}^{N_E} R_j(t) \tag{8.3}$$

8.4.1 RELATING SERVICE FLOWS AND EQUIPMENT FLOWS

Because most services traffic will flow through multiple network elements, (8.3) most likely to be a strict inequality. We need to construct a relationship between the service traffic, $R^{(k)}(t)$, and network element throughputs, $R_j(t)$. Consider a specific service traffic flow $R^{(k)}$. This flow will most likely propagate through several network elements as it travels from its source to its destination. Assume that some or all this traffic flow travels through the j th network element. We represent the traffic flow if the k th service through the j th network element by $R_j^{(k)}(t)$. Then we can write

$$R_j^{(k)}(t) = \alpha_j^{(k)}(t) R^{(k)}(t) \tag{8.4}$$

The alpha matrix, $\alpha_j^{(k)}$ represents the proportion of service traffic $R^{(k)}$ that propagates through network element j. The total traffic routed through the j th network element, $R_j(t)$, is given by

$$R_j(t) = \sum_{k=1}^{N^{(svc)}} R_j^{(k)}(t) = \sum_{k=1}^{N^{(svc)}} \alpha_j^{(k)} R^{(k)}(t) \qquad (8.5)$$

If all of the k th service traffic travels through the j th network element, then $R^{(k)} = R_j^{(k)}$ for the j th network element and so $\alpha_j^{(k)} = 1$. For all network elements, j' which do not carry any traffic of the k th service we have $R_{j'}^{(k)} = 0$; that is $\alpha_{j'}^{(k)} = 0$. We can have $0 \leq \alpha_j^{(k)} \leq 1$. For example, if the traffic of the k th service is split over two paths with $0.3R^{(k)}$ allocated to the j th network element and $0.7R^{(k)}$ allocated to the j' th network element, then $\alpha_j^{(k)} = 0.3$ and $\alpha_{j'}^{(k)} = 0.7$.

For a service k, we have $R^{(k)} \leq \sum_j R_j^{(k)}$ because the traffic of service k is most likely to travel through multiple network elements and may include distinct protection paths.

The parameters $\alpha_j^{(k)}$ represent a relationship between the physical layer (via the network element j-index) and the service layer (via the service k-index). The $\alpha_j^{(k)}$ describes the path(s) a service follows through a network (including protection paths, if allocated). The service paths are set up between edge routers which may aggregate multiple service flows onto high capacity bearers (e.g. via SDH, SONET, OTN, Ethernet, MPLS).

Provided that the network traffic is relatively well spread across the network nodes, we can use the $\alpha_j^{(k)}$ to relate the power consumption of the network equipment to the power consumption of the services and hence of the network. (By "relatively widely spread across the network nodes", we mean that a substantial number of the $\alpha_j^{(k)}$ are non-zero. That is, the substantial number of network elements have non-zero utilisation, and different service flows typically follow different paths through the network.)

To provide a concrete example of the $\alpha_j^{(k)}$, consider the small network depicted in Figure 8.2. In this network there are three service flows, $R^{(1)}$, $R^{(2)}$ and $R^{(3)}$. Flow $R^{(1)}$ is split into two paths from element $j = 1$ to element $j = 6$. One path carries $0.3R^{(1)}$ from element $j = 1$ through elements $j = 3, j = 4$ to element $j = 6$. The remaining 70% of flow $R^{(1)}$ travels from element $j = 1$ through elements $j = 2$, $j = 3$ and $j = 5$ to $j = 6$. All of the flow $R^{(2)}$ travels from element $j = 1$ through elements $j = 2$ and $j = 4$ to element $j = 6$. Flow $R^{(3)}$ originates from element $j = 7$, is split at element $j = 3$ equally into two paths. One path is through $j = 2$ and $j = 4$ where it is recombined with the second path that is through $j = 5$ and $j = 4$. The flow for $R^{(3)}$ then travels from $j = 4$ to element $j = 7$. The flows in this network can be represented with the $\alpha_j^{(k)}$ values given in Table 8.3.

8.4.2 TIME SCALES AND AVERAGING

The $R^{(k)}(t)$'s and $R_j(t)$'s traffic levels vary over the diurnal cycle but in many cases the hour-by-hour data on all these traffic flows is not readily available to the modeller. In such situations often only the averaged traffic levels over one or more diurnal cycles are available. For example, weekly or monthly average traffic levels.

This gives two possible situations for developing the power models:

(a) If all of the diurnal-cycle data is available for all of the service traffic flows, $R_j^{(k)}(t)$, and all of the network element throughputs, $R_j(t)$, then we can relate them for each time, t, over the cycle. In this case, the $\alpha_j^{(k)}$ values may include a time dependence and are defined by

$$R_j(t) = \sum_{k=1}^{N^{(svc)}} R_j^{(k)}(t) = \sum_{k=1}^{N^{(svc)}} \alpha_j^{(k)}(t) R^{(k)}(t) \qquad (8.6)$$

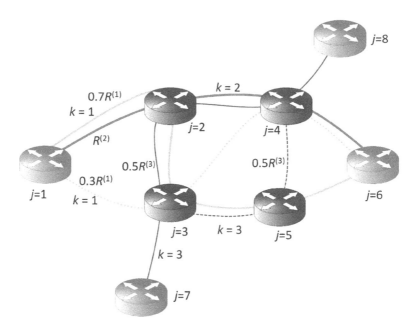

FIGURE 8.2   Network with service flows $R^{(1)}$ $R^{(2)}$ and $R^{(3)}$ and network elements 1 to 6. The flow $R^{(1)}$ is shown with a yellow line and is split over two paths with 30% travelling through network elements $j = 3, j = 4$ (dashed line) and 70% through network elements $j = 2, j = 3$ and $j = 4$ (solid line). Service traffic $R^{(2)}$ is not split, shown in red. Service traffic $R^{(3)}$ is split and shown with the solid and dashed blue lines. The $\alpha_j^{(k)}$ matrix is shown in Table 8.1.

TABLE 8.1
The $\alpha_j^{(k)}$ Values for the Network Depicted in Figure 8.2

$k \backslash j$	1	2	3	4	5	6	7	8
1	$\alpha_1^{(1)} = 1$	$\alpha_2^{(1)} = 0.7$	$\alpha_3^{(1)} = 1$	$\alpha_4^{(1)} = 0.3$	$\alpha_5^{(1)} = 0.7$	$\alpha_6^{(1)} = 1$	$\alpha_7^{(1)} = 0$	$\alpha_8^{(1)} = 0$
2	$\alpha_1^{(2)} = 1$	$\alpha_2^{(2)} = 1$	$\alpha_3^{(2)} = 0$	$\alpha_4^{(2)} = 1$	$\alpha_5^{(2)} = 0$	$\alpha_6^{(2)} = 1$	$\alpha_7^{(2)} = 0$	$\alpha_8^{(2)} = 0$
3	$\alpha_1^{(3)} = 0$	$\alpha_2^{(3)} = 0.5$	$\alpha_3^{(3)} = 1$	$\alpha_4^{(3)} = 1$	$\alpha_5^{(3)} = 0.5$	$\alpha_6^{(3)} = 0$	$\alpha_7^{(3)} = 1$	$\alpha_8^{(3)} = 1$

(b) However if the traffic data is averaged over a time duration T, we construct a relationship between the averaged traffic, R_j, through the j th network element to the averaged service traffic flows. In this the $\alpha_j^{(k)}$ are defined via the averaged values:

$$\left\langle R_j \right\rangle_T = \sum_{k=1}^{N^{(svc)}} \left\langle R_j^{(k)} \right\rangle_T = \sum_{k=1}^{N^{(svc)}} \alpha_j^{(k)} \left\langle R^{(k)} \right\rangle_T \qquad (8.7)$$

where

$$\left\langle X \right\rangle_T = \frac{1}{T} \int_T X(t) \, dt \qquad (8.8)$$

In situation a) above, the $\alpha_j^{(k)}$ typically only change on time scales much longer than a diurnal cycle. This is because most networks are static over that time scale. However, with networks that are dynamically reconfigurable (such as those operating MPLS, GMPLS, ASON, OpenFlow, SDN and NFV [21] protocols) [22–26] the $\alpha_j^{(k)}$ may vary on the diurnal time scale.

If one adopts model option a) above, then the R_j, $R^{(k)}$ in the equations below can be understood to include the time dependence t unless otherwise stated. In contrast, if one adopts model option b) then the R_j, $R^{(k)}$ in the equations below are taken to refer to the time averages defined by (8.8).

In other words, in the equations below, the quantities R_j and $R^{(k)}$ include an inferred time dependence if we choose model option a) and time averaged if we choose model option b). The choice of model option depends on the amount of information we have about the data flows - for the network and the services it provides.

Below we implement multiple levels of averaging: averaging over network elements and averaging over service traffic flows. We simplify the equations by expressly displaying the time dependence or time averaging only when these are needed for clarity.

To construct the relationships between the $\alpha_j^{(k)}$ and network power, we derive some general relationships for $\alpha_j^{(k)}$ that will be of use.

To do this we note the following: For two independent random variables, x_i and y_i, $i = 1,...,M$, and $M \gg 1$ we have:

$$\langle x \rangle = \frac{1}{M}\sum_{i=1}^{M} x_i \quad \text{and} \quad \langle xy \rangle = \frac{1}{M}\sum_{i=1}^{M} x_i y_i \approx \langle x \rangle \langle y \rangle \tag{8.9}$$

For two independent random variables and $M \gg 1$, we also have:

$$\sum_{i=1}^{M} x_i y_i \approx M \langle x \rangle \langle y \rangle \approx \langle x \rangle \sum_{i=1}^{M} y_i \tag{8.10}$$

In the calculations below, there are averages across the number of network elements (j subscript index) as well as averages across the number of services (k superscript index). We need to distinguish between averaging over these distinct sets of indices and hence denote these different averages as follows:

$$\frac{1}{N^{(svc)}}\sum_{k=1}^{N^{(svc)}} x_j^{(k)} = \left\langle x_j \right\rangle^{(svc)} \quad \text{average over services for given network element } j$$

$$\frac{1}{N_E}\sum_{j=1}^{N_E} x_j^{(k)} = \left\langle x^{(k)} \right\rangle_E \quad \text{average over network elements for given service } k \tag{8.11}$$

$$\frac{1}{N_E N^{(svc)}}\sum_{j=1}^{N_E}\sum_{k=1}^{N^{(svc)}} x_j^{(k)} = \left\langle x \right\rangle \quad \text{average over both network elements and services}$$

We define the mean traffic per service, $\langle R \rangle^{(svc)}$, and mean traffic per network element, $\langle R \rangle_E$, respectively, by

$$\left\langle R \right\rangle^{(svc)} = \frac{1}{N^{(svc)}}\sum_{k=1}^{N^{(svc)}} R^{(k)} = \frac{R_{Ntwk}}{N^{(svc)}}$$

$$\left\langle R \right\rangle_E = \frac{1}{N_E}\sum_{j=1}^{N_E} R_j \tag{8.12}$$

Note: For model option a) these equations include a diurnal-cycle time dependence. For model option b) these equations do not have a cyclic time dependence in the traffic values R_j and $R^{(k)}$. In both cases N_E and $N^{(svc)}$ do not have a time dependence.

8.4.3 α Matrix Relationships

The matrix $\alpha_j^{(k)}$ and its averages can provide information that is of use in modelling the power consumption and energy efficiency of a network. This matrix relates the service traffic flows, $R^{(k)}$, to the network element throughput traffic, R_j, which determine the power consumption of the network elements. Therefore, we can use the $\alpha_j^{(k)}$ to relate services to their corresponding power consumption and the total network traffic (which is the sum over all components of the service traffic) to overall network power consumption.

The purpose of this section it to provide the modeller with a range of approximations that may assist if they do not have access to detailed information regarding the network architecture and service traffic flows through the network. Different modellers may have access not different information regarding the network and service traffic flows. For example, many models are based on only knowing the average number of hops for a typical service. The results in this section provide relationships between that value and other parameters that may be of use in constructing approximations for service and/or network power consumption.

If no service traffic flow is split across multiple paths, then $\alpha_j^{(k)} = 1$ or 0 for every network element j and service, k. In this case, the number of services routed through the j th network element, $N_j^{(svc)}$, is given by:

$$N_j^{(svc)} = \sum_{k=1}^{N^{(svc)}} \alpha_j^{(k)} \qquad \text{where } \alpha_j^{(k)} = 1 \text{ or } 0 \tag{8.13}$$

We can generalise this equality to include cases where service traffic is split over multiple paths between its source and destination, resulting in some of the $\alpha_j^{(k)}$ being less than 1 such as in Figure 8.2 and the corresponding $\alpha_j^{(k)}$ matrix in Table 8.1. With this generalisation, some of the $N_j^{(svc)}$ may be non-integer.

We assume that the route through the network for a service can be considered independent of the service data rate. This is reasonable because typically we expect $R_j >> R^{(k)}$ for all j and k; hence the choice of a path through the network for a service is unlikely to be network element dependent. Assuming that the $\alpha_j^{(k)}$ and $R^{(k)}$ are independent we can apply (8.10) to approximate (8.5):

$$R_j = \sum_{k=1}^{N^{(svc)}} \alpha_j^{(k)} R^{(k)} \approx \langle R \rangle^{(svc)} \sum_{k=1}^{N^{(svc)}} \alpha_j^{(k)} \approx \langle \alpha_j \rangle^{(svc)} \sum_{k=1}^{N^{(svc)}} R^{(k)} = \langle \alpha_j \rangle^{(svc)} R_{Ntwk} \tag{8.14}$$

where $\langle R \rangle^{(svc)}$ is the mean of the traffic flows taken across all the services. In addition, generalising (8.13) we also have:

$$R_j = \sum_{k=1}^{N_j^{(svc)}} R_j^{(k)} = \sum_{k=1}^{N_j^{(svc)}} \alpha_j^{(k)} R^{(k)} \approx N_j^{(svc)} \langle R \rangle^{(svc)} \tag{8.15}$$

Using (8.14) and (8.15) we also have for the number of services routed through the j th network element:

$$N_j^{(svc)} = \sum_{k=1}^{N^{(svc)}} \alpha_j^{(k)} = N^{(svc)} \left\langle \alpha_j \right\rangle^{(svc)} \approx R_j / \left\langle R \right\rangle^{(svc)} \qquad (8.16)$$

and

$$\sum_{k=1}^{N^{(svc)}} \alpha_j^{(k)} R^{(k)} \approx N^{(svc)} \left\langle R \right\rangle^{(svc)} \left\langle \alpha_j \right\rangle^{(svc)} \approx R_{Ntwk} \left\langle \alpha_j \right\rangle^{(svc)} \qquad (8.17)$$

Returning to the situation in which the k th service flow is not split across multiple paths and kept as a single flow, then the $\alpha_j^{(k)}$ take on values 0 or 1. Summing over the j index we get the number of network elements through which the k th service is routed:

$$N_E^{(k)} = \sum_{j=1}^{N_E} \alpha_j^{(k)} \quad \text{where} \quad \alpha_j^{(k)} = 1 \text{ or } 0 \qquad (8.18)$$

With the paths not split, the number of network elements through which the service k flows, $N_E^{(k)}$, satisfies:

$$\sum_{j=1}^{N_E} R_j^{(k)} = N_E^{(k)} R^{(k)} \qquad (8.19)$$

We generalise (8.18) and (8.19) to include the case when service traffic is split over multiple paths giving non-integer values of $\alpha_j^{(k)}$. We have the relationship between $\alpha_j^{(k)}$ and number of network elements in the path of the k th service, $N_E^{(k)}$, is

$$\sum_{j=1}^{N_E} \alpha_j^{(k)} R^{(k)} = N_E^{(k)} R^{(k)} \qquad (8.20)$$

With this generalisation, the number of network elements through which that service flows, $N_E^{(k)}$, is given by:

$$N_E^{(k)} = \sum_{j=1}^{N_E} \alpha_j^{(k)} = N_E \left\langle \alpha^{(k)} \right\rangle_E \qquad (8.21)$$

We note that $N_E^{(k)}$ equals the number of hops in the path of service k plus 1, that is $N_E^{(k)} = N_{hops}^{(k)} + 1$. The number of hops of a service is used for estimates of service power consumption and energy efficiency [6, 7, 10]. From (8.16) and (8.21):

$$\left\langle \alpha^{(k)} \right\rangle_E = \frac{N_E^{(k)}}{N_E}, \qquad \left\langle \alpha_j \right\rangle^{(svc)} \approx \frac{R_j}{R_{Ntwk}} \qquad (8.22)$$

That is, $\left\langle \alpha^{(k)} \right\rangle_E$ is the proportion of network elements that deal with the k th service flow and $\left\langle \alpha_j \right\rangle^{(svc)}$ is the fraction of network traffic through the j th network element.

Using (8.19) and (8.21), we also have the following relationships:

$$N_E^{(k)} = N_{hops}^{(k)} + 1 = \sum_{j=1}^{N_E} \alpha_j^{(k)} = N_E \left\langle \alpha^{(k)} \right\rangle_E = \sum_{j=1}^{N_E} R_j^{(k)} \bigg/ R^{(k)} = N_E \left\langle R^{(k)} \right\rangle_E \bigg/ R^{(k)} \tag{8.23}$$

Using their generalised definitions, depending upon the network architecture $N_j^{(svc)}$ and $N_E^{(k)}$ may not be integers. Using (8.16) and (8.23), and because the summations are finite, we also have the following relationships involving α:

$$\sum_{j=1}^{N_E} \sum_{k=1}^{N^{(svc)}} \alpha_j^{(k)} = N_E N^{(svc)} \left\langle \alpha \right\rangle = \sum_{j=1}^{N_E} \left(\sum_{k=1}^{N^{(svc)}} \alpha_j^{(k)} \right) = N_E \left\langle N^{(svc)} \right\rangle_E$$

$$= \sum_{k=1}^{N^{(svc)}} \left(\sum_{j=1}^{N_E} \alpha_j^{(k)} \right) = N^{(svc)} \left\langle N_E \right\rangle^{(svc)} = N^{(svc)} \left(\left\langle N_{hops} \right\rangle^{(svc)} + 1 \right) \tag{8.24}$$

In (8.24), $\langle N^{(svc)} \rangle_E$ is the mean number of services per network element and $\langle N_E \rangle^{(svc)}$ is the mean number of network elements in the path of a service (= mean number of hops for that service + 1). Therefore, the mean number of hops for a service, over all services, is given by:

$$\left\langle N_{hops} \right\rangle^{(svc)} + 1 = \left\langle N_E \right\rangle^{(svc)} = N_E \left\langle N^{(svc)} \right\rangle_E \bigg/ N^{(svc)} \tag{8.25}$$

Multiplying the numerator and denominator by the average traffic per service $\langle R \rangle^{(svc)}$ and noting that $N^{(svc)} \langle R \rangle^{(svc)} \approx R_{Ntwk}$ we get:

$$\left\langle N_{hops} \right\rangle^{(svc)} + 1 = \left\langle N_E \right\rangle^{(svc)} \approx \frac{N_E \left\langle N^{(svc)} \right\rangle_E \left\langle R \right\rangle^{(svc)}}{N^{(svc)} \left\langle R \right\rangle^{(svc)}} = \frac{N_E \left\langle R \right\rangle_E}{R_{Ntwk}} = \frac{\displaystyle\sum_{j=1}^{N_E} R_j}{\displaystyle\sum_{k=1}^{N^{(svc)}} R^{(k)}} \tag{8.26}$$

$$= \frac{\text{Total network element throughput}}{\text{Network traffic}}$$

Also, we have:

$$\sum_{j=1}^{N_E} R_j = N_E \left\langle R \right\rangle_E = \sum_{j=1}^{N_E} \sum_{k=1}^{N^{(svc)}} \alpha_j^{(k)} R^{(k)} \approx N_E \sum_{k=1}^{N^{(svc)}} R^{(k)} \left\langle \alpha^{(k)} \right\rangle_E$$

$$= \approx N_E N^{(svc)} \left\langle R \right\rangle^{(svc)} \left\langle \alpha \right\rangle \approx R_{Ntwk} N_E \left\langle \alpha \right\rangle \tag{8.27}$$

Therefore

$$\left\langle \alpha \right\rangle \approx \frac{\left\langle R \right\rangle_E}{R_{Ntwk}} \approx \frac{\text{mean network element traffic}}{\text{Total network traffic}} \tag{8.28}$$

From this we can interpret $\langle \alpha \rangle$ as the proportion of network traffic per network element.

Also, from (8.24):

$$
\langle \alpha \rangle = \frac{\langle N_E \rangle^{(svc)}}{N_E} = \frac{\langle N_{hops} \rangle^{(svc)} + 1}{N_E} =
$$

$$
= \frac{\left(\begin{array}{c} \text{mean number of network elements} \\ \text{in path of a service} \end{array} \right)}{\text{Number network elements}} = \frac{\text{mean hops for a service} + 1}{\text{Number network elements}} \tag{8.29}
$$

Assuming the network traffic to be evenly distributed across all network elements, this expression for $\langle \alpha \rangle$ also corresponds to the average proportion of network elements in the path of a service.

To check the accuracy of these approximations, a series of 1400 network simulations was set up to calculate the exact and approximated values defined above. The simulations ranged from 150 to 2000 service flows through mesh networks consisting of between 20 and 200 elements. The service paths were randomly generated to travel through between 4 and 16 elements. A proportion of the data flows was split into two paths and recombined, with each of the paths having a different length. The service flow over these split paths was randomly split across the two paths in the ratio x and $(1 - x)$ where $0.1 \leq x \leq 0.9$. Therefore, for these paths, the α 's have non-integer values. The data rates for the services were randomly set to be between 1 Gb/s and 40 Gb/s. This set-up resulted in the mean throughput of the network elements, $\langle R \rangle_E$, ranging from approximately 220 Gb/s to approximately 1.7 Tb/s and the total network traffic, R_{Ntwk}, ranging from approximately 500 Gb/s to approximately 40 Tb/s.

Over the full 1400 simulation data set, the correlation values are listed in Table 8.2. We can see that the forms derived above provide very good approximations for the actual values.

8.4.4 NETWORK POWER EQUATION

With these preliminaries completed, we approximate the power consumption of the network, (8.1), with the total of the power consumption of the flows using the $\alpha_j^{(k)}$ parameters; that is:

$$
P_{Ntwk}(t) = M_{OH} \sum_{j=1}^{N_E} \left(P_{idle,j} + E_j R_j(t) \right) \approx M_{OH} \sum_{j=1}^{N_E} \left(P_{idle,j} + E_j \sum_{k=1}^{N^{(svc)}} \alpha_j^{(k)} R^{(k)}(t) \right) \tag{8.30}
$$

TABLE 8.2
Correlation Values for Derived Approximations

Equation	Parameters	Correlation (R^2 Value)
(8.15), (8.16)	$\sum_{k=1}^{N^{(svc)}} \alpha_j^{(k)}$ and $R_j / \langle R \rangle^{(svc)}$	0.9
(8.24)	$\sum_{j=1}^{N_E} \sum_{k=1}^{N^{(svc)}} \alpha_j^{(k)}$, $N_E \langle N^{(svc)} \rangle_E$ and $N^{(svc)} \langle N_E \rangle^{(svc)}$	1
(8.26)	$\langle N_E \rangle^{(svc)}$ and $N_E \langle R \rangle_E / R_{Ntwk}$	≥ 0.9
(8.27)	$\langle R_E \rangle$ and $\langle R \rangle_{Ntwk} \langle \alpha \rangle$	≥ 0.9
(8.28)	$\langle \alpha \rangle$ and $\langle R \rangle_E / R_{Ntwk}$	≥ 0.99
(8.29)	$\langle \alpha \rangle$ and $\langle N_E \rangle^{(svc)} / N_E$	1

Assuming that the incremental energy per bit of the network elements, E_j, the service traffic, $R^{(k)}$, and the routing of the service traffic, $(\alpha_j^{(k)})$, can be treated as independent random variables and the traffic is widely spread throughout the network, we can use (8.9) and (8.10) on the double summation. We then use (8.24) and (8.29) to get:

$$
\begin{aligned}
P_{Ntwk}(t) &\approx M_{OH}\left(N_E \left\langle P_{idle} \right\rangle_E + N_E \left\langle E \right\rangle_E \left\langle \alpha \right\rangle R_{Ntwk}(t) \right) \\
&\approx M_{OH}\left(N_E \left\langle P_{idle} \right\rangle_E + \left\langle N_E \right\rangle^{(svc)} \left\langle E \right\rangle_E R_{Ntwk}(t) \right) \\
&\approx M_{OH}\left(N_E \left\langle P_{idle} \right\rangle_E + \left(\left\langle N_{hops} \right\rangle^{(svc)} + 1 \right)\left\langle E \right\rangle_E R_{Ntwk}(t) \right)
\end{aligned}
\tag{8.31}
$$

Note that the averages in (8.30) are over the network elements (j-index as indicated by a subscript E) and services (k-index indicated by a superscript (svc)), not the diurnal cycle (time).

Ideally, all the data required to calculate the total network power using the summations in (8.30) will be available. However, this will often not be the case. In such circumstances, the averaged values and approximations in (8.31) can be used. Also available are one or more of the approximations listed in Table 8.2. For example, (8.31) can be rearranged to involve the quotient $\langle N_E \rangle^{(svc)}/N_E$ which can be approximated with $\langle \alpha \rangle$ by using (8.28). Which approximations are useful will depend upon the data available for the model.

The same 1400 network simulations used to validate the approximations derived above were extended to include idle power, $P_{idle,j}$ and energy per bit, E_j, for the equipment. The idle power and energy per bit values were randomly selected ranging from 1.5kW to 11kW and 0.2nJ/bit to 0.8nJ/bit, respectively [7]. The correlation between (8.30) and (8.31) was greater than 0.95. From this we can see that, with the use of averaging and several simplifications, the approximation in (8.31) provides a good estimate for network power consumption without the need for detailed information about each item of equipment in the network.

To calculate network power using (8.31), the traffic $R_{Ntwk}(t)$ is determined by the diurnal cycle, $R_2(t,\tau)$, and annual traffic growth trends, $R_3(\tau)$. The approach for calculating the parameters in (8.31) for various network growth scenarios will now be considered.

8.5   MODELLING POWER CONSUMPTION GROWTH

The amount of network equipment deployed to deal with network traffic over a diurnal cycle is dimensioned to accommodate the peak load, R_{peak}, accounting for "maximum utilisation", μ_{max} ($0 \le \mu_{max} \le 1$) [27, 28]. The utilisation parameter is typically set by the network operator by considering allowances for:

- Network traffic growth.
- "Flash crowds", sudden short bursts of traffic due to special events.
- Ensure no more than a fraction $x\%$ of packets are lost over $y\%$ of the diurnal cycle. The values of x and y will be determined by network operation policies.
- Maximum or average allowed packet delay and delay jitter.
- Other relevant operator policies.

For the j th network element with maximum capacity, $C_{max,j}$, if the peak traffic through that network element starts to exceed $\mu_{max}C_{max,j}$ repeatedly on a daily basis, then the network operator will deploy additional equipment to accommodate the increased traffic load.

To calculate the number of deployed network elements, N_E, that appears in (8.31), we need to consider the traffic levels at the peak traffic time, t_{peak}. To do this we set $R_{Ntwk}(t_{peak}) = R_{peak}$, and the traffic through the network elements at t_{peak} to be $R_j(t_{peak}) = \mu_{max} C_{max,j}$. Using these values in (8.26) and solving for the number of network elements, N_E, required to accommodate network traffic at peak time, we get:

$$N_E = \left\lceil \frac{\left(\langle N_{hops} \rangle^{(svc)} + 1\right) R_{Ntwk}\left(t_{peak}\right)}{\langle R\left(t_{peak}\right)\rangle_E} \right\rceil = \left\lceil \frac{\left(\langle N_{hops} \rangle^{(svc)} + 1\right) R_{peak}}{\mu_{max}\langle C_{max}\rangle_E} \right\rceil \tag{8.32}$$

Where $\lceil x \rceil$ = smallest integer greater than x and we define the mean maximum throughput of the network elements, $\langle C_{max}\rangle_E$, by

$$\langle C_{max}\rangle_E = \frac{1}{N_E}\sum_{j=1}^{N_E} C_{max,j} \tag{8.33}$$

Using these results in (8.31), the power consumption can be written in the form:

$$P_{Ntwk}\left(t\right) \approx M_{OH}\left(\langle N_{hops} \rangle^{(svc)} + 1\right)\left(\frac{\langle P_{idle}\rangle_E}{\mu_{max}\langle C_{max}\rangle_E} R_{peak} + \langle E\rangle_E R_{Ntwk}\left(t\right)\right) \tag{8.34}$$

Variants of the form in (8.34) have been widely used to estimate the power consumption of networks and services [6–10, 12]. In most cases, these variants have not accounted for the P_{idle} contribution to network element power consumption and have included only the $E = P_{max}/C_{max}$ contribution. The peak power consumption, corresponding to peak traffic, R_{peak}, is given by

$$P_{Ntwk}\left(t_{peak}\right) \approx M_{OH}\left(\langle N_{hops} \rangle^{(svc)} + 1\right)\left(\frac{\langle P_{idle}\rangle_E}{\mu_{max}\langle C_{max}\rangle_E} + \langle E\rangle_E\right) R_{peak} \tag{8.35}$$

The average power, calculated over a diurnal cycle, T, is

$$\langle P_{Ntwk}\rangle_T \approx M_{OH}\left(\langle N_{hops} \rangle^{(svc)} + 1\right)\left(\frac{\langle P_{idle}\rangle_E}{\mu_{max}\langle C_{max}\rangle_E} R_{peak} + \langle E\rangle_E \langle R_{Ntwk}\rangle_T\right) \tag{8.36}$$

Using (7.9) we can write (8.34) as

$$P_{Ntwk}\left(t,\tau\right) \approx M_{OH}\left(\langle N_{hops} \rangle^{(svc)} + 1\right) R_{peak}\left(\tau\right)\left(\frac{\langle P_{idle}\rangle_E}{\mu_{max}\langle C_{max}\rangle_E} + \langle E\rangle_E f\left(t,\tau\right)\right) \tag{8.37}$$

Using (3.11) and setting $P_{idle} = yP_{max}$, $(0 < y \leq 1)$ this, in turn, can be written as:

$$P_{Ntwk}\left(t,\tau\right) \approx M_{OH}\frac{\left(\langle N_{hops} \rangle^{(svc)} + 1\right)\langle P_{max}\rangle_E R_{peak}\left(\tau\right)}{\langle C_{max}\rangle_E}\left(\frac{y}{\mu_{max}} + \left(1-y\right)f\left(t,\tau\right)\right) \tag{8.38}$$

In (8.38) the y/μ_{max} term represents the network element idle power consumption (i.e. element power when $R_j = 0$) summed over all the deployed elements, which is determined by the amount of equipment that must be deployed to deal with the peak traffic. The $f(t,\tau)$ dependent term represents the incremental power that is determined by the hourly and yearly evolution of the diurnal cycle, $f(t,\tau)$. The corresponding forms for the peak and average power are

$$P_{Ntwk}\left(t_{peak}\right) \approx M_{OH}\left(\langle N_{hops}\rangle^{(svc)}+1\right)\frac{\langle P_{max}\rangle_E}{\langle C_{max}\rangle_E}R_{peak}\left(\frac{y}{\mu_{max}}+\left(1-y\right)\right) \tag{8.39}$$

$$\langle P_{Ntwk}\rangle_T \approx M_{OH}\left(\langle N_{hops}\rangle^{(svc)}+1\right)\frac{\langle P_{max}\rangle_E}{\langle C_{max}\rangle_E}R_{peak}\left(\frac{y}{\mu_{max}}+\left(1-y\right)\langle f(\tau)\rangle_T\right) \tag{8.40}$$

Forms (8.38), (8.39) and (8.40) represent the power consumption for a given traffic profile and collection of network elements. They do not account for the deployment of multiple generations of equipment over successive years to accommodate the growth in network traffic. To provide a realistic estimation of network power consumption, we need to account for the fact that a network will typically include the deployment of several generations of equipment over the years of operation of the network. We now consider this issue and how to modify (8.38) accordingly.

8.5.1   Network Growth and Technology Trends

Over the years almost every term in (8.38) will likely change. The technology parameters, P_{min}, C_{max} and y will change as new generations of equipment are deployed in the network. Design parameters μ_{max} and N_{hops} will change as the network operator modifies its design practices with the evolution of traffic demands on its network. Finally, $R_{peak}(\tau)$ and $f(t,\tau)$ will change as customer demand grows and changes over the years. If we can model or approximate these changes, we can use (8.38) to estimate the evolution of network power consumption over the years of network operation.

The energy efficiency (typically measured in Joules/bit) of network equipment has been improving over the years. This has been made manifest through their power consumption per unit throughput (energy/bit) reducing year on year [29–31]. Focussing on (8.38), several publications have tracked the evolution of the parameters in this equation. For example, Neilson [29], Schien and Preist [3] and Van Heddeghem and Idzikowski [9] have published trends for $\langle P_{max}\rangle/\langle C_{max}\rangle$. All these publications represent the annual technology improvements in the form of this quotient reducing year on year. That is, expressing this ratio for year n in terms of year 0 we have:

$$\frac{\left\langle P_{max}^{[n]}\right\rangle_E}{\left\langle C_{max}^{[n]}\right\rangle_E}=\frac{\left\langle P_{max}^{[0]}\right\rangle_E}{\left\langle C_{max}^{[0]}\right\rangle_E}\gamma^n \tag{8.41}$$

where $0 \leq \gamma \leq 1$.

We have $\langle P_{max}^{[n]}\rangle \geq \langle P_{min}^{[n]}\rangle$ for all n; therefore, we set the relationship $\langle P_{min}^{[n]}\rangle = y^{[n]}\langle P_{max}^{[n]}\rangle$ to apply for all years. In this equation $y^{[n]}$ quantifies the relationship between idle and peak network element power in the nth year accounting for newer generations of equipment. Different approaches to reducing power are available [9, 32]. For example, improvements in electronic technologies may result in both P_{idle} and P_{max} reducing proportionately, which is accounted for by γ alone. Another approach is to introduce sleep states, which will reduce P_{idle} more significantly than P_{max}. This approach will be represented by reducing $y^{[n]}$ relative to $y^{[0]}$.

It is important to note that the index, γ, in (8.41) represents annual reduction of the ratio $\langle P_{max}^{[n]}\rangle_E/\langle C_{max}^{[n]}\rangle_E$. Therefore, if there are q years between two generations of equipment released onto the market, the change in the ratio between the two generations of equipment will be γ^q.

Trends for network traffic growth can be described using (7.18) and (7.19) [33, 34]. To accommodate traffic growth, network operators will schedule the deployment of additional equipment, based upon expected or projected traffic growth.

Traffic surveys and forecasts indicate network traffic is growing year on year [34]. We assume this trend will continue and consider the power consumption due to network equipment added or updated in the Nth year based upon the growth trends given in (7.18) and (7.19). Considering the intervening years (0 to N) the diurnal-cycle traffic at the end of the mth year is set equal to the traffic at the start of year $(m+1)$, $R^{[m+1]}(t)$. This will be given by the diurnal-cycle traffic at the start of the mth year, denoted by $R^{[m]}(t)$, plus the increase in diurnal-cycle traffic during that year, which we denote as $\Delta R^{[m]}(t)$. Therefore for $m \geq 0$ we write:

$$R^{[m+1]}(t) = R_{peak}^{[m+1]} f^{[m+1]}(t) = R^{[m]}(t) + \Delta R^{[m]}(t) = R_{peak}^{[m]} f^{[m]}(t) + \Delta R^{[m]}(t) \qquad (8.42)$$

From this we have for $m \geq 0$:

$$\Delta R^{[m]}(t) = R_{peak}^{[m+1]} f^{[m+1]}(t) - R_{peak}^{[m]} f^{[m]}(t) = R_{peak}^{[0]} \sigma^m \left(\sigma f^{[m+1]}(t) - f^{[m]}(t) \right) \qquad (8.43)$$

and

$$\Delta R_{peak}^{[m]} = R_{peak}^{[m+1]} - R_{peak}^{[m]} = R_{peak}^{[0]} \sigma^m (\sigma - 1) \qquad (8.44)$$

We can write for traffic at the start of the Nth year $(N > 0)$:

$$R^{[N]}(t) = R^{[0]}(t) + \sum_{m=0}^{N-1} \Delta R^{[m]}(t) = \sigma^N R_{peak}^{[0]} f^{[N]}(t) \qquad (8.45)$$

Taking the diurnal-cycle average of (8.45) and using (7.19) gives

$$\left\langle R^{[N]} \right\rangle_T = \beta^N R_{peak}^{[0]} \left\langle f^{[0]} \right\rangle_T \qquad (8.46)$$

8.5.2 Network Power Consumption Forecasting

We now use the equations above to construct the power consumption forecasts for three network deployment scenarios. Each scenario can accommodate annual traffic growth, but differ in their equipment deployment strategies. In all scenarios, the average number of hops in a service path, $\langle N_{hops} \rangle^{(svc)}$, the ratio y, the maximum utilisation, μ_{max}, and overhead multiplier, M_{OH}, are assumed to remain constant over the years.

8.5.2.1 No Equipment Improvement – Scenario (i)

This scenario assumes traffic growth continues to follow current growth trends but there is no improvement in network technology; that is $\gamma = 1$. This scenario is a "worst case" and provides an upper bound to network power consumption because the network traffic is being dealt with by equipment that has seen no improvement in energy efficiency. In this case, the equipment parameters $\langle P_{max}^{[N]} \rangle_E / \langle C_{max}^{[N]} \rangle_E = \langle P_{max}^{[0]} \rangle_E / \langle C_{max}^{[0]} \rangle_E$ for all N while $R_{peak}^{[N]}$ and $f^{[N]}(t)$ evolve. Several authors have used this scenario for the growth in power consumption of the Internet to indicate the importance of improving the energy efficiency of telecommunications networks [6, 10, 35, 36]. In this scenario, the approach that is modelled is deploying equipment in year N to accommodate the traffic in that year [1, 4, 6]. Although this scenario is far from representing "real world" network planning, it is considered because it provides a set of "worst case" results corresponding to an upper limit for power consumption.

For this case, there is no improvement of the equipment and the technology is fixed at year 0. With annually increasing traffic, we use (8.45) in (8.38) giving the instantaneous network power consumption at time t in year N as

$$P_{Ntwk}^{[N]}(t) \approx M_{OH}\left(\left\langle N_{hops}\right\rangle^{(svc)} + 1\right)\frac{\left\langle P_{max}^{[0]}\right\rangle_E}{\left\langle C_{max}^{[0]}\right\rangle_E} R_{peak}^{[N]}\left(\frac{y^{[0]}}{\mu_{max}} + \left(1 - y^{[0]}\right)f^{[N]}(t)\right) \tag{8.47}$$

Using (8.46) with (8.47), the mean power consumption over a diurnal cycle, T_2, is given by

$$\left\langle P_{Ntwk}^{[N]}\right\rangle_T \approx M_{OH}\left(\left\langle N_{hops}\right\rangle^{(svc)} + 1\right)\frac{\left\langle P_{max}^{[0]}\right\rangle_E}{\left\langle C_{max}^{[0]}\right\rangle_E} R_{peak}^{[0]}\left(\frac{y^{[0]}}{\mu_{max}}\sigma^N + \left(1 - y^{[0]}\right)\beta^N\left\langle f^{[0]}\right\rangle_T\right) \tag{8.48}$$

Noting that $f^{[N]}(t_{peak}) = 1$, from (8.47) the network power consumption at peak traffic time, t_{peak}, is

$$P_{Ntwk}^{[N]}(t_{Peak}) \approx M_{OH}\left(\left\langle N_{hops}\right\rangle^{(svc)} + 1\right)\frac{\left\langle P_{max}^{[0]}\right\rangle_E}{\left\langle C_{max}^{[0]}\right\rangle_E} R_{peak}^{[0]}\sigma^N\left(y^{[0]}\left(\frac{1}{\mu_{max}} - 1\right) + 1\right) = \sigma^N P_{Ntwk}^{[0]}(t_{peak}) \tag{8.49}$$

8.5.2.2 Latest Generation Equipment across Entire Network – Scenario (ii)

In this scenario, all the network traffic is dealt with by the latest generation of equipment (i.e. all network equipment is upgraded each year including additional equipment to accommodate annual traffic growth). This scenario provides a "best case" and provides a lower bound to network power consumption. This scenario is not feasible because it represents a situation in which all the equipment in the network is replaced annually with the latest generation of equipment and every year a new generation of improved equipment is made commercially available. Such a situation is not possible due to financial constraints and the fact that new generation equipment is seldom released on an annual basis.

Despite this, several publications have provided network power forecasts based on this approach. As stated above, these forecasts provide a lower limit to future network power consumption [6, 8, 11].

To model this scenario, all the equipment in the network has the annual improvement included as described in (8.41) along with the annual increase in traffic. Using (8.41) and (8.45) in (8.38), the instantaneous network power consumption at time t is given by

$$P_{Ntwk}^{[N]}(t) \approx M_{OH}\left(\left\langle N_{hops}\right\rangle^{(svc)} + 1\right)\frac{\left\langle P_{max}^{[0]}\right\rangle_E}{\left\langle C_{max}^{[0]}\right\rangle_E} R_{peak}^{[0]}\left(\sigma\gamma\right)^N\left(\frac{y^{[N]}}{\mu_{max}} + \left(1 - y^{[N]}\right)f^{[N]}(t)\right) \tag{8.50}$$

The mean network power consumption over a diurnal cycle T is

$$\left\langle P_{Ntwk}^{[N]}\right\rangle_T \approx M_{OH}\left(\left\langle N_{hops}\right\rangle^{(svc)} + 1\right)\frac{\left\langle P_{max}^{[0]}\right\rangle_E}{\left\langle C_{max}^{[0]}\right\rangle_E} R_{peak}^{[0]}\gamma^N\left(\frac{y^{[N]}}{\mu_{max}}\sigma^N + \left(1 - y^{[N]}\right)\beta^N\left\langle f^{[0]}\right\rangle_T\right) \tag{8.51}$$

Network power consumption at t_{peak} is

$$P_{Ntwk}^{[N]}\left(t_{peak}\right) \approx M_{OH}\left(\left\langle N_{hops}\right\rangle^{(svc)}+1\right)\frac{\left\langle P_{max}^{[0]}\right\rangle_E}{\left\langle C_{max}^{[0]}\right\rangle_E} R_{peak}^{[0]}\left(\sigma\gamma\right)^N\left(y^{[N]}\left(\frac{1}{\mu_{max}}-1\right)+1\right) \tag{8.52}$$

8.5.2.3 Equipment Improvement and Replacement – Scenario (iii)

The above two scenarios are useful in that they provide an upper and lower bound to network power consumption over the forecast years. However, they do not represent actual practice in the planning and deployment of a telecommunications network. A more realistic representation of the approach adopted by network operators is have a schedule for planning ahead to deploy sufficient equipment to cater for a few years of network traffic growth. The "few years" being a matter of operator policy." In this scenario, we need to keep track of traffic growth during each year of the planning cycle. We will index the years covered by the planning period with index m with $0 \leq m \leq N$ and the start year for the forecast is $m = 0$. Therefore, the m index will range over years 0 to N, where N is the number of years into the future the plan covers. During the planning stage for the next set of network upgrades, the network operator will decide upon the number of years between deploying network upgrades. We denote this number of years by p. That is, if there was a network upgrade deployed in year m, the next planned upgrade will be in year $m + p$.

If we decide to deploy a network upgrade in year m, to determine how much equipment is required for that upgrade a forecast is made for the traffic beyond year m out to the year $(m + p)$. Using this approach, the amount of equipment deployed in the year m is dimensioned so that the anticipated peak traffic in the network corresponds to the maximum allowed utilisation of the equipment, μ_{max}, in year $m + p$. Between the years m and $m + p$, no additional equipment is added and the utilisation of the equipment (deployed in year m) increases as the traffic in the network increases. The equipment deployed in year m to deal with traffic growth over the years m to $m + p$ will be based on the latest technology available in year m.

Starting at year $m = 0$, following this approach, equipment deployment will be in years p, $2p$, $3p$, ... etc. (Generally, the years in which additional equipment is deployed is given by the non-zero values $\lfloor m/p \rfloor p$, where $\lfloor x \rfloor$ is the largest integer less than or equal to x. For $0 \leq m \leq N$, the non-zero values given by $\lfloor m/p \rfloor p$ are p, $2p$, $3p$, ...)

The equipment deployed will be the latest generation available at the time of deployment; however, each new generation of equipment is available every q years which need not align with the deployment schedule of each p years. Again, starting at $m = 0$, the appearance of a new generation of equipment will be in years q, $2q$, $3q$, ..., or more generally $\lfloor m/q \rfloor q$.

As the network traffic increases, in addition to deploying latest-generation equipment to deal with traffic growth, network operators typically take the opportunity to also remove old (depreciated) equipment. The depreciated equipment will be replaced with latest generation equipment and the traffic previously carried by the replaced equipment will be transferred to the new equipment. Therefore, in this scenario, we include the impact on network power consumption of removing old depreciated equipment and replacing it with the latest generation equipment.

To include this in the model we adopt the approach that equipment replacement only occurs at the time of a network upgrade (i.e. at multiples of p years). Therefore, we set a parameter r such that rp is the number of years a piece of equipment operates in the network. After rp years from its initial deployment, that equipment is replaced with the latest generation equipment. The value of r will be set by the network operator's equipment depreciation policy.

Again, starting at year $m = 0$ this replacement process will start at year $m = rp$. From that time on, if m is a year in which new equipment is deployed (i.e. m/p is an integer), then equipment deployed in the year $(\lfloor m/p \rfloor - r)p$ will be replaced. Further, all traffic carried by equipment deployed in the year

($\lfloor m/p \rfloor$-r)p will be transferred across to the latest generation equipment. The generation of the new equipment will be $\lfloor m/q \rfloor$ and the equipment to be replaced with this new generation of equipment was $\lfloor (m-rp)/q \rfloor$ generation equipment.

For example, if $p = 2$ and $r = 3$, we introduce new equipment every $p = 2$ years and replace equipment after $rp = 6$ years of operation. If $q = 4$, then in year $m = 8$ the operator will be deploying generation $\lfloor m/q \rfloor = \lfloor 8/4 \rfloor = 2$ equipment and replacing generation $\lfloor (m-rp)/q \rfloor = \lfloor (8-6)/4 \rfloor = 0$ equipment (deployed in year $\lfloor m/p \rfloor - r)p = (\lfloor 8/2 \rfloor - 3)2 = 2$) with generation 2 equipment. The traffic that was carried by the depreciated (generation 0) equipment will be transferred across to that new generation 2 equipment.

To calculate the amount of new equipment that must be deployed in year m requires a forecast of the traffic growth over duration of years m to $m + p$. Based on the published reports [34], the traffic growth in the network is assumed to be exponential. The growth in peak traffic is depicted in Figure 8.3.

Using the approach in (8.42) and (8.43), the increase in traffic over the years between two successive deployments, $\lfloor m/p \rfloor p$ and $(\lfloor m/p \rfloor + 1)p$, denoted by $\Delta R^{[m,p]}(t)$ is

$$\Delta R^{[m,p]}(t) = R_{peak}^{[(\lfloor m/p \rfloor +1)p]} f^{[(\lfloor m/p \rfloor +1)p]}(t) - R_{peak}^{[\lfloor m/p \rfloor p]} f^{[\lfloor m/p \rfloor p]}(t) \tag{8.53}$$

The increase in peak traffic from year $\lfloor m/p \rfloor p$ to year $(\lfloor m/p \rfloor +1)p$ is given by

$$\Delta R_{peak}^{[m,p]} = R_{peak}^{[(\lfloor m/p \rfloor +1)p]} - R_{peak}^{[\lfloor m/p \rfloor p]} = R_{peak}^{[0]} \sigma^{\lfloor m/p \rfloor p}(\sigma^p -1) \tag{8.54}$$

To construct an expression for network power in year N, we need to keep track of the equipment deployed during each upgrade and the generation of that equipment, as well as the equipment that is to be removed and replaced by the newer generation equipment. To make these distinctions clearer in the resulting equation, we express the relationship between network traffic growth and network

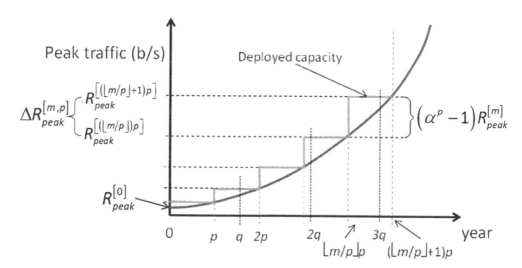

FIGURE 8.3 Network peak traffic growth over years. Network equipment is deployed each p years. The years when additional network equipment is deployed are indicated by dashed vertical lines. The equipment in year $\lfloor m/p \rfloor$ p must accommodate traffic growth until the next round of new deployment in year $(\lfloor m/p \rfloor + 1)p$. The new generation of equipment is released each q years, which may or may not coincide with a year of deployment. The green line shows the capacity of the equipment deployed as a function of the year of operation of the network

power in two parts. The first part represents the equipment deployed in years up to year N (accounting for the newer generations of equipment) *without* any replacement of depreciated equipment. The second part represents the equipment that has been removed from the network (i.e. the depreciated equipment) by subtracting out the contributions of the depreciated equipment. Then adding in the contributions resulting from the deployment of the latest generation equipment that has been dimensioned to deal with the traffic that the depreciated equipment previously carried.

The resulting expression is

$$
\begin{aligned}
P_{Ntwk}^{[N]}(t) = M_{OH} \Bigg(& \left\langle N_{hops} \right\rangle^{(svc)} + 1 \Bigg) \frac{\left\langle P_{max}^{[0]} \right\rangle_E}{\left\langle C_{max}^{[0]} \right\rangle_E} R_{peak}^{[0]} \left(\frac{y^{[0]}}{\mu_{max}} + \left(1 - y^{[0]}\right) f^{[0]}(t) \right) \\
& + \sum_{m=0}^{N} \gamma^{\lfloor m/q \rfloor q} \Bigg(\frac{y^{[m]}}{\rho} \sigma^{\lfloor m/p \rfloor p} \left(\alpha^p - 1\right)\left(\lfloor m/p \rfloor - \lfloor (m-1)/p \rfloor\right) \\
& \qquad + \left(1 - y^{[m]}\right) \sigma^m \left(\sigma f^{[m+1]}(t) - f^{[m]}(t)\right) \Bigg) \\
& + \sum_{m=0}^{N} \theta\left(\lfloor m/p \rfloor - r\right) \left(\gamma^{\lfloor m/q \rfloor q} - \gamma^{\lfloor (m-rp)/q \rfloor q}\right) \Bigg(y^{[m]} \frac{\sigma^{(\lfloor m/p \rfloor - r)p} \left(\sigma^p - 1\right)}{\mu_{max}} \left(\lfloor m/p \rfloor - \lfloor (m-1)/p \rfloor\right) \\
& \qquad + \sigma^{m-rp} \left(\sigma f^{[m+1-rp]}(t) - f^{[m-rp]}(t)\right)\left(1 - y^{[m]}\right) \Bigg) \Bigg)
\end{aligned}
\tag{8.55}
$$

where θ is the unit step function defined by $\theta(x) = 0$ for $x < 0$ and $\theta(x) = 1$ for $x \geq 0$. The term $(\lfloor m/p \rfloor - \lfloor (m-1)/p \rfloor)$ equals 1 in years when new equipment is added to the network (i.e. multiples of p) and zero otherwise. It is only applied to the idle power contribution (i.e. $y^{[m]}/\mu_{max}$ terms) because it represents the stepwise increases in idle power when new equipment is introduced. The other terms represent the continuous increase in power consumption as traffic increases.

In (8.55) terms the first summation term represents the first part of the expression: that is, the power consumption of the network equipment without replacement of old generation equipment. The second summation, which includes $\theta(\lfloor m/p \rfloor - r)$, is the second part of the expression representing the removal of the old equipment, its replacement with the newest generation equipment, and the transfer of traffic from the replaced equipment to the new equipment. In this part of the expression, the term $(\gamma^{\lfloor m/q \rfloor q} - \gamma^{\lfloor (m-rp)/q \rfloor q})$ represents the equipment energy efficiency improvement that is applied to the power consumption contribution arising from the traffic that was carried by the replaced equipment.

For average network power we get:

$$
\begin{aligned}
\left\langle P_{Ntwk}^{[N]} \right\rangle_T = M_{OH} \Bigg(& \left\langle N_{hops} \right\rangle^{(svc)} + 1 \Bigg) \frac{\left\langle P_{max}^{[0]} \right\rangle_E}{\left\langle C_{max}^{[0]} \right\rangle_E} R_{peak}^{[0]} \left(\frac{y^{[0]}}{\mu_{max}} + \left(1 - y^{[0]}\right) \left\langle f^{[0]} \right\rangle_T \right) + \\
& + \sum_{m=0}^{N} \gamma^{\lfloor m/q \rfloor q} \Bigg(\frac{y^{[m]}}{\rho} \sigma^{\lfloor m/p \rfloor p} \left(\sigma^p - 1\right)\left(\lfloor m/p \rfloor - \lfloor (m-1)/p \rfloor\right) \\
& \qquad + \left(1 - y^{[m]}\right) \beta^m \left(\beta - 1\right)\left\langle f^{[0]} \right\rangle_T \Bigg) \\
& + \sum_{m=0}^{N} \theta\left(\lfloor m/p \rfloor - r\right) \left(\gamma^{\lfloor m/q \rfloor q} - \gamma^{\lfloor (m-rp)/q \rfloor q}\right) \Bigg(\frac{y^{[m]}}{\mu_{max}} \sigma^{(\lfloor m/p \rfloor - r)p} \left(\sigma^p - 1\right)\left(\lfloor m/p \rfloor - \lfloor (m-1)/p \rfloor\right) \\
& \qquad + \beta^{m-rp} \left(\beta - 1\right)\left\langle f^{[0]} \right\rangle_T \left(1 - y^{[m]}\right) \Bigg) \Bigg)
\end{aligned}
\tag{8.56}
$$

For peak network power we have:

$$
\begin{aligned}
P_{Ntwk}^{[N]}\left(t_{peak}\right) = M_{OH}\Bigg(\!\left(\langle N_{hops}\rangle^{(svc)}+1\right)\frac{\left\langle P_{max}^{[0]}\right\rangle_E}{\left\langle C_{max}^{[0]}\right\rangle_E}R_{peak}^{(0)}\left(y^{[0]}\left(\frac{1}{\mu_{max}}-1\right)+ \right. \\
+\sum_{m=0}^{N}\gamma^{\lfloor m/q\rfloor q}\left(\frac{y^{[m]}}{\rho}\sigma^{\lfloor m/p\rfloor p}\left(\sigma^p-1\right)\left(\lfloor m/p\rfloor-\lfloor(m-1)/p\rfloor\right)\right. \\
\left.+\left(1-y^{[m]}\right)\sigma^m\left(\sigma-1\right)\right) \\
+\sum_{m=0}^{N}\theta\left(\lfloor m/p\rfloor-r\right)\left(\gamma^{\lfloor m/q\rfloor q}-\gamma^{\lfloor(m-rp)/q\rfloor q}\right)\left(\frac{y^{[m]}}{\mu_{max}}\sigma^{(\lfloor m/p\rfloor-r)p}\left(\sigma^p-1\right)\left(\lfloor m/p\rfloor-\lfloor(m-1)/p\rfloor\right)\right. \\
\left.\left.+\sigma^{m-rp}\left(\sigma-1\right)\left(1-y^{[m]}\right)\right)\Bigg)
\end{aligned}
\tag{8.57}
$$

8.5.3 Normalising the Results

Viewing the equations for $P_{Ntwk}^{[N]}(t)$, $\langle P_{Ntwk}^{[N]}\rangle_T$ and $P_{Ntwk}^{[N]}(t_{peak})$ we see a common term in all of them is

$$
M_{OH}\left(\langle N_{hops}\rangle^{(svc)}+1\right)\frac{\left\langle \dot{P}_{max}^{[0]}\right\rangle_E}{\left\langle C_{max}^{[0]}\right\rangle_E}R_{peak}^{[0]}
\tag{8.58}
$$

The terms $\langle P_{max}^{[0]}\rangle_E$, $\langle C_{max}^{[0]}\rangle_E$ and $R_{peak}^{[0]}$ are constants and we set M_{OH} and $\langle N_{hops}\rangle^{(svc)}$ to be constant over the years of the estimate. The form in (8.58) has been used as a basis to approximate network power consumption [6, 7, 9, 10, 12].

Dividing out the common term identified in (8.58) from the power equations for each scenario provides a "normalised" (dimensionless) measure of the evolution of the network power consumption, for a given annual growth in traffic and technology improvement independent of the specific details of the equipment. All that is required to determine the normalised results are

- the equipment improvement rate, γ;
- the annual growth rates for peak traffic σ and mean traffic β;
- the ratio of equipment idle to peak power y (typically greater than 0.8),
- network design parameter μ_{max}
- the year zero mean value $\langle f^{[0]}\rangle_T$.

There are published values for σ and β [34], γ [3, 9, 29, 31], y [37, 38] and μ_{max} [27, 28].

To determine a value for $\langle f^{[0]}\rangle_T$ we can use data collected from several published papers, which have provided diurnal-cycle traffic data [39–41] or data taken from traffic logs of the network operator [42–44]. Typical values for these parameters are listed in Table 8.3.

After using (8.58) to normalise the peak and average power consumption expressions above, if we know the average or peak power consumption of the network in year 0 and assuming the average number of hops for traffic, $\langle N_{hops}\rangle^{(svc)}$, the overhead factor, M_{OH}, and power ratio y are constant over the years of the forecast, then the forms above can be used to estimate proportional increase in future power consumption for a given deployment scenario (i.e. values of p, q and r) irrespective of the fine details of the network.

The results for the three deployment scenarios using the parameters in Table 8.3 with $p = 2$, $q = 3$ and $r = 2$ are graphed in Figure 8.4. We see that, during the first few years, Scenario (iii) consumes

---

TABLE 8.3

Parameter Values for Network Normalised Power and Energy per Bit Calculations

Parameter	Value
σ	1.37
β	1.30
γ	0.85
y	0.9
μ_{max}	0.5
$\langle f^{[0]} \rangle_T$	0.63

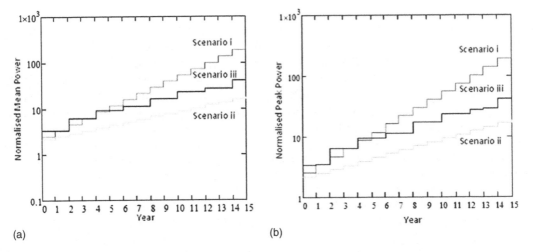

(a) (b)

FIGURE 8.4 Normalised mean, (a), and peak, (b), power evolution for Scenarios; (i) "no equipment improvement", (ii) "latest generation equipment" and (iii) "traffic forecasting with equipment replacement".

most power. This is a consequence of installing equipment to accommodate several (p) years of future traffic growth. However, after several years, Scenario (i) consumes most power because it is not upgrading to the latest generation of equipment, yet having to deal with the growth in traffic. Although Scenario (ii) consumes least power, however replacing all network equipment each year would most likely be prohibitively expensive and logistically challenging.

Using the normalised forms of equations (8.56) and (8.57), a network planner can test various network upgrading scenarios by varying p, q and r. The planner can also investigate the impact of changing the value of $y^{[n]}$ over the years to see of sleep states provide enough power savings to justify investment in such equipment with this functionality.

8.6 TECHNICAL SUMMARY

The results presented in this chapter are oriented towards medium-to-large networks that have at least 10's of network elements and 100's of service traffic flows. The results are based on several assumptions. We assume the following:

- The network has a sufficiently large number of network elements so that using averaged network element parameters (P_{idle}, C_{max}, P_{max}) provides a reasonable estimate of total network power consumption and approximations (8.9) and (8.10) can be applied. (If this assumption is not satisfied, it is highly likely that the amount of equipment in the network is small enough to allow direct calculation of power consumption by simply adding equipment powers.)

- There is a sufficiently large number of traffic flows through the network that using averaged traffic flows through equipment will provide a reasonably accurate estimate of total equipment power consumption and approximations (8.9) and (8.10) can be applied. (Again, if the number of flows is small, a simple direct summation of equipment powers should be feasible.)
- The traffic is sufficiently well spread over the network elements that almost no network elements are idle or at very low utilisation. This should be a reasonable assumption for several reasons. Installing equipment and then not utilising it would be a waste of resources. Further, most traffic in edge and core networks have diverse protection paths so that, in the event of a network failure, the traffic will still reach its destination. Factors such as these favour the spreading of traffic across network equipment.
- The energy per bit, E_j, of a network element is independent of that network element's throughput, R_j.

If these assumptions are valid for the network under consideration, then the total network equipment power consumption, $P_{Ntwk}(t)$, at time t in a diurnal cycle is approximated by (8.34). The peak network equipment power is given by (8.35) and the average network equipment power over duration T is given by (8.36). In these expressions, the overhead power factors such as environmental control (cooling), power supplies and regulation are cumulatively represented by the single multiplier M_{OH} [6, 8, 10].

When using this model to forecast future network power consumption we focus on peak and average network power. The peak power forecast provides an indication of the requirements for total power provisioning for the network. The average power forecast can be used to estimate the energy and carbon footprint of the network. To construct these forecasts we recast (8.35) and (8.36) into more convenient forms by introducing the ratio $y = P_{max}/P_{idle}$. With this and using $E = (P_{max} - P_{idle})/C_{max}$ we get (8.39) and (8.40).

These forms can then be used to forecast the peak and average network power consumption in future years. To do this we need to introduce the year index, superscript $[n]$, introduced in Chapter 7, which represents the changes in the parameters due to evolving technologies and traffic growth. From Chapter 7, in year n we have for the peak traffic, $R_{peak}^{[n]}$, and mean traffic, $\langle R^{[n]} \rangle_T$, in terms of their values in year 0:

$$R_{peak}^{[n]} = R_{peak}^{[0]}\sigma^n$$
$$\left\langle R^{[n]} \right\rangle_{T_2} = \beta^n \left\langle R^{[0]} \right\rangle_{T_2} = \sigma^n R_{peak}^{[0]} \left\langle f^{[n]} \right\rangle_{T_2} = \beta^n R_{peak}^{[0]} \left\langle f^{[0]} \right\rangle_{T_2} \tag{8.59}$$

General technology improvement trends can be represented using (8.41). In addition to these, we note that the ratio P_{max}/P_{idle} can also change with new generations of equipment with the introduction of technologies such as sleep states. To account for this, we include a time dependence for this parameter: $y^{[n]} = P_{max}^{[n]}/P_{idle}^{[n]}$.

We consider three scenarios. The first two are not realistic but provide an upper and lower bound to future network power growth. The third scenario represents a more realistic approach for planning future network growth.

Scenario (i): Growing traffic but no improvement in network equipment. This gives an upper bound to total network power consumption over future years. The upper bounds for average and peak network power consumption in year N are given in (8.48) and (8.49), respectively. In this scenario, if $y^{[N]}$ is approximately constant over the years, which has been the trend to date, we get:

$$P_{Ntwk}^{[N]}\left(t_{peak}\right) \approx \sigma^N P_{Ntwk}^{[0]}\left(t_{peak}\right) \tag{8.60}$$

Scenario (ii): Growing network traffic and all network equipment is the latest generation. This gives a lower bound to total network power consumption over future years. The lower bounds for average and peak network power consumption in year N are given by (8.51) and (8.52), respectively.

Scenario (iii): Growing network traffic with periodic network upgrades to latest generation equipment to accommodate traffic growth and replacement of old (depreciated) equipment with the latest generation equipment. In this case, the traffic previously carried by the depreciated equipment is transferred to the latest generation equipment that replaces it. This scenario represents a more realistic network planning by including:

- The deployment of additional equipment to deal with traffic growth at pre-set intervals;
- The inclusion of the impact of new generations of network equipment with improved energy efficiency;
- The removal of depreciated equipment, its replacement with latest generation equipment and the transfer of the traffic handled by the depreciated equipment across to latest generation equipment.

Scenario (iii) can provide network operators with a reasonable forecast for future network power consumption. In this forecast, we have the following additional parameters: p is the years between network deployments, q is the years between the availability of new generation equipment, and rp is the number of years between when a network element is deployed and when it is replaced by the latest generation of equipment. That is, equipment deployed in year m will be replaced by the latest generation of equipment in the year $m + rp$. The traffic previously carried by the depreciated equipment is transferred across to the latest generation equipment that takes its place.

In this case, the average and peak network power are given by (8.56) and (8.57), respectively. In these equations, the contributions to network power are split into two parts. The first part consists of the $y^{[0]}$ term plus the first summation, and represents the equipment deployed in years up to year N (accounting for the newer generations of equipment) without any replacement of depreciated equipment. The second part, which is the second summation, represents the equipment that has been removed from the network (i.e. the depreciated equipment) by subtracting out the contributions of that depreciated equipment. Then adding in the contributions resulting from the deployment of the latest generation equipment (to replace the depreciated equipment, which has been dimensioned to deal with the traffic that the depreciated equipment previously dealt with, plus growth.

Looking at the results for all three Scenarios, we note that they all have a common multiplier out front, this being:

$$M_{OH} \left(\langle N_{hops} \rangle^{(svc)} + 1 \right) \frac{\left\langle P_{max}^{[0]} \right\rangle_E}{\left\langle C_{max}^{[0]} \right\rangle_E} R_{peak}^{[0]} \tag{8.61}$$

Assuming that the number of hops averaged across the services, $\langle N_{hops} \rangle^{(svc)}$, remains constant over the forecast period, this factor can be divided out of all the scenario power equations, leaving a set of "normalised" network power consumption forecasts. These equations forecast the proportional increase in network power consumption over the forecast period for a given network deployment strategy, removing the need to estimate network power consumption in year [0].

Using the equations (8.48), (8.49), (8.51), (8.52), (8.56) and (8.57) and dividing out the term in (8.61), Table 8.4 shows the network power forecasts and the information required for each forecast. The power forecasts are for proportional power increases relative to year 0 values given in (8.61).

TABLE 8.4

Scenarios for Forecasting Average and Peak Network Powers

Scenario	Forecast Provided	Inputs Required
Scenario (i): (8.48) and (8.49)	Upper bounds for proportional increase in diurnal-cycle average and paeak network powers relative to year 0	Peak hour traffic growth rate, σ Average traffic growth rate, β
Scenario (ii): (8.51) and (8.52)	Lower bounds for proportional increase in diurnal-cycle average and peak network powers relative to year 0	Peak hour traffic growth rate, σ Average traffic growth rate, β Equipment improvement rate, γ P_{idle}/P_{max} ratio for each year, $y^{[n]}$
Scenario (iii): (8.56) and (8.57)	Approximate value for proportional change in diurnal-cycle average and peak network powers relative to year 0	Peak hour traffic growth rate, σ Average traffic growth rate, β Equipment improvement rate, γ P_{idle}/P_{max} ratio for each year, $y^{[n]}$ Years between new equipment deployments, p Years between new generation equipment releases, q Equipment depreciation lifetime, rp

8.7   EXECUTIVE SUMMARY

Forecasting communications network peak and average power consumption provides important guidance for comparing future deployment strategies. Peak power provides an indication of the requirements for power provisioning for the network to operate over the entire diurnal cycle, including peak traffic times. Average power forecasts provide an indication of future network energy and carbon footprints. These will be important for determining carbon offsets and other carbon abatement strategies.

Highly accurate forecasting will require a significant amount of information on network equipment and deployment architectures. By adopting a more statistically based approach, the amount of required information is reduced and the modelling can be significantly simplified. Although the accuracy of the forecasts will also be reduced, the model still provides information useful for future network planning.

Another advantage of the approximate model described in this chapter is that it can be applied to providing year-on-year forecasts of proportional network power increases. With the knowledge of peak and average traffic growth as well as technology improvement trends (which are available), this model can enable network planners to determine upper and lower bounds for future network consumption growth. With additional information on scheduling the replacement of old equipment with new generation equipment, more detailed peak and average power forecasts can be constructed.

REFERENCES

[1] K. Ishii *et al.*, "Unifying Top-Down and Bottom-Up Approaches to Evaluate Network Energy Consumption," *Journal of Lightwave Technology*, vol. 33, no. 21, p. 4395, 2015.

[2] D. Schien and C. Preist, "*A Review of Top-Down Models of Internet Energy Intensity,*" in *2nd International Conference on ICT for Sustainability, (ITC4S)*, 2014.

[3] D. Schien and C. Preist, "Approaches to the Energy Intensity of the Internet," *IEEE Communications Magazine*, vol. 52, no. 11, p. 130, 2014.

[4] J. Aslan *et al.*, "Electricity Intensity of Internet Data Transmission: Untangling the Estimates," *Journal of Industrial Ecology*, vol. 22, no. 4, p. 785, 2017.

[5] V. Coroama and L. Hilty, "Assessing Internet Energy Intensity: A Review of Methods and Results," *Environmental Impact Assessment Review*, vol. 45, p. 63, 2014.

[6] J. Baliga *et al.*, "Energy Consumption of Optical IP Networks," *Journal of Lightwave Technology*, vol. 27, no. 13, p. 2391, 2009.

[7] U. Mandal *et al.*, "Energy-Efficient Networking for Content Distribution Over Telecom Network Infrastructure," *Optical Switching and Networking*, vol. 10, p. 393, 2013.

[8] W. Van Heddeghem *et al.*, "Power Consumption Modeling in Optical Multilayer Networks," *Photonic Network Communications*, vol. 24, no. 2, p. 86, 2014.

[9] W. Van Heddeghem and *et al.*, "A Quantitative Survey of the Power Saving Potential in IP-over-WDM Backbone Networks," *IEEE Communications Surveys & Tutorials*, vol. 18, no. 1, p. 706, 2016.

[10] D. Kilper *et al.*, "Power Trends in Communication Networks," *IEEE Journal of Selected Topics In Quantum Electronics*, vol. 17, no. 2, p. 275, 2011.

[11] B. Skubic *et al.*, "Energy-Efficient Next Generation Optical Access Networks," *IEEE Communications Magazine*, vol. 50, no. 1, p. 122, 2012.

[12] A. Taal *et al.*, "*Storage to Energy: Modelling the Carbon Emissions of Storage Task Offloading between Data Centres,*" in *IEEE 11th Annual Consumer Communications and Networking Conference*, 2014.

[13] S. Lambert *et al.*, "Worldwide Electricity Consumption of Communications Networks," *Optics Express*, vol. 20, no. 26, p. 513, 2012.

[14] A. Andrae and T. Edler, "On Global Electricity Usage of Communication Technology: Trends to 2030," *Challenges*, vol. 6, p. 117, 2015.

[15] S. Lanzisera *et al.*, "Data Network Equipment Energy Use and Savings Potential in Buildings," *Energy Efficiency*, vol. 5, p. 149, 2012.

[16] C. Weber *et al.*, "The Energy and Climate Change Implications of Different Music Delivery Methods," *Journal of Industrial Ecology*, vol. 14, p. 754, 2010.

[17] J. Malmodin and D. Lunden, "The Energy and Carbon Footprint of the Global ICT and E&M Sectors 2010-2015," *Sustainability*, vol. 10, p. 3027, 2018.

[18] P. Corcoran and A. Andrae, "Emerging Trends in Electricity Consumption for Consumer ICT," 2013.

[19] J. Koomey *et al.*, "Network Electricity Use Associated with Wireless Personal Digital Assistants," *Journal of Infrastructure Systems*, vol. 10, no. 3, p. 131, 2004.

[20] B. Raghavan and J. Ma, "*The Energy and Emergy of the Internet,*" in *Proceedings of the 10th ACM Workshop on Hot Topics in Networks, Hotnets'11*, Cambridge, 2011.

[21] M. Veeraraghava *et al.*, "Network Function Virtualization: A Survey," *IEICE Transactions in Communications*, Vols. E100-B, no. 11, p. 1978, 2017.

[22] M. Chamania and A. Jukan, "Dynamic Control of Optical Networks," in *Springer Handbook of Optical Networks*, B. Mukherjee *et al.*, Eds., Springer, 2020, p. 535.

[23] J. Amazona *et al.*, "*A Critical Review of OpenFlow/SDN-based Networks,*" in *International Conference on Information Technology Converge Services (ICTON 2014)*, 2014.

[24] Infinera, "Infinera Virtualized Control Plane for Automatically Switched Optical Network (vASON)," Infinera Corporation, 2019.

[25] R. Masoudi and A. Ghaffari, "Sofware Defined Networks: A Survey," *Journal of Network and Computer Applications*, vol. 67, p. 1, 2016.

[26] H. Fardady *et al.*, "Softward-Defined Networking: A Survey," *Computer Networks*, vol. 81, p. 79, 2015.

[27] International Telecommunications Union, "Telecom Network Planning for evolving: Network Architectures Reference Manual; Document NPM/4.1," International Telecommunications Union.

[28] Cisco, "Best Practices in Core Network Capacity Planning," Cisco, 2013.

[29] D. Neilson, "*Power Dissipation Limitations to the Scalability of Network Elements,*" in *European Conference on Optical Communications*, 2011.

[30] R. S. Tucker, "Optical Communications—Part I: Energy Limitations in Transport," *IEEE Journal on Selected Topics in Communications*, vol. 17, no. 2, p. 245, 2011.

[31] C. Lange *et al.*, "Energy Consumption of Telecommunication Networks and Related Improvement Options," *IEEE Journal of Selected Topics In Quantum Electronics*, vol. 17, no. 2, p. 285, 2011.

[32] B. Buype *et al.*, "Power Reduction Techniques in Multilayer Traffic Engineering," in *International Conference on Information Technology Converge Services (ITCON)*, 2009.

[33] S. Korotky, "Semi-empirical Description and Projections of Internet Traffic Trends Using a Hyperbolic Compound Annual Growth Rate," *Bell Laboratories Technical Journal*, vol. 18, no. 3, p. 5, 2013.

[34] Cisco, "The Zettabyte Era: Trends and Analysis," Cisco White Paper, June 2017.

[35] T. Klein, "Future of Network Energy," in *Future X Network: A Bell Labs Perspective*, M. Weldon, Ed., CRC Press, 2015, p. 439.

[36] R. S. Tucker, "*Green Optical Communications – Part 2: Energy Limitations in Networks,*" *IEEE Journal of Selected Topics in Quantum Electrontronics*, vol. 17, no. 2, p. 261, 2011.

[37] X. Fan *et al.*, *"Power Provisioning for a Warehouse-sized Computer,"* in *Proceedings of the 34th annual International Symposium on Computer Architecture*, 2007.

[38] A. Vishwanath and *et al.*, "Modeling Energy Consumption in High-Capacity Routers and Switches," *Journal of Selected Areas in Communications*, vol. 32, no. 8, p. 1524, 2014.

[39] J. Yang *et al.*, *"Characterizing Internet Backbone Traffic from Macro to Micro,"* in *IEEE International Conference on Network Infrastructure and Digital Content*, 2009.

[40] M. Kihl *et al.*, *"Traffic Analysis and Characterization of Internet User behaviour,"* in *International Conference on Ultra-modern Telecommunications and Control Systems*, 2010.

[41] G. Maier *et al.*, *"On Dominant Characteristics of Residential Broadband Internet Traffic,"* in *Proceedings of 9th ACM SIGCOMM Conference on Internet Measurement*, 2009.

[42] Netnod, "IX Statistics," Netnod, 2020. [Online]. Available: https://www.netnod.se/ix/statistics. [Accessed 12 July 2020].

[43] Cenic, "CENIC Network Statistics," Cenic, [Online]. Available: http://cricket.cenic.org/. [Accessed 12 July 2020].

[44] London Internet Exchange, "Linx," London Internet Exchange, 2020. [Online]. Available: https://www.linx.net/. [Accessed 12 July 2020].

9 Data Centres

Power consumption in a data centre is a major operational cost for data centre operators. Consequently, there are many ongoing efforts to understand power consumption and improve the energy efficiency of data centres. An outcome of this effort is the development of many sophisticated and detailed power models for data centres. For an extensive survey of these models, readers are referred to Dayarathna *et al.* [1]. Our focus is on providing relatively simple and approximate power models for telecommunications systems and services and hence we will concentrate our efforts on describing simpler data centre power models.

Data centres are designed to provide a diversity of services to end users. As a user, be it corporate or individual consumer, our experience of data centres is the specific services they provide. Therefore, although the total power consumption of a data centre is important, from the perspective of a user, it is the power consumption of the services provided by the data centre that is of most direct interest. For example, a financial institution may use one or more data centres to store its transaction data and other business records. If that institution wishes to undertake a carbon footprint audit, its interest will not be on the total power consumption of the data centres storing the institution's data, rather it will be on the power consumption of the services they access when using the data centres.

The main difference between total data centre and data centre service power models is that the service model only considers the resources (CPU, memory, input/output, disk, etc.) required by the specific service under study. This will only constitute a fraction of the overall available data centre resources. Therefore, the challenge is more about how to ascertain what fraction of data centre resources a service requires.

Rather than construct a detailed data centre power model, this chapter will provide an overview of a generic data centre power model. It will then develop a server power model that can provide an estimate of the power consumption of several types of services provided by data centres. These models can then be used with later chapters to construct power models of services provided by a data centre.

9.1 DATA CENTRE POWER MODELLING

A major problem with power consumption modelling of data centres is the lack of detailed information of the power consumption of the sub-systems in a data centre. Depending upon the amount of this information to which the modeller has access, there are several approaches that can provide a working estimate of data centre power consumption that will be described in the following sections.

9.1.1 WHOLE OF DATA CENTRE MODELS

If the modeller has almost no information available on the specific data centre of interest, one approach for constructing a "ball-park" estimate is to use publicly available information to construct a Joule/bit value for a total data centre. This approach provides an estimate for the energy/bit of a data centre rather than an estimate of total data centre power consumption. However, this model can be used to construct an estimate for the "energy efficiency" of a data centre. Furthermore, in many cases the modeller is more interested in the power consumption of a data centre due to a specific service and this model provides such an estimate.

DOI: 10.1201/9780429287817-9

This approach was used by Jalali *et al.* [2] to compare the power consumption of using nano-data centres and centralised data centres. The Joule/bit value in that paper was derived from a published itemised cost model of a data centre [3, 4]. Using that information and adopting typical industry values for the data centre equipment, Jalali *et al.* calculated a data centre energy per bit value of 4 to 7 microJ/bit.

To provide guidance on how such an energy per bit model can be built, we now describe an approach to deriving an approximate power model of a Facebook server. This example shows how, in some cases, several different sources of information are brought together to build the model. A problem with using diversity of sources is that it assumes the authors of those sources have adopted similar definitions for the parameters used to construct their model. For example, "utilisation" of equipment and/or a data centre has a variety of possible interpretations.

This model provides an approximate value for the energy per bit (Joules/bit) for a Facebook server. The value is derived from Facebook's publicly posted total data centre energy consumption, which was reported as 3.24 TWh in 2018 [5]. The average Power Use Efficiency (PUE) for Facebook data centres in 2018 was reported as 1.11. Therefore, the Facebook IT equipment energy consumption in 2018 was 2.91 TWh = 10.5×10^{15} Joules.

To get an estimate of Facebook traffic (in bit/s), we use several other sources. Cisco publishes periodic reports on Internet traffic; the Cisco VNI published in 2018 [6] gives total Internet traffic in 2018 as 156 Exabytes/month. According to the Sandvine September 2019 "Global Internet Phenomenon Report" [7], Facebook was responsible for 7.79% of total Internet traffic. Therefore, an estimate of total Facebook traffic for 2018 is 156 Exabytes/month $\times$ 0.0779 $\times$ 12 $\times$ 8 = 1.16 $\times$ 10^{21} bits/year. This gives the Joules per bit for Facebook data centres as $10.5 \times 10^{15}/1.16 \times 10^{21}$ = 9.0 microJ/bit. Of course, this is an average value across all Facebook data centres. Specific servers and possibly specific data centres will give different values.

Despite the somewhat different approach to making this estimation, the values reported by Jalali *et al.*, 4 to 7 microJ/bit, and this estimate of 9 microJ/bit are in good agreement. Thus we should be able to use this value with some confidence that it will provide an overall estimate of power or energy consumption for a generic data centre.

9.1.2 Data Centre Sub-System-Based Models

Adopting a preference for simple data centre models, the model described in this section will follow the approaches described in [8–10]. These models break down the power consumption of a data centre into power for its sub-systems, focusing on modelling the dominant contributors to data centre power consumption. Listing these sub-systems, we have [8]:

1. Servers and storage systems;
2. Power supply equipment including power backup;
3. Cooling and air flow equipment;
4. Data centre networking equipment;
5. Control, security, lighting, etc.

Dayarathna *et al.* [1] provide an extensive survey of models applied to these sub-systems within a data centre. These models range from very sophisticated to relatively simple.

The relative power consumption of the sub-systems 1) to 5) listed above varies across data centres. Typical values are shown in Table 9.1. Because data centre architectures and technologies have been evolving over time, the table shows the percentage contributions and the corresponding reference with its year of publication.

From this we see that the dominant contributors to overall data centre power consumption are as follows: (1) the servers and storage equipment; (2) the cooling and air flow equipment; and

TABLE 9.1

Percentage Power Contributions to Data Centre Power Consumption

Reference, Year	Servers & Storage	Data Centre Networking	Cooling & Air Flow	Power Supply	Control, Security, etc.
[8] 2009	56%	5%	30%	8%	1%
[11] 2013	68%	5%	15%	8%	4%
[12] 2010	26%	10%	50%	11%	3%
[13] 2011	51%		33%	13%	3%
[14] 2011	46%	4%	38%	8%	4%

(3) power supply – altogether typically accounting for more than 80% of overall data centre power consumption. From this, we can understand the focus of the data centre power models in [8–10] being on server utilisation, cooling and power supply. In particular, these power models provide an estimate of data centre power consumption based upon two parameters:

(1) The aggregate load presented to the server infrastructure, represented by the utilisation at time t, $U(t)$, of the servers in the data centre. (Typically, $U(t)$ is normalised so that $0 \leq U(t) \leq 1$.)
(2) The temperature difference between the server room of the data centre and the outside temperature which affects the cooling efficiency.

9.2 SERVER POWER MODELS

The data centre utilisation, $U(t)$, is dimensionless, aggregated across all servers in the data centre and normalised to range between 0 and 1. Utilisation $U(t) = 0$ represents the situation when all the servers are idle. Utilisation $U(t) = 1$ represents when all the servers are operating at their maximum. For individual servers, there are multiple published models for the dependence of the power consumption of a server on its individual utilisation, $u(t)$ [1, 15, 16]. For the purposes of simplicity and noting that a linear server power model is a relatively good approximation for servers [17, 18], we adopt the following form for the power consumption of in individual server:

$$P_{svr}\left(u(t)\right) = P_{svr,idle} + \left(P_{svr,max} - P_{svr,idle}\right)u(t) \tag{9.1}$$

where $P_{svr,idle}$ is the idle power of the server (when $u(t) = 0$) and $P_{svr,max}$ is the maximum power consumption of the server (when $u(t) = 1$). One can also model the idle and maximum server powers [19]. The value of $u(t)$ can also be expressed in terms of compute capability using generic measures such as a Millions of Instructions Per Second (MIPS) [15], or a more task-oriented measure, such as Intel's EDA MIPS [16], or the SpecPower benchmark [9].

When applied across all the servers in a data centre, the utilisation for individual servers may vary due to differing workloads, load types, server types and consolidation of workloads within the data centre. Therefore, we define the data centre server utilisation, $U(t)$, as the average utilisation of all the servers in the data centre. That is:

$$U(t) = \left\langle u(t) \right\rangle_{svr} = \frac{1}{N_{svr}} \sum_{i=1}^{N_{svr}} u_i(t) \tag{9.2}$$

where N_{svr} is the number of servers in the data centre. To simplify the notation, in the equations below, we shall drop the t-dependence and express the quantities as functions of U. This is done with the understanding that U remains a function of time as its value changes over a diurnal cycle.

The total data centre server power consumption is the sum of server power consumptions and, using (9.2) in (9.1), we see that with each server operating at $\langle u \rangle_{svr}$, we get total server power consumption:

$$P_{svr,tot}\left(U\right) = N_{svr}P_{svr,idle} + \left(P_{svr,max} - P_{svr,idle}\right)N_{svr}U$$
$$= N_{svr}P_{svr,max}\left(y + \left(1-y\right)U\right) \tag{9.3}$$

where $y = P_{svr,idle}/P_{svr,max}$. Looking at (9.3), we see that spreading the load evenly over all servers (i.e. each server operating at utilisation $\langle u \rangle_{svr}$) will result in every server expending its idle power, even if the data centre utilisation, U (hence $\langle u \rangle_{svr}$), is close to zero. This is considered a waste of power and so server "consolidation" strategies have been developed to power down some servers and focus the load on a reduced number of servers [8, 9]. The process of consolidation can be represented by a parameter, c, representing the degree of consolidation. For $c = 0$, the data centre's workload is packed into the minimum number of active servers needed to service that workload, with as many servers as required operating at utilisation of unity. The remaining servers are powered down to save on their idle power. For $c = 1$, the data centre workload is perfectly balanced across all servers, each operating at utilisation $\langle u \rangle_{svr}$. With degree of consolidation c, the number of active servers, $N_{svr,act}$, is

$$N_{svr,act}\left(U\right) = \left(U + \left(1-U\right)c\right)N_{svr} \tag{9.4}$$

With maximum consolidation, $c = 0$, $N_{sver,act}(U) = UN_{svr}$ with each active server operating at utilisation $u_{c=0} = 1$. When $c = 1$, all N_{svr} servers are active with each at utilisation $u_{c=1} = u$. The average utilisation, $\langle u_c \rangle_{svr}$, of the active servers is

$$u_{c\,svr} = \frac{U}{\left(U + \left(1-U\right)c\right)} \tag{9.5}$$

with the remaining servers in a low-power "sleep" state. With this, the total power consumption of the data centre servers is

$$P_{svr,tot}\left(U\right) = N_{svr,act}\left(U\right)P_{svr,max}\left(y + \langle u_c \rangle_{svr}\left(1-y\right)\right) + \left(N_{svr} - N_{svr,act}\left(U\right)\right)P_{sleep}$$
$$= N_{svr}P_{svr,max}\left(U + \left(1-U\right)yc + y_{sleep}\left(1-U\right)\left(1-c\right)\right) \tag{9.6}$$

where P_{sleep} is the low-power state of the inactive servers and $y_{sleep} = P_{sleep}/P_{svr,max}$. If we focus only on the incremental power of the servers, $\langle u_c \rangle_{svr}(P_{svr,max} - P_{svr,idle})$, we see that this does not change with consolidation. That is:

$$N_{svr,act}\left(U\right)\langle u_c \rangle_{svr}\left(P_{svr,max} - P_{svr,idle}\right) = N_{svr}U\left(P_{svr,max} - P_{svr,idle}\right) \tag{9.7}$$

The power savings are attained by placing servers into a low energy state and hence saving on idle power. The overall incremental power consumption due to the workload is the same, irrespective of consolidation. This independence occurs with the linear server power model given in (9.1).

With nonlinear power models, there may be a savings in the incremental power consumption component that can be secured using consolidation. Further, where a data centre includes different types of server, consolidation of tasks onto the more efficient servers is an advantage.

9.3   COOLING AND AIR FLOW

There is a diversity of methods for data centre cooling and environmental control [20]. A review of data centre cooling technologies is provided by Capozzolia and Primiceria [21]. There is a variety of published power models for the environmental conditioning in data centres to cool servers [1], ranging from relatively simple generic models [22] through to sophisticated Computational Flow Dynamic (CFD) models [23]. For the purposes of simplicity, by and large we adopt the approach of Pelley *et al.* [8] and Rahmani *et al.* [9].

Practitioners distinguish between two main types of server room air cooling equipment: Computer Room Air Handler (CRAH) and Computer Room Air Conditioner (CRAC). A CRAH system uses chillers to provide a supply of low-temperature coolant (e.g. water) that is then used by a CRAH to provide cool air circulation around the server room, as indicated in Figure 9.1. A CRAC uses a (direct expansion) refrigeration cycle to cool air that is then circulated around the server room. The refrigerant is first compressed, raising its temperature, after which it is cooled in the condenser. This is often with the aid of cold water from the cooling tower, but the cooling operation is much simpler than that of the chiller of the CRAC system. The compressed refrigerant is then allowed to expand in the evaporator, absorbing heat from the data centre air flow as it does so. A block diagram of such plant is shown in Figure 9.1. CRAH and chiller systems are generally more efficient than CRAC and compressor because of the compressor power consumption [24]. As a rule of thumb, CRAH power consumption is approximately 70% of the total peak load being supported, whereas CRAC systems require closer to 100% of total peak load being supported [24].

This section provides a relatively straightforward set of equations for the power consumption of CRAH and CRAC and the pumps used to circulate the water coolant. Some of the models are empirically based on particular equipment and so may not reflect the specific details of equipment in other facilities. Consequently, the models presented here may only be approximate if one is seeking to model a specific environmental control system. However, they will provide an overall understanding of power modelling of data centre cooling systems.

As a rough "rule of thumb", the power consumption of data centre cooling systems can be estimated by noting that a 2003 study by Hewlett-Packard researchers reported that for every 1 Watt of power utilised for operation of servers, an additional 0.5 Watts is required to remove the dissipated heat [25]. Although this reference is relatively old, the ratio of cooling power to server power of 0.5 to 1 has been restated more recently [1, 26]. If one is seeking the simplest possible model for data centre cooling, a value in the range of $0.5 N_{svr,tot} P_{svr,max}$ to $N_{svr,tot} P_{svr,max}$ could be used.

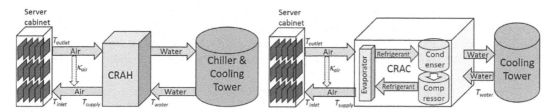

FIGURE 9.1   CRAH and CRAC cooling models. The cold water temperature, T_{water}, supply, T_{supply}, inlet, T_{inlet}, and outlet, T_{outlet}, air temperatures used in the models are indicated. Also, the "air confinement factor", κ_{air}, is used to represent mixing of the cold and hot air flows in the server room.

9.4 COMPUTER ROOM AIR HANDLING (CRAH) AND CHILLER POWER MODELS

Because the servers are the dominant generators of heat in a data centre, it follows that server utilisation, $U(t)$, is a key variable for quantifying the operation of the environmental control systems. Adopting this viewpoint, the authors of [8] and [9] utilise the US Department of Energy's "DOE2" chiller model to derive a chiller power consumption equation that is a function of utilisation $U(t)$. Pelley et al. [8] use a specific quadratic curve-fit equation relating utilisation to the power consumption of the chillers. As an example, for a chiller designed to remove 8 MW of heat at a peak chiller power consumption of 3200 kW, an outside atmospheric temperature of 29.4°C (85°F), and a chilled water supply to the CRAH of 7.22°C (45°F) – and complete consolidation ($c = 0$) – Pelley et al. give the chiller power consumption as a function of U:

$$P_{chiller}(U) = 742.8U^2 + 1844.6U + 538.7 \text{ (kWatts)} \qquad (9.8)$$

Pelley et al. also provide the chiller power consumption profile (as a function of data centre utilisation, U) for outside temperatures of 12.7°C (55°F), 18.3°C (65 °F), 23.9 °C (75 °F) and 35°C (95 °F). Using this model for the chiller will require more than one unit if the total power consumption of the servers is greater than 8 MW. Therefore, the number of chillers required, $N_{chiller}$, will be:

$$N_{chiller} = \left\lceil \frac{N_{svr}P_{svr,max}}{8MW} \right\rceil \qquad (9.9)$$

where $\lceil x \rceil$ is the smallest integer larger than x.

Rahmani et al. [9] provide a more generic form for the chiller power consumption given by

$$P_{chiller}(U) = 0.7N_{svr}P_{svr,\max}\left(0.32U^2 + 0.11U + 0.63\right) \qquad (9.10)$$

where the constants 0.32, 0.11 and 0.63 were obtained by curve fitting to measurements from several examples of real data centre data in which the water temperature was kept steady at 7.22°C. Factor $N_{svr}P_{svr,max}$ is the maximum total server power consumption. The factor of 0.7 arises from the rule-of-thumb dimension of chiller plant power to correspond to 70% of the total maximum server power consumption [24]. It is noted in [8] that the power consumption of the chiller can be highly dependent upon the chilled water temperature. When the load on the chiller is low (U much smaller than unity), the chiller power consumption is relatively insensitive to the chilled water temperature. However, as U increases toward unity, the power increase required to lower the chilled water temperature becomes substantial.

Turning our attention to the power consumption of the CRAH unit, several power models have been published. These include simple models with CRAH power linearly related to server utilisation, $U(t)$ [9], models in which the CRAH power is related to the cube of fan speed [8, 27], models that include optimisation with respect to server power consumption [28], advanced Computational Flow Dynamics (CFD) models [29, 23], thermodynamic models [30], and total server room models [31]. In this book, we will discuss the simpler models because they require less detailed information, making them more amenable for a wider audience.

Typically, a single CRAH unit can accommodate the removal up to a maximum amount of heat energy to keep the server equipment at the desired temperature. Therefore, if the total amount of heat generated by the servers is greater than a single CRAH unit can accommodate, multiple units will be required. The cooling capacity of CRAH units is dependent on several factors, including the air flow speed, the temperature difference between the inflow and outflow air, and the air flow design of the data centre. The simplest expression for the rate of heat energy removal, dQ_{heat}/dt (Joules/s), by a CRAH or CRAC unit is dependent upon the air flow, f_{air} (m³/s) [32, 33]:

$$\frac{d}{dt}Q_{heat} = specific\_gravity \times specific\_heat \times \Delta T \times f_{air}$$

$$= S_{air}C_{air}\Delta T f_{air} \tag{9.11}$$

where S_{air} is the *specific gravity* (1.2 kg/m³) and C_{air} is the *specific heat* (1.0 kJ/kg°C) are the parameters for air cooling at the operating temperature of the computer room and ΔT is the temperature difference between the exit and entry air flow. For the selected air flow operating temperatures, the amount of heat removed is controlled by the air flow rate. This formula can be modified to allow for imperfect containment of the air flow, where the cooling air can become mixed with warmer air. The "containment factor", κ_{air}, represents the fraction of cold air supplied to the server relative to the cold air supplied from the CRAH and $0 \le \kappa_{air} \le 1$. A value of $\kappa_{air} = 1$ corresponds to perfect containment in that there is no mixing of hot and cold server room air; hence, the temperature of the air flow into the server is identical to that supplied by the CRAH. In contrast, κ_{air} close to zero means the air flow into the server is close to that leaving the server, meaning there is significant mixing of the air flow into and leaving the server housing. The containment factor is given by [28]:

$$\kappa_{air} = \frac{T_{outlet} - T_{inlet}}{T_{outlet} - T_{supply}} \tag{9.12}$$

Referring to Figure 9.1, T_{supply} is the temperature of the cold air emerging from the CRAH, T_{inlet} is the temperature of the air inlet into the server housing and T_{outlet} is the temperature of the air outlet from the server housing. Including this, the rate of heat removal becomes [8, 28]:

$$\frac{d}{dt}Q_{heat} = \kappa_{air}S_{air}C_{air}\left(T_{outlet} - T_{inlet}\right)f_{air} \tag{9.13}$$

More sophisticated models are presented in Pelley *et al.* [8] and Meinser & Wenisch [28] which include more details of the air flow of CRAH or CRAC operation in a data centre. They model the rate of heat energy removal by a CRAH unit, dQ_{heat}/dt, using a modified effectiveness-Number of Transfer Units (NTU) method [8]:

$$\frac{d}{dt}Q_{heat}\left(f_{air,norm}\right) = \eta_{heat}\kappa_{air}m_{CRAH}C_{air}f_{air,norm}^{0.7}\left(\kappa_{air}T_{outlet} + \left(1-\kappa_{air}\right)T_{inlet} - T_{water}\right) \tag{9.14}$$

where m_{CRAH} is the mass flow rate of the CRAH unit (kg/s) and $f_{air,norm}$ is the normalised fan air flow rate $f_{air,norm} = f_{air}/f_{CRAH,max}$, where f_{air} is the fan volume flow rate of the CRAH and $f_{CRAH,max}$ is the maximum CRAH fan volume flow rate. The CRAH fan mass flow rate can be expressed as $m_{CRAH} = f_{CRAH,max}S_{air}$. $\eta_{heat} \approx 0.5$ is the heat transfer efficiency of the CRAH.

Rahmani *et al.* [9] provide a CRAH power model based on a linear relationship between the CRAH power consumption and the volume flow rate of air, f_{air}, required to flow in the server room. Including an idle power for the CRAH and representing the linear air flow relationship relative to the server utilisation, $f_{air} = f_{air,max}U$, Rahmani *et al.* give the CRAH power consumption as a function of U:

$$P_{CRAH}\left(U\right) = P_{CRAH,idle} + 1.33 \times 10^{-5}\frac{P_{svr,tot}\left(1\right)}{\eta_{heat}}f_{air,max}U \tag{9.15}$$

where $P_{svr,tot}(1) = P_{svr,tot}(U = 1)$ is the maximum total server power consumption (in kW), given by (9.6) with $U = 1$ and $f_{air,max}$ is the maximum air flow rate (set to 14000 m^3/hour). The CRAH idle power, $P_{CRAH,idle}$, is set to be 7% to 10% of $P_{svr,tot}(1)$ [9].

A variable speed drive CRAH unit has an idle power consumption, $P_{CRAH,idle}$, corresponding to the fans operating at minimum flow rate, $f_{CRAH,min}$. With variable speed drive fans, the incremental power consumption of the CRAH power is dominated by the fan power [8]. Variable speed drive fan power consumption increases with the cube of the air flow rate [8, 34]. This gives a power consumption equation:

$$P_{CRAH}\left(f_{air,norm}\right) = P_{CRAH,idle} + \left(P_{CRAH,max} - P_{CRAH,idle}\right) f^3_{air,norm}$$
$$= P_{CRAH,idle} + \Delta P_{CRAH} f^3_{air,norm} \tag{9.16}$$

where $\Delta P_{CRAH} = P_{CRAH,max} - P_{CRAH,idle}$.

Pelley et al. note that, as the volume flow rate of the CRAH increases, both the mass available to transport and the efficiency of the heat exchange increase, which reduces the cubic dependence to between quadratic (i.e. f_{air}^2) and cubic. They also note that a CRAH's ability to remove heat depends on several other factors beyond temperature difference between T_{water}, T_{inlet}, T_{outlet} and κ_{air}. To include these effects requires a more sophisticated model than that considered here. For example, see Dayarathna et al. [1], Zhang et al. [29] and Zhabelova et al. [31].

To get numerical values from this model, we need parameter values for $P_{CRAH,idle}$, $P_{CRAH,max}$, and $f_{CRAH,max}$. As stated above, Rahmani et al. [9] give a value for total CRAH power consumption, $N_{CRAH}P_{CRAH,idle}$ (where N_{CRAH} is the number of CRAH units), of around 7% to 10% of $N_{svr}P_{svr,max}$, and cite an example of a CRAH unit with $f_{CRAH,max} = 14000$ m^3/hour and having $P_{CRAH,max} = 7.5$ kW. Pelley et al. [8] cite a CRAH model with $f_{CRAH,max} = 11723$ m^3/hour (6900 Cubic Feet/Minute), $P_{CRAH,max} = 3$ kW and $P_{CRAH,idle} = 0.1$ kW.

In The Green Grid White Paper [34], multiple 30 Ton, 3 fan CRAH units were capable of operating the fans between 50% and 100% of maximum speed. At $f_{air} = 0.571 f_{CRAH,max}$ (57% maximum speed) the CRAH power was 1.34 kW. At $f_{air} = f_{CRAH,max}$ (maximum speed) the CRAH power was $P_{CRAH,max} = 7.44$ kW. At maximum fan speed the CRAH unit heat removal load was 106 kW. Curve fitting the CRAH data provided in The Green Grid White Paper gives $P_{CRAH}(f_{air,norm}) = 7.441 f^{2.998}_{air,norm}$. This gives $P_{CRAH,idle} = 0$.

Product specification sheets are available that provide the relationship between, dQ_{heat}/dt, T_{water}, T_{inlet}, T_{outlet} and P_{CRAH}. For example, Vertiv Liebert System Design Manuals can provide values for these parameters [35]. In this case, we approximate P_{CRAH} by the sum of the fan powers.

Heat removal is designed to match the heat generated by the servers operating at utilisation $U(t)$ to the heat removing capacity of one or more CRAH or CRAC units. The heat removing capacity of a single CRAH unit is controlled by its air flow rate, f_{air}. Note that the operational air flow rate may be set below the maximum possible air flow rate for maintenance and management reasons [32]. For example, Rahmani et al. [9] and Ito et al. [32] give as an example a CRAH with maximum airflow of $f_{air,max} = 14000$ m^3/hour and they consider scenarios where the unit is operated at airflow rates less than $f_{air,max}$.

We need to calculate the value of U to insert into (9.15) or $f_{air,norm}$ to insert into (9.16). This is found by first calculating the number of CRAH units, N_{CRAH}, and then determining how to spread the cooling load across that number of units. In doing this, the modeller can decide whether or not to minimise N_{CRAH} and operate all the units at a high capacity or include additional units each operating at a lower capacity [34]. With the cubic dependence of the CRAH power on fan speed, there can be power savings by operating more CRAHs at a lower capacity [34]. The choice of N_{CRAH} should ensure each CRAH unit should be operating at or below maximum capacity.

The number of CRAH units required to accommodate the maximum power consumption of the servers, N_{CRAH}, is set by ensuring the total CRAH heat removal rate at least matches the maximum total server power:

$$N_{CRAH} \geq \left\lceil N_{svr} P_{svr,max} \middle/ \left(\frac{d}{dt} Q_{heat} \left(f_{air,max} \right) \right) \right\rceil \tag{9.17}$$

where $\lceil x \rceil$ is the smallest integer larger than x and $f_{air,max}$ represents the maximum flow rate of the CRAH. Exactly matching the maximum CRAH heat removal with total server power corresponds to an equality. One may adopt a "safety margin" by using $dQ_{heat}(f)/dt$ with f appropriately less than its maximum value in (9.17). That is, to remove the server room heat with server utilisation $U(t)$, we require the number of active CRAH units, N_{CRAH}, to satisfy for all values of U between 0 and 1:

$$N_{CRAH} \frac{d}{dt} Q_{heat} \left(f \right) \geq N_{svr} P_{srv,max} \left(U + \left(1 - U \right) yc + y_{sleep} \left(1 - U \right) \left(1 - c \right) \right) \tag{9.18}$$

where we have used (9.6) and the value of c corresponds to the minimal amount of server consolidation. With the number of CRAH units determined, there are several possible scenarios for CRAH power consumption:

1. The CRAHs are fixed airflow units and all CRAHs are permanently active. In this situation, the CRAH power is just the sum of the CRAH unit powers.
2. The CRAHs are fixed airflow units and can be independently operated [32]. In this case, the total CRAH power is the sum of the active CRAH units. The number of active CRAH units will be set by the requirement to satisfy (9.18) for varying $U(t)$ over the diurnal cycle. To minimise CRAH power consumption, one would expect to use the minimum number of CRAHs that satisfy that inequality.
3. The CRAH units have "variable speed drive" fans which enable the air flow to vary from a minimum, $f_{air,min}$, to a peak volume flow rate, $f_{air,max}$, [8, 34]. In this case the power model is more complicated as will now be shown.

We can either assume the heat load is evenly carried by each CRAH unit or allocate a different heat load to the various units [32]. Assuming equal sharing of the heat load, using the Rahmani *et al.* model in (9.15), the total CRAH power consumption is:

$$P_{CRAH,tot} \left(U \right) = N_{CRAH} P_{CRAH} \left(U \right) \tag{9.19}$$

Similarly, with the Pelley *et al.* model, we solve (9.14) for $f_{air,norm}$ given the appropriate heat load. For example, using (9.18) and then solving (9.14) gives:

$$f_{air,norm} \left(U \right) = \left(\frac{N_{svr} P_{svr} \left(U + \left(1 - U \right) yc + y_{sleep} \left(1 - U \right) \left(1 - c \right) \right)}{N_{CRAH} \eta_{heat} \kappa_{air} m_{CRAH} C_{air} \left(\kappa_{air} T_{outlet} + \left(1 - \kappa_{atr} \right) T_{inlet} - T_{water} \right)} \right)^{1/0.7} \tag{9.20}$$

This value is then placed into (9.16) and the total CRAH power is given by summing over the number of CRAH units:

$$P_{CRAH,tot} \left(U \right) = N_{CRAH} \left(P_{CRAH,idle} + \Delta P_{CRAH} f_{air,norm}^3 \left(U \right) \right) \tag{9.21}$$

If the CRAH units do not share the heat load equally, then this approach is applied individually to each of the groups of servers that are shared by each CRAH unit.

Note however that the CRAH unit represents only the air flow portion of this data centre cooling plant; an overall assessment of the cooling system needs also to include the power consumption of the water (or coolant) chiller plant.

9.5 COMPUTER ROOM AIR CONDITIONING (CRAC) AND CONDENSER POWER MODELS

A CRAC system, introduced in Section 9.3, differs from a CRAH primarily in that it uses a compressor, evaporator, and condenser rather than a chiller.

There are several relatively simple power models published for CRAC systems. Zhan & Reda [36] provide a power model based on a "Coefficient of Performance" (CoP). To cool the servers, they apply a CRAC unit to cool a set number of servers. With this the total heat removed by a CRAC is given by $B_{CRAC} = \sum_k P_{svr,k}$, where the k th server consumes $P_{svr,k}$ power. The power consumption of the m th CRAC unit, $P_{CRAC,m}$, that is removing heat from that group of servers is given by

$$P_{CRAC,m} = \frac{\sum\limits_{k} P_{svr,k}}{CoP} \tag{9.22}$$

where the CoP is defined as the ratio of heat removed per second (dQ_{heat}/dt) and the power required to remove that heat, P_{rem}, [37]:

$$CoP = \left(\frac{d}{dt}Q_{heat}\right)\bigg/ P_{rem} \tag{9.23}$$

Moore et $al.$ [37] have provided a numerical relationship between the CoP and the CRAC water supply temperature to the CRAC in °C. Based on data from the HP Utility Data Centre:

$$CoP(T_{water}) = 0.0068T_{water}^2 + 0.0008T_{water} + 0.458 \tag{9.24}$$

Moore et $al.$ do not provide clear guidance as to the range of validity of the form for $CoP(T_{water})$. For temperatures, T_{water}, below around 10°C, the $CoP < 1$, meaning the CRAC is consuming more power than the servers it is cooling.

Rahmani et $al.$ [9] model a CRAC noting that the amount of heat that has to be removed is the same as with a CRAH, but the method of dispersing the heat differs. The CRAC uses the evaporator to remove heat from the data centre air flow, then together the compressor and condenser dump that heat into the outside environment or cooling tower. Consequentially, they provide a linear model for the CRAC system based on the incremental (U dependent) term in (9.15) with the air-cooled condenser represented by its CoP. The CRAC power is modelled by the following equation [9]:

$$P_{CRAC}(U) = P_{CRAC,idle} + 1.33 \times 10^{-5}(1 + CoP_{condenser})\frac{P_{svr,tot}(1)}{\eta_{heat}} f_{air,max}U \tag{9.25}$$

where $CoP_{condenser}$ is the CoP of the condenser and is usually in the range 3 to 4.5 for a simple water assisted compressor cooler, and the range 4.5 to 6 for an air cooled condenser. Rahmani et $al.$ state that the idle power component, $P_{CRAC,idle}$, is typically in the range $0.1P_{svr,tot}(1)$ to $0.3P_{svr,tot}(1)$.

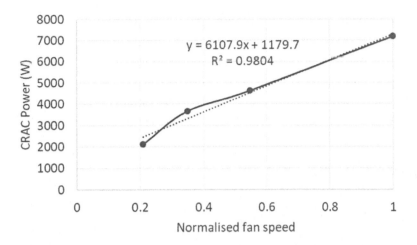

FIGURE 9.2 Plot of CRAC power for range of normalised fan speed from data collected by Ito et al. [32]. As shown, a linear curve fit gives a value $R^2 = 0.98$.

Both cooling technologies include a series of fans to distribute cool air, and their power consumption increases with the cube of the fan speed. However for both technologies, the heat removal systems, chiller for CRAH and compressor-condenser for CRAC, add substantially to the overall power consumption of the cooling system, often being dominant. For example, as shown in Figure 9.2, power consumption measurements by Itoh *et al.* [32] for a CRAC system show a linear fit with fan speed rather than cubic fit:

$$P_{CRAC}\left(f_{air}\right) \approx 1179.7 + 531.1 f_{air}$$
$$P_{CRAC}\left(f_{air,norm}\right) \approx 1179.7 + 6107.9 f_{air,norm} \qquad (9.26)$$
$$= P_{CRAC,idle} + \Delta P_{CRAC} f_{air,norm}$$

where $f_{air,norm} = f_{air}/f_{air,max}$. The CRAC used for these measurements has a maximum air volume flow rate of 14000 m³/hour and can remove up to 27 kW of heat [32]. From this, we have $P_{CRAH,idle} \approx 1180$ W and $\Delta P_{CRAH} \approx 6108$ W.

We need to relate the total CRAC power to data centre utilisation, U. As with the CRAH units, the total number of required CRAC units, N_{CRAC}, is calculated using the same approach as with calculating the number of CRAH units. Using (9.17) or (9.18), we have:

$$N_{CRAC} \geq \left\lceil N_{svr}P_{svr,max} \middle/ \left(\frac{d}{dt}Q_{heat}\left(f_{air,max}\right)\right) \right\rceil \qquad (9.27)$$

or

$$N_{CRAC} \frac{d}{dt}Q_{heat}\left(f\right) \geq N_{svr}P_{svr,max}\left(\left(U + \left(1-U\right)yc\right) + y_{sleep}\left(1-U\right)\left(1-c\right)\right) \qquad (9.28)$$

Adopting the approach of Rahmani *et al.*, we assume a linear relationship $U = f_{air}/f_{air,max} = f_{air,norm}$, which gives:

$$P_{CRAC,tot}\left(U\right) = N_{CRAC}\left(P_{CRAC,idle} + \Delta P_{CRAC}U\right) \qquad (9.29)$$

Adopting the approach of Pelley *et al.*, we use (9.20):

$$P_{CRAC,tot}\left(U\right) = N_{CRAC}\left(P_{CRAC,idle} + \Delta P_{CRAC} f_{air,norm}\left(U\right)\right) \tag{9.30}$$

In some publications and white papers, CRAH and CRAC units are dimensioned using the weight of the unit [34, 23]. A reasonable rule of thumb for the relationship between CRAH or CRAC tonnage and cooling is 3.5 kW requires 1 ton of cooling equipment. This rule has been applied for CRAH units [34] and CRAC units [23].

9.6 POWER SUPPLY EQUIPMENT

The Power Distribution Units (PDUs) are used to transform the high voltage supply provided to the data centre down to voltage levels appropriate for the servers and other data centre equipment. Uninterruptable Power Supply (UPS) provides temporary power supply redundancy in the case of short-term power failures. Both of these systems incur losses, which can be modelled as follows [8, 9, 38]. The PDU power consumption has the form:

$$P_{PDU}\left(U\right) = P_{PDU,idle} + \rho_{PDU}\left(P_{svr,tot}\left(U\right)\right)^{2} \tag{9.31}$$

The idle PDU power consumption, $P_{PDU,idle}$, is approximately 2% of the total power for a 10 MW data centre [8, 9]. At peak load, PDU power consumption is approximately 6% of the power for a 10 MW data centre [8, 9]. If the total power of a data centre is known, then, assuming linearity, the values of $P_{PDU,idle}$ and ρ_{PDU} can be directly calculated. In this case, we have:

$$\begin{aligned} P_{PDU,idle} &\approx 0.02 P_{DC}\left(1\right) \\ \rho_{PDU} &\approx 0.04 P_{DC}\left(1\right) \Big/ \left(P_{svr,tot}\left(1\right)\right)^{2} \end{aligned} \tag{9.32}$$

In these expressions, $P_{DC}(1) = P_{DC}(U = 1)$ is the total data centre power consumption at maximum utilisation, which will be the maximum data centre power consumption.

In some cases, the model for the total data centre may be being built up from its various components. In that case, we may not have a value for $P_{DC}(1)$ to use to calculate $P_{PDU,idle}$ and ρ_{PDU}. In this case, we can approximate the total power consumption of the data centre by noting from Table 9.1 that the total power consumption of the servers and cooling are around 80% of total data centre power. Therefore, we can use the sum of server and cooling power to provide estimates for $P_{PDU,idle}$ and ρ_{PDU}. In this case, we have:

$$\begin{aligned} P_{PDU,idle} &\approx 0.02 \frac{\left(P_{svr,tot}\left(1\right) + P_{cooling}\left(1\right)\right)}{0.8} \\ \rho_{PDU} &\approx 0.04 \frac{\left(P_{svr,tot}\left(1\right) + P_{cooling}\left(1\right)\right)}{0.8\left(P_{svr,tot}\left(1\right)\right)^{2}} \end{aligned} \tag{9.33}$$

where $P_{cooling}(1) = P_{cooling}(U = 1)$ is the maximum power consumption of the server cooling system. For chiller and CRAH, it is $P_{chiller}(U = 1) + P_{CRAH,tot}(U = 1)$. For a CRAC, it is $P_{CRAC,tot}(U = 1)$.

For the UPS, power consumption can be modelled as approximately linear with respect to utilisation [9]:

$$P_{UPS}\left(U\right) = P_{UPS,idle} + \Delta P_{UPS} U \tag{9.34}$$

The idle UPS power consumption, $P_{UPS,idle}$, for a 10 MW data centre is approximately 4% of total data centre power peak consumption [8, 9]. At peak load it is approximately 8% of total data centre peak power consumption [8, 9]. As with $P_{PDU}(U)$, if we know the maximum data centre power consumption, $P_{DC}(1)$, then:

$$P_{UPS,idle} \approx 0.04 P_{DC}\left(1\right)$$
$$\Delta P_{UPS} \approx 0.04 P_{DC}\left(1\right)$$

(9.35)

Alternatively, if the value of $P_{DC}(1)$ is not available:

$$P_{UPS,idle} \approx 0.04 \frac{\left(P_{svr,tot}\left(1\right) + P_{cooling}\left(1\right)\right)}{0.8}$$
$$\Delta P_{UPS} \approx 0.04 \frac{\left(P_{svr,tot}\left(1\right) + P_{cooling}\left(1\right)\right)}{0.8}$$

(9.36)

9.7   PUMPS AND MISCELLANEOUS POWER CONSUMPTION

The pumps required to circulate the chiller water around the data centre are reported to consume between 6% [8] and 16% [9] of the total data centre power. As with the PDU and UPS powers, depending on the information available:

$$P_{pumps} \approx 0.06 P_{DC}\left(1\right) \quad to \quad 0.16 P_{DC}\left(1\right)$$

(9.37)

or

$$P_{pumps} \approx 0.06 \frac{\left(P_{svr,tot}\left(1\right) + P_{cooling}\left(1\right)\right)}{0.8} \quad to \quad 0.16 \frac{\left(P_{svr,tot}\left(1\right) + P_{cooling}\left(1\right)\right)}{0.8}$$

(9.38)

As shown in Table 9.1, the power consumption of the remaining operations such as networking in a data centre are reported to be around 6% of peak power consumption [8, 9]. Again, depending upon the information available, we can use:

$$P_{misc} = 0.06 P_{DC}\left(1\right)$$

(9.39)

or

$$P_{misc} = 0.06 \frac{\left(P_{svr,tot}\left(1\right) + P_{cooling}\left(1\right)\right)}{0.8}$$

(9.40)

9.8   BRINGING THE MODEL TOGETHER

The sections above described power models for the dominant components of a data centre. Bringing them all together will provide an overall model that should be able to provide a reasonable approximation for the power consumption of a data centre. The key parameters are the maximum power consumption of the servers within the data centre, $N_{svr}P_{svr,max}$, the data centre utilisation, $U(t)$, the degree of consolidation of the servers, c, and the sleep state power, P_{sleep}. With these known, the generic values of the other parameters, which are based on data collected by various researchers and industry participants, should provide reasonable guidance as to the power consumption of the other dominant contributors to total data centre power consumption.

Assuming that we have access to reasonable estimates for these values or we adopt the values given in the references cited in this text, then the total data centre power is given by

$$P_{DC}(U) = P_{svr,tot}(U) + \begin{Bmatrix} P_{chiller}(U) + P_{CRAH,tot}(U) \\ P_{CRAC,tot}(U) \end{Bmatrix} + P_{PDU}(U) + P_{UPS}(U) + P_{pumps} + P_{misc} \quad (9.41)$$

where the upper term in the {.} brackets is used when the data centre uses chillers and CRAH units and the bottom term is used when the data centre uses CRAC units.

A widely used energy efficiency measure for data centres is the Power Use Effectiveness (PUE) defined in terms of either power or energy by [1, 39, 40]:

$$PUE = \frac{\text{Total facility power (energy) consumption}}{\text{Total IT power (energy) consumption}} \quad (9.42)$$

In this definition, "IT" equipment refers to the servers and the data centre communications network that interconnects servers and connects the data centre to the Internet. Using the terms in (9.41) if we use power to define the PUE, we can write for a given utilisation:

$$PUE = \frac{P_{data\_centre}(U)}{N_{svr} P_{svr,max}(U + (1-U)yc + y_{sleep}(1-U)(1-c))} \quad (9.43)$$

Alternatively, using energy, we need to integrate the power consumption over a set duration, T, (for example a year) to give:

$$PUE = \frac{\int_T P_{data\_centre}(U(t)) dt}{\int_T P_{svr,tot,max}\left((U(t) + (1-U(t))yc) + P_{sleep}N_{svr}(1-U(t))(1-c)\right) dt}$$

$$= \frac{\langle P_{data\_centre}(U) \rangle_T}{P_{svr,tot,max}\left(\langle U \rangle_T + (1-\langle U \rangle_T)yc\right) + P_{sleep}N_{svr}(1-\langle U \rangle_T)(1-c)} \quad (9.44)$$

where $\langle X \rangle_T$ is the average value of $X(t)$ over the time duration T.

The energy efficiency of data centres became of focus of interest in the late 1990s and early 2000s. At that time, PUE values ranged between 2 and 3. PUE values such as these indicate that more power is being consumed by non-IT equipment in the facility than IT equipment. Although data centres with PUE values around 2 or more are still operating [41], over recent years, data centre operators have undertaken significant actions to reduce the PUE [41] and today some report PUE values around 1.1 [5, 42, 43] and even closer to unity [44]. A range of "best practices" being adopted to attain these low PUE values are described in Shehabi et al. [45]

Plots for a 10 MW server power consumption data centre are presented in Figure 9.3, using a chiller and CRAH, and in Figure 9.4, using a CRAC, for two diurnal cycles. The details of the model are given in Table 9.2. The PUEs for these modelled data centres are PUE = 2.5 for the chiller and CRAH cooled data centre model and PUE = 2.8 for the CRAC cooled data centre model. These values are relatively high by current standards; however, the model's simplicity fails to include many of the modern details for highly optimised data centres [31, 41] and some the parameter

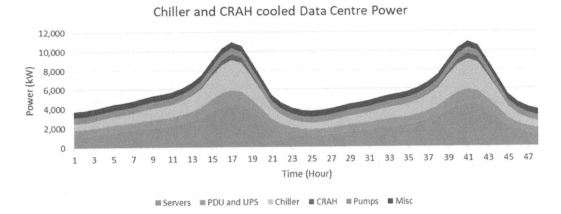

FIGURE 9.3 Power model for a data centre with cooling using a chiller and Computer Room Air Handler, CRAH.

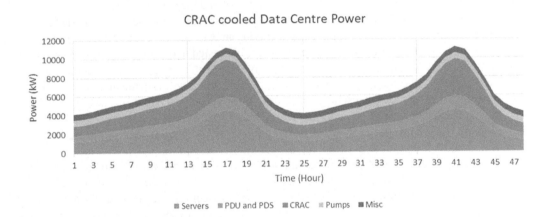

FIGURE 9.4 Power model for a data centre with cooling using Computer Room Air Conditioning, CRAC.

TABLE 9.2
Parameter Values for the Data Centre Power Model for Figures 9.3 and 9.4

Component	Equation	Parameter Values
Servers	(9.6)	$P_{svr,max} = 300W$, $y = 0.44$, $P_{sleep} = 0.5y$, $N_{svr} = 15000$
Chiller	(9.8)	
CRAH	(9.21), (9.20) and (9.17)	$P_{CRAH,idle} = 100W$ [28], $P_{CRAH,max} = 4667$ W [32], $T_{water} = 10°C$, $T_{supply} = 20°C$, $T_{inlet} = 20.5°C$, $T_{outlet} = 26°C$
CRAC	(9.22) and (9.24)	$T_{water} = 10°C$
PDU	(9.31) and (9.33)	
UPS	(9.34) and (9.36)	
Pumps	(9.38)	6%
Misc	(9.40)	6%

values are taken from publications that are several years old. Unfortunately, it is difficult to find current publicly available power data on this equipment. This, again, displays the ongoing problem of accessing data that corresponds to current technologies.

Over recent times, it has been noted that the power consumption trend for US data centres has seen relatively flat growth since around 2015 [45]. This flat trend has been explained by improvements in cooling, power provisioning, servers, storage and use of virtualisation [46]. In contrast, other researchers have stated that power consumption of data centres has grown over recent years and that the power consumption trends for data centres are not clear [47].

9.9   SERVER POWER MODELS

As the use of data centres has grown over recent years, models for data centre servers have also been developed to assist in understanding the power consumption of the services provided by data centres. Constructing server power models is non-trivial for several important reasons. Firstly, server technology has been improving over the years, as one would expect. In addition, the development of virtualisation and evolution of data centre architectures has changed the load and character of server utilisation. The diversity of applications that are now run in data centres has dramatically grown, as more and more organisations move their IT services into data centres.

Depending upon the requirement of the server power model and the amount of detailed information about the server(s) to which the modeller has access, one of several types of server power model may be considered most appropriate. The following sections provide an overview of several server model types.

As discussed in the first section of this chapter, we are interested in developing a power model for services. These services will typically involve the use of data centre resources, in particular servers and their sub-systems. The challenge is in developing a model that will provide an estimate for the power consumption of the server resources allocated to providing a specific service. Because the servers in a data centre may well be providing many, possibly thousands, of simultaneous services the complication when constructing a service power model is to provide a meaningful representation of the power consumption of the service of interest.

For the power model of a service to be useful, it must provide an estimate that is justifiable and of worth to the modeller. Almost all users of data centre services are geographically remote from the data centre and have minimal knowledge of how an application is dealt with by data centre equipment. Even if we know the total data exchanged between a user and the servers in a data centre, this does not provide much, if any, information about how that application utilises the server sub-systems. Knowing the application "type" may assist us in determining which components of the server will dominate its power consumption when servicing the user request. Some server power models have adopted this approach and endeavoured to construct models that can accommodate different application types [15, 16, 48].

In many cases, although we may know the type of application, this still is not enough to provide an accurate model of a service's power consumption in a data centre. This is because such a model will require detailed information on the server components' power consumption and utilisation as they respond to service requests. Beyond the server's details, there is also a variety of data centre networks used to interconnect the servers and connect them to the Internet [49]. The use of distributed servers, where the service may be provided by devices not located in the same data centre, let alone the same device or rack within a data centre, further complicates the issue [1].

9.9.1   SINGLE VALUE SERVER POWER MODELS

The simplest models provide a single power per server value. The power of a server will depend upon its type and generation. For example, in the Schneider Electric White Paper, "Allocating Data Center Energy Costs and Carbon to IT Users" [14], servers are categorised into five representative

---

TABLE 9.3

Indicative Server Average or Maximum Power Consumption for Several "Server Classes"

Server Class	Power (Watts)	Year	Reference
1U app server	250 (average)	2011	[14]
Virtual server	90 (average)	2011	[14]
Web blade	200 (average)	2011	[14]
ERP blade	200 (average)	2011	[14]
Mainframe	4000 (average)	2011	[14]
3U to 10U server	2000 (average)	2011	[14]
1 Socket	118 (maximum)	2020	[50]
2+ Socket	365 (maximum)	2020	[50]
Volume Server	330 (maximum)	2020	[50]
Mid-range	1860 (average)	2020	[50]
High-end	20000 (average)	2020	[50]

The "Year" indicates the year of the server generation for which that power value can be applied. See the corresponding Reference for details.

"Server Classes", each with an expected power consumption, as listed in Table 9.3. The document "United States Data Center Energy Usage Report" [50] also lists maximum power consumption values for a variety of Server Classes. These are also listed in Table 9.3.

An example of using server power to provide an estimate for the energy consumption of a service is provided by Baliga *et al.* [51]. In that paper the calculation of energy consumption of "Processing as a Service" calculates the contribution of the server to the service energy as the product:

$$E_{proc} = t_{proc} P_{svr} PUE \tag{9.45}$$

where t_{proc} is the time taken to complete the processing, P_{svr} is the power consumption of the processing server and *PUE* is the Power Use Effectiveness of the data centre in which the server is located.

When using equipment power values, care must be taken because of the following factors: 1) Depending upon the Server Class, the server power may have reduced over subsequent generations of that Server Class as the technology used for that Type has changed. Therefore, using server power values from some years ago may not be appropriate today. 2) Even for the Server Classes that have not changed power consumption over the years (see Ref. [50] for details), the power consumption for a given workload most likely has changed as the servers have improved their energy efficiency. In other words, the improvement in technology has been realised by increasing processing capability per Watt rather than reducing the overall server power consumption.

Having power data on the total server power may not be enough to determine how much power a specific service consumes. Determining the server power consumed by a specific service is non-trivial. For example, in the Google White Paper, "Google's Green Computing: Efficiency at Scale", published in 2011 [52], the server "power per user" of less than 0.25 Watts was calculated for its Gmail service. To construct this estimate, Google stated in an endnote [52]:

> This is using a detailed estimate of all the energy necessary to host Gmail divided by our 7-day active users as our user count.

There are many more Gmail accounts than 7-day active users, but this represents those accounts in regular use.

This type of calculation requires detailed information not publicly available, a common problem when attempting to construct power models today.

To apply the Gmail power-per-user value in a power model developed for 2020 or beyond, we will require that service to be similar to Gmail and will have to implement appropriate scaling of the power per user to reflect the energy efficiency improvements since 2011. This scaling will need to account for factors such as any reduction in server power consumption and consolidation of tasks or virtualisation in data centre servers.

A 17-times improvement in server power efficiency (operations/Watt) between 2007 and 2016 has been reported [53]. As a rough guide, if we assume a similar ratio applies between 2011 and 2020 then the server power per user for an email service in 2020 (based on the Gmail value) will be less than 0.015 Watts. Given that Google makes a significant effort to maximise the efficiency of its data centres, it would be reasonable to adopt this value as an approximate value (rather than an upper limit) for a typical data centre mail server.

An example of using this approach for modelling server power consumption when providing a service is in Vishwanath *et al.* [54], where an estimate of power consumption of several cloud-based services was presented.

Another example of a power model on a per-user basis is in Baliga *et al.* [51], in which a power model of "Software as a Service" was developed using the power consumption of an example server split over the number of users of that server. In this case, we need information on the number of users sharing a given server. This is information that is unlikely to be easily available. In the case of the Baliga *et al.* "Software as a Service" model, this problem was addressed by considering several scenarios in which different numbers of users shared a given server.

These approaches for constructing a (rough) power approximation have become increasingly common in network and Internet service power modelling over recent years. This is because most telecommunications and Internet service companies have become increasingly reticent to publish much detail of their internal workings as the competitive environment has become more aggressive. For example, the value for its Gmail power-per-user calculation was published by Google in 2011, yet is still referred to in its 2019 Environmental Report rather than being updated [42]. This practice is not uncommon and, consequently, values used in power modelling may require rescaling from earlier years.

A major drawback of models based on a single value is their inability to distinguish between different types of applications. For example, an application that merely transfers simple text (such as a simple text-only email service) will require significantly less server resources than an application that involves significant computing and/or data transfer (such as advanced numerical modelling). As the range of cloud services has broadened, researchers have developed more sophisticated server power models that endeavour to accommodate this diversity of service types.

9.9.2 WHOLE SERVER POWER MODELS

When more detailed modelling than a single value is required, utilisation-based server power models are a common approach. As with data centres, there is a diverse range of utilisation-based server power models. Some models treat a server as a single unit and provide a power consumption model based only on the utilisation of the server CPU [55, 56]. This approach has typically been based on measurements across multiple servers in a data centre and thus the model represents a generic machine rather than a specific server. Fan *et al.* [55] provide two CPU-based models, one linear and the second nonlinear:

$$P_{svr}(u) = P_{svr,idle} + \left(P_{svr,max} - P_{svr,idle}\right)u(t)$$
$$P_{svr}(u) = P_{svr,idle} + \left(P_{svr,max} - P_{svr,idle}\right)\left(2u(t) - u(t)^r\right)$$

(9.46)

where $u(t)$ is the server CPU utilisation and the exponent r of $u(t) = 1.4$. To use this model we require values for $P_{svr,idle}$ and $P_{svr,max}$. As we can see from Table 9.3, server power consumption values can vary significantly depending upon the Server Class and model. Therefore, again we encounter the paucity of published relevant data on devices. As noted by Fan $et\ al.$ [55], the "name plate" power consumption value (i.e. the power supply value listed in the server manual) can be up to 1.7 times greater than the actual maximum server power consumption. Therefore, unless we have access to the specific power consumption details and corresponding utilisation of the server(s) being modelled, a more generic approach is required.

Here we will describe one such approach to give the reader an idea of how to construct an esti-mate of server power consumption. From Table 9.3 we have a set of server power values for a range of different servers. We will consider two cases. The first case is a "2+ Socket" server, which has a maximum power consumption of 365 Watts. For the second case, we consider a "mid-range" server. According to Table 9.3, a mid-range server has an average power consumption of 1860 Watts. For both cases, we need values for $P_{svr,idle}$ and $P_{svr,max}$. For the 2+ Socket server we immediately have $P_{svr,max} = 365$ Watts. To get $P_{svr,idle}$, we use the report by Shehabi $et\ al.$, "United States Data Centre Energy Usage Report" [50], which provides estimates for the ratio of $P_{svr,idle}/P_{svr,max}$ (referred to as the Dynamic Range, DR, in that report). According to Figure 8 of Shehabi $et\ al.$, the average Dynamic Ratio for a "volume server" (i.e. a server used in a large service provider data centre) in 2020 is 0.44. From this we have for the 2+ Socket server, $P_{svr,idle} = 160$ Watts.

For the mid-range server Table 9.3 provides the average power. We assume this corresponds to the power consumption for average utilisation, u. Therefore, to find the idle and maximum power values we require a value for the average utilisation. An estimate of utilisation is also contained in Shehabi $et\ al.$ According to Table 1 of that report, the average server utilisation for a service provider server in 2020 can be approximated as 25%. With this and using the Dynamic Range value of 0.44, we have two equations for $P_{svr,idle}$ and $P_{svr,max}$:

$$P_{svr,idle}/P_{svr,\max} = 0.44$$
$$P_{svr,ave} = 1860W = P_{svr,idle} + \left(P_{svr,\max} - P_{svr,idle}\right)0.25$$
$$or \tag{9.47}$$
$$P_{svr,ave} = 1860W = P_{svr,idle} + \left(P_{svr,\max} - P_{svr,idle}\right)\left(2 \times 0.25 - 0.25^{1.4}\right)$$

where we have included both the linear and nonlinear power profiles given in (9.46). Solving these, we get for the two models for a mid-range server:

$$P_{svr,idle} = 1411\ Watts, \quad P_{svr,max} = 3207\ Watts \quad linear\ model$$
$$P_{svr,idle} = 1279\ Watts, \quad P_{svr,max} = 2906\ Watts \quad non\ linear\ model \tag{9.48}$$

The linear model values can also be used as the server component in (9.41). When constructing estimates such as these, it is important to keep track of the annual improvements in technologies as Shehabi $et\ al.$ [50] have done.

Zhang $et\ al.$ [56] found that a linear model does not apply to all the Server Types they studied and so considered CPU utilisation models based on linear, quadratic and cubic curve fits:

$$P_{svr}\left(u(t)\right) = a_1 + b_1 u(t)$$
$$P_{svr}\left(u(t)\right) = a_2 + b_2 u(t) + c_2 u(t)^2$$
$$P_{svr}\left(u(t)\right) = a_3 + b_3 u(t) + c_3 u(t)^2 + d_3 u(t)^3 \tag{9.49}$$

TABLE 9.4

Server Model Parameters for Two Servers Studied by Zhang *et al.* [56]

Server	Linear Fit	Quadratic Fit	Cubic Fit
Fujitsu PRIMERGY BX920 S3	$a_1 = 762.1$	$a_2 = 1268$	$a_3 = 1067$
	$b_1 = 35.92$	$b_2 = 2.156$	$b_3 = 34.2$
		$c_2 = 0.338$	$c_3 = -0.504$
			$d_3 = 0.005624$
Colfax International CX2266-N2	$a_1 = 193.7$	$a_2 = 177.43$	$a_3 = 168.33$
	$b_1 = 0.9385$	$b_2 = 2.0432$	$b_3 = 3.5294$
		$c_2 = -0.0112$	$c_3 = -0.0051$
			$d_3 = 0.000271$

Their results showed that the cubic model based on CPU utilisation only provided a good fit with a low average error in power estimation for applications that required significant memory and disk access. From (9.49) we see that the model's idle power, $P_{svr,idle}$, corresponds to the a parameters and the maximum power, $P_{svr,max}$, corresponds to the sums of the a, b, c and d parameters. Zhang *et al.* provided values for these parameters for two specific server models, as listed in Table 9.4.

The models in (9.49) and values in Table 9.4 can also be adopted as a server power model, assuming the servers being modelled are similar to the servers named in the table.

9.9.3 SERVER SUB-SYSTEM POWER MODELS

Beyond CPU utilisation only models, a range of approaches for server power modelling have been published [1, 15, 16, 57]. The majority of models typically represent server power as consisting of contributions from the following sub-systems [15, 16, 58]:

1. Server Central Processing Unit (CPU) processing
2. Memory access
3. Disk access
4. Network Interface Card (NIC) operation.

There is a range of model types for each of these sub-systems [1, 15, 48, 56] and, keeping with the principle of simplicity, we will adopt linear models for all four sub-systems. Furthermore, we will focus on integrated servers which include all of these functions within a single unit, although many data centres employ arrays of compact blade servers that are housed separately from shared disk arrays. The same modelling principles however apply to these two configuration types.Thus the power consumption model for a server used in this book can be written in the form:

$$
\begin{aligned}
P_{svr}\left(u_{CPU}(t), R_{mem}(t), R_{disk}(t), R_{NIC}(t)\right) &= P_{CPU,idle} + R_{CPU}(t)E_{CPU} + P_{mem,idle} + R_{mem}(t)E_{mem} \\
&\quad + P_{disk,idle} + R_{disk}(t)E_{disk} + P_{NIC,idle} + R_{NIC}(t)E_{NIC}
\end{aligned} \tag{9.50}
$$

where $P_{X,idle}$ are the idle powers of sub-system X (being CPU, memory, disk, and Network Interface Card). $R_{CPU}(t)$ is the operations per second of the CPU when processing data, $R_{mem}(t)$ is the memory accesses per second, $R_{disk}(t)$ is the disk reads or writes per second, and $R_{NIC}(t)$ is the input or output data rate (bit/s) for the server. Also, E_{CPU} is the energy per CPU operation, E_{mem} is the energy per memory access, E_{disk} is the energy per disk read or write, and E_{NIC} is the energy per bit for the

NIC. In some papers, the server power consumption is expressed in terms of utilisations. In this case, we have [48]:

$$P_{svr}\left(u_{CPU}\left(t\right),u_{mem}\left(t\right),u_{disk}\left(t\right),u_{NIC}\left(t\right)\right) = P_{CPU,idle} + u_{CPU}\left(t\right)\Delta P_{CPU} + P_{mem,idle} + u_{mem}\left(t\right)\Delta P_{mem}$$
$$+ P_{disk,idle} + u_{disk}\left(t\right)\Delta P_{disk} + P_{NIC,idle} + u_{NIC}\left(t\right)\Delta P_{NIC} \quad (9.51)$$

where $P_{X,idle}$ and ΔP_X are the idle and incremental powers of the sub-system X (being CPU, memory, disk, and Network Interface Card), respectively. We have:

$$\Delta P_X = P_{X,max} - P_{X,idle} \quad (9.52)$$

with $P_{X,max}$ the maximum power consumption of the sub-system X, which occurs at utilisation of unity.

The utilisations $u_X(t)$ are defined by

$$a) \quad u_{CPU} = \frac{\text{operations per second}}{\text{maximum operations per second}}$$

$$b) \quad u_{mem} = \frac{\text{memory accesses per second}}{\text{maximum memory accesses per second}}$$

$$c) \quad u_{disk} = \frac{\text{disk reads / writes per second}}{\text{maximum disk reads / writes per second}}$$

$$d) \quad u_{NIC} = \frac{\text{NIC input / output bits per second}}{\text{maximum NIC input / output bits per second}}$$

$$(9.53)$$

The relationships between the rates, $R_x(t)$, and utilisations, $u_X(t)$, and between the E_X and ΔP_X are as follows:

$$R_X\left(t\right) = u_X\left(t\right)R_{X,max}$$
$$E_X = \frac{P_{X,max} - P_{X,idle}}{R_{X,max}} = \Delta P_X / R_{X,max} \quad (9.54)$$

The primary difficulty with these models is the need to determine the various ΔP_X values as well as the quantities required to provide the various utilisations, $u_X(t)$. Again we face the problem that different servers will have different values for each of these parameters and different applications will require different utilisations of the various server sub-systems. Further, it is very unlikely that server manufacturers will make these parameters publicly available. Thus we need to simplify the model to minimise the amount of data needed to construct the power model. To do this we start by noting that several research groups have published data that we can use. For example, [15, 48, 59] provide quantitative server models based on the form of (9.51). Further, some authors have reported that a number of the parameters E_X can be set to zero. For example, Economu et al. [48] report that for the Itanium server u_{NIC} can be set to zero. Zhou et al. [15] undertook a Principal Component Analysis of power measurements on a Dell PowerEdge R720 server to find the dominant contributors to its power consumption. Based on the server running three different types of applications, one CPU intensive (SPEC CPU 2006 [60]), one memory/disk intensive (Iozone [61]) and one NIC intensive (HP LoadRunner [62]), they found the contributions to the server's power consumption are as listed in Table 9.5. Although the quantities considered by these authors included several parameters that require direct access to the server, their analysis shows that the dominant contributors can be reduced to around three and still account for around 95% of server power consumption.

TABLE 9.5

Percentage Contributions of Different Factors to Server Power Consumption for Three Application Types [15]

Parameter	Application Contribution to Server Power Consumption (%)		
	CPU Intensive	NIC/Server Intensive	Memory/Disk Intensive
Processor time	62	63	53
Disk Bytes/s	19	21	27
Disk time	14	11	15
Page Faults/s	4	3	4
Memory used	1	1	1
Bytes/s	0	1	0
NIC bandwidth	0	0	0

Entries in this Table are self-explanatory except for the page fault. A page fault occurs when a page of data that the CPU requires is not available in memory and must be transferred from virtual memory or the disk to RAM.

Zhou *et al.* also analysed the dependence of server power on six of these parameters (NIC bandwidth having no impact) using the following regression analyses: multivariate linear regression, power regression, exponential regression, and polynomial regression. Focussing on the linear regression model, further analysis of their data shows that the CPU intensive application can be modelled with 98% correlation with measured server power by including only the CPU utilisation and the disk Bytes/s access; for the NIC/server intensive application, a 91% correlation with measured server power including only Memory Used and Disk Time; finally, for the Memory/disk intensive application, an 89% correlation with server power using only Page Faults per second and Disk Time. The forms for these linear regression models are as follows:

$$P_{svr,CPU}\left(u_{CPU}, R_{disk\_B/s}\right) = 220u_{CPU} - 1.53 \times 10^{-5} R_{disk\_B/s} + 112 \; Watts$$
$$P_{svr,NIC}\left(u_{mem}, u_{disk\_time}\right) = 2590u_{mem} + 255u_{disk\_time} - 7.3 \; Watts \quad\quad (9.55)$$
$$P_{svr,mem/disk}\left(R_{page\_fault}, u_{disk\_time}\right) = 2.3 \times 10^{-4} R_{page\_fault} - 61.3u_{disk\_time} + 130 \; Watts$$

In these equations, u_{CPU} is CPU utilisation (taken to be the fraction of elapsed time the CPU spends executing a non-idle thread), $R_{disk\_B/s}$ is the total number of bytes sent to and retrieved from the disk per second (i.e. disk reads and writes per second), u_{mem} is the proportion of memory used by the application, u_{disk_time} is the proportion of disk time the server disk was busy servicing the read and write requests, and R_{page_fault} is the number of memory page faults per second.

Although these simplifications reduce the number of parameters required to approximate the power consumption of the server, these parameters still require specific measurements on the server, measurements that are very unlikely be available to a typical modeller.

Attempting to model the server power when restricting the parameter set to just the three dominant factors listed in Table 9.5 leads to correlations with the measured power consumption of 70% or less. Depending upon the accuracy required, this may be too low to be considered acceptable.

9.9.4 MINIMAL INFORMATION UTILISATION SERVER POWER MODELS

It often occurs that the most appropriate server power model must include its utilisation. That is, the single value power models described in Section 9.9.1 may not be adequate. However, the vast majority of published server power models (being sub-system based) require information that may not be publicly available. In most cases the data required to construct an accurate model will only be available to people who have direct access to, or are well acquainted with, the data centre server that provides the service of interest. As we found previously, the lack of specific or detailed information is not an uncommon situation when trying to construct power models of telecommunications networks and services. Therefore, we have to construct a variety of "work-arounds" to counter this paucity of directly usable data on equipment.

Although the CPU utilisation-based whole server models described in Section 9.9.2 are widely considered to be less accurate than the sub-system-based models [1, 63], they do not require such detailed information as the sub-system-based models. On the other hand, CPU utilisation-based models provide more information than the single value models discussed in Section 9.9.1.

In this section, we will present several examples of utilisation-based server power models that have been constructed assuming minimal specific information on servers in a data centre. By "minimal information" we mean these models require much less detailed information than those presented in Section 9.9.3 and do not refer to any specific type of server at all. Rather, the model is a very approximate version for a generic server. Recognising the constraints on the information available, these approximate models will be whole-of-server models and forgo the accuracy that comes with the more detailed models. The models described in this section can provide guidance regarding the power consumption of the equipment and are based on only a minimal amount of information.

The example models described here are not the only approach available; however, they require only information that is likely to be publicly available and not require the modeller to have any special relationship with the cloud service provider.

As a starting point, we assume the modeller has access to the amount of data (or data rate and duration) their service exchanges with the data centre. That is the total number of bits of data transmitted to and received from the data centre. This information can be accessed using a network packet analyser such as Wireshark [64].

We will discuss several approaches to constructing models that require minimal specific information. In some cases, we will assume the application of interest can be roughly categorised in a manner that will provide some information regarding the utilisation of the server's CPU. For example, simple email will not require significant CPU processing. In contrast, an application that requires sophisticated numerical processing will demand significant utilisation of the CPU. Alternatively, the application of interest may be able to be placed into a minimal set of representative apps common across users. For example, simple transactions (email, banking, etc.), web browsing, messaging, image storage (Facebook), video streaming (Netflix), etc. This categorisation of applications is a common approach [60, 65–67].

Therefore, for these models, we assume we have access only to information such as:

1. We know the data rate and duration of the service being served by a data centre.
2. We can categorise or represent the service type, as described below.
3. We have access to publicly available information required to estimate server parameters that can be used to construct a server power model. Although some modellers have access to more detailed information, in keeping with the general tenor of this book, we assume minimal specialised information.

The first minimalist model example is taken from Baliga *et al.* [51]. That paper provides a simple power model specifically for a "Storage as a Service" application.

9.9.4.1   Storage as a Service Server Power Model

Storage as a Service is a file storage and backup service (e.g. Dropbox, photo storage), where most of the processing is performed on the user's computer and the data centre is used merely to store the user data. Files can be downloaded from the data centre for viewing and editing on the user's computer and then uploaded back to the data centre for storage. The model is constructed specifically for data storage on a server with the per-user power consumption of the storage service $P_{st}(R_D,B_d)$, calculated as a function of file size, B_d, and file transfers per hour, R_D, being:

$$P_{st}\left(R_D,B_d\right) = B_d\left(\frac{R_D}{3600}\frac{P_{st,max}}{C_{st,max}} + X_{red}\frac{P_{storage}}{B_{storage}}\right)PUE \tag{9.56}$$

where $P_{st,max}$ is the maximum power consumption of the storage server (Watts), $C_{st,max}$ is the maximum throughput capacity of the storage server (bit/s), $P_{storage}$ is the power consumption of the data storage system (e.g. hard disk or solid-state disk), $B_{storage}$ (bits) is the capacity of the data storage system. The parameter X_{red} accounts for any redundancy used in the system in case of failure ($X_{red} = 2$ for the case where each storage system has one backup), PUE is the Power Use Effectiveness for the data centre. Finally, the 3600 converts downloads per hour to downloads per second. In (9.56) it is assumed that the storage server does minimal processing of the data so that its power consumption can be well represented by its throughput.

Data on the parameters that appear in (9.56) is available in the report by Shehabi *et al.* [50].

9.9.4.2   Generic Application-Specific Server Power Model

As discussed in Section 9.9.3, different applications will place a greater load on different sub-systems of the server. For example, a CPU-bound application may drive the CPU to maximum utilisation without placing a correspondingly high load on any of the other sub-systems. In contrast, an NIC-bound application will drive the network card to maximum utilisation well before the CPU is highly utilised. Therefore, when constructing a server power model, we must be aware of this detail. It is also important to note that, even though the utilisation of the server sub-systems may be limited by the fact that an application is bound by one sub-system, this sub-system may not be the dominant contributor to server power consumption. For example, we can have applications that are NIC bound but the power consumption of the server is dominated by the CPU.

It is possible to construct an application-specific server model that avoids the need to have detailed information on all the server sub-systems. However, this approach is typically limited to a range of application types because it assumes a direct relationship between the data exchanged between the user and server (R_{NIC} in (9.50) or u_{NIC} in (9.51)) and the utilisation of the server. Such a relationship is unlikely to be available for all application types. Examples of application-specific servers are the video servers discussed in Section 9.9.4.3.

In the generic application-specific server model described here, the server NIC data rate, R_{NIC}, is mapped onto the CPU utilisation, u_{CPU}, with the mapping being dependent upon the type of application. With this relationship determined, a whole server power model as described in Section 9.9.2 is used to construct an estimate of server power consumption based on the user data flow. We use a whole of server power model to avoid the need for details regarding the power consumption of the other server sub-systems.

To construct a relationship between user data rate and CPU utilisation we use the results from Koller *et al.* [63]. In that publication the authors display the measured relationship between NIC utilisation and the corresponding CPU utilisation for three applications. The relationships, along with linear approximations, are shown in Figure 9.5. The applications are [63]:

1. Lotus Domino: This application provides enterprise email, messaging, directory services, web services and application services. In this case the Domino app services 500 users and is disk bound.
2. Domino- (Domino minus): This is a variant of Domino, in that it services 10 users and is CPU bound.
3. TPC-W: This is a transactional application that is CPU bound.

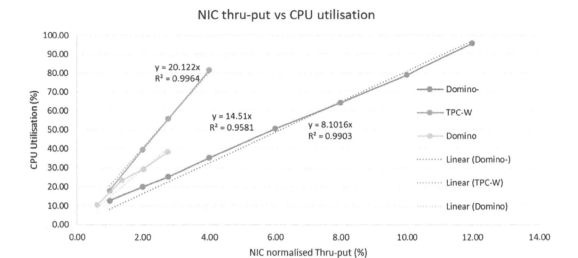

FIGURE 9.5   Relationships between NIC throughput and CPU utilisation for applications studied in [63].

Therefore, this model is applicable only to applications types that are similar to these three.

Note that, in Figure 9.5, the Domino application only reaches approximately 40% utilisation of the CPU and around 3% NIC utilisation. This is because it is bound by disk usage. In contrast, Domino minus is CPU bound and so can achieve quite high CPU utilisation while having only relatively low NIC utilisation. Application TCP-W is CPU and memory bound and so can attain reasonably high CPU utilisation but is limited to quite low NIC utilisation. None of these applications are bound by the NIC, hence none of them will reach 100% utilisation of the NIC, because they are bound by sub-systems other than the NIC.

Using linear fits to the measured data, the relationship between the NIC utilisation and CPU utilisation is well approximated with:

$$
\begin{aligned}
u_{CPU}\left(u_{NIC}\right) &= 8.101 u_{NIC} & \text{Domino} & \ \left(\text{CPU bound}\right) \\
u_{CPU}\left(u_{NIC}\right) &= 20.122 u_{NIC} & \text{TPC} & \ \text{W}\left(\text{CPU, memory bound}\right) \\
u_{CPU}\left(u_{NIC}\right) &= 14.51 u_{NIC} & \text{Domino}&\left(\text{disk bound}\right)
\end{aligned}
\tag{9.57}
$$

Using these forms in (9.46) or (9.49) provides a direct relationship between the data rate into the server, corresponding to u_{NIC}, and the server power consumption for applications that are similar to these applications. The fact that applications may be bound by different server sub-systems means that the maximum value for u_{NIC} that can be used in (9.57) for an application is set by the corresponding maximum value of u_{CPU} in the plots for the application shown in Figure 9.5. The maximum allowed value of u_{NIC} for Domino is 3% (corresponding to maximum u_{CPU} of 40%), for Domino minus 12% (corresponding to maximum u_{CPU} of 96%), and for TPC-W, the allowed maximum value for u_{NIC} is 4% (corresponding to maximum u_{CPU} of 80%).

Although the measurements used to determine (9.57) were undertaken for a specific server, as is often done with this type of modelling, we use that server as an example machine to provide this relationship for generic modelling.

To use (9.57) we note that the coefficients appearing in that equation relate to a server that is based on 2008 technology. Therefore, we need to rescale these equations to account for the improvements in technology since then. The CPU and NIC utilisations are given in (9.53) a) and d), respectively. Assume the improvement rate per year for the maximum number of operations per second for the CPU is β_{CPU} and for the maximum capacity improvement rate per year for the NIC is β_{NIC}. Then, we have at year n relative to year 0:

$$u_{CPU}^{[n]} = \frac{ops_{usr}}{ops_{CPU}^{[0]}\left(1+\beta_{CPU}\right)^n} = \frac{u_{CPU}^{[0]}}{\left(1+\beta_{CPU}\right)^n}$$

$$u_{NIC}^{[n]} = \frac{R_{usr}}{C_{NIC}^{[0]}\left(1+\beta_{NIC}\right)^n} = \frac{u_{NIC}^{[0]}}{\left(1+\beta_{NIC}\right)^n}$$

(9.58)

where $ops_{CPU}^{[n]}$ is the maximum number of CPU operations per second in year n and $C_{NIC}^{[n]}$ is the NIC maximum capacity in year n.

Annual improvements in CPU processing generally follow Moore's Law [68], which states a 60% annual improvement in computing ability: hence $\beta_{CPU} = 0.6$. For NIC ports, over the years 1982 to 2018 Ethernet port speed has increased from 10 Mbit/s to 40 Gbit/s [71]. This corresponds to an annual improvement of 25%: therefore $\beta_{NIC} = 0.26$. Including these improvements in (9.57) we have for equipment released in year n (relative to 2008):

$$u_{CPU}^{[n]} = 8.101 u_{NIC}^{[n]} \left(\frac{1.26}{1.6}\right)^{n-2008} \quad CPU \;\; bound$$

$$u_{CPU}^{[n]} = 20.122 u_{NIC}^{[n]} \left(\frac{1.26}{1.6}\right)^{n-2008} \quad CPU, \;\; memory \, bound$$

(9.59)

$$u_{CPU}^{[n]} = 14.51 u_{NIC}^{[n]} \left(\frac{1.26}{1.6}\right)^{n-2008} \quad disk \, bound$$

where $u_{CPU}^{[n]}$ and $u_{NIC}^{[n]}$ correspond to the utilisations in the year n, the year for which the model is to be applied. The exponent $(n-2008)$ is applied because the server for which (9.57) was constructed was released in 2008. A further issue is the range of allowed values of utilisation in the year n, $u_{NIC}^{[n]}$ that can be used in (9.59). Recall that this range was set by the maximum value of u_{CPU} in Figure 9.5 due to the server being bounded by one of its sub-systems. Over the years of technology improvement, the absolute value of this bound may change. Without information on the magnitude of this change over the years, we assume that this bound remains the same in relative terms. In other words, without further information to the contrary, we assume the maximum value of CPU utilisation in year n is the same as in year 0; that is maximum $u_{CPU}^{[n]}$ = maximum $u_{CPU}^{[0]}$. Of course, even if there is information available regarding changes in maximum values utilisation of all the sub-systems over the years, none of the utilisation values can be greater than unity. Hence we cannot have $u_{CPU}^{[n]} > 1$ for any of the equations in (9.59).

Another complication with using (9.57) is that we need to know the maximum NIC port capacity, C_{NIC}, in (9.58) for the server to determine u_{NIC}. This information is not likely to be easily available and there are most likely multiple types of NIC port speeds used in the various servers in a data centre. This is particularly so if the modeller is unsure as to the location and age of the data centre that is servicing the user. Therefore, we introduce another approximation into the model in an endeavour to account for this lack of information.

If the modeller does not have a value for C_{NIC}, one way around this problem is to use an average value of C_{NIC} for installed network ports across the current population of installed ports. This information is available in Shehabi *et al.* [50]. This report provides data on the number of network ports with maximum speeds of 100 Mbit/s, 1 Gbit/s, 10 Gbit/s and 40 Gbit/s installed in US data centres during the years 2006 to 2020 and their power consumption. Using this data, the average (maximum) port speed for ports deployed in US data centres in 2020 is 18 Gb/s. Therefore, without any other source of more detailed information for 2020, we can set $C_{NIC} = 18$ Gb/s in (9.58).

9.9.4.3   Video Server Power Model

Some organisations continue to post details of their power consumption that can be used to construct power models of the equipment they deploy. For example, Netflix has posted the details of several of their video servers [69]. With the information provided, we can construct an approximate model for several types of video server.

The Netflix web page has details on three server types, summarised in Table 9.6:

1. Storage Appliances: "Storage appliances are 2U servers that are focused on reliable dense storage and cost-effective throughput. This appliance is used to hold the Netflix catalogue in many IX locations around the world and embedded at our larger ISP partner locations." [69]
2. Global Appliances: "Global appliances are 1U lower cost appliances that we use for smaller ISP partners and emerging markets. This appliance is designed for low-cost 10GE attached content delivery." [69]
3. Flash Appliances: "For large deployments, we might also include flash-based appliances. These 1U appliances help scale network delivery for large sites up to Terabits per second as required." [69]

We will now use this information to construct some quantitative comparisons of some of the models described above. The approach described in the following paragraphs is indicative of the kinds of stratagems that may be required to construct quantitative estimates of power consumption. To use the Netflix information, we can take either of two approaches:

1. Simple peak power model. This uses the peak power without any consideration of the fact that most servers are not operating at peak utilisation.
2. Utilisation-based model that includes an idle power estimate.

To do this, we need to make a series of assumptions to construct the power model. First, we have to choose a power model for the server. We shall adopt the "whole of server" linear models given in (9.46). Next, we need an estimate for P_{idle} to express power as a function of the server utilisation. To estimate P_{idle}, we use the Dynamic Range reported by Shehabi $et\ al.$ [50] for 2020, which gives $P_{idle} = 0.44P_{max}$. Bringing these together gives for the overall server linear power models:

$$P_{storage}\left(u_{CPU}\right) = P_{idle} + u_{CPU}\left(P_{\max} - P_{idle}\right) = 220 + u_{CPU}280 \quad Watts$$
$$P_{global}\left(u_{CPU}\right) = 110 + u_{CPU}140 \quad Watts \tag{9.60}$$
$$P_{flash}\left(u_{CPU}\right) = 132 + u_{CPU}168 \quad Watts$$

TABLE 9.6

Details for Three Types of Content Server Deployed by Netflix [69]

Server Type	Peak Power (P_{max})	Operational Throughput	NIC Port Capacity	Storage	Storage Technology
Storage	~ 500 Watts	36 Gb/s	60 or 100 Gb/s	288 TByte	HDD & SSD
Global	~ 250 Watts	8.5 Gb/s	10 Gb/s	100 TByte	HDD & SSD
Flash	~ 300 Watts	35 Gb/s	40 Gb/s	14 TByte	SSD

To estimate the operational power consumption of the servers we need utilisations, u_{CPU}. We use the average utilisation of the NIC port given in Table 9.6 and set u_{NIC} = (Operational throughput)/ (NIC port capacity). This gives

$$u_{NIC,storage} = 36/60 = 0.6$$
$$u_{NIC,global} = 8.5/10 = 0.85 \qquad\qquad (9.61)$$
$$u_{NIC,flash} = 35/40 = 0.875$$

We now need to relate the NIC utilisation, u_{NIC}, to the CPU utilisation, u_{CPU}, of the server. We could simply assume $u_{CPU} = u_{NIC}$ and substitute the values in (9.61) directly for the values of u_{CPU} in the corresponding equation in (9.60). Alternatively, one could select one of (9.59) that is considered most appropriate for a video server and equate the server utilisation in (9.60) with the appropriate u_{CPU} in (9.59). Given the Netflix data is relatively current, to update the relationships between u_{NIC} and u_{CPU}, we can set the year $n = 2020$. Using (9.61) this gives for the three server types:

$$u_{CPU,storage}^{[2020]} = 8.101 \times 0.6 \left(\frac{1.26}{1.6}\right)^{(2020-2008)} = 0.27$$

$$u_{CPU,global}^{[2020]} = 8.101 \times 0.85 \left(\frac{1.26}{1.6}\right)^{(2020-2008)} = 0.39 \qquad\qquad (9.62)$$

$$u_{CPU,flash}^{[2020]} = 8.101 \times 0.875 \left(\frac{1.26}{1.6}\right)^{(2020-2008)} = 0.40$$

Repeating this for the CPU, memory-bound and disk-bound applications in (9.59), we get $u_{CPU} > 1$ for all three server types. This means those forms are unacceptable. This situation means that, with the assumption that (9.59) well approximates the relationship between u_{NIC} and u_{CPU} for these servers, then they could not attain an NIC utilisation as high as 60% or more because the server would be bounded by its disk or memory access. Given that these servers are operating at these NIC utilisations, it would appear the content server application is not disk or memory access bound. In simplistic terms, we would expect these servers to primarily read the video data off the disk, recode it for transmission via the Internet and then output the data stream. Given the data rate is most likely limited by a bottleneck outside the data centre, we would not expect the application to be either disk or memory limited.

Using the u_{CPU} values in (9.62) in (9.60) we get:

$$P_{storage} \approx 296 \; Watts$$
$$P_{global} \approx 165 \; Watts \qquad\qquad (9.63)$$
$$P_{flash} \approx 200 \; Watts$$

Using the nonlinear model from (9.46) gives slightly different results.

It is instructive to compare these estimates with those resulting from using the approach of Jalali *et al.* [2], described above in Section 9.1.1. Those models produced an estimate of energy per bit for a data centre of between 4 and 9 microJ/bit. Using the operational throughput from Table 9.6, and the power consumption values in (9.63), we get energy/bit values, E, for the servers of $E_{storage}$ = 8.2 nJ/bit, E_{global} = 19.4 nJ/bit and E_{flash} = 5.1 nJ/bit. These values are several orders of magnitude less than the values estimated for data centres. There are several reasons for this. Most importantly, the video server itself is very simple compared with a typical data centre installation, with no significant compute capability, scheduling, nor many of the other functions of a server in a data centre. A further important reason for this difference is the fact that the estimate in Section 9.1.1 accounts for the power consumption of all the servers and traffic between them that occurs when the data centre is servicing a request. The Cisco Global Cloud Index, 2016-2021 has reported that, of all the traffic in a data centre, only around 15% is between the data centre and the user, approximately 13% is between data centres, and 72% of traffic occurs within the data centre [72]. Therefore, the traffic generated by users causes almost 6 times as much traffic within and between data centres. This traffic generates activities by other servers within one or more data centres. None of these factors are included in $E_{storage}$, E_{global} or E_{flash}. Also, the energy per bit values for the servers do not include any other infrastructure in the data centre, whereas the estimates in Section 9.1.1 include other IT infrastructure in the data centre.

This example highlights a difference between server power consumption and data centre power consumption that must be appreciated when modelling the power consumption of ICT. This difference is important when modelling the power consumption of services provided by data centres.

9.9.4.4   Blockchain Server Power Model

Cryptocurrencies, mentioned in Chapter 1, are an increasingly important new approach to transferring value (i.e. currency) between parties, via the Internet. The most well-known example is "Bitcoin" which is one of many cryptocurrencies now in operation.

A crucial component of the operation of cryptocurrency transactions is blockchain technology which provides a decentralised validation of the transaction. (In a traditional transaction this is provided either by a financial institution or the government.) The blockchain technology is implemented using highly specialised blockchain servers that undertake intensive calculations to provide a "Proof of Work" (PoW) as part of the validation process. Details of cryptocurrency and the blockchain technology are beyond the scope of this work, there is a plethora of books and other sources that provide an overview of blockchain technologies and their use.

The reason for this section is to introduce the approach used to model the power consumption of blockchain servers. As mentioned in Chapter 1, there is concern that the environmental footprint of cryptocurrencies is rapidly growing due to the electrical power requirements of blockchain servers. These servers are highly specialised machines built specifically for the purpose of blockchain calculations and are typically located in dedicated facilities that only undertake blockchain calculations [73]. These calculations ultimately produce a "hash value" to provide a unique identity to each block in the blockchain.

To estimate the power consumption (either globally, for specific cryptocurrencies or on a per-transaction basis) the blockchain server power consumption is modelled using an "energy per operation" approach. For blockchain servers, the specific model is "energy per hash". With this, the power consumption of a blockchain server is given by $P(W)$ = (Joules/hash)x(hashes generated per second) [74, 75]. Tables of Joules/hash or Watts/hash/s are available [74, 75, 76]. It is important to note that the Joules/hash values provided do not include environmental control, power backup or other redundancies and overheads [75].

The specifications for these servers are impressive, in terms of both the rate at which they can process hashes and their computation energy efficiencies, being a fraction of a joule per GigaHash. However putting those in perspective, the probability that a randomly selected number (called a "nonce") will meet the requirement of determining a Blockchain hash code is of the order 1 in 6×10^{22}, which explains the enormous computation and energy cost of a blockchain application using the PoW algorithm, such as Bitcoin. More recent blockchain products are starting to use less demanding algorithms, requiring computation resources lowered by many orders of magnitude. [74] Readers should be aware that this is a rapidly changing technology area.

9.10   VIRTUALISATION

When constructing server power models, it is important to appreciate that, in modern data centres, a high degree of consolidation/virtualisation is often used to improve energy efficiency. (See Section 9.2 for details.) With this technology, rather than running different applications on different servers, multiple applications are run within virtual machines on a single server. The degree of virtualisation (i.e. number of virtual servers operating on a given physical server) can impact the power consumption of an application due to the hypervisor overhead for a given degree of virtualisation. The impact of virtualisation on a server running multiple applications is dependent upon the type of applications being run on that physical server [63, 70].. Metri *et al.* have shown that the energy efficiency of applications running on a virtual machine is dependent upon the application type, machine utilisation, virtual machine size and physical server type [66].

This somewhat complicates modelling server power consumption using only the server utilisation. There are several possible approaches to accommodate the impact of virtualisation on server power consumption. Possible approaches are as follows:

1. Just ignore it. This could be acceptable provided the utilisation of the servers is not too high. For example, Koller *et al.* [63] have reported that, at low utilisation, running multiple virtual machines on a server does not significantly increase server power consumption. Also, some application types are not impacted by virtualisation: for example, applications that do not have significant I/O operations or have a large working set of operations.
2. Include a power overhead factor to account for virtualisation. This approach was adopted by Shehabi *et al.* [50], who gave a consolidation overhead value of 5%. However, it should be noted that Koller *et al.* have reported that the virtualisation overhead is application dependent [63].
3. Include a virtualisation overhead dependent on the number of virtual machines and application types. This approach has been validated by Koller *et al.*; however, it requires information on the number of virtual machines running on a server and application types [63]. One positive aspect of this approach is that, after accounting for these two factors, a linear utilisation-based power model still provides a good estimate.

The models developed in this section will play a crucial role in estimating the power consumption of Internet-based services. However, as we will see in Chapter 10, we will also need to address several other key contributors and conceptual issues when constructing models for Internet service power consumption.

9.11   TECHNICAL SUMMARY

To construct a data centre power model, a common approach is to segment the operational parts of a data centre into the following:

1. Servers and storage systems
2. Power supply equipment including power backup;
3. Cooling and air flow equipment;
4. Data Centre networking equipment;
5. Control, security, lighting, etc.

A power model for each is then constructed. The overall power consumption of the server equipment is the primary determinant of overall data centre power consumption because it is a major contributor and the power consumption of several of the other segments of data centre operation is determined by the power consumption of the servers. In particular, the power supply equipment, the cooling and air flow and networking are directly related to total server power consumption.

A common and useful power model for a server is linearly dependent upon the server's "utilisation". This model is given in (9.1) where the utilisation is typically taken to be that of the server CPU. The server and storage system components of data centre power consumption can be based on an "average" server whose utilisation is given by an average utilisation defined in (9.2). Using the average server approach, the total power consumption of all the servers in the data centre can be expressed as the number of servers multiplied by the power consumption of the average server, as in (9.3).

In modern data centres, server "consolidation" is used to reduce total server power consumption. Consolidation enables servers with low or zero utilisation to be placed into a lower power "sleep" state and the applications that are running are brought together onto a reduced number of servers. The degree of consolidation is quantified with a parameter c, which ranges from $c = 0$ (for maximal consolidation where all operating servers are operating at maximum utilisation) to $c = 1$ (for minimal consolidation where all servers are operating with the same utilisation). Consolidation primarily saves on server idle power, $P_{svr,idle}$, by placing idle servers into a sleep state with power $P_{sleep} \ll P_{svr,idle}$. With consolidation the total server power is given by (9.6).

The temperature of the server room, where the servers are located, is typically controlled by either or both of CRAH and CRAC. CRAH units use chilled water provided by an external cooling system, whereas CRAC units include their own refrigeration systems for controlling the server room air temperature.

For a CRAH system, the power consumption of the chillers and CRAH unit can both be expressed in terms of the server utilisation, $U(t)$, and are given by (9.10) and (9.16), respectively. The power consumption of a CRAC system is similarly dependent upon $U(t)$ and is given by (9.30).

The power supply to the servers can be represented as consisting of two systems. The Power Distribution Unit (PDU) transforms the mains input power into the data centre to the appropriate voltage level for the racks of equipment. The Uninterruptable Power Supply (UPS) ensures short-term continuity of power supply should the main supply to the data centre fail. The power consumptions of these two systems are given in (9.31) and (9.34), respectively.

In addition to these systems, a data centre also includes a range of activities that consume power. For example, the internal data centre communications network that provides connections between servers and to the Internet are an essential part of a data centre; however, this network's power consumption is significantly less than the contributions discussed above. This plus the administration, data centre lighting, control and security operations contribute only a few per cent of the overall data centre power consumption and do not vary much with respect to utilisation, $U(t)$ [8]. The power consumption of these systems can be aggregated and can be approximated by (9.39).

Bringing these components together enables an approximation for total data centre power consumption to be constructed and is given in (9.41).

A widely used energy efficiency metric for data centres is the Power Usage Effectiveness (PUE), which is the ratio of total data centre power consumption to its IT equipment power consumption. The PUE gives an indication of the amount of power consumed by the data centre that is not used for providing data services to customers. An expression for the PUE, based on the power model described above, is given in (9.43) (in terms of power) and (9.44) (in terms of energy).

Although data centre power models play an important role when considering the sustainability of ICT and Internet services, in many cases the interest is more on the power or energy consumption of the services and applications provided by a data centre. To construct power models of applications we require more detailed power models of servers. This presents most modellers with a problem, because there is a paucity of publicly available detailed information on the power consumption of most servers. Therefore, construction of a server power model inevitably requires multiple assumptions and the bringing together of information from a diversity of sources.

The simplest approach for modelling the power consumption of a server is to use a single value. An example of this is given in (9.45) or in published power values, as shown in Table 9.3. Another approach to constructing a single value power model is to use published information to construct an energy/bit model, as described in Section 9.9.1.

In some cases, a more detailed model is required, such as one that includes a dependency on server utilisation. There are two approaches to utilisation-based models. The first is whole server models, as described in Section 9.9.2. These models express the total server power consumption as a function of the CPU utilisation. These include models which express server power consumption in terms of linear and nonlinear equations in CPU utilisation. Examples of this approach are given in Section 9.9.2. An advantage of these models is that they require less information than the sub-system-based models.

Sub-system-based models express server power consumption as the sum of its sub-systems' power consumptions, as described in Section 9.9.3. These sub-systems are as follows:

1. Server Central Processing Unit (CPU) processing;
2. Memory access;
3. Disk access;
4. Network Interface Card (NIC) operation.

The total server power consumption is then expressed in the form (9.50) or (9.51). A disadvantage of these models is the need to access information on each sub-system, as well as the utilisations of those sub-systems, due to the particular applications running on the server. The parameter values for this type of model can be determined by measurement; however, this typically requires collecting enough data to enable curve fitting because of the number of variables involved. In some cases, it has been found that some parameters can be ignored; however, this can only be determined after measurement.

It is highly unlikely the parameters that apply to one server model will apply to all others. This means that a modeller will have to assume that the model parameters to which they have access will be applicable to the server being modelled.

A significant challenge for most modellers is likely to be the fact that they only have direct access to the amount of user data that is exchanged between the user and the server, which is most likely to be located in a data centre inaccessible to the modeller and service user. The modeller may also have an appreciation of the type of service application under consideration, such as video delivery, data storage, email, or financial analysis. The type of application will influence which sub-system of the server limits the server's performance and/or is most intensively used by the application. It is likely the modeller will not know the details of the servers that process the application of interest.

When only this information is available, the modeller has limited options on how to construct a power model. With knowledge of the type of application, the modeller may assume the server is optimised for the type of application on which they are focused. For example, a video streaming service is most likely using servers designed for that purpose. This will allow the modeller to narrow down the typical server parameters for their power model.

If a more generic model is required, the approach described in Section 9.9.4.2 provides a possible option. However, this model has several caveats that need to be observed; in particular, the parameter values used in the equations may need adjusting to allow for technology advances.

Virtualisation is very widely used in modern data centres and does impact the power consumption of servers running applications. Several approaches to dealing with virtualisation have been presented in published literature. These are reviewed in Section 9.10.

9.12 EXECUTIVE SUMMARY

Without detailed information about the data centre systems, it is unlikely that an accurate model for total data centre power consumption can be constructed. However, approximate models have been published. These models typically model total data centre power consumption by segmenting the data centre into its dominant power consuming systems, which are as follows:

1. Servers and storage systems;
2. Power supply equipment including power backup;

3. Cooling and air flow equipment;
4. Data centre networking equipment;
5. Control, security, lighting, etc.

The dominant contributor to data centre power consumption is item 1), the servers. The total server power consumption can be modelled using the "utilisation" of the servers, $U(t)$, at time t. This parameter can also be used to estimate the power consumption of the other segments of the data centre. Using published generic parameters, an approximate power model for a data centre over the time, t, can be constructed.

From this model, the PUE can be calculated over the diurnal cycle or to provide an average value. The PUE is a widely used metric for ascertaining the energy efficiency of data centres.

In many cases, the issue of interest is the power consumption of the services or applications provided by a data centre. To construct these models, we require power models for the servers. There is a diverse range of such models. The simplest use a single number to represent the server power consumption. Such models just use publicly available information on the average power of servers and models based on "energy per bit" derived from global data collected from multiple sources.

If a more detailed model is required, there are published "utilisation" based models. However, these require more detailed information regarding servers and their sub-systems. These sub-systems are as follows:

1. Server Central Processing Unit (CPU) processing;
2. Memory access;
3. Disk access;
4. Network Interface Card (NIC) operation.

This may require measurement of the servers under various loads to determine the parameter values used in the power model equations. Unfortunately, such detailed information is not easily available from the manufacturers.

In some cases, the only information that is readily available to the modeller is the amount of data exchanged between the user and the data centre and the type of application. Examples of different application types include video delivery, data storage, web browsing, email, and financial analysis. The type of application will influence which sub-system of the server limits the server's performance and/or is most intensively used by the application. It is likely the modeller will not know the details of the servers that process the application of interest.

Despite this limited amount of information, it is still possible to construct models that provide some guidance on the power consumption of services provided by a data centre. These models will only be approximate because they unavoidably use parameter values that are based on general server characteristics or correspond to server models that do not represent to the actual machines providing the service.

REFERENCES

[1] M. Dayarathna *et al.*, "Data Center Energy Consumption Modelling: A Survey," *IEEE Communications Surveys and Tutorials*, vol. 18, no. 1, p. 732, 2016.
[2] F. Jalali *et al.*, "Fog Computing May Help to Save Energy in Cloud Computing," *IEEE Journal on Selected Areas in Communications*, vol. 34, no. 5, p. 1728, 2016.
[3] J. Hamilton, "Perspectives data center cost and power," [Online]. Available: http://mvdirona.com/jrh/TalksAndPapers/PerspectivesDataCenterCostAndPower.xls. [Accessed 17 July 2020].
[4] J. Hamilton, "Perspectives: Overall data center costs," [Online]. Available: http://perspectives.mvdirona.com/2010/09/18/OverallDataCenterCosts.aspx. [Accessed 17 July 2020].
[5] Facebook, "Facebook Sustainability Data 2019," Facebook 2020. [Online]. Available: https://sustainability.fb.com/case-studies/. [Accessed 17 July 2020].

[6] Cisco, "Cisco Visual Networking Index: Forecast and Trends 2017–2022," Cisco White Paper 2018.

[7] Sandvine, "The Global Internet Phenomena Report," Sandvine, October 2018.

[8] S. Pelley *et al.*, "*Understanding and Abstracting Total Data Center Power,*" in *Workshop on Energy Efficient Design (WEED09)*, 2009.

[9] S. Rahmani *et al.*, "A Complete Model for Modular Simulation of Data Centre Power Load," arXiv.org, 2018. [Online]. Available: https://arxiv.org/abs/1804.00703. [Accessed 16 July 2020].

[10] S. Livieratos *et al.*, "A New Proposed Energy Baseline Model for a Data Center as a Tool for Energy Efficiency Evaluation," *International Journal of Power and Energy Research*, vol. 3, no. 1, p. 1. 2019.

[11] L. Barroso *et al.*, *The Datacenter as a Computer: An Introduction to the Design of Warehouse-Scale Machines, H. D. Mark*, Ed., Morgan & Claypool, 2013.

[12] Info-Tech, "Top 10 energy-saving tips for a greener data center," Info-Tech Research Group, 2007.

[13] D. Bouley, "Estimating a Data Center's Electrical Carbon Footprint," Schneider Electric, 2011.

[14] N. Rasmussen, "Allocating Data Center Energy Costs and Carbon to IT Users," Schneider Electric, 2011.

[15] Z. Zhou *et al.*, "Analysis of Energy Consumption Model in Cloud Computing Environments," in *Advances on Computational Intelligence in Energy. Green Energy and Technology*, T. Herawan *et al.*, Eds., 2019.

[16] S. Rivoire *et al.*, "*A Comparison of High-Level Full-System Power Models,*" in *Proceedings of the 2008 conference on Power aware computing and systems, HotPower'08*, 2008.

[17] R. Buyya *et al.*, "Energy-Efficient Management of Data Center Resources for Cloud Computing: A Vision, Architectural Elements, and Open Challenges," arXiv, 2010. [Online]. Available: https://arxiv.org/abs/1006.0308. [Accessed 16 July 2020].

[18] R. A. Giri and A. Vanchi "Increasing Data Center Efficiency with Server Power Measurements," Intel Information Technology2010

[19] H. Cheunga *et al.*, "A Simplified Power Consumption Model of Information Technology (IT) Equipment in Data Centers for Energy System Real-Time Dynamic Simulation," *Applied Energy*, vol. 222, p. 329, 2018.

[20] T. Evans, "The Different Technologies for Coolilng Data Centers," Schneider Electric, 2012.

[21] A. Capozzolia and G. Primiceria, "Cooling Systems in Data Centers: State of Art and Emerging Technologies," *Energy Procedia*, vol. 83, p. 484, 2015.

[22] N. Rasmussen, "Electrical Efficiency Modeling of Data Centers," APC, 2006.

[23] M. Iyengar *et al.*, "*Reducing Energy Usage in Data Centers through Control of Room Air Conditioning Units,*" in *12th IEEE Intersociety Conference on Thermal and Thermomechanical Phenomena in Electronic Systems*, 2010.

[24] R. Sawyer, "Calculating Total Power Requirements for Data Centers," American Power Conversion, 2004.

[25] C. Patel *et al.*, "*Smart Coolling of Data Centers,*" in *Proceedings of IPACK'03 International Electronic Packaging Technical Conference and Proceedings*, 2003.

[26] Z. Wang *et al.*, "Opportunities and Challenges to Unify Workload, Power, and Cooling Management in Data Centers," *ACM SIGOPS Operating Systems Review*, vol. 44, no. 3, p. 41, 2010.

[27] A. Uchechukwu, *et al.*, "Energy Consumption in Cloud Computing Data Centers," *International Journal of Cloud Computing and Services Science*, vol. 3, no. 3, p. 145, 2014.

[28] D. Meisner and W. Wenisch, "*Does Low-Power Design Imply Energy Efficiency for Data Centers?,*" in *IEEE International Symposium on Low Power Electronics and Design (ISLPED)*, 2011.

[29] X. Zhang *et al.*, "Power Consumption Modeling of Data Center IT Room with Distributed Air Flow," *International Journal of Modeling and Optimization*, vol. 6, no. 1, p. 33, 2016.

[30] R. Schmidt and M. Iyengar, "Thermodynamics of Information Technology Data Centers," *IBM Journal of Research and Development*, vol. 53, no. 3, p. Paper 9, 2009.

[31] G. Zhabelova *et al.*, "*Towards an Open Model for Data Center Research: from CPU to Cooling Tower,*" in *44th Annual Conference of the IEEE Industrial Electronics Society (IECON 2018)*, 2018.

[32] S. Itoh *et al.*, "*Power Consumption and Efficiency of Cooling in a Data Center,*" in *11th IEEE/ACM International Conference on Grid Computing (GRID)*, 2010.

[33] K. Dunlap, "Cooling Audit for Ide notifying Potential Cooling Problems in Data Centers, Revision 3," Schneider Electric, 2014.

[34] The Green Grid, "Cases Study: The ROI of Cooling System Energy Efficiency Upgrades," The Green Grid, 2011.

[35] Liebert Corporation, "https://www.vertiv.com/en-us/," 2020. [Online]. [Accessed 7 August 2020].

[36] X. Zhan and S. Reda, *"Techniques for Energy-Efficient Power Budgeting in Data Centers,"* in *50th ACM/ EDAC/IEEE Design Automation Conference (DAC)*, 2013.

[37] J. Moore et al., *"Making Scheduling "Cool": Temperature-Aware Workload Placement in Data Centers,"* in *USENIX Annual Technical Conference*, 2005.

[38] R. Sawyer, "Making Large UPS Systems More Efficient, Revision 3," Schneider Electric, 2012.

[39] D. Minoli, *Designing Green Networks and Network Operations*, CRC Press, Taylor & Francis Group, 2011.

[40] J.-M. Pierson, Ed., *Large-Scale Distributed Systems and Energy Efficiency: A Holistic View*, Wiley & Sons, 2015.

[41] J. Cho and Y. Kim, "Improving Energy Efficiency of Dedicated Cooling System and Its Contribution towards Meeting and Energy-Optimised Data Center," *Applied Energy*, vol. 165, p. 697, 2016.

[42] Google, "Google Environmental Report 2019," Google, 2019. [Online]. Available: https://services. google.com/fh/files/misc/google_2019-environmental-report.pdf. [Accessed 17 July 2020].

[43] Microsoft, "Microsoft's Cloud Infrastructure Data Centers and Network Fact Sheet," Microsoft, 2015

[44] Intel, "Data Center Strategy Leading Intel's Business Transformation," Intel, 2019.

[45] B. Guenin, "Thermal Facts & Fairy Tales: Whatever Happened to the Predicted Data Center Energy Consumption Apocalypse?," 30 May 2019. [Online]. Available: https://www.electronics-cooling. com/2019/05/thermal-facts-fairy-tales-whatever-happened-to-the-predicted-data-center-energy-consumption-apocalypse/. [Accessed 7 August 2020].

[46] E. Masanet et al., "Recalibrating Global Data Center Energy-Use Estimates," *Science*, vol. 367, no. 6481, p. 984, 2020.

[47] R. Hintemann, "Efficiency Gains Are Not Enough: Data Center Energy Consumption Continues to Rise Significantly," 2018. [Online]. Available: https://www.borderstep.de/. [Accessed 8 August 2020].

[48] D. Economou et al., *"Full-System Power Analysis and Modeling for Server Environments,"* in *Workshop on Modeling Benchmarking and Simulation (MoBS)*, 2006.

[49] T. Chen et al., "The Features, Hardware, and Architectures of Data Center Networks: A Survey," *Journal of Parallel and Distributed Computing*, vol. 96, p. 45, 2016.

[50] A. Shehabi et al., "United States Data Center Energy Usage Report," Ernest Orlando Lawrence Berkeley National Laboratory, 2016.

[51] J. Baliga et al., "Green Cloud Computing: Balancing Energy in Processing, Storage and Transport," *Proceedings of the IEEE*, vol. 99, no. 1, p. 149, 2011.

[52] Google, "Google's Green Computing: Efficiency at Scale," Google, 2011.

[53] S. Sharma, *"Trends in Server Efficiency and Power Usage in Data Centers,"* in *SPEC Asia Summit*, 2016.

[54] A. Vishwanath et al., "Energy Consumption Comparison of Interactive Cloud-Based and Local Applications," *IEEE Journal on Selected Areas in Communications*, vol. 33, no. 4, p. 616, 2015.

[55] X. Fan et al., *"Power Provisioning for a Warehouse-Sized Computer,"* *Proceedings of the 34th Annual International Symposium on Computer Architecture*, 2007.

[56] X. Zhang et al., "A High-Level Energy Consumption Model for Heterogeneous Data Centers," *Simulation Modelling Practice and Theory*, vol. 39, p. 41, 2013.

[57] P. Garraghan et al., *"A unified Model for Holistic Power Usage in Cloud Datacenter Servers,"* in *9th IEEE/ACM International Conference on*, 2016.

[58] W. Lin et al., "A Cloud Server Energy Consumption Measurement System for Heterogeneous Cloud Environments," *Information Sciences*, vol. 468, p. 47, 2018.

[59] J. Arjona, *"A Measurement-based Analysis of the Energy Consumption of Data Center Servers,"* in *e-Energy 14, Proceedings of the 5th International Conference on Future Energy Systems*, 2014.

[60] Standard Performance Evaluation Corporation, "SPEC CPU 2006," Standard Performance Evaluation Corporation, 9 January 2018. [Online]. Available: https://www.spec.org/cpu2006/. [Accessed 17 July 2020].

[61] W. D. Norcott, "Iozone Filesystem Benchmark."

[62] Hewlett-Packard Development Company, "An Introduction to HP LoadRunner software," 2011.

[63] R. Koller et al., *"WattApp: An Application Aware Power Meter for Shared Data Centers,"* in *7th International Conference on Autonomic Computing and Communications (ICAC'10)*, 2010.

[64] Wireshark, "Wireshark," [Online]. Available: https://www.wireshark.org/. [Accessed 17 July 2020].

[65] Z. Li et al., "A Survey on Modeling Energy Consumption of Cloud Applications: Deconstruction State of the Art and Trade-off Debates," *IEEE Transactions on Sustainable Computing*, vol. 2, p. 255, 2017.

[66] G. Metri *et al.*, "*Experimental Analysis of Application Specific Energy Efficiency of Data Centers with Heterogeneous Servers*," in *IEEE Fifth International Conference on Cloud Computing*, 2012.

[67] A. Quintiliani *et al.*, "*Understanding "workload-related" Metrics for Energy Efficiency in Data Center*," in *20th International Conference on System Theory, Control and Computing (ICSTCC)*, 2016.

[68] J. Chovan and F. Uherek, "Photonic Integrated Circuits for Communication Systems," *RadioEngineering*, vol. 27, no. 2, p. 357, 2018.

[69] Netflix, "Open Connect Appliances," 2020. [Online]. Available: https://openconnect.netflix.com/en/appliances/#software. [Accessed 17 July 2020].

[70] F. Chen *et al.*, "*Experimental Analysis of Task-based Energy Consumption in Cloud Computing Systems*," in *Proceedings of the 4th ACM/SPEC International Conference on Performance Engineering (IPCE'13)*, 2013.

[71] Shehabi *et al.* "Data center growth in the United States: decoupling the demand for services from electricity use," *Environment Research Letters*, vol. 13, no. 12, 124030 (2018).

[72] Cisco, "Cisco Global Cloud Index: Forecast and Trends 2016–2021," Cisco White Paper 2018.

[73] Sedlmeir *et al.* "Recent Developments in Blockchain Technology and their Impact on Energy Consumption" arXiv:2102.07886v1 [cs.CR] 15 Feb 2021.

[74] Gallersdorfer *et al.* "Energy Consumption of Cryptocurrencies Beyond Bitcoin", *Joule*, vol. 4, p. 1843, 2020.

[75] M. Krause, T. Tolaymay, "Quantification of energy and carbon costs for mining cryptocurrencies", *Nature Sustainability*, vol. 1, p. 711, 2018.

[76] Cambridge Centre for Alternative Finance, "Cambridge Bitcoin Electricity Consumption Index: Methodology" [Online]. Available: https://cbeci.org/cbeci/methodology. [Accessed 18 June 2021].

10 Service Transport Power Consumption Models

Services provided by means of telecommunications networks take on many forms. To many consumer households, telecommunications services might mean access to daily news, services like Instagram or Facebook for exchange of personal news and photographs, e-mail, information searching, online banking, or video streaming. At a business level, other services such as office applications, collaboration, information storage and revival, online conferencing, cloud computing, serving of advertising and promotional material add to the list of services. These and other such services form the primary focus of users of the Internet and Information and Communications Technologies (ICT).

Such services are generally provided by data centres; these are warehouse-scale computer and storage facilities. Data centres and their power consumption were discussed in Chapter 9.

However, the data centre providing a service will very commonly be located in a different city, or even a different continent, to the service user. End users are generally unaware of where their service originates, or of the networks that enable and support these services, or their complexity. Nonetheless, the energy consumed by these supporting networks is significant and a full accounting of the energy consumption and carbon footprint of ICT services should where possible include the network energy consumption. This has been a matter of study in the context of the Greenhouse Gas Protocol, and appropriate methods and principles are being developed [1]

In addition to the end-user services already mentioned, the major network operators also provide transport and routing services to smaller operators, corporates, and Internet Service Providers. This is beneficial in providing economies of scale in terms of both cost and total energy consumption, but adds to the complexity of breaking down the energy or carbon footprint of a service. In addition, smaller operators may provide communication services to Internet Service Providers, which in turn connect to end customer's premises either directly or through another party. Thus a "service" may describe one of several different contexts, but the basic principles and equations are generic.

With this in mind, this chapter complements the discussion of data centres and their typical services in Chapter 9, by focusing on the construction of power consumption models for the telecommunications networks supporting these services. These models will heavily leverage the work presented in the earlier chapters; however, there is a range of additional issues that need to be considered when constructing a model of a service rather than an entire network. For example, the data traffic arising from a service represents only a fraction of the total traffic in a network. Thus we need some methodology for identifying a proportion of total network power that can be justifiably allocated to the service of interest. There are several ways of undertaking this task and we need to develop an approach that is both implementable and justifiable to stakeholders involved in the telecommunications and Internet services.

10.1 ATTRIBUTIONAL AND CONSEQUENTIAL MODELS

Life Cycle Analysis (LCA) is the general term for a range of methods that provide an assessment or estimate of the environmental impact of a product or service. In most cases, this assessment covers a product's or service's life cycle including extraction and processing of the raw materials, manufacturing, distribution, use phase, recycling, and final disposal. In our case, we are focussing on the use phase only, in particular, the power consumption of a service provided via the Internet or telecommunications network.

There are two widely adopted approaches for LCA, namely Consequential and Attributional [2]:

- Consequential LCA provides "an estimate of how the production and use of the study object affect the global environmental burdens" [3]. Focussing on Internet services, Consequential LCA provides an estimate of the change in total emissions as a result of change in the level of consumption of an Internet service [4].
- Attributional LCA provides "an estimate of what part of the global environmental burdens belongs to the study object" [3]. Focussing on Internet services, Attributional LCA provides an estimate of the total emissions that can be attributed to an Internet service [4].

This book focuses on the use phase of Internet services and the impact of the traffic resulting from the introduction and growth of an Internet service. We find that, due to the existence of idle power consumption for network elements, the methods of consequential and attributional assessments of Internet service power consumption differ in the scenarios in which they should be applied, as well as their outcome. Further, determining the attributional energy consumption of an Internet service requires some consideration.

10.2   A SIMPLE "FIRST-CUT" MODEL

The models developed in the later sections of this chapter involve a significant degree of complexity. In some cases, a very simple "first-cut" model may suffice. In this section, we describe such a model. This type of model estimates the power consumption of a service by first calculating the energy per bit of the equipment or network dealing with the service. This is then used with the data rate of the service to provide the power estimate. This approach has been used to estimate power consumption of several Internet services [5, 6]. The general approach of using energy per bit has been widely adopted, frequently in top-down models [1, 7] and some bottom-up models [8].

We start by considering a single node in a network. Within this node, there are N network elements (routers or switches) dealing with traffic. We assume the traffic only propagates through a single network element and that each network element has the typical power dependence on throughput, namely $P(t) = P_{idle} + \mu(t)(P_{max} - P_{idle})$, where $\mu(t) = R(t)/C_{max}$ is the utilisation of the network element with $R(t)$ the throughput traffic of a network element and C_{max} its maximum capacity.

The power consumption of a service is estimated by first calculating the energy per bit, $E = \Delta P/\Delta R$, of the node given by approximating the slope of the throughput vs power relationship for many network elements ($N_E \gg 1$), as depicted in Figure 10.1. The power consumption of a service propagating through that node is then given by $P^{(svc)} \approx ER^{(svc)}$, where $R^{(svc)}$ is the traffic due to the service of interest. The total power consumption of that service is then approximated by $P_{Tot}^{(svc)}\left(R^{(svc)}\right) \approx N_E^{(svc)}P^{(svc)}$ where $N_E^{(svc)}$ is the number of nodes in the network through which the service propagates.

Viewing Figure 10.1, we see that, as the total capacity, $C_{Tot,Node}$, of the node is increased by adding more network elements, the power consumption, $P_{Tot,Node}$, has a step-wise profile due to the idle power of the added equipment.

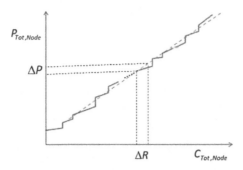

FIGURE 10.1 Power profile of a network node as network elements are added to the node to deal with increased traffic through the node. We wish to approximate the slope, $\Delta P/\Delta R$, of the node to estimate the power consumption of a service's traffic through the node.

Averaging P_{idle}, P_{max} and C_{max} over the N network elements in the node, we have for a linear approximation for the slope in Figure 10.1, which is the energy per bit, E, of the node:

$$E \approx \frac{N\left(\langle P_{idle}\rangle + \mu\left(\langle P_{max}\rangle - \langle P_{idle}\rangle\right)\right) - \langle P_{idle}\rangle}{N\mu\langle C_{max}\rangle} \tag{10.1}$$

$$= \frac{P_{idle}\left(\frac{1}{\mu}-1\right) + P_{max}}{C_{max}}$$

From this, the total power consumption, $P^{(svc)}$, of the service can be approximated by

$$P^{(svc)}(R^{(svc)}) = N_E^{(svc)}P^{(svc)} = N_E^{(svc)}ER^{(svc)} \tag{10.2}$$

Although this is an approximation, it provides a very direct and relatively simple approach to estimating the power consumption of a service through a network.

10.3 UNSHARED AND SHARED EQUIPMENT

We now consider more sophisticated models. The power consumption of a service arises from the power consumption of the network equipment over which that service operates. Therefore, when a network element carries traffic from several services, the power consumption of a specific service depends on the number of other services and their traffic. In the home Customer Premises Equipment (CPE), it is highly likely that very few services are simultaneously operating on that equipment. When developing a model for CPE, we need to account for this fact. Measurements on some CPE have shown their power consumption has a non-negligible dependence on service traffic [6, 9]. In this section we describe a service power model that can be applied to equipment, such as CPE, that is supporting only one or a few services.

We need a model that can allocate equipment power consumption to a specific service, among possibly several services, that an end user may be utilising as part of their online activities. The problem with making this assessment is that, at any specific time of the day, the CPE may be "on" or "operational" whether or not the end user is accessing the service of interest. Therefore, we need an approach that can distinguish between those times the user is accessing the service and those times is they are not.

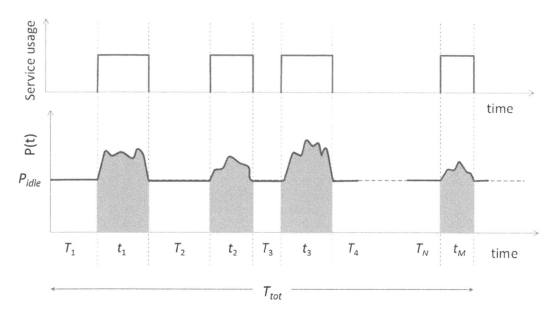

FIGURE 10.2 Example use and power consumption as a function of time for user CPE that is operating with a single service. The regions shaded in pink correspond to the equipment being active with the user's service.

Consider a user accessing a single service via their CPE (ONU, modem, etc.). Let the user's use of the service be represented by the plots in Figure 10.2. In this case, the user is "active" in that they are accessing the service only during times t_m, $m = 1, \ldots, M$ and **not** accessing the service for times T_m, during which the CPE is idle as indicated in Figure 10.2. The total time under consideration is T_{tot}. The equipment is operational over time T_{tot} in that it is energised and ready for use for all of T_{tot}.

Detailed monitoring the times of use of the service and the consequential power consumption of the CPE requires specialised power monitoring equipment. It is not likely this monitoring is readily available; therefore, we will construct a set of simplified, averaged, energy and power consumption equations that only require a minimal amount of information about the service and the CPE.

We have:

$$\sum_{m=1}^{M} T_m + t_m = T_{tot} \tag{10.3}$$

For the total energy consumption, Q_{tot}, of the equipment over the operational duration T_{tot}, we have:

$$Q_{tot} = \int_{0}^{T_{tot}} P(t)dt = \sum_{m=1}^{M}\left(T_m P_{idle} + \int_{t_m} P(t)dt \right) \tag{10.4}$$

where $\int_{t_m} f(t)dt$ is the integral of $f(t)$ over the time interval t_m. Note that in this instance there is only one service and we have allocated all the equipment idle power during the active times, t_m, to that service. We return to this issue later. Because we are focused on the service that is only active during time durations t_m, we define the average power of the service, $\langle P^{(svc)} \rangle$, by:

$$\left\langle P^{(svc)} \right\rangle = \sum_{m=1}^{M}\int_{t_m} P(t)dt \Bigg/ \sum_{m=1}^{M} t_m \tag{10.5}$$

We also define average service active and inactive times $\langle t \rangle$ and $\langle T \rangle$, respectively, by:

$$\langle t \rangle = \frac{1}{M}\sum_{m=1}^{M} t_m \qquad \langle T \rangle = \frac{1}{M}\sum_{m=1}^{M} T_m \qquad (10.6)$$

With this we have:

$$\sum_{M=1}^{M}\int_{t_M} P(t)\,dt = M\left\langle P^{(svc)}\right\rangle\langle t \rangle \qquad (10.7)$$

The power consumption of an equipment unit is typically well approximated with the linear form:

$$P(t) = P_{idle} + ER(t) \qquad (10.8)$$

With this, the energy consumed by the service during the mth time period of its use, Q_m, is:

$$Q_m = \int_{t_m} P(t)\,dt = P_{idle}t_m + E\int_{t_m} R(t)\,dt = P_{idle}t_m + EB_m \qquad (10.9)$$

In this equation, B_m is the number of bits transferred by the unit in the user's mth active time, t_m, and E is the differential energy per bit of the device. The B_m are given by:

$$B_m = \int_{t_m} R(t)\,dt \qquad (10.10)$$

We allocate all the energy expended during the service's active times to the service; therefore, the energy consumption of the service is given by

$$Q^{(svc)} = MP_{idle}\langle t \rangle + E\sum_{m=1}^{M} B_m = \frac{\langle t \rangle}{\langle t \rangle + \langle T \rangle}T_{tot}P_{idle} + EB^{(svc)} = \rho T_{tot}P_{idle} + EB^{(svc)} \qquad (10.11)$$

In this equation, $B^{(svc)} = \sum_m B_m$ is the total data transferred for the service of interest over duration T_{tot}, we have used $M(\langle T \rangle + \langle t \rangle) = T_{tot}$. The ratio of active time, $M\langle t \rangle$ to operational time $M(\langle t \rangle + \langle T \rangle)$ is written as:

$$\rho = \frac{\langle t \rangle}{\langle t \rangle + \langle T \rangle} \qquad (10.12)$$

This ratio is the proportion of time in which the user is actively accessing the service.
We can also define the average data rate for the service under study:

$$\langle R \rangle = \sum_{m=1}^{M}\int_{t_m} R(t)\,dt \left/ \sum_{m=1}^{M} t_m \right. \qquad (10.13)$$

Using this and noting that t_m and $R(t)$ can be considered independent random variables, we have:

$$B^{(svc)} = M \langle R \rangle \langle t \rangle = \rho T_{tot} \langle R \rangle \tag{10.14}$$

Again, this gives for the overall energy consumption of the service, as in (10.11):

$$Q^{(svc)} = \rho T_{tot} \left(P_{idle} + E \langle R \rangle \right) = \rho T_{tot} P_{idle} + EB^{(svc)} \tag{10.15}$$

If there is a reason for not allocating all the idle power, P_{idle}, to the service under study when that service is active (i.e. during the periods t_m), then we can introduce a factor ξ to represent the proportion of idle power allocated to that service. This may be appropriate if the equipment is simultaneously running management or operationally required functions in the background during times when the service of interest is active. In this case, (10.11) and (10.15) become:

$$Q^{(svc)} = \rho T_{tot} \left(\xi P_{idle} + E \langle R \rangle \right) = \rho T_{tot} \xi P_{idle} + EB^{(svc)} \tag{10.16}$$

Because the equipment is only intermittently actively running the service of interest, we can define the average power of the service either of two ways. Firstly, we can define the average power as the energy expended over the active time, ρT_{tot}, of the service, giving:

$$\left\langle P^{(svc)} \right\rangle_{\rho T_{tot}} = \xi P_{idle} + E \langle R \rangle \tag{10.17}$$

Secondly, the average power of the service can be defined over the total duration the equipment is operating, T_{tot}, giving:

$$\left\langle P^{(svc)} \right\rangle_{T_{tot}} = \rho \left(\xi P_{idle} + E \langle R \rangle \right) \tag{10.18}$$

From these equations, we can see that the service energy, $Q^{(svc)}$, and average power $\langle P^{(svc)} \rangle$ take on relatively simple and intuitive forms that do not require detailed information of each service interaction with the user.

We now consider the situation in which the service of interest is one of several services being provided via the CPE during the operational time T_{tot}. Times between the service of interest utilising the CPE, T_m, may no longer be idle. Rather the CPE may be utilised providing one of the other services. It is relatively easy to generalise this approach to the situation where the user is accessing other services, one at a time, during the times T_m. In this case, Figure 2 will be replaced by something like Figure 10.3.

In Figure 10.3 the time slices in the T_m periods represent different services being accessed by the user. We continue to focus on the service represented by the time periods t_m. We note that the times T_m, during which the service of interest is not accessed, can be expressed as:

$$T_m = \sum_{n=1}^{N_m} T_{n,m} \tag{10.19}$$

where $T_{n,m}$ is the time within time duration T_m during which the nth service (which does not include the service of interest to us) is accessed. The sum in (10.19) is over the sub-intervals $T_{n,m}$ of T_m during which the nth service is accessed. In total there are $N + 1$ activities, the service of interest plus N other activities, which may include the equipment being idle. Within duration T_m, if the nth service

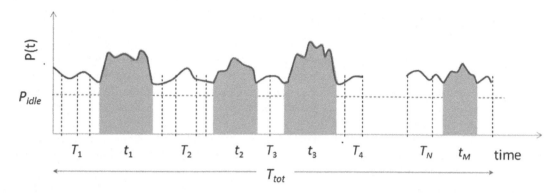

FIGURE 10.3 Example of power consumption as a function of time for user CPE that is sequentially providing multiple services over a given time duration.

is not active, then $T_{n,m} = 0$. With this, we see that all the equations above still stand and we only need change the definition of ρ to be based on the total average time the device is not dealing with the service of interest:

$$\langle T \rangle = \left\langle \sum_{n=1}^{N_M} T_{n,m} \right\rangle = \frac{1}{M} \sum_{m=1}^{M} \sum_{n=1}^{N_m} T_{n,m} \qquad (10.20)$$

For some CPE, the device operates at a maximum capacity line rate whenever it is actively exchanging data with the network. Let this fixed capacity line rate of the device be C_{line}, this being the line rate at which the equipment exchanges data with the network. With this, we have $R(t) = C_{line}$ for all active times including the t_m during which the equipment is providing the service of interest. Then we have:

$$C_{line} \sum_{m=1}^{M} t_m = B^{(svc)} \qquad (10.21)$$

Referring to (10.14) and noting that, with a fixed-line rate, we have $\langle R \rangle = C_{line}$, we have over the times during which the device is providing the service of interest:

$$B^{(svc)} = \rho T_{tot} C_{line} \qquad (10.22)$$

We can also define the effective data rate, $R^{(svc)} = B^{(svc)}/T_{tot}$, which is the overall average data transfer rate for the service of interest when we take the ratio of total bits delivered into the network divided by the total time taken to deliver them. We have:

$$R^{(svc)} = \frac{B^{(svc)}}{T_{tot}} = \rho C_{line} \qquad (10.23)$$

The quantity $1/\rho = C_{line}/R^{(svc)}$ is related to the burstiness of the service traffic [10, 11]. Using (10.23) in (10.15), we get:

$$P^{(svc)} = \frac{Q^{(svc)}}{T_{tot}} = \left(\frac{P_{idle}}{C_{line}} + E \right) R^{(svc)} \qquad (10.24)$$

10.4   GENERALISED SERVICE POWER MODEL AND DEALING WITH IDLE POWER

In the discussion above, allocating the incremental power $ER^{(svc)}$ to a service with throughput $R^{(svc)}$ came about rather naturally. In contrast, allocating idle power requires more careful consideration. Recall a brief mention in Section 10.3 that in some cases we may decide to allocate only a fraction of idle power during active service times to that service. This was encapsulated in (10.16) by introducing the parameter ξ to represent the proportion of idle power allocated to the service. The issue of allocating idle power to a service is more nuanced than that of allocating incremental power because it involves deciding if the service power model will be Attributional or Consequential. We now consider this issue in detail.

Our endeavour in this chapter is to construct a methodology that provides a quantitative value for the power consumption of a service. This endeavour is complicated by the fact that most network elements consume idle power irrespective of their throughput. The question then arises: "How is idle power, which is consumed irrespective of the traffic load, to be allocated to individual service flows as the magnitude and number of those flows varies over time?" This is where the distinction between Consequential and Attributional analyses comes into play.

To appreciate this, consider the power footprint of a new Internet service. This service will be introduced onto a network that is already carrying a plethora of other services and the additional traffic generated by the new service will be easily accommodated by the existing network. For simplicity, we consider a single network element, which has idle power, P_{idle}, and incremental energy per bit, E. Let the new service generate additional traffic, $R^{(svc)}(t)$, which adds to the existing traffic already being dealt with by the network element. The additional power consumption due to the traffic resulting from the new service is $R^{(svc)}(t)E$. Because idle power is constant, independent of the load on the network element, there is no increase in idle power consumption due to the new service. Given the definitions above, we can identify the increase in power $R^{(svc)}(t)E$ as "Consequential", because it represents a change in power footprint due to the new service.

If the provider of this new service is required to offset their service's power footprint, we may ask: "If we only consider the increase in power consumption, $R^{(svc)}(t)E$, who will be held responsible to offset the idle power, P_{idle}, of the network element?" Or: "What overall power consumption should we attribute to the new service, given the functions that contribute to the network element idle power are essential for that element to deal with the traffic arising from the new service?" These questions align with assessing the Attributional power footprint.

Most would agree that some form of attributional footprint, which includes allocation of idle power consumption, is more appropriate if we are to allocate a power footprint to all the services that are being provided via a telecommunications network. Accepting this approach, an important issue is how to make this allocation. We now consider several approaches to this.

The power consumption of a network element can be written in the form:

$$P(t) = P_{idle} + R(t)E \tag{10.25}$$

where P_{idle} is the idle power, E the incremental energy per bit and $R(t)$ is the throughput in bit/s.

We are considering the situation in which the total traffic, $R(t)$, will consist of multiple flows. Given we are focussing on the power consumption of a service, it is most convenient to separate the total traffic into separate data flows such that each data flow is identified with one of $N^{(svc)}$ services.

We are specifically interested in allocating a power consumption to the k th traffic flow, $R^{(k)}(t)$, which is one of $N^{(svc)}(t)$ at time t. The power consumption of this flow arising from the incremental energy per bit component can be straightforwardly set as $R^{(k)}(t)E$. The allocation of a component of the idle power to the k th flow is not as intuitively obvious.

We assume the total data flowing through the network element is the sum of multiple service flows. Therefore, we have:

$$R(t) = \sum_{k=1}^{N^{(svc)}} R^{(k)}(t) \qquad (10.26)$$

We wish to construct a relationship between the flow, $R^{(k)}(t)$, and a portion of total network element power, $P(t)$, which can be allocated to the k th flow. Most likely this power allocation will be used to calculate the environmental footprint of the services whose data flows are being dealt with by the network element. Therefore, we would expect the allocations to sum to the total power, $P(t)$, of the network element to ensure the allocation covers its entire footprint. Therefore, to make this construction we adopt the requirement:

1) Sum of the power allocated to the flows, $P^{(k)}(t)$, equals the total element power:

$$\sum_{k=1}^{N^{(svc)}} P^{(k)}(t) = P(t) \qquad (10.27)$$

With this requirement, we adopt a simple form for power allocation:

$$P^{(k)}(t) = h^{(k)}(t) P_{idle} + ER^{(k)}(t) \qquad (10.28)$$

where $h^{(k)}$ is a function that allocates an amount of idle power to the k th service. We require:

$$\sum_{k=1}^{N^{(svc)}(t)} h^{(k)}(t) = 1 \qquad (10.29)$$

An obvious candidate for $h^{(k)}(t)$ is:

$$h^{(k)}(t) = \frac{R^{(k)}(t)}{R(t)} \qquad (10.30)$$

Another option is:

$$h^{(k)}(t) = \frac{1}{N^{(svc)}(t)} \qquad (10.31)$$

There are other options (e.g. a slight variant of (10.30) is discussed below). However, in terms of simplicity, (10.30) and (10.31) represent the simplest approaches to allocating idle power over multiple services.

Comparing the two, we note that (10.31) equally shares idle power across all the flows. However, it is highly unlikely that all flows will be of equal magnitude. Service providers with relatively small data rate requirements are likely to object to being allocated the same idle power footprint as service providers who have significantly larger data rate requirements.

Therefore, we will also adopt a second requirement:

2) The power consumption allocated to a flow is dependent upon the flow magnitude:

$$P^{(k)}(t) = P^{(k)}\left(R^{(k)}(t)\right) \tag{10.32}$$

In this equation, $R^{(k)}(t)$ may not be the actual flow magnitude at time t; rather it may be the data rate reserved for the service that is provided by the flow. For example, if the service uses an MPLS tunnel, then the data rate committed to that tunnel will be used. In contrast, if the service is a best-effort service, then there is no committed bit rate and $R^{(k)}(t)$ will be whatever is sustainable for that service at time t.

With large core routers and switches, there may be many thousands of flows. Therefore, keeping track of the number of separate flows, $N^{(svc)}(t)$, may not be easily implemented. In contrast, the traffic parameters required to implement (10.30) are most likely available via network element management. Therefore, we adopt (10.30) on the basis that a service that demands a larger flow should be attributed a larger proportion of the total power consumption.

The Greenhouse Gas Protocol, "ICT Sector Guidance built on the GHG Protocol Product Life Cycle Accounting and Reporting Standard", has adopted the form (10.30) for allocation of power over multiple services through a network element [1]. In addition (10.30) is in accord with (10.24) in that it uses a ratio of data rates rather than the number of users.

Adopting (10.30) gives:

$$P^{(k)}(t) = P_{idle}\frac{R^{(k)}(t)}{R(t)} + ER^{(k)}(t) = \left(\frac{P_{idle}}{R(t)} + E\right)R^{(k)}(t) \tag{10.33}$$

Equation (10.33) thus describes the power consumption $P^{(k)}(t)$ attributable to a service k on the basis of the traffic $R^{(k)}(t)$ generated by the service. By extension, provided that is small compared with the total traffic $R(t)$, we can write:

$$\delta P^{(k)}(t) = \left(\frac{P_{idle}}{R(t)} + E\right)\delta R^{(k)}(t) \tag{10.34}$$

where $\delta R^{(k)}$ represents the increment in $R^{(k)}(t)$ and $\delta P^{(k)}$ the corresponding change in power allocated to the k th service. This would be applicable in adding a single user service to the traffic level in a high capacity link or router, however if dealing with situations where the traffic level $R^{(k)}(t)$ is significant compared with $R(t)$, or where other service traffic levels are varying, the analysis needs to be more sophisticated.

Another advantage of the form (10.30) is that a commonly used energy efficiency metric, $H(t)$, for a network element is the ratio Power/throughput. That is:

$$H(t) = \frac{P(t)}{R(t)} = \frac{P_{idle}}{R(t)} + E \tag{10.35}$$

Adopting this metric, the power consumption of that network element due to the k th service has the simple form:

$$P^{(k)}(t) = H(t)R^{(k)}(t) \tag{10.36}$$

Adopting (10.30), even if $E = 0$, we will have a time-dependent power consumption for the k th service, as its throughput, $R^{(k)}(t)$, changes over time. This means the power consumption allocated to a service can vary over time, even if the network element's power consumption is constant.

In the above, we have adopted an approach in which the network element power is well represented by the simple linear function in (10.25). However, for some network elements the power consumption is not a linear function of the throughput in bit/s. In Chapter 4, it was shown that mobile base station power consumption is a linear function of the resource elements transmitted in each symbol duration. This is related to the data throughput (in bit/s) by the spectral efficiency (bit/s/Hz) of the transmissions to users.

Although the form (10.33) has several advantages, it has to be used with care, as we shall now show. We can write (10.33) in the form:

$$P^{(k)} = \left(\frac{P_{idle}}{R} + E \right) R^{(k)} \tag{10.37}$$

where we have suppressed the time dependence for simplicity. Because the total throughput traffic, R, is a function of the service traffic flows, any change in the traffic of any service (including the k th) will result in a change in R. Therefore, the change in power consumption allocations that occur when the k th service traffic changes needs to be carefully calculated. Let the k th service traffic level change by a small amount $\delta R^{(k)}$. Applying differential analysis, we have for the corresponding change in power allocated to the k th service, $\delta P^{(k)}$:

$$\delta P^{(k)} = \frac{dP^{(k)}}{dR^{(k)}} \delta R^{(k)} = \left(\frac{\partial P^{(k)}}{\partial R} \frac{dR}{dR^{(k)}} + \frac{\partial P^{(k)}}{\partial R^{(k)}} \right) \delta R^{(k)} \tag{10.38}$$

Using (10.36), this gives:

$$\delta P^{(k)} = \left(-P_{idle} \frac{R^{(k)}}{R^2} + \frac{P_{idle}}{R} + E \right) \delta R^{(k)} = \left(\frac{P_{idle}}{R} \left(1 - \frac{R^{(k)}}{R} \right) + E \right) \delta R^{(k)} \tag{10.39}$$

In some network elements, the differential energy $E \approx 0$; consequentially, the idle power allocation alone determines the power change due to a change in service throughput in the machine.

We also note that R being present in (10.37) means a change in throughput of the p th service ($p \neq k$) can result in a non-zero change in power allocation, $P^{(k)}$, to the k th service. That is, we have the change in power consumption of the k th service due to a change in throughput of the p th service. Using (10.37) we have for the change in the power consumption allocated to the k th service due to change in traffic, $\delta R^{(p)}$, of the p th service:

$$\delta P^{(k,p)} = \frac{\partial P^{(k)}}{\partial R^{(p)}} \delta R^{(p)} = \frac{\partial}{\partial R^{(p)}} \left(\frac{P_{idle}}{R} + E \right) R^{(k)} \delta R^{(p)} = \frac{-P_{idle}}{R^2} R^{(k)} \delta R^{(p)} \tag{10.40}$$

From this we see that, if we share the idle power across all services in proportion to their current traffic throughput as described in (10.30), the power consumption allocated to the j th service will change when the throughput of any other service changes. In other words, adopting this approach for allocating service power consumption will result in a change in traffic of the p th service causing a change in the power consumption footprint allocated to the k th service. It is unlikely that the provider of service k will consider this to be reasonable, because they may be penalised when another service provider reduces their service traffic through the network element. The decision of the p th service provider will almost certainly be beyond the control of the k th service provider. We consider this issue in Section 10.4.1.

If the k th service provider's service traffic is much smaller than the total traffic through the network element, we have $R^{(k)}/R \ll 1$. In this circumstance, we can ignore terms of the form $R^{(k)}/R$ in (10.38) and (10.39), giving:

$$\delta P^{(k)} \approx \left(\frac{P_{idle}}{R} + E \right) \delta R^{(k)} \qquad\qquad R^{(k)} \ll R$$
$$\delta P^{(k,p)} \approx 0 \qquad\qquad R^{(k)} \ll R \tag{10.41}$$

From this we see that only in the limit of very small service traffic ($R^{(k)} \ll R$) does the proportional allocation of idle power proposed in (10.30) provide a power footprint allocation that is independent of the other services using that network element.

From these results, we see that, although the principle behind attributing idle power to users via (10.30) may appear to be sensible and equitable, it is problematic for service providers who wish to determine a power consumption allocation for their service. We now consider a modified approach that endeavours to addresses this issue.

10.4.1 BRINGING IN THE NETWORK OPERATOR

Viewing the equations above, we note that the reason for the power allocation of a flow being dependent upon the magnitude of other flows is the change in utilisation that occurs when any service flow changes. This can be avoided by adopting a fixed value for total traffic, $R(t)$, while still allowing the service traffic flows, $R^{(k)}(t)$, to vary. A reason for using an Attributional model for service traffic flows is to enable service providers to assess strategies they may have for reducing their power footprint (and hence carbon footprint). Apart from service providers, there is another stakeholder in the overall quest to reduce power consumption, that being the network operator. Just as service providers have an interest in reducing the power footprint of their service, network operators also have an interest in reducing the power footprint of their network. Therefore, making an allocation of idle network element power to the network operator is in accord with the principles of improving energy efficiency. However the network operator needs to deal with several other issues which affect the energy efficiency of the services delivered.

Whereas the equations developed in section 10.4 are based on the total actual network traffic R, it is tempting to use the total traffic average as a basis for idle power allocation. However in practice the network is invariably over-dimensioned compared with the average traffic, for several reasons. The various components of traffic show marked diurnal cycles, for example business traffic peaking during the day and domestic traffic, especially entertainment, peaking during the evening. Within the busy hour for each traffic class, traffic is subject to short-duration fluctuations potentially creating congestion. Special events also often create increased demand. Furthermore, network links need to include redundancy, in order to carry offloaded traffic arising from faults in other parts of the network. All of these factors have an impact on the Quality of Service, which in many countries is monitored by the telecommunications regulator. The operator would also, when upgrading a network, make some capacity provision to accommodate short-term traffic growth. All of these lead to a network providing significantly greater traffic capacity than is needed for most of each day, and thus more infrastructure with higher power consumption than the average traffic level would demand. It is suggested in [12] that average traffic levels typically amount to about 50% of network capacity. That is the network capacity is "over-provisioned" such that average traffic levels are half of the network capacity. The level of over-provisioning is a decision made by the network owner.

This over-provisioning is included in the energy consumption of a service in some work (such as [1]), but this has the effect of passing the energy cost through to the end user of a service. However as the network operator is largely in control, from building the network to balancing Quality of Service requirements against cost and energy consumption, it is appropriate for the operator to incur part of that additional energy cost.

The approach we adopt is to implement a bandwidth allocation to the operator of the network element. This allocation is reflects the over-provisioning set by the network operator. In this approach, in place of (10.26), we define a fixed traffic level for the network element, given by:

$$R_{fixed} = \sum_{k=1}^{N^{(svc)}} R^{(k)}(t) + R^{(op)}(t) = R(t) + R^{(op)}(t) \tag{10.42}$$

where R_{fixed} is set at a predetermined value greater than the maximum of $R(t)$, and $R^{(op)}(t)$ is the difference between the total traffic of all the services at time t and R_{fixed}. The traffic value $R^{(op)}(t)$ is allocated to the network operator, just as $R^{(k)}(t)$ is allocated to the k th service.

Typically the value of R_{fixed} would be the maximum capacity of the network element, $R_{fixed} = C_{max}$, however there may be reasons for adopting a different value. In this case, the operator of the network element takes responsibility for all the unused capacity through the element. This means in place of (10.33) we have:

$$P^{(k)}(t) = \left(\frac{P_{idle}}{R_{fixed}} + E \right) R^{(k)}(t) \tag{10.43}$$

In (10.43), the first term in the brackets is the attributed idle power consumption of the service and the second term is the incremental power consumption. Because R_{fixed} = constant, in place of (10.39) and (10.40) we get, respectively:

$$\Delta P^{(k)} = \left(\frac{P_{idle}}{R_{fixed}} + E \right) \Delta R^{(k)} \tag{10.44}$$

$$\Delta P^{(k,p)} = 0$$

Therefore, under this approach, the power consumptions of the services are independent.

Even though the operator of the equipment is not allocated any incremental power consumption of the network element, the operator is allocated an amount of the idle power:

$$P^{(op)}(t) = P_{idle} \frac{R^{(op)}(t)}{R_{fixed}} = P_{idle} \frac{R_{fixed} - R(t)}{R_{fixed}} = P_{idle} \left(1 - \frac{R(t)}{R_{fixed}} \right) \tag{10.45}$$

This allocation can be justified on the basis that the operator of the network element chooses the make and model, configuration and R_{fixed} of the element to deal with the traffic it must carry. Therefore, the operator needs to take on some of the responsibility of the power (and hence carbon) footprint of the machine.

The desire to reduce the resulting (OPEX) cost of that energy/carbon could be seen as a driver to reduce the unused capacity of the machine (i.e. decrease R_{fixed}) and/or reduce the idle power consumption of the machine.

In cases of a network operator providing carriage and switching services to large volume users, e.g. corporates, second-tier carriers or Internet Service Providers, it has been common practice to provide services on the basis of a fixed capacity. For example, a service user might purchase a 10 Gbit/s stream, which although it may not be fully used at times, would incur the capacity-based financial and idle energy cost. That is appropriate given that the bulk of the energy consumption arises from the idle power consumption of router line cards and ports which are dedicated to that user.

There is however a trend toward dynamic bandwidth adjustment techniques, which often exploit the capabilities of software defined networking [13–19]. These are claimed to better utilise the underlying physical infrastructure in the network, compared with established fixed capacity allocation models, and would thus lead to reduced network energy consumption. In turn, more flexible network service power allocation models would be used. These models would broadly follow the diurnal power consumption models to follow in Section 10.7.

10.4.2 MOBILE NETWORKS

From Chapter 4, the power consumption of a mobile base station can be expressed in the form:

$$P_{BS}\left(\mu_{BS}\right) = P_{idle} + \mu_{BS}\left(P_{max} - P_{idle}\right) = P_{idle} + \mu_{BS}\Delta_P P_{max,out} \tag{10.46}$$

where P_{max} and P_{idle} are the power consumed by the base station under maximum utilisation ($\mu_{BS} = 1$) and zero utilisation ($\mu_{BS} = 0$), respectively; $P_{max,out}$ is the maximum transmit power from the base station antennas; and Δ_P is a scaling factor relating transmit power to base station consumed power.

Constructing an estimate for the power or energy consumption of a service provided through a mobile network is more complicated than the models described above. Power consumption of a base station is expressed as a function of utilisation, μ_{BS}, which is defined in terms of the number of PDSCH REs used to transport data to the user equipment:

$$\mu_{BS}\left(t\right) = \frac{N_{PDSCH}\left(t\right)}{N_{max,PDSCH}} \tag{10.47}$$

In this equation, $N_{PDSCH}(t)$ is the number of PDSCH REs allocated at time t to transmit data to all active users and $N_{max,PDSCH}$ is the maximum number of PDSCH REs. The utilisation is a dimensionless parameter, which is different to (10.25) which expresses equipment power consumption as a function data throughput, $R(t)$, which has dimensions of bit/s.

With a mobile network, the utilisation of the p th user can be expressed as (4.52):

$$\mu_{BS}^{(p)} = \frac{N_{RF,PDSCH}^{(p)}\left(t\right)}{N_{max,RF,PDSCH}} \tag{10.48}$$

Where $N_{RF,PDSCH}^{(p)}\left(t\right)$ is the number of PDSCH REs allocated to transmit data to the p th user in a Radio Frame. We have for the utilisation of the base station (4.53):

$$\mu_{BS}\left(t\right) = \sum_{p=1}^{N^{(usr)}} \mu_{BS}^{(p)}\left(t\right) \tag{10.49}$$

The data rate for transmission of data to the individual p th user is related to that user's utilisation via (4.56):

$$R^{(p)}\left(t\right) = \mu_{BS}^{(p)}\left(t\right) SE\left(z^{(p)},t\right) BW \sigma_{PDSCH} \tag{10.50}$$

where $SE(z^{(p)},t)$ is the spectral efficiency (bit/s/Hz) at the location of the p th user and $BW\sigma_{PDSCH}$ is the base station bandwidth available for data transmission to active users.

The utilisation definition in (10.48) for the users can also be generalised to define the utilisation of the k th service. Doing so, we recast (10.49) to give the relationship between the utilisation of the services to the overall base station utilisation, to give:

$$\mu_{BS}(t) = \sum_{k=1}^{N^{(svc)}} \mu_{BS}^{(k,svc)}(t) \tag{10.51}$$

where $N^{(svc)}$ is the number of services being dealt with by the base station and, $\mu_{BS}^{(k,svc)}$ is the utilisation for the k th service (not k th user), given by:

$$\mu_{BS}^{(k,svc)} = \sum_{p=1}^{N^{(k,svc)}} \mu_{BS}^{(k,p)} \tag{10.52}$$

In this equation, $N^{(k,svc)}$ is the number of users of the k th service that are active in the mobile base station service area, and $\mu_{BS}^{(k,p)}$ is the p th user's utilisation of the k th service.

Unlike utilisation, the total data rate of a service is not a simple recasting of (10.50) because the users of a given service are at locations, $z^{(p)}$, scattered around the base station service area. This means the spectral efficiency is likely to vary across the users of a service. Therefore, the total throughput of the k th service, $R^{(k)}$, will have the form:

$$R^{(k)}(t) = BW\sigma_{PDSCH} \sum_{p=1}^{N^{(k,svc)}} \mu_{BS}^{(k,p)} SE\left(z^{(k,p)},t\right) \tag{10.53}$$

where $SE(z^{(k,p)},t)$ is the spectral efficiency for the p th user of the k th service. We consider the issue of throughput for a service in greater detail in Chapter 10. For now, we focus on the power consumption of a service as part of the traffic provided by a mobile base station.

With this in mind, referring back to (10.46), the attribution of incremental power to the k th service is intuitive, being $\mu_{BS}^{(k,svc)} \Delta_P P_{max,out}$. For the allocation of idle power, we adapt the approach used above, based on "ICT Sector Guidance built on the GHG Protocol Product Life Cycle Accounting and Reporting Standard" [1] to utilisation. With this, we modify (10.33) by replacing the data rates, $R(t)$ and $R^{(k)}(t)$, with μ_{BS} and $\mu_{BS}^{(k,svc)}$, respectively. This gives the base station power attributed to the k th service as:

$$P^{(k,svc)}\left(\mu_{BS}^{(k,svc)}\right) = \left(\frac{P_{idle}}{\mu_{BS}} + \Delta_P P_{max.out}\right)\mu_{BS}^{(k,svc)} \tag{10.54}$$

With this allocation, we will find the same issues as above with allocation in (10.33) in terms of the cross-effects of one service changing its utilisation impacting the utilisation of all the other services. Similarly, we can address this by introducing the network operator via a fixed utilisation, $\mu_{fixed,BS}$, such that the network operator picks up the energy footprint corresponding to (10.45), which in this case will be:

$$P_{BS}^{(op)}(t) = P_{idle}\left(1 - \frac{\mu_{BS}(t)}{\mu_{fixed,BS}}\right) \tag{10.55}$$

10.5 SELECTING A MODEL

Focussing on the load-dependent component of network element power consumption, the derivation of service power consumption based on "time sharing" of equipment over multiple applications, as depicted in Figure 10.3 and equation (10.3), gives the power consumption equations (10.17) and (10.18). In contrast, derivation of service power based on "throughput sharing", as depicted in (10.26), gives form (10.33) or (10.43). This raises the question of which model should be used when considering a given situation. When should one use the "time-sharing" service power model (based on $\langle t \rangle$) or the "throughput-sharing" service power model (based on $R^{(k)}$)?

There are multiple factors that influence the type of model we use to assess the energy consumption of a service. When estimating the energy consumption of a network, we would typically adopt models based upon the total traffic flow through the network element. In contrast, estimating the energy consumption of a service requires a model that will select out the specific traffic flow for that service. Several factors that should be considered when deciding between time-sharing and throughput-sharing models are as follows:

1. Measurement: It may not be possible to easily undertake the measurements required for one or the other of these two approaches. For example, the traffic-based model is much easier to implement with core routers, because the management system typically provides a traffic log recording the packets/s, bit/s throughput of the network element. These management systems do not provide a time log for every flow through the machine. When undertaking (power) measurements on edge and core network elements, the input into these elements is often provided by a traffic generator. Traffic generators typically measure the flows they generate in bits or packets per second. This is better suited to a throughput-sharing model.

   In contrast, a time log is often relatively easy to implement with a customer device such as a tablet, notebook, smartphone or other CPE. In this case, a time-sharing approach is likely to be easier to implement as has been done for several publications on power consumption of cloud services [9, 20, 21]. Constructing a traffic log requires extracting and processing data from a network packet analyser to calculate the packet/bit rate of the service.

2. The use cases of equipment can be organised into the following:
   a. A network element that simultaneously deals with many data streams. For example, large capacity routers can be processing a large number, $N_{shared} \gg 1$, of parallel or simultaneous service flows. For example, large core routers and switches having many simultaneously active input and output ports.
   b. A network element, such as CPE, that deals with a single service at a time. In this case, there is only one flow. The device may deal with multiple services, but will do so sequentially in time, not simultaneously.

3. The number of input/output ports for the machine: For CPE there is typically only one access network facing port and the device deals with different services sequentially in time. In contrast, for aggregation, edge and core routers, there are likely to be many ports and the network element deals with many services simultaneously.

4. The number of functions the machine undertakes. For example, a laptop's overall power consumption may be the cumulative effect of running several applications apart from exchanging packets with the network to which it is connected. In contrast, the incremental power consumption of a core router is primarily due to its routing of packets and little else. The allocation of power consumption of a laptop to a service may need to account for the power consumed by the other applications. Typically, this is done on a time-sharing basis. A router's power allocated to the different packet flows is easiest done on a throughput-sharing basis.

The forms for the energy/power equations for time-sharing and throughput-sharing power models are summarised in Table 10.1.

TABLE 10.1

Guidance for the Choice of Power Model for a Network Element

	Time-Sharing Model	Throughput-Sharing Model
Power/Energy equation	$P^{(svc)}(t) = \left(\dfrac{P_{idle}}{C_{line}} + E \right) R^{(svc)}(t)$	$P^{(k)}(t) = \left(\dfrac{P_{idle}}{R(t)} + E \right) R^{(k)}(t)$
Number of simultaneous service data flows	1	$N_{shared} \gg 1$
Number of input/output ports	1	Many

Equipment that has only a single service flow at any instant in time (but may allocate different times to different services) can be modelled using the "time-sharing" approach. Equipment that deals with many services simultaneously can be modelled using the "throughput-sharing" approach.

In this section, we focussed on the power consumption of a service in a single device, such as user customer equipment or a router or switch that has many service flows. Almost all service traffic flows travel through multiple devices as they make their way through a network. Using the models derived above, we now consider the issue of power consumption of transporting service traffic through a network.

10.6 SERVICE TRANSPORT POWER CONSUMPTION IN A NETWORK

When considering the power consumption of a service transported via a network, we distinguish between "consequential" and "attributional" assessments. As the popularity of Internet services has grown, there has been an ever-increasing collection of new services becoming available to end users. Also, many existing services have experienced continued increases in popularity. This raises questions such as: "What is the additional power consumption of the Internet due to the appearance and/or growth of a given service?" Questions such this can be addressed using a consequential assessment of the service's power consumption.

In contrast, one may ask: "What is the overall power consumption of a given service?" In this case, we are not looking at the increase (as in the previous paragraph), but some form of allocation of total power consumption of a service. This is a question that is better addressed using an "attributional" assessment.

This raises the question of how best to address consequential and attributional questions and, therefore, what kind of power consumption modelling is best suited for each of these types of questions.

We will consider attributional power assessment first. As discussed in Section 10.4, attributional assessment includes allocating a portion of equipment idle power to a service. Despite the issues discussed in Section 10.4, because the Greenhouse Gas Protocol "ICT Sector Guidance built on the GHG Protocol Product Life Cycle Accounting and Reporting Standard" [1] is currently the best candidate for an international standard on calculating the power footprint of a service, we shall allocate the idle power using to (10.30) in accordance with that standard.

To construct an equation for the overall network power of a service, we start with (8.30). This equation provides an estimate of total network power consumption by summing over services and network elements. To construct an equation for total network power of a service, we remove the sum over services to pick out only the k th service contributions. This gives:

$$P_{Ntwk}^{(k)}(t) = M_{OH} \sum_{j=1}^{N_E} \left(\frac{P_{idle,j}}{R_j(t)} + E_j \right) \alpha_j^{(k)} R^{(k)}(t) \tag{10.56}$$

Using (8.11) to express the sum using its average and (8.31), we get:

$$
\begin{aligned}
P_{Ntwk}^{(k)}(t) &\approx M_{OH} N_E^{(k)} \left(\langle P_{idle} \rangle_E \left\langle \frac{1}{R(t)} \right\rangle_E + \langle E \rangle_E \right) R^{(k)}(t) \\
&\approx M_{OH} \left(N_{hops}^{(k)} + 1 \right) \left(\langle P_{idle} \rangle_E \left\langle \frac{1}{R(t)} \right\rangle_E + \langle E \rangle_E \right) R^{(k)}(t)
\end{aligned}
\tag{10.57}
$$

where we recall that $\langle X(t) \rangle_E$ is the average of $X_j(t)$ over the network elements and $N_{hops}^{(k)}$ is the number of hops in the path of the k th service. The form for $P_{Ntwk}^{(k)}(t)$ is complicated by the presence of the $\langle 1/R(t) \rangle_E$ term. It should be noted that, except for special situations, this does not have the same value as $1/\langle R(t) \rangle_E$. An approach to deal with this term is discussed in Section 10.7.

Moving our attention to consequential power allocation, the growth of the Internet is driven by the increasing popularity of the services for which it provides a platform. When considering the power or carbon footprint of services provided via the Internet, there are two issues that have generated interest:

1. The power footprint of a single user accessing a service. For example, over recent years there has been on ongoing debate regarding the energy consumption of a Google search [22, 23]. Estimates for the carbon footprint of a Google search range from 7-10 gm at the high end to 0.2 gm (Google estimate) at the low end [16]. Other examples include the calculation of the power consumption of specific instances of using cloud-computing services [6, 20, 25, 26].

2. The total power consumption of a network service. For example, the network energy consumption of photo-sharing via Facebook has been estimated to be approximately 60% of the total power consumption of all the IT equipment in Facebook data centres [6]. More recently, global annual energy consumption of "smart home applications" has been estimated [20].

We can undertake these calculations using expressions for total network power (8.31) and (8.34), reproduced here:

$$
\begin{aligned}
P_{Ntwk}(t) &\approx M_{OH} \left(N_E \langle P_{idle} \rangle_E + N_E \langle E \rangle_E \langle \alpha \rangle R_{Ntwk}(t) \right) \\
&\approx M_{OH} \left(\langle N_{hops} \rangle^{(svc)} + 1 \right) \left(\frac{\langle P_{idle} \rangle_E}{\mu \langle C_{max} \rangle_E} R_{peak} + \langle E \rangle_E R_{Ntwk}(t) \right)
\end{aligned}
\tag{10.58}
$$

For simplicity we will assume the network operator adopts the same network element utilisation policy across the transport (edge and core) networks: i.e. μ is uniform across all network elements. We consider two cases:

1. The increase in service traffic, ΔR, is small enough for it to be accommodated by allowing an increase in utilisation of already deployed equipment without deploying additional equipment [8]. To avoid the deployment of additional equipment, the additional traffic, ΔR, needs to be smaller than the average spare capacity per network element, $\langle C_{max} \rangle_E - \langle R \rangle_E$. We need to account for the fact that, on average, the traffic of a service travels through $\langle N_E \rangle^{(svc)} = (\langle N_{hops} \rangle^{(svc)} + 1)$ elements in the network. Therefore, the total additional capacity that must be accommodated by the network is $(\langle N_{hops} \rangle^{(svc)} + 1)\Delta R$. Adopting the uniform utilisation policy, we assume the network traffic is relatively evenly distributed over the N_E elements of the network and we note from (8.28) that the additional traffic per network element with increase in network traffic from R_{Ntwk} to $R_{Ntwrk} + \Delta R$, will be $\langle \alpha \rangle \Delta R$. This must be less than the average spare capacity

per network element. Therefore, this case corresponds to traffic increase ΔR satisfying the condition:

$$\langle \alpha \rangle \Delta R = \frac{\langle N_{hops} \rangle^{(svc)} + 1}{N_E} \Delta R < \langle C_{max} \rangle_E - \langle R \rangle_E \ll R_{Ntwk} \qquad (10.59)$$

In this scenario, the number of network elements, N_E, is kept constant and the increase in network traffic is $\Delta R_{Ntwk} = \Delta R$. Using the second line of (10.58), the increase in network power consumption will be:

$$\Delta P_{Ntwk}\big|_{N_E} = M_{OH} \left(\langle N_{hops} \rangle^{(svc)} + 1 \right) \langle E \rangle_E \Delta R \qquad (10.60)$$

where $\Delta P_{Ntwk}|_{NE}$ represents the change in network power due to a small change in traffic, ΔR.

2. This case considers the change in power when the increase in service traffic, ΔR, is too large to be accommodated by merely increasing the throughput per network element as in case 1) above. That is, the increase in traffic is large enough to require the deployment of additional equipment. In other words, ΔR is too large to be accommodated by merely increasing the utilisation of the network elements. That is:

$$\langle \alpha \rangle \Delta R > \langle C_{max} \rangle_E - \langle R \rangle_E \qquad (10.61)$$

To construct this estimate of the change in power, we assume the average operational utilisation of the network elements, given by $\mu_{op} = \langle R \rangle_E / \langle C_{max} \rangle_E$, is kept constant. In this case, we apply the differentiation chain rule to the first line of (10.58). We have:

$$\Delta P_{Ntwk}\big|_{\rho_{op}} = \frac{\partial P_{Ntwk}}{\partial N_E} \frac{\partial N_E}{\partial R_{Ntwk}} \Delta R \qquad (10.62)$$

To evaluate this expression, we apply the first partial derivative, $\partial P_{Ntwk}/\partial N_E$ to the first line in (10.56). For the second partial derivative, $\partial N_E/\partial R_{Ntwk}$, we note from (8.32) that the number of network elements can be expressed as:

$$N_E \approx \frac{\left(\langle N_{hops} \rangle^{(svc)} + 1 \right) R_{Ntwk}}{\mu_{op} \langle C_{max} \rangle_E} \qquad (10.63)$$

Using these results, we get:

$$\begin{aligned}
\Delta P_{Ntwk}\big|_{\mu_{op}} &= M_{OH} \left(\langle P_{idle} \rangle_E + \langle E \rangle_E \langle \alpha \rangle R_{Ntwk} \right) \frac{\langle N_{hops} \rangle^{(svc)} + 1}{\mu_{op} \langle C_{max} \rangle_E} \\
&= M_{OH} \frac{\langle N_{hops} \rangle^{(svc)} + 1}{\mu_{op}} \left(\frac{\langle P_{idle} \rangle_E}{\langle C_{max} \rangle_E} + \mu_{op} \langle E \rangle_E \right) \Delta R
\end{aligned} \qquad (10.64)$$

In case 1), we note that the incremental total network power will be negligible if the incremental energy per bit for all the equipment is small (i.e. $\langle E \rangle_E \approx 0$). Also, note that the results (10.60) and (10.64) are an estimate of the increase in overall network power due to the introduction of additional traffic generated by a service. This is not necessarily the same as the power attributed to a service, because that depends on how the idle power is attributed to services as discussed in Section 10.4.

10.7 DIURNAL CYCLE SERVICE POWER CONSUMPTION MODELLING

In many cases we are interested in the total energy consumption of a service over a duration, T, such as a diurnal cycle or year. Network element traffic and service traffic both exhibit diurnal-cycle variation. However, it may occur that the peak and minimum traffic times of one does not correspond to that of the other. For example, the traffic generated by a data back-up service may have its peak time in the early hours of the morning (say 2am to 3am). If this traffic is dealt with by a router that carries video streaming traffic and typically has its peak traffic time during the middle evening (say 8pm to 11pm), then the data back-up service will be "out-of-phase" with the peak traffic of the router. This phase relationship can have a significant impact on the attributional power consumption of a service. By "phase relationship" we mean the time delay between the peak service traffic and peak network element traffic. Most diurnal cycles have a characteristic shape, in that they typically vary in a roughly cyclical manner between the maximum (peak) traffic value and the minimum.

In this section, we use this characteristic traffic profile to construct several simplified models for service traffic flow, relative to total network element traffic, that enables the calculation of an approximate attributional energy consumption of a service.

Note that because this section largely concerns the relationship between the timing of certain services and the timing of the total network traffic, the actual total traffic at any time $R(t)$ has been included in these equations, not the form R_{fixed} as introduced in Section 10.4.1. In cases where service capacity can be varied or reallocated according to demand levels using dynamic bandwidth adjustment or similar techniques this is appropriate.

In situations where a service allocations are fixed or the balance of unused network capacity and energy cost is assigned to the network operator, R_{fixed} should replace the variables $R(t)$ and the abbreviated form R. This constant value for capacity leads to greatly simplified equations and their handling.

From (10.33) and (10.57), we see that the attributional power consumption of the k th service can be expressed using the form:

$$P^{(k)}(t) = \left(\frac{P_{idle}}{R(t)} + E \right) R^{(k)}(t) \tag{10.65}$$

To calculate the energy consumption or average power consumption of the service, we integrate over a set duration, T:

$$
\begin{aligned}
Q^{(k)}(T) &= P_{idle} \int_0^T \frac{R^{(k)}(t)}{R(t)} dt + E \int_0^T R^{(k)}(t) dt \\
&= P_{idle} \int_0^T \frac{R^{(k)}(t)}{R(t)} dt + E B^{(k)}(T)
\end{aligned} \tag{10.66}
$$

where $B^{(k)}(T)$ is the total number of bits for the service over time duration T. The mean power consumption $\langle P^{(k)} \rangle_T = Q^{(k)}(T)/T$ is given by:

$$
\begin{aligned}
\left\langle P^{(k)} \right\rangle_T &= \frac{P_{idle}}{T} \int_0^T \frac{R^{(k)}(t)}{R(t)} dt + E \frac{B^{(k)}(T)}{T} \\
&= P_{idle} \left\langle \frac{R^{(k)}}{R} \right\rangle_T + E \left\langle R^{(k)} \right\rangle_T
\end{aligned} \tag{10.67}
$$

Note that the average power involves the average of a quotient. This cannot be split into a quotient of averages unless $R(t)$ or R are constant.

A way around this is to focus on the term $R^{(k)}(t)/R(t)$ and its average value. This term is the traffic through the network element due to the service relative to the total traffic through the network element. We have:

$$0 \leq \frac{R^{(k)}(t)}{R(t)} \leq 1 \tag{10.68}$$

Hence:

$$0 \leq \left\langle \frac{R^{(k)}}{R} \right\rangle \leq 1 \tag{10.69}$$

The total throughput of the element $R = R^{(k)} + \sum_{p \neq k} R^{(p)} = R^{(k)} + \hat{R}$, where $\hat{R} = \sum_{p \neq k} R^{(p)}$ is all the other traffic through the element that is not due to the k th service. We now consider several profiles for $R^{(k)}(t)$ relative to $R(t)$, the total element throughput. We consider four cases in which service traffic $R^{(k)}(t)$ is expressed as a function of total traffic $R(t)$. That is:

$$R^{(k)}(t) \approx R^{(k)}\big(R(t)\big) \tag{10.70}$$

We consider these four cases.

1. The k th service traffic is approximately constant over the diurnal cycle. Because the service traffic must always be less than or equal to $R(t)$ over the diurnal cycle, we require $R^{(k)}(t) \leq R(t_{min})$, where t_{min} is the time in the diurnal cycle of minimum total traffic. That is, t_{min} is such that $R(t_{min}) = \min\{R(t)\}$. In this case, $R^{(k)}(t) \approx \gamma R(t_{min})$ where $0 \leq \gamma \leq 1$. Therefore:

$$\frac{R^{(k)}}{R(t)} \approx \frac{\gamma R(t_{min})}{R(t)} \tag{10.71}$$

which gives:

$$\left\langle \frac{R^{(k)}}{R} \right\rangle_T \approx \left\langle \frac{1}{R} \right\rangle_T \gamma R(t_{min}) \tag{10.72}$$

Using (10.66), the total energy consumption of the k th service is:

$$Q^{(k)}(T) = P_{idle} \gamma T \left\langle \frac{1}{R} \right\rangle_T R(t_{min}) + EB^{(k)}(T) \tag{10.73}$$

2. The k th service traffic is in-phase with the diurnal cycle of $R(t)$. Here we consider two cases. The first is rather special in that the k th service traffic is approximately proportional to the total network element traffic. The second is more general in that the two traffic flows are in-phase but not directly proportional.

a. If the service has a data rate that is approximately proportional to the total data rate, then $R^{(k)}(t) \approx \gamma R(t)$ with a constant, $0 \leq \gamma \leq 1$.

$$\frac{R^{(k)}(t)}{R(t)} \approx \gamma \qquad (10.74)$$

which gives

$$\left\langle \frac{R^{(k)}}{R} \right\rangle_T \approx \gamma \qquad (10.75)$$

Using (10.66) the energy consumption of the k th service is:

$$Q^{(k)}(T) = P_{idle}\gamma T + EB^{(k)}(T) \qquad (10.76)$$

b. Even if the k th service has a diurnal cycle that is in-phase with $R(t)$, that does not necessarily mean the two traffic flows are directly proportional. More generally, even though they have very similar shapes, they will most likely have independent maxima and minima. In that case, we can rescale the total traffic, $R(t)$, to approximate the k th service's traffic.

Let the maximum and minimum values of the element traffic, $R(t)$, be R_{max} and R_{min}, respectively. Similarly, let the maximum and minimum values of the k th service traffic be $R_{max}^{(k)}$ and $R_{min}^{(k)}$, respectively. Then, on the basis that the k th service traffic has approximately the same shape as the total network traffic, we can express the k th service traffic in the form:

$$R^{(k)}(t) \approx R_{min}^{(k)} + \frac{\left(R(t) - R_{min}\right)}{\Delta R} \Delta R^{(k)} \qquad (10.77)$$

where $\Delta R^{(k)} = R_{max}^{(k)} - R_{min}^{(k)}$, and $\Delta R = R_{max} - R_{min}$. This gives for the energy consumption when $R(t)$ and $R^{(k)}(t)$ are "in-phase":

$$Q_{in-phase}^{(k)}(T) \approx P_{idle}T\left(\frac{\Delta R^{(k)}}{\Delta R} + \left\langle \frac{1}{R} \right\rangle_T \left(\frac{R_{min}^{(k)}\Delta R - R_{min}\Delta R^{(k)}}{\Delta R}\right)\right) + EB^{(k)}(T) \qquad (10.78)$$

The subscript "in-phase" for the energy consumption is included in this equation because we use this expression below. Also, we have for $B^{(k)}(T)$:

$$B^{(k)}(T) \approx \frac{\left(R_{max}^{(k)} - R_{min}^{(k)}\right)}{\left(R_{max} - R_{min}\right)} B(T) = \frac{\left(R_{max}^{(k)} - R_{min}^{(k)}\right)}{\left(R_{max} - R_{min}\right)} \int_T R(t)dt \qquad (10.79)$$

3. The k th service traffic is 180 degrees out-of-phase relative to $R(t)$. For example, a data backup service that does most of its data transferring at times of low network traffic. Again, we consider two cases.

a. In this first case, which is likely to occur only in rather unique situations, we approximate the service traffic as approximately inversely proportional to the total traffic:

$$R^{(k)}(t) \propto \frac{1}{R(t)} \qquad (10.80)$$

Again, we require the maximum service traffic, $\max\{R^{(k)}(t)\}$, to be less than $\min\{R(t)\}$. From (10.80), we see that at the time t_{min} such that $R(t_{min}) = \min\{R(t)\}$ we have that $R^{(k)}(t_{min})$ will be at its maximum. Therefore, we have with an appropriate constant, γ:

$$R^{(k)}\left(t_{min}\right) \approx \frac{\gamma}{R\left(t_{min}\right)} \leq R\left(t_{min}\right) \tag{10.81}$$

which fixes the value of γ as $0 < \gamma \leq R(t_{min})^2$. In this case, we have:

$$\frac{R^{(k)}(t)}{R(t)} \approx \frac{\gamma}{R(t)^2} \tag{10.82}$$

Giving:

$$\left\langle \frac{R^{(k)}}{R} \right\rangle_T \approx \gamma \left\langle \frac{1}{R^2} \right\rangle_T \tag{10.83}$$

In this case, the energy consumption of the k th service is:

$$Q^{(k)}\left(T\right) = P_{idle}\gamma T \left\langle \frac{1}{R^2} \right\rangle_T + EB^{(k)}\left(T\right) \tag{10.84}$$

b.   More generally, if the diurnal cycle of $R^{(k)}(t)$ has approximately the same shape but may be 180 degrees out-of-phase with that of $R(t)$, we can express the diurnal cycle of $R^{(k)}(t)$ as:

$$R^{(k)}\left(t\right) \approx R_{max}^{(k)} - \frac{\left(R(t) - R_{min}\right)}{\Delta R} \Delta R^{(k)} \tag{10.85}$$

This gives for the energy consumption of the k th service when it and $R(t)$ are 180 degrees out-of-phase:

$$Q_{out-of-phase}^{(k)}\left(T\right) \approx P_{idle}T\left(\left\langle \frac{1}{R} \right\rangle_T \left(\frac{R_{max}^{(k)}\Delta R + R_{min}\Delta R^{(k)}}{\Delta R}\right) - \frac{\Delta R^{(k)}}{\Delta R}\right) + EB^{(k)}\left(T\right) \tag{10.86}$$

4. If the k th service has a data rate that is negligible relative to the other traffic through the element, then $R^{(k)}(t)/R(t) \ll 1$. In this case, we could write:

$$\frac{R^{(k)}(t)}{R(t)} \approx 0 \tag{10.87}$$

Giving:

$$\left\langle \frac{R^{(k)}}{R} \right\rangle_T \approx 0 \tag{10.88}$$

which gives for the energy consumption of the k th service over duration T is just the product of the energy per bit, E, and the total number of bits for that service over the duration T, $B^{(k)}(T)$. That is:

$$Q^{(k)}(T) \simeq EB^{(k)}(T) \qquad (10.89)$$

From this, we see that, by considering the traffic of the service relative to the total traffic, which will be dependent upon the service type, we can use various approximations of the diurnal cycle profile to gain an appreciation of the energy consumption of that service. An advantage of adopting one of the above four approximations is that we do not require the details of the service traffic. Rather, we use the total traffic, $R(t)$, to provide an approximation for the service traffic, $R^{(k)}(t)$. This reduces the amount of information we require about the service traffic. Rather than knowing the total time profile of the service traffic diurnal cycle, we only need to determine either the constant γ, or the service's maximum and minimum traffic (assuming we know $R(t)$). However, it is important to note that, as a consequence of this simplification, we will lose accuracy in the energy consumption calculation.

Using (10.78) and (10.86), we can look at the issue of when is it more energy conserving to transport data through a network. For example, let the k th service be a data backup service and assume the service provider wishes to minimise the energy consumption of the service. One way to do so is to carefully choose the time the backup data is transported through the network from the primary data location to the backup location. To simplify the issue, the question we consider is whether or not the backup service is operated at a time that is in-phase or out-of-phase with the overall traffic through the network. This reduces the question to the whether or not $Q^{(k)}_{out\text{-}of\text{-}phase} - Q^{(k)}_{in\text{-}phase}$ is positive or negative. We consider a service operated over a time interval T (much shorter than a diurnal cycle) during which the overall diurnal cycle traffic, $R(t)$, is relatively constant. By this we mean that the service can be scheduled to commence at time t in the diurnal cycle, such that T satisfies the condition:

$$\frac{|R(t) - R(t + \Delta T)|}{\langle R(t) \rangle_{\Delta T}} << 1 \qquad (10.90)$$

Where $\langle R(t) \rangle_{\Delta T}$ is the average of $R(t)$ over time t to $t + T$. In other words, the service can be scheduled to operate entirely "in-phase" or "out-of-phase" with the overall traffic diurnal cycle and not overlapping into the transition between these two phases.

Subtracting (10.78) from (10.86) gives:

$$Q^{(k)}_{out\text{-}of\text{-}phase} - Q^{(k)}_{in\text{-}phase} = P_{idle} T \frac{\Delta R^{(k)}}{\Delta R} \left(\left\langle \frac{1}{R} \right\rangle_T (R_{max} + R_{min}) - 2 \right) \qquad (10.91)$$

From this, we see that the value of $\langle 1/R \rangle_T$ is the crucial parameter. We have $(1/R_{max}) < \langle 1/R \rangle_T < 1/R_{min}$, therefore:

$$1 + \frac{R_{min}}{R_{max}} < \left\langle \frac{1}{R} \right\rangle_T (R_{max} + R_{min}) < 1 + \frac{R_{max}}{R_{min}}$$
$$\text{and} \qquad (10.92)$$
$$1 + \frac{R_{min}}{R_{max}} < 2 < 1 + \frac{R_{max}}{R_{min}}$$

This means that the sign of $Q^{(k)}_{out\text{-}of\text{-}phase} - Q^{(k)}_{in\text{-}phase}$ can be either positive or negative. If $\langle 1/R \rangle_T \sim 1/R_{min}$, then $Q^{(k)}_{out\text{-}of\text{-}phase} > Q^{(k)}_{in\text{-}phase}$ and more energy consumption can be attributed to transporting the backup data if it is conducted during off-peak periods. A diurnal cycle that has $\langle 1/R \rangle_T \sim 1/R_{min}$ has a relatively short peak traffic period or a peak-to-average traffic ratio that is much greater than unity. This type of traffic profile is seen in networks that have a significant proportion of video download traffic that occurs primarily during the mid-evening peak viewing time [27]. As video traffic becomes more dominant and assuming it continues to only occur during several hours (typically mid to late evenings), then transporting the backup data across a network at times away from this peak will result in a larger proportion of network energy being attributed to that service. Assuming the backup service operator endeavours to minimise the energy consumption attributed to their service, they will be motivated to undertake the backups during peak hours. This will just exacerbate the high load on the network during peak times. Consequentially, it may be worthwhile constructing a price or regulatory regime that encourages all services that are not time-of-day critical to shift their service times to off-peak. If enough such services do so, then the diurnal cycle will be flattened out.

If $\langle 1/R \rangle_T \sim 1/R_{max}$, then the diurnal cycle is flattened. In this case the diurnal cycle has long durations of heavy traffic with only a relatively short duration "off-peak" period or the diurnal cycle is relatively shallow with the off-peak traffic not significantly less than the peak traffic. In this case we have $Q^{(k)}_{out\text{-}of\text{-}phase} < Q^{(k)}_{in\text{-}phase}$ and a backup service has lower attributed energy consumption when it operates during the relatively brief off-peak times.

From these results, we see that there is a definite advantage to "flattening" the diurnal cycle. It reduces the amount of additional equipment required to accommodate the peak traffic relative to the mean traffic. It also enables service providers to shift their peak traffic times away from the network peak time without incurring an energy penalty.

If we bring in the operator, as described in Section 10.4.1, then the appropriate equation is (10.43) and the attributed power consumption of a service becomes independent of the time of day. However, viewing (10.45), we see that the network operator sees variation in their attributed power consumption over the diurnal cycle.

10.8   TECHNICAL SUMMARY

The methods described in this chapter provide estimates for the power consumption of service traffic when it is one of possibly many services being transported by a network element or network. These models provide a transparent and justifiable approach to attributing a power footprint to a service's traffic in a network that carries many services. These models range from relatively simple "energy per bit" models through to models that include an allocation to the network operator for their network utilisation policies and approaches for accounting for the traffic diurnal cycle of a service.

The simple model only requires estimates for the typical maximum power, P_{max}, minimum power, P_{min}, capacity, C_{max}, of the equipment transporting the service, and the total throughput data rate, $R(t)$. The power consumption of the service is then given by (10.2), where $N_E^{(svc)}$ is the number of nodes through which the service travels, E is the energy per bit of the network equipment (given by (10.1)) and $R^{(svc)}$ is the data rate of the service.

Some network equipment, such as CPE, typically has a single user and the services being accessed by that user are dealt with on the basis of time sharing. In this case the power consumption is modelled by tracking times when the service is active. This model is based on a "time-sharing" approach and the resulting power for a service is dependent upon the proportion of time it is active. For a service that is active for a portion ρ of total duration T_{Tot}, the energy consumption, $Q^{(svc)}$, of that service is given in (10.15). During other times in $(1-\rho)T_{Tot}$, the equipment will be either idle or undertaking tasks not associated with the service of interest.

For equipment that is dealing with many service traffic flows simultaneously, such as an edge or core router, we need to consider how to attribute idle power of the network equipment. In this case we adopt a "throughput-sharing" model in which power is allocated on the basis of service traffic data rate (bit/s) rather than the service active time described above.

To attribute power consumption to a service, we need to consider how the network element idle power will be allocated. A simple approach is to allocate idle power to services in proportion to their data rate, $R^{(k)}(t)$, relative to the total network element throughput, $R(t)$, which gives a power allocation to the k th service as shown in (10.33).

Because the k th service traffic, $R^{(k)}(t)$, contributes to the total element traffic, $R(t)$, a small change, $\delta R^{(k)}$, in the traffic of the k th service changes its power consumption by $\delta P^{(k)}$, given in 10.39). Further, because all the traffic flows through the network element contribute to the total element throughput, $R(t)$, a change in the traffic flow of any service will have an impact on the power consumption attributed to all other services. A small change, $\delta R^{(p)} << R$, in the p th service will result in a change $\delta P^{(k,p)}$ attributed to the k th service, given by (10.40).

These results show that, although a proportional allocation of idle power to services provides an equitable approach to attributing network element power to the services it is dealing with, we have a problem in that any change in one service will impact the power allocated to all other services. This is unlikely to be considered acceptable to the providers of those other services.

An approach to resolving this issue is to replace the $R(t)$ value in the denominator of (10.33) with a fixed value, R_{fixed}, as shown in (10.43). The value of R_{fixed}, given in (10.42), is greater than the maximum of $R(t)$ over the diurnal cycle. It could be set to C_{max}, the maximum capacity of the network element, or μC_{max}, the maximum operational value of the network element throughput (where μ is the maximum operational utilisation of the network element). The component of idle power given in (10.45) is taken to be the idle power footprint of the network operator for the network element. This can be justified because it gives the network operator an incentive to also improve the energy efficiency of their network and equipment.

If the unused throughput capacity is allocated to the network operator, then the power attributed to the k th service provider is somewhat simplified and the impact on the k th service of a change in other services' traffic is zero, as shown in (10.44). In this case, the network operator is attributed a power footprint given by (10.45).

When applying the concept of attributional power consumption to a service provided in a mobile network, the definition of attributed power needs to be modified to accommodate the fact that mobile base station power is expressed in terms of base station utilisation, $\mu_{BS}(t)$, as in (10.46). This can be relatively easily done by replacing the total throughput, R, and service throughput, $R^{(k)}$, with the total base station utilisation, μ_{BS}, and the service utilisation, $\mu_{BS}{}^{(k,svc)}$, as shown in (10.54). Similarly, the interdependence of different service utilisations can be removed by allocating some base station utilisation to the network provider, as in (10.55).

When considering the power consumption of a service, we often are interested in two scenarios:

1. The power consumption of an individual's use of a service. In this case, the traffic, $\delta R^{(svc)}$, represents the traffic of a single user. For example the power consumption of a single photo shared via a social network [6] or the power consumption of a single "smart" house [20]. We expect $\delta R^{(svc)}$ to satisfy the constraint that it is small enough to not require the deployment of any additional network equipment for it to be accommodated by the network. That is, N_E remains constant. Rather, this additional traffic is accommodated by allowing the utilisation of the existing network equipment to slightly increase. In this case the additional network power consumption that can be allocated to that service is given by (10.60).
2. The additional traffic due to the service, $\Delta R^{(svc)}$, is so great that extra equipment has to be deployed. This case corresponds to the situation where we are considering the cumulative power consumption of a service resulting from many or all the users of that service. For example, the global power consumption of photo sharing via social networks [6] or that of smart houses nationwide [20]. In this case, we assume the amount of added equipment is such that the utilisation of the equipment, μ_{op}, remains the same as before the additional traffic. This gives the additional power consumption of the network due to the service, in (10.64).

An issue of significant interest is the overall energy consumption of a service over some time duration, T, which may be a week, month or year. Attributing power to a service using (10.33), we find the relative phases of the service traffic to total traffic through the network element has a significant impact on the attributed energy footprint. By "relative phase" we mean the relationship between the times of peak service traffic and peak total traffic. There are two cases of particular interest: When the two peak times align; and when the peak of one aligns with the minimum of the other. Ideally, we would have access to the total diurnal cycle data of both the total traffic and the service traffic. However, this may not be always possible. An approximation is to assume the service traffic diurnal cycle has a similar shape to that of the total traffic, but of smaller magnitude. If we know the diurnal-cycle total traffic and can ascertain just the maximum and minimum traffic of the service (rather than its values at all hours over the entire cycle), we can construct an approximation of the service energy for the two cases of aligned peak traffic times (i.e. in-phase) and when they are 180 degrees out-of-phase.

For the case when the peak traffic times are in-phase the k th service energy consumption, $Q^{(k)}(T)$, over duration T, can be approximated by (10.78). For the case when the service traffic, $R^{(k)}(t)$, is 180 degrees out-of-phase with the total traffic, $R(t)$, we get (10.86). Comparing the values of $Q^{(k)}(T)$ for the in-phase and 180 degrees out-of-phase traffic, we find that, depending upon the shape of the total traffic diurnal cycle, there may be energy consumption advantages of scheduling the traffic of a service to have its peak away from the time of peak total traffic.

10.9 EXECUTIVE SUMMARY

When constructing an estimate for the power or energy footprint of a transporting a service through a network, it must be recognised that there may be a significant difference between the "consequential" and "attributional" footprints. The consequential footprint is a measure of the additional energy consumption of transporting a service in addition to the current network energy consumption. In many cases, this may be relatively small, particularly if the incremental energy (energy/bit) of network elements is close to zero.

However, the attributional footprint is an allocation of a proportion of the total energy footprint of a network that can justifiably be attributed to the transport of that service. This requires the allocation of an amount of network equipment idle power to a service, because network element idle power is an integral part of providing services via the network.

Although the attribution of incremental network element power is relatively intuitive, given by $R^{(k)}(t)E$, the attribution of idle power is more nuanced. One approach is to adopt the "ICT Sector Guidance built on the GHG Protocol Product Life Cycle Accounting and Reporting Standard" [1]. With this, each service will be attributed an amount of idle power given by $R^{(k)}(t)P_{idle}/R(t)$: that is, in proportion to the service traffic relative to total traffic through the network element.

Although this approach is transparent and equitable, in that it allocates power in proportion to the magnitude of service traffic, it has the side effect that every service provider's attributed footprint changes whenever any other service provider changes their traffic. One way of addressing this is to allocate the unused traffic capacity (i.e. the difference between the maximum allowable traffic and total services traffic) to the network operator. With this, the operator is attributed some of the idle power footprint and it is then also in their interest to minimise the overall energy footprint of the network element or network.

When allocating power consumption to a service provided via a mobile network, a slightly different approach is required. In a mobile network, the power consumption of a base station has a simple linear form in terms of the base station "utilisation", not the base station throughput (bit/s). The utilisation is the ratio of radio resources allocated to transmit data to users to the total amount of radio resources available. This change from throughput to utilisation is relatively easily carried across to attributing base station power to the services it is transporting to users.

There are two scenarios of particular interest when considering the energy footprint of a service. The first is the energy consumption of a single user of the service. (For example, "What is the carbon footprint of a web search?" or "What is the energy footprint of sharing a photo via a social network?") With this calculation, we can assume the additional traffic (i.e. from the single web search or single photo upload) does not require any additional network equipment to be deployed. The second scenario of interest is the global energy footprint of a service. ("What is the global carbon footprint of all web searching worldwide?" or "What is the energy footprint of photo sharing via social networks worldwide?") To consider this scenario, we need to recognise that to accommodate the global traffic generated by a service will most likely require the addition of equipment to the network. Therefore, a different approach for calculating the energy footprint is used.

Effectively all services and networks display a traffic diurnal cycle in which the service or network has a peak traffic time and a minimum traffic time. It can occur that the peak traffic time for a service is well away from the peak traffic time of the network element. In this case, energy consumption that can be attributed to that service may be quite different to the situation where the peak traffic time of the service and network element coincide.

An estimate for the attributional energy footprint of a service can be estimated using the diurnal cycle of the total network element traffic. To construct this estimate, we assume the diurnal cycle of the service of interest is roughly the same shape as that of the total element traffic. (Most diurnal cycles have the same periodicity and roughly common times between peak and minimum traffic.) We also require an estimate for the peak and minimum traffic of the service. With this, there are several approaches to estimating the energy footprint of a service. These estimates can include the situation in which the peak traffic of the network element and service are at the same time or in which the peak traffic times are 12 hours apart (i.e. 180 degrees out-of-phase).

Some services are not tied to the local diurnal cycle (e.g. data backup services). Depending upon the shape of the total traffic diurnal cycle, such services may be able to reduce their attributed energy footprint by operating with their peak traffic out-of-phase with the overall network element traffic.

REFERENCES

[1] GeSI, Carbon Trust, "ICT Sector Guidance built on the GHG Protocol Product Life Cycle Accounting and Reporting Standard," 2017.

[2] G. Finnveden, "Recent developments in Life Cycle Assessment," *Journal of Environmental Management*, vol. 91, no. 1, p. 1, 2009.

[3] T. Ekvall, "Attributional and Consequential Life Cycle Assessment," in *Sustainability Assessment at the 21st Century*, M. J. Bastante-Ceca, Ed., IntechOpen, 2020, p. Chapter 4.

[4] M. Brander *et al.*, *Consequential and Attributional Approaches to LCA: A Guide for Policy Makers with Specific Reference to Greenhouse Gas LCA of Biofuels*, Ecometrica Press, 2008.

[5] F. Jalali *et al.*, "Energy Consumption Comparison of Nano and Centralized Data Centers," *ACM SIGMETRICS Performance Evaluation Review*, vol. 43, no. 3, p. 49, 2014.

[6] F. Jalali *et al.*, "*Energy Consumption of Photo Sharing in Online Social Networks*," in *14th IEEE/ACM International Symposium on Cluster, Cloud and Grid Computing*, 2014.

[7] C. Chan *et al.*, "Methodologies for Assessing the Use-Phase Power Consumption and Greenhouse Gas Emissions of Telecommunications Network Services," *Environmental Science and Technology*, vol. 47, no. 1, p. 485, 2013.

[8] V. Coroama *et al.*, "The Direct Energy Demand of Internet Data Flows," *Journal of Industrial Ecology*, vol. 17, no. 5, p. 680, 2013.

[9] F. Jalali *et al.*, "Fog computing may help to save energy in cloud computing," *IEEE Journal on Selected Areas in Communications*, vol. 34, no. 5, p. 1728, 2016.

[10] F. Hung-Ming *et al.*, "On Burstiness of Self Similar Traffic Models," in *ATM, Networks and LANs*, D. Faulkner and A. Hamer, Eds., IOS Press, 1996, p. 146.

[11] R. van de Meent *et al.*, "*Burstiness Predictions Based on Rough Network Traffic Measurements*," in *Proceedings of the 19th World Telecommunications Congress (WTC/ISS 2004)*, 2004.

[12] Cisco, "Best Practices in Core Network Capacity Planning, White Paper," Cisco, 2013.

[13] D. Moltchanov, "Automatic Bandwidth Adjustment for Content Distribution in MPLS Networks," *Advances in Multimedia*, vol. 2008, p. 624941, 2008.

[14] L. Altmanova *et al.*, "Deliverable DS1.1.1,2: Final GÉANT Architecture," GÉANT, 2011.

[15] J. Kuri *et al.*, *"On the Resource Efficiency of Virtual Concatenation in Next-Generation SDH Networks,"* in *2nd International Conference on Broadband Networks*, 2005.

[16] Y. Zhou *et al.*, "Supporting Dynamic Bandwidth Adjustment Based on Virtual Transport Link in Software-Defined IP Over Optical Networks," *Optical Communications and Networking*, vol. 10, no. 3, p. 125, 2018.

[17] V. Lopez and L. Velasco, Eds., *Elastic Optical Networks: Architectures, Technologies and Control*, Springer International, 2016.

[18] A. Monge and K. Szarkowicz, MPLS in the SDN Era; Interoperable Scenarios to Make Networks Scale to New Services, O'Reilly Media, 2015.

[19] E. Osborne and A. Simha, Traffic Engineering with MPLS, Cisco, 2003.

[20] C. Gray *et al.*, "Smart' Is Not Free: Energy Consumption of Consumer Home Automation Systems," *IEEE Transactions on Consumer Electronics*, vol. 66, no. 1, p. 87, 2020.

[21] C. Gray *et al.*, "Energy-Efficient Network Protocols for Domestic IoT Application Design," *Journal of Telecommunications and the Digital Economy*, vol. 7, no. 2, p. 50, 2019.

[22] Full Fact, "How Energy intensive Is a Google Search?," 15 August 2019. [Online]. Available: https://fullfact.org/environment/google-search/. [Accessed 14 September 2020].

[23] Direct Energy, "Powering a Google Search: The Facts and Figures," 2020. [Online]. Available: https://business.directenergy.com/blog/2017/november/powering-a-google-search. [Accessed 14 September 2020].

[24] Google, "Powering a Google Search," 11 January 2009. [Online]. Available: https://googleblog.blogspot.com/2009/01/powering-google-search.html. [Accessed 14 September 2020].

[25] J. Baliga *et al.*, "Green Cloud Computing: Balancing Energy in Processing, Storage and Transport," *Proceedings of the IEEE*, vol. 99, no. 1, p. 149, 2011.

[26] A. Vishwanath *et al.*, "Energy Consumption Comparison of Interactive Cloud-Based and Local Applications," *IEEE Journal on Selected Areas in Communications*, vol. 33, no. 4, p. 616, 2015.

[27] Cisco, "Cisco Visual Networking Index: Forecast and Trends, 2017–2022," Cisco White Paper, 2018.

11 Energy Efficiency

Energy efficiency has come to the attention of many organisations and researchers over recent years. The growing appreciation that we live on a planet with finite resources has encouraged a trend to securing more "output per Joule" (which is often converted to "output per gram of CO_2") as one approach to moving towards a more sustainable society [1].

Energy efficiency in the ICT sector and the Internet are now attracting significant industry attention and research efforts [2–6]. These efforts include modelling and quantifying the power consumption and the energy efficiency (or energy intensity) of communications equipment, networks and services [7–10]. Energy efficiency metrics play a crucial role in identifying, comparing and reducing power consumption of the systems and sub-systems that make up the telecommunications infrastructure underpinning Internet services. The adage "no improvement without measurement" applies equally to network energy efficiency as any other process improvement activity [11]. As the energy demands by ICT and the Internet grow and with the spreading adoption of carbon accounting across industries and jurisdictions, the need to understand and quantify energy efficiency of equipment, networks and services is increasing.

The selection of an energy efficiency metric is an important aspect in any endeavour to reduce or control power consumption of telecommunications equipment, networks or services. This is because improvement strategies (i.e. plans for continuing improvement of a chosen aspect of an organisation's operation) are substantially influenced by metrics adopted to ascertain the success or otherwise of the strategies for improvement. The focus of this book is mitigating use-phase power consumption footprint. With this, the question we are interested in answering is as follows: "What is an appropriate metric for developing strategies to reduce the use-phase power consumption footprint of a telecommunications network element, network or service?"

If we just choose total power consumption, $P_{Ntwk}(t)$, as the metric, then any growth in the network (either in terms of traffic, geographical spread or customers served) will be considered retrograde because the power consumption will increase. Adopting the power as a metric will suggest improvement is attained by reducing the size, coverage and traffic flow; in other words, by minimising the network. This clearly is not an acceptable strategy.

Therefore, a metric must accommodate the fact that the network should be allowed to grow to satisfy customer demand. Typically, the greater the customer demand the more traffic is dealt with by the network. This naturally leads us to a metric based on power per unit data throughput or energy per bit. This type of metric will allow network growth on the condition that the network power does not grow "out of proportion" with the amount of traffic it is transporting.

More sophisticated metrics are available. For example, a power footprint metric for submarine networks is "energy per bit per kilometre" [12]. For a mobile network, a candidate metric is "energy per bit per unit area" [13–15] and energy per user per unit area [16]. Surveys of efficiency metrics for networks have been provided by Alsharif *et al.* [17], Fang *et al.* [18] and Wu *et al.* [19]. A review of energy efficiency in optical networks is provided in Kilper *et al.* [20].

Metrics such as energy efficiency (Joules/bit or Watts/bit/s) and energy intensity (bits/Joule or bit/s/Watt) are often used to compare the energy efficiency of different equipment and networks [3] because they effectively "normalise" the metric with respect to the traffic capacity of the equipment, network or service. For example, the GreenTouch international consortium provided a technology roadmap that would improve the energy efficiency of the Internet by a factor of 1000 between the years 2010 and 2020. The success of this undertaking was measured by comparing the bits/Joule of a 2010 "state of the art" network with that of a GreenTouch designed network with forecast 2020

DOI: 10.1201/9780429287817-11

traffic [21]. That is, the GreenTouch strategy was based on applying the metric of "energy per bit" to the 2010 and 2020 network technologies and architectures.

This approach typifies almost all methods for "improving" a system. That is: adopt a metric that provides a quantitative measure of "improvement" and then develop strategies to modify the system to produce the desired changes in the value of the metric. We adopt this principle here and in this chapter will discuss several options for metrics and compare those metrics. In this chapter, we focus on several "bottom-up" energy/bit metrics by investigating the information they provide and strategies they suggest to improve energy efficiency.

The energy/bit metric has been used to compare the energy efficiency of networks [2, 3] as well as to estimate network power consumption [22–25]. There exist a variety of energy/bit metrics, each derived using a different approach. This means using them to make direct comparisons of network energy efficiency requires some care [8, 9].

11.1 NETWORK POWER AND ENERGY CONSUMPTION MODELS

Energy consumption models used to calculate energy efficiency metrics are typically classified as either "top-down" or "bottom-up" [8, 9]. A detailed discussion of these models was provided in Chapter 7. In summary, "top-down" can be described as [26]:

> top-down analyses are based on two estimates: 1) the overall energy demand of either the entire Internet or a part of it (e.g. a country or a continent), and 2) the total Internet traffic of that region.

Examples of "top-down" models include Koomey *et al.* [27] and Lanzisera *et al.* [28].

A succinct description of "bottom-up" is given by Coroama *et al.* [29]:

> bottom-up approaches model parts of the Internet (i.e. deployed number of devices of each type) based on network design principles. Such a model combined with manufacturers' consumption data on typical network equipment leads to an estimate of the overall energy consumption, which is then related to an estimate of the corresponding data traffic.

Examples of "bottom-up" models include Baliga *et al.* [30, 31] and Kilper *et al.* [32].

As a general observation, the network technology community shows a preference for bottom-up models, whereas environmental/sustainability workers tend to construct top-down models. However, there are exceptions to this.

11.2 RECAP ON EQUIPMENT, NETWORK AND SERVICE POWER CONSUMPTION MODELS

The metrics we consider in this work are all based upon power/bit/s or energy/bit. To assist the reader, in this section we provide a brief recap of the equations for power consumption that were developed on previous chapters which we will require to construct and investigate energy efficiency metrics.

11.2.1 EQUIPMENT POWER CONSUMPTION MODEL

From Section 3.4, the generic form for the traffic load-dependent power model for the j th network equipment is given by (3.11)

$$P_j(t) = P_{idle,j} + \frac{(P_{max,j} - P_{idle,j})}{C_{max,j}} R(t) = P_{idle,j} + E_j R(t) \qquad (11.1)$$

where $C_{max,j}$ is the maximum capacity (throughput) of the network element, $P_{max,j}$ is the maximum power consumption of the element and $R(t)$ is the element throughput (bit/s). We refer to the quantity E_j as the "incremental energy per bit" for the j th network element. The situation in which the idle power is zero, $P_{idle,j} = 0$ (often referred to as "load" or "energy proportional"), is of particular interest because it provides significant energy savings when the network element is lightly loaded (i.e. $R_j(t) \ll C_{max,j}$), relative to network elements with a high idle power [33]. Load proportional equipment is widely considered as energy efficient because its power consumption is minimal at low load. In contrast, equipment for which $E_j \approx 0$ is considered energy inefficient because its power consumption remains approximately the same for both high and low load. A significant number of strategies for improving energy efficiency have been proposed to attain energy proportionality [33–35].

11.2.2 Network Power Model

Using the bottom-up approach for networks, we have (8.1), (8.31) and (8.34) (recalling that $\langle N_E \rangle^{(svc)} = (\langle N_{hops} \rangle^{(svc)} + 1)$):

$$
\begin{aligned}
P_{Ntwk}(t) &= M_{OH} \sum_{j=1}^{N_E} \left(P_{idle,j} + E_j R_j(t) \right) \\
&\approx M_{OH} \left(N_E \langle P_{idle} \rangle_E + \langle N_E \rangle^{(svc)} \langle E \rangle_E R_{Ntwk}(t) \right) \\
&\approx M_{OH} \langle N_E \rangle^{(svc)} \left(\frac{\langle P_{idle} \rangle_E}{\mu_{max} \langle C_{max} \rangle_E} R_{peak} + \langle E \rangle_E R_{Ntwk}(t) \right)
\end{aligned}
\tag{11.2}
$$

where the total network traffic R_{Ntwk} is given by (8.2)

$$
R_{Ntwk}(t) = \sum_{k=1}^{N^{(svc)}} R^{(k)}(t)
\tag{11.3}
$$

and the traffic through the j th network element, $R_j(t)$, is as follows (8.5):

$$
R_j(t) = \sum_{k=1}^{N^{(svc)}} \alpha_j^{(k)} R^{(k)}(t) = \sum_{k=1}^{N^{(svc)}} R_j^{(k)}(t)
\tag{11.4}
$$

Where $R_j^{(k)}(t) = \alpha_j^{(k)} R^{(k)}(t)$ is the proportion of the k th service traffic through the j th network element. Also, recall the notation $\langle X \rangle_E$ is the average over network elements:

$$
\langle X \rangle_E = \frac{1}{N_E} \sum_{j=1}^{N_E} X_i
\tag{11.5}
$$

11.2.3 Service Power Consumption Model

In Chapter 10, we found that to model the power consumption of a service requires consideration of whether or not the equipment is being shared by many services and, if so, how to allocate idle power to the service or services with which the network element or network are dealing.

In Section 10.3, we discussed power modelling of equipment (such as CPE) that is either not shared or shared only by a few services. From (10.24), we have for the power consumption of a service in equipment that is not shared:

$$P^{(svc)} = \frac{Q^{(svc)}}{T_{tot}} = \left(\frac{P_{idle}}{C_{line}} + E \right) R^{(svc)} \tag{11.6}$$

where we have $R^{(svc)} = B^{(svc)}/T_{tot}$ is the effective data rate of the service of interest over the duration T_{tot} and C_{line} is the line rate of the connection between the CPE and the access network. Because C_{line} is a constant, this power allocation is similar to that proposed in Section 10.4.1, where the idle power attributed to a service is based upon a fixed data rate, R_{fixed}, as in (10.48).

To consider the energy efficiency of a service that is one of many being dealt with by a network element, we recall Section 10.4, where the issue of allocating network element idle power was discussed. The reader may recall that two approaches to allocating the idle power were considered in detail. The first approach adopted a rule of proportional allocation of idle power to the k th service, based on the ratio, $R^{(k)}(t)/R_j(t)$, of the service throughput relative to the total throughput of the j th network element. This resulted in the power allocation, $P_j^{(k)}(t)$, to the k th service being dealt with by the j th network element being (10.33):

$$P_j^{(k)}(t) = P_{idle,j} \frac{R_j^{(k)}(t)}{R_j(t)} + E_j R_j^{(k)}(t) = \left(\frac{P_{idle,j}}{R_j(t)} + E_j \right) R_j^{(k)}(t) \tag{11.7}$$

A disadvantage of this approach is that, should any other service ($p \neq k$) change its throughput of the j th network element, $R_j^{(p)}(t)$, then the power allocation to the k th service will also change. This means a service provider will find its power allocation may be impacted by decisions by other service providers and beyond its control.

The second approach, discussed in Section 10.4.1, avoids this cross-impact issue by allocating network element capacity to the network operator, so that idle power consumption allocated to the k th service is based on a fixed traffic throughput and given by (10.49). The complication with (10.49) is that it requires the cooperation of and coordination between all the service providers and the network operator.

The Greenhouse Gas Protocol accounting and reporting standard for ICT equipment published by The Carbon Trust has adopted (11.7) for allocation of service power [36]. Therefore, we will focus our attention on energy efficiency metrics that use this approach and discuss their properties.

The model for service power consumption over a network is provided by (10.62) and (10.63), reproduced here:

$$P^{(k)}(t) = M_{OH} \sum_{j=1}^{N_E} \left(\frac{P_{idle,j}}{R_j(t)} + E_j \right) \alpha_j^{(k)} R^{(k)}(t)$$
$$\approx M_{OH} N_E^{(k)} \left(\langle P_{idle} \rangle_E \left\langle \frac{1}{R(t)} \right\rangle_E + \langle E \rangle_E \right) R^{(k)}(t) \tag{11.8}$$

11.3   STANDARDISED ENERGY EFFICIENCY METRICS

With the relationships for power consumption determined, we now discuss several energy efficiency metrics that have been defined in the literature and standards documents. The ITU has described an energy-efficiency metric in ITU-T Rec. L.1310 as [3]:

> The energy efficiency metric is typically defined as the ratio between the functional unit and the energy necessary to deliver the functional unit.

This definition results in a metric with units "bits/Joule". ITU-T Rec. L.1310 also recognises that: "The inverse metric, energy divided by functional unit, could be used as an alternative." [3]

In this text, we shall adopt "energy per bit" metrics rather than "bit per energy" metrics. This is because energy per bit metrics (defined by $H(t) = P(t)/R(t)$) provide for simpler mathematics. For example, if we have two network elements, NE_1 and NE_2 in series (i.e. nodes in a long haul link), then, if the energy per bit of NE_1 is H_1 and of NE_2 is H_2, the energy consumption, Q_{1+2}, of B bits through NE_1 and NE_2 in series is $Q_{1+2} = B(H_1 + H_2)$.

For a situation in which a stream of B bits is split into two parallel paths with amount αB bits passing through NE_1 and amount $(1 - \alpha)B$ passing through NE_2, then the energy consumption of the overall flow of B bits can be expressed as $Q_{1\&2} = B(\alpha H_1 + (1 - \alpha)H_2)$. Bit per energy metrics do not provide this simple mathematics.

There is a diverse range of energy efficiency metrics in standards and other publications. A review of the more widely used metrics is provided by Minoli [37] and Hamdoun et al. [38]. The majority of energy efficiency metrics currently used in standards documents are based on the ratio of power consumed to traffic throughput for a set of pre-defined traffic load levels of the equipment. For example, the "Energy Consumption Ratio-Variable Load", ECR-VL, is defined by the ratio [38]:

$$ECR = \sum_m a_m \times P_m \bigg/ \sum_m a_m \times R_m \qquad (11.9)$$

where $\sum_m a_m = 1$, and P_m and R_m are the power and throughput values, respectively, at utilisations indexed by m. The values of P_m and R_m are specified and depend upon the type of network element.

For the j th network element, using (11.1), this metric gives:

$$ECR_j = \frac{P_{idle,j}}{C_{max,j} \sum_m a_m b_m} + E_j \qquad (11.10)$$

where $b_m = R_{m,j}/C_{max,j}$ with $C_{max,j}$ the maximum throughput capacity of the network element, $R_{m,j}$ the throughput required to give the ratio b_m and E_j is the incremental energy per bit of the network element. Example values for b_m are as follows: $b_1 = 1$ (full load), $b_2 = 0.5$ (half load), $b_3 = 0.3$, $b_4 = 0.1$ and $b_5 = 0$ (idle) with corresponding weights $a_1 = 0.35$, $a_2 = 0.4$, $a_3 = 0.25$, $a_4 = 0$ and $a_5 = 0$ [37].

There are several problems with this and similarly defined metrics (TEER [39], EER [40] and TEEER [41]). First, although the definition includes an average over loads, it effectively corresponds to a single traffic load and is unlikely to correspond to the actual load on the element over all times of the day. Therefore, the ECR value does not incorporate the impact of traffic variation over the diurnal cycle.

Another issue with these metrics is seen by considering two routers with the (approximately) same C_{max} but different values for P_{idle} and E (as defined in (11.1)). Let the values for the routers be $P_{idle,1}$, E_1 and $P_{idle,2}$ E_2, respectively, and $P_{idle,2} = xP_{idle,1}$, where x is constant. Setting the ECR metric to have the same value for both machines, $ECR_1 = ECR_2$, and using (11.1), the relationship between the incremental energy per bit, E_1 and E_2, is as follows:

$$E_2 = E_1 + \frac{P_{idle,1}}{C_{max} \sum_m a_m b_m}(1 - x) \qquad (11.11)$$

For the values above, we have $\sum_m a_m b_m = 0.44$.

Of particular interest is when $x = 0$. This means router 2 is load proportional (i.e. $P_{idle,2} = 0$), which is viewed as desirable in terms of energy efficiency. However, it has the same ECR as router 1, which may have large idle power.

On the other hand, setting

$$x = 1 + \left(\frac{E_1 C_{max}}{P_{idle,1}} \right) \sum_m a_m b_m \qquad (11.12)$$

gives $E_2 = 0$ which is often considered energy inefficient because, as discussed in Section 11.2.1, the element's power consumption remains the same independent of load. Therefore, depending upon the actual throughput of the two routers, one may have a higher energy per bit than the other, despite both having the same ECR value.

This means that metrics based on the form (11.9) need to be used with care. Just because two network elements have the same ECR value does not mean they are similarly energy efficient in terms of power consumed for a given throughput. Also, these types of metrics may not provide reliable guidance on energy efficiency over a diurnal cycle or independent of the diurnal cycle shape.

Although these metrics have units of "Joules/bit" they are, in fact, not a ratio of "energy per bit". This can be seen in several ways. Firstly, energy consumption and the number of bits are only meaningful over a time duration (i.e. energy is the integral of power over time and bits are the integral of throughput over time). The ratio of power per bit-rate provided by metrics of the form (11.10) is an instantaneous measure corresponding to a specific traffic load, which may only occur at a specific time of the day. Secondly, as we will see below, the quantities "energy per bit" and "power per bit-rate" may have similar dimensions, but they have very different properties.

Another problem with the type of metric defined in (11.9) is that it does not lend itself to being applied to networks or services. It is not clear how to apply this metric to a collection of interconnected network elements. Even less apparent is how to apply this metric to a service that may be one of many propagating through a given network element.

A comprehensive survey of energy efficiency metrics and measurement tools is provided in Riekstin *et al.* [42] and Wu et al. [19].

When considering the energy efficiency, particularly for services, traffic variations over the diurnal cycle are important because, as shown in Section 10.7 and also discussed below, the shape of the diurnal cycle significantly impacts the power consumption of equipment that is allocated to a service. This, in turn, significantly impacts an energy efficiency metric.

11.4 CONSTRUCTING ENERGY EFFICIENCY METRICS

11.4.1 ENERGY PER BIT METRICS

In this section, we will study several metrics that have been proposed or applied [8, 21, 22, 25, 29]. We will implement them in a manner that is applicable to network equipment, networks and services. We will focus on simple quotient-based metrics (such as power/user, energy/bit) rather than more complicated function-based metrics [43]. In many cases, the precise value of the metric requires a significant amount of detailed information about network equipment, service flows and equipment throughput. Therefore, in most cases one or more approximate forms will also be presented for those situations in which only limited or averaged data is available. As one would expect, the fewer data available, the more approximate the value provided by the metric. Although these forms only provide approximations to the actual value of the metric, they should be able to provide guidance on the development of improvement strategies.

The forms of simple energy efficiency metrics employed in the literature include:

(a) Instantaneous power per throughput (Watts/bit/s), $^1H(t)$. Using (11.1) for network elements, (11.2) for networks and (11.8) for services, respectively, we get:

$$a)\ ^1H_j(t) = \frac{P_{idle,j}}{R(t)} + E_j$$

$$^1H_{Ntwk}(t) = \frac{P_{Ntwk}(t)}{R_{Ntwk}(t)} = \frac{M_{OH}}{R_{Ntwk}(t)} \sum_{j=1}^{N_E} \left(P_{idle,j} + E_j \sum_{k=1}^{N^{(svc)}} \alpha_j^{(k)} R^{(k)}(t) \right)$$

$$\approx \frac{M_{OH} N_E}{R_{Ntwk}(t)} \left(\langle P_{idle} \rangle_E + \langle N^{(svc)} \rangle_E \langle E \rangle_E \right)$$

(11.13)

$$b)\ \approx M_{OH} \langle N_E \rangle^{(svc)} \left(\frac{\langle P_{idle} \rangle_E}{\mu_{max} \langle C_{max} \rangle_E} \frac{R_{peak}}{R_{Ntwk}(t)} + \langle E \rangle_E \right)$$

$$^1H^{(k)}(t) = \frac{P^{(k)}(t)}{R^{(k)}(t)} = M_{OH} \sum_{j=1}^{N_E} \left(\frac{P_{idle,j}}{R_j(t)} + E_j \right) \alpha_j^{(k)}$$

$$c)\ \approx M_{OH} N_E^{(k)} \left(\langle P_{idle} \rangle_E \left\langle \frac{1}{R(t)} \right\rangle_E + \langle E \rangle_E \right)$$

In (11.13) and later equations in this chapter, specific lines of the equations have been labelled "a)", "b)" and "c)". These refer to the line of the equation on which they occur and will be used below to refer to those specific expressions of that energy efficiency metric.

In the form for $^1H^{(k)}(t)$, $\langle 1/R(t) \rangle_E$ is the average of $1/R_j(t)$ over the network elements $j = 1$ to N_E. The ease of evaluating terms such as $\langle 1/R(t) \rangle_E$ depends on the types of information available. In some cases, we can use the results derived in Chapter 8 to re-express these in terms of other parameters that may be more easily accessed.

For example, using (8.11), (8.6), (8.17) and (8.16) we can express the average of $1/R_j(t)$ over network elements in terms of average over service traffic flow volumes, $1/R^{(k)}(t)$ and average of number of service flows through the network elements, $1 / N_j^{(svc)}$:

$$\left\langle \frac{1}{R(t)} \right\rangle_E = \frac{1}{N_E} \sum_{j=1}^{N_E} \frac{1}{R_j(t)} = \frac{1}{N_E} \sum_{j=1}^{N_E} \frac{1}{\sum_{k=1}^{N^{(svc)}} \alpha_j^{(k)} R^{(k)}(t)}$$

$$\approx \frac{1}{R_{Ntwk}(t) N_E} \sum_{j=1}^{N_E} \frac{1}{\langle \alpha_j \rangle^{(svc)}} \approx \frac{N^{(svc)}}{R_{Ntwk}(t) N_E} \sum_{j=1}^{N_E} \frac{1}{N_j^{(svc)}}$$

(11.14)

$$\approx \frac{1}{\langle R(t) \rangle^{(svc)}} \left\langle \frac{1}{N^{(svc)}} \right\rangle_E$$

where we have used $\langle R(t) \rangle^{(svc)} = R_{Ntwk}(t)/N^{(svc)}$.

The approach taken to determine the values of these metrics will be highly dependent upon what information we have available regarding the network equipment and service traffic flows.

In some cases, we may have the average, $\langle X \rangle$, of a quantity but not the average of its reciprocal, $\langle 1/X \rangle$. Under certain conditions, we can approximate $\langle 1/X \rangle$ in terms of $1/\langle X \rangle$ and the variance of X, $\mathrm{var}(X)$. For all cases in which we are interested, $X_j > 0$; therefore, provided the condition $\max\{X_j\}\,/\langle X \rangle < 2$, is satisfied we have the following: We express the value of X_j relative to the mean of the X_j, that is; $X_j = \langle X \rangle + \Delta X_j$. With this:

$$\left\langle \frac{1}{X} \right\rangle = \frac{1}{N}\sum_{j=1}^{N}\frac{1}{X_j} = \frac{1}{N\langle X \rangle}\sum_{j=1}^{N}\frac{1}{1+\left(\Delta X_j/\langle X \rangle\right)} \approx \frac{1}{N\langle X \rangle}\sum_{j=1}^{N}1 - \frac{\Delta X_j}{\langle X \rangle} + \left(\frac{\Delta X_j}{\langle X \rangle}\right)^2$$
$$= \frac{1}{\langle X \rangle}\left(1 + \frac{\mathrm{var}(X)}{\langle X \rangle^2}\right)$$

(11.15)

where we have used $\sum_j \Delta X_j = 0$ and $\Delta X_j/\langle X \rangle < 1$ for all j to truncate the power series at $N = 2$. For example, if the variation away from the mean is less than 40%, then $\Delta X_j < 0.4\langle X \rangle$ for all j. In this case, we get $1/\langle X \rangle \approx 0.952\langle 1/X \rangle$. That is a 5% error using $1/\langle X \rangle$ to approximate $\langle 1/X \rangle$. This approximation may be useful if we only know the mean and variance of the X_j.

Because $^1H(t)$ metrics are based on instantaneous power and throughput, they will vary over the diurnal cycle of traffic and the time of calculation will also be required [44]. For example, Baliga *et al.* [22] chose to use $^1H(t)$ at the time of peak traffic. Adopting this will give a lower value for the metric relative to other times within the diurnal cycle. Depending upon the depth of the diurnal cycle, the choice of time for evaluating the metric may significantly impact the resulting value.

(b) "Energy per bit", $^2H(T)$, is defined by the quotient of total energy expended over duration T divided by total throughput (bits) over duration T:

$$^2H(T) = \frac{\int_T P(t)\,dt}{\int_T R(t)\,dt} = \frac{\langle P \rangle_T}{\langle R \rangle_T} = \frac{Q(T)}{B(T)}$$

(11.16)

where

$$\langle X \rangle_T = \frac{1}{T}\int_T X(t)\,dt$$

(11.17)

In this equation the time integral is over a predetermined duration, T; therefore, $Q(T)$ is the energy consumed over time T and $B(T)$ is the total number of bits transferred over that time. An example of the use of this metric is the GreenTouch consortium "GreenMeter" [21]. The GreenMeter assessed the improvement in network energy efficiency using $1/(^2H(T))$ (i.e. bits per Joule) based on a bottom-up model with $T = 1$ year for the years 2010 and 2020. The GreenTouch mission was to develop a roadmap of technologies that would provide a target improvement of a factor of 1000. In terms of the ^2H metric, the target improvement was quantified by the ratio $^2H(\text{year }2010)/^2H(\text{year }2020) = 1000$.

Metrics of the form $^2H(T)$ have has also been employed in ITU-T energy efficiency standards L.1330 [45] and L.1332 [46] for equipment and networks.

For equipment, a network and a service, we have, respectively:

$$a) \; ^2H_j\left(T\right) = \frac{P_{idle,j}}{\left\langle R_j \right\rangle_T} + E_j$$

$$^2H_{Ntwk}\left(T\right) = \frac{M_{OH}}{\left\langle R_{Ntwk} \right\rangle_T} \sum_{j=1}^{N_E} \left(P_{idle,j} + E_j \left\langle R_j \right\rangle_T \right)$$

$$b1) \approx M_{OH} \left(\frac{N_E \left\langle P_{idle} \right\rangle_E}{\left\langle R_{Ntwk} \right\rangle_T} + \left\langle N_E \right\rangle^{(svc)} \left\langle E \right\rangle_E \right)$$

$$b2) \approx M_{OH} \left\langle N_E \right\rangle^{(svc)} \left(\frac{\left\langle P_{idle} \right\rangle_E}{\mu_{max} \left\langle C_{max} \right\rangle_E} \frac{R_{peak}}{\left\langle R_{Ntwk} \right\rangle_T} + \left\langle E \right\rangle_E \right)$$

$$^2H^{(k)}\left(T\right) = \frac{M_{OH}}{\left\langle R^{(k)} \right\rangle_T} \sum_{j=1}^{N_E} \left(P_{idle,j} \left\langle \frac{R_j^{(k)}}{R_j} \right\rangle_T + E_j \left\langle R_j^{(k)} \right\rangle_T \right) \tag{11.18}$$

$$c1) \approx M_{OH} N_E^{(k)} \left(\frac{\left\langle P_{idle} \right\rangle_E}{\left\langle R^{(k)} \right\rangle_T} \left\langle \left\langle \frac{R^{(k)}}{R} \right\rangle_T \right\rangle_E + \left\langle E \right\rangle_E \right)$$

$$c2) \approx M_{OH} N_E^{(k)} \left(\left\langle P_{idle} \right\rangle_E \left\langle \left\langle \frac{1}{R} \right\rangle_T \right\rangle_E + \left\langle E \right\rangle_E \right)$$

In this the form recall that the term $\langle X \rangle_E$ is an average over the network elements, $j = 1, 2,$ ..., N_E:

$$\left\langle X \right\rangle_E = \frac{1}{N_E} \sum_{j=1}^{N_E} X_j \tag{11.19}$$

With this, $\langle\langle X \rangle_E \rangle_T$ is the average of X over the network elements $j = 1$ to N_E and over time duration T:

$$\left\langle \left\langle X \right\rangle_E \right\rangle_T = \frac{1}{T N_E} \int_T \sum_{j=1}^{N_E} X_j\left(t\right) dt \tag{11.20}$$

In (11.18)c the kth service diurnal cycle flow is considered adequately independent of the jth network element diurnal cycle flow to allow the approximation $\langle\langle R^{(k)}/R \rangle_E \rangle_T \approx \langle R^{(k)} \rangle_T \langle\langle 1/R \rangle_E \rangle_T$. In general, provided the number of independent services comprising the traffic flow through the jth network element is significantly greater than one, this approximation should apply. As a further step, provided the conditions to use (11.15) are satisfied, then $^2H^{(k)}(T)$ simplifies further to:

$$^2H^{(k)}\left(T\right) \approx M_{OH} N_E^{(k)} \left(\frac{\left\langle P_{idle} \right\rangle_E}{\left\langle \left\langle R \right\rangle_E \right\rangle_T} + \left\langle E \right\rangle_E \right) \tag{11.21}$$

In some publications the relationships $\langle N_E \rangle^{(svc)} = \langle N_{hops} \rangle^{(svc)} + 1$ and $N_E^{(k)} = N_{hops}^{(k)} + 1$ are used to express the energy efficiency in terms of the number of hops in the path of the service traffic route [22, 32].

This metric can also be defined using top-down network power models. In those cases, $Q(T)$ is determined from information such as equipment deployment inventory data and energy consumption over duration T and $B(T)$ is an assessment of the total network traffic over duration T, typically set to 1 year [27, 28].

This type of metric is proposed in ITU-T Y.3022 (08/2014) [47]. In that document, two types of metrics are described. The first is "power efficiency metrics ($P_{metrics}$) are formulated as follows:

$$P_{metrics} = \text{Power efficiency metrics}\left(\frac{bit/s}{Watt}\right)$$
$$= \frac{\text{Average data rate during measurement time of } T}{\text{Average power consumption during measurement time of } T} \tag{11.22}$$

and "energy efficiency metrics ($E_{metrics}$) are formulated as follows":

$$E_{metrics} = \text{Energy efficiency metrics}\left(\frac{bits}{Joule}\right)$$
$$= \frac{\text{Total length of data bits during measurement time of } T}{\text{Accumulation of power consumption during measurement time of } T} \tag{11.23}$$

Although $P_{metrics}$ and $E_{metrics}$ appear to be different, they are actually identical over a given time duration T. We can see this by noting the definition of "average" for these quantities, as shown in (11.17), corresponds to "(total over duration T)/(duration T)". Consequentially the ratios in (11.22) and (11.23) will be identical over a given duration T.

This brings us to a subtle yet important distinction between "power" metrics of the form $^1H(t)$ and "energy" of the form $^2H(T)$. Note that $^1H(t)$ is defined at time instant t, whereas $^2H(T)$ is defined over a duration T. This distinction means that, although the dimensions of both $^1H(t)$ and $^2H(T)$ are the same (energy/bit), the metrics themselves are quite different. The quantities that make up $^1H(t)$ are "Watts" and "bit/s"; both of these are "instantaneous" (sometimes called "intensive" [48, 49]). The metric $^1H(t)$ is intensive in that it is defined at a specific instant in time. In contrast, the quantities that make up $^2H(T)$, "energy" and "number of bits", both of which are extensive [48] in that they are only meaningful when defined over an extended duration of time, T. Unless the system is unchanging in time, only in the limit of $T \rightarrow 0$ for $^2H(T)$ are these two metrics equal.

(c) "Average power per throughput", $^3H(T)$. This metric is the average of $^1H(t)$ over time duration T, defined by

$$^3H(T) = \frac{1}{T}\int_T \frac{P(t)}{R(t)} dt = \left\langle \frac{P}{R} \right\rangle_T = \left\langle {}^1H(t) \right\rangle_T \tag{11.24}$$

Although this metric has not been proposed in the standards, it does provide information not available with $^2H(T)$. In particular, the metric $^2H(T)$ does not provide information about the impact of diurnal cycle shape. Inspecting the definition of $^2H(T)$, we note that two different systems which have diurnal cycles for $P(t)$ and $R(t)$ with different hourly values but the same mean values for $\langle P \rangle_T$ and $\langle R \rangle_T$, respectively, will give the same value for $^2H(T)$. However, there may be times within the diurnal cycle when one system is much less energy efficient, as given by $^1H(t)$, than the other. In this case, there may be efficiency improvement strategies that can "even out" the traffic over the diurnal cycle so that, according to $^1H(t)$, times of severe energy inefficiency are avoided. The $^2H(T)$ metric will not expose such situations. In other words, metric $^3H(T)$ can distinguish between systems that have durations of high and low instantaneous energy efficiency. Applying this metric, we have:

$$a)\ ^3H_j\left(T\right)=\left\langle\frac{P_j}{R_j}\right\rangle_T=\frac{1}{T}\int_T\frac{P_{idle,j}}{R_j\left(t\right)}dt+E_j=P_{idle,j}\left\langle\frac{1}{R_j}\right\rangle_T+E_j$$

$$^3H_{Ntwk}\left(T\right)=\left\langle\frac{P_{Ntwk}}{R_{Ntwk}}\right\rangle_T=M_{OH}\sum_{j=1}^{N_E}\left(P_{idle,j}\left\langle\frac{1}{R_{Ntwk}}\right\rangle_T+E_j\sum_{k=1}^{N^{(svc)}}\alpha_j^{(k)}\left\langle\frac{R^{(k)}}{R_{Ntwk}}\right\rangle_T\right)$$

$$b1)\approx M_{OH}\left(N_E\left\langle P_{idle}\right\rangle_E\left\langle\frac{1}{R_{Ntwk}}\right\rangle_E+\left\langle N_E\right\rangle^{(svc)}\left\langle E\right\rangle_E\right)$$

$$b2)\approx M_{OH}\left\langle N_E\right\rangle^{(svc)}\left(\frac{\left\langle P_{idle}\right\rangle_E}{\mu_{max}\left\langle C_{max}\right\rangle_E}R_{peak}\left\langle\frac{1}{R_{Ntwk}}\right\rangle_T+\left\langle E\right\rangle_E\right)$$

$$c1)\ ^3H^{(k)}\left(T\right)=\left\langle\frac{P^{(k)}}{R^{(k)}}\right\rangle_T=M_{OH}\sum_{j=1}^{N_E}\left(P_{idle,j}\left\langle\frac{1}{R_j}\right\rangle_T+E_j\right)\alpha_j^{(k)}$$

$$c2)\approx M_{OH}N_E^{(k)}\left(\left\langle P_{idle}\right\rangle_E\left\langle\left\langle\frac{1}{R}\right\rangle_E\right\rangle_T+\left\langle E\right\rangle_E\right)$$

(11.25)

11.4.2 Power per User Metrics

Another common approach to quantifying energy efficiency is power per user [22, 32, 50–53]. This metric tends to be more frequently used when considering networks for which the number of users can be relatively easily assessed: for example, access networks with a known number of access lines, or wireless access networks where the user density (per unit area) is known or specified by the modeller.

To implement a power per user metric, we need the number of users at time t, $N^{(usr)}(t)$. Depending upon the information available, different approaches for determining this parameter may be used. For example, in Baliga *et al.* [22], $N^{(usr)}$ was taken to be a constant given by the peak number of users. Kilper *et al.* [32] set the number of users as the number of households connected to the network. If a user behaviour model is available, such as discussed in Chapter 7, we may have access to the average bandwidth per user, $\langle R(t)\rangle^{(usr)}$, in which case we could use [22, 23]:

$$N^{(usr)}\left(t\right)=\frac{R_{Tot}\left(t\right)}{\left\langle R\left(t\right)\right\rangle^{(usr)}}$$

(11.26)

where $R_{Tot}(t)$ is the total (network element, network or service) traffic at time t and $\langle R(t)\rangle^{(usr)}$ is taken to be the average user traffic at time t. If detailed information on user behaviour types over the diurnal cycle is available, we can use the average over user types:

$$\left\langle R\left(t\right)\right\rangle^{(usr)}=\frac{1}{N^{(usr)}\left(t\right)}\sum_{k=1}^{N^{(usr)}\left(t\right)}R^{(k)}\left(t\right)=\frac{1}{N^{(usr)}\left(t\right)}\sum_{p=1}^{N^{(type)}}N^{(type,p)}(t)\left\langle R^{(usr)}\left(t\right)\right\rangle^{(type,p)}$$

(11.27)

In (11.27) we have included two possible approaches. The summation over $N^{(usr)}(t)$ on the right-hand side assumes we know the user data rate, $R^{(k)}(t)$, for each user $k = 1,\ldots, N^{(usr)}(t)$ at time t. The summation over $N^{(type)}$ applies to the situation where we only know there are several user "types" that are connected to the network, such as the type 1 and type 2 users discussed in Section 7.5. In the summation over user types, $N^{(type)}$, there are $N^{(type,p)}(t)$ users of the p th user type online at time t and the average data rate for the p th type of user is $\langle R^{(usr)}(t)\rangle^{(type,p)}$.

Just as with the H metrics above, several approaches are possible for defining a power per user energy efficiency metric:

a) "Instantaneous Power per user", $^1G(t)$. This metric has been used by several authors for quantifying energy efficiency [21, 30, 53]. For a network element, network and service, respectively, we have:

$$a)\ ^1G_j\left(t\right)=\frac{P_j\left(t\right)}{N_j^{(usr)}\left(t\right)}$$

$$^1G_{Ntwk}\left(t\right)=\frac{P_{Ntwk}\left(t\right)}{N_{Ntwk}^{(usr)}\left(t\right)}$$

$$\approx\frac{M_{OH}}{N_{Ntwk}^{(usr)}\left(t\right)}\left(N_E\left\langle P_{idle}\right\rangle_E+\left\langle N_E\right\rangle^{(svc)}\left\langle E\right\rangle_E R_{Ntwk}\left(t\right)\right)\tag{11.28}$$

$$b)\ \approx\frac{M_{OH}}{N_{Ntwk}^{(usr)}\left(t\right)}\left(\left\langle N_{hops}\right\rangle^{(svc)}+1\right)\left(\frac{\left\langle P_{idle}\right\rangle_E}{\mu_{max}\left\langle C_{max}\right\rangle_E}R_{peak}+\left\langle E\right\rangle_E R_{Ntwk}\left(t\right)\right)$$

$$c)\ ^1G^{(k)}\left(t\right)=\frac{P^{(k)}\left(t\right)}{N^{(usr,k)}\left(t\right)}\approx\frac{M_{OH}N_E^{(k)}}{N^{(usr,k)}\left(t\right)}\left(\left\langle P_{idle}\right\rangle_E\left\langle\frac{1}{R(t)}\right\rangle_E+\left\langle E\right\rangle_E\right)R^{(k)}\left(t\right)$$

where $N_j^{(usr)}\left(t\right)$ is the number of users of the j th network element at time t, $N_{Ntwk}^{(usr)}\left(t\right)$ is the number of users of the network at time t, and $N^{(usr,k)}(t)$ is the number of users of the k th service. In some publications, the metric has been evaluated at a specific time within the diurnal cycle [22].

One can also define power per user metrics analogous to $^2H(T)$ and $^3H(T)$.

b) "Average power per average number of users", $^2G(T)$. This metric is the quotient of the average power over duration T and the average number of users over the time T. This metric can also be interpreted as the total energy over duration T divided by the total number of user-hours over duration T.

$$a)\ ^2G_j\left(T\right)=\frac{\left\langle P_j\right\rangle_T}{\left\langle N_j^{(usr)}\right\rangle_T}=\frac{P_{idle,j}}{\left\langle N_j^{(usr)}\right\rangle_T}+E\frac{\left\langle R_j\right\rangle_T}{\left\langle N_j^{(usr)}\right\rangle_T}$$

$$^2G_{Ntwk}\left(T\right)=\frac{\left\langle P_{Ntwk}\right\rangle_T}{\left\langle N_{Ntwk}^{(usr)}\right\rangle_T}=\frac{M_{OH}}{\left\langle N_{Ntwk}^{(usr)}\right\rangle_T}\sum_{j=1}^{N_E}\left(P_{idle,j}+E_j\left\langle R_j\right\rangle_T\right)$$

$$\approx\frac{M_{OH}}{\left\langle N_{Ntwk}^{(usr)}\right\rangle_T}\left(N_E\left\langle P_{idle}\right\rangle_E+\left\langle N_E\right\rangle^{(svc)}\left\langle E\right\rangle_E\left\langle R_{Ntwk}\right\rangle_T\right)$$

$$b)\ \approx\frac{M_{OH}\left\langle N_E\right\rangle^{(svc)}}{\left\langle N_{Ntwk}^{(usr)}\right\rangle_T}\left(\frac{\left\langle P_{idle}\right\rangle_E}{\mu_{max}\left\langle C_{max}\right\rangle_E}R_{peak}+\left\langle E\right\rangle_E\left\langle R_{Ntwk}\right\rangle_T\right)\tag{11.29}$$

$$^2G^{(k)}\left(T\right)=\frac{\left\langle P^{(k)}\right\rangle_T}{\left\langle N^{(usr,k)}\right\rangle_T}=\frac{M_{OH}}{\left\langle N^{(usr,k)}\right\rangle_T}\sum_{j=1}^{N_E}\left(P_{idle,j}\left\langle\frac{R_j^{(k)}}{R_j}\right\rangle_T+E_j\left\langle R_j^{(k)}\right\rangle_T\right)$$

$$c)\ \approx\frac{M_{OH}N_E^{(k)}}{\left\langle N^{(usr,k)}\right\rangle_T}\left(\left\langle P_{idle}\right\rangle_E\left\langle\left\langle\frac{R^{(k)}}{R}\right\rangle_E\right\rangle_T+\left\langle E\right\rangle_T\left\langle R^{(k)}\right\rangle_T\right)$$

The ^2G type metrics have been used to estimate network power consumption [30, 32, 52] and the power consumption of a service [54–56].

c) "Average power per user", $^3G(T)$. This metric is the average of $^1G(t)$ over time duration T, defined by

$$a)\ ^3G_j(T)=\left\langle\frac{P}{N_j^{(usr)}}\right\rangle_T=P_{idle,j}\left\langle\frac{1}{N_j^{(usr)}}\right\rangle_T+E\left\langle\frac{R_j}{N_j^{(usr)}}\right\rangle_T$$

$$^3G_{Ntwk}(T)=\left\langle\frac{P_{Ntwk}}{N_{Ntwk}^{(usr)}}\right\rangle_T$$

$$\approx M_{OH}\left(N_E\langle P_{idle}\rangle_E\left\langle\frac{1}{N_{Ntwk}^{(usr)}}\right\rangle_T+\langle N_E\rangle^{(svc)}\langle E\rangle_E\left\langle\frac{R_{Ntwk}}{N_{Ntwk}^{(usr)}}\right\rangle_T\right) \qquad (11.30)$$

$$b)\approx M_{OH}\left(\langle N_{hops}\rangle^{(svc)}+1\right)\left(\frac{\langle P_{idle}\rangle_E}{\mu_{max}\langle C_{max}\rangle_E}R_{peak}\left\langle\frac{1}{N_{Ntwk}^{(usr)}}\right\rangle_T+\langle E\rangle_E\left\langle\frac{R_{Ntwk}}{N_{Ntwk}^{(usr)}}\right\rangle_T\right)$$

$$c)\ ^3G^{(k)}(T)=\left\langle\frac{P^{(k)}}{N^{(usr,k)}}\right\rangle_T\approx M_{OH}N_E^{(k)}\left(\langle P_{idle}\rangle_E\left\langle\left\langle\frac{1}{R}\right\rangle_E\frac{R^{(k)}}{N^{(usr,k)}}\right\rangle_T+\langle E\rangle_E\left\langle\frac{R^{(k)}}{N^{(usr,k)}}\right\rangle_T\right)$$

11.5   COMPARING METRICS

As discussed above, metrics are most frequently used for either improvement (i.e. developing strategies to change the value of a system's metric) or benchmarking (comparing the value of a metric for the systems being compared by the benchmark). When used for improvement, the choice of metric will directly influence the strategies adopted for "improvement". When used for benchmarking, the choice of metric will determine what we mean when we say one system is "better" than another. Therefore, the choice of metric is important. Different metrics may not provide identical information when applied to the same situation. A major reason for this is that equipment, networks and services are subject to daily cyclic variation in traffic, which is incorporated into the various metrics in different ways. The analysis presented in this section is based on the work by Hinton and Jalali [57].

Viewing the forms for ^2H, ^3H, ^2G and ^3G metrics we can see that they all include terms of the form $\langle 1/X\rangle_T$ and $\langle Y/X\rangle_T$, where $X(t)$ and $Y(t)$ represent data rates, $R(t)$, or numbers of users, $N(t)$, that have been averaged over a diurnal cycle. To understand how these metrics can provide different information about energy efficiency, we will take a closer look at the metrics $^1H(t)$, $^2H(T)$ and $^3H(T)$. After that, we will use the similarities between the H and G metrics to extend our analysis to the G metrics.

To compare these metrics, it is helpful to have closed forms for each of them. Because the $^2H(T)$ and $^3H(T)$ metrics involve averages over a diurnal cycle, we shall use a sinusoid to approximate the traffic diurnal cycle of a service, $R^{(k)}(t)$, to provide closed-form expressions for the metrics. Likewise, to take a close look at the G metrics, we will also approximate the diurnal cycle for the number of users, $N_{user}(t)$, with a sinusoid. Despite the fact that most real-world diurnal cycles are only very roughly approximated by a sinusoid, we show below that this approximation provides useful information about these metrics.

11.5.1   TRAFFIC DIURNAL CYCLES

Diurnal traffic cycles result from the fact that many users are typically "off line" and "on line" during common times over each 24-hour period. This results in a daily, approximately, cyclic variation in network traffic within a (local) region. Examples of diurnal cycles for several commercially deployed routers are shown by the solid lines in Figure 11.1. Also shown are a pure sinusoidal

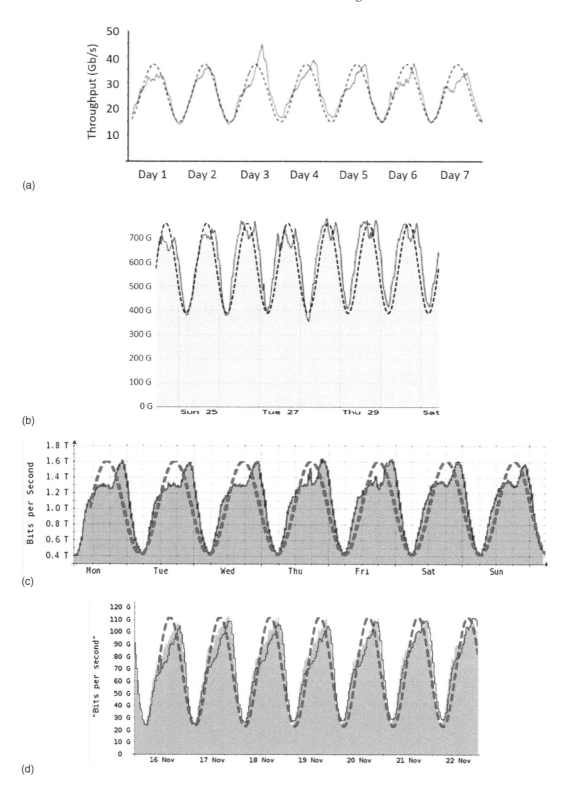

FIGURE 11.1 Examples of 7-day diurnal cycles from several commercially deployed routers (solid lines) and a sinusoidal approximation (dashed lines). The routers are from Internet Exchanges in: (a) Perth Australia, (b) New York (DE-CIX) [58], USA, (c) Hong Kong (HKIX) [59], (d) Copenhagen (Netnod) [60].

approximations (dashed lines) of the 24 hour diurnal cycle. The sinusoidal approximations are set so that they cycle every 24 hours and their amplitude approximates that of the router data. As can be seen, an appropriate sinusoid provides a reasonable approximation to the cycle.

For the purposes of comparing the general characteristics of the metrics, we shall use a simple approximation for the diurnal cycle of the k th service's traffic flow (in bit/s) of the form:

$$R^{(k)}(t) = R_{avg}^{(k)} + \Delta R^{(k)} \cos\left((2\pi t/T) - \phi^{(k)}\right) \tag{11.31}$$

with $T = 24$ hours (the duration of a diurnal cycle), $R_{avg}^{(k)}$ the average traffic of the k th service flow over duration T (given by $\langle R^{(k)} \rangle_T$), and $\Delta R^{(k)}$ the amplitude of the variation away from the average for the k th flow (given by $\Delta R^{(k)}(t) = R^{(k)}(t) - R_{avg}^{(k)}$) The phase $\phi^{(k)}$ accounts for the fact that the diurnal cycles of the individual services may not be synchronised (i.e. different services will have a different time of peak traffic). All of the time dependence of the diurnal cycle is via the "cos" term. We will use (11.31) to calculate closed forms for the metrics $^2H(T)$ and $^3H(T)$.

From (11.3), the time dependence of the total network traffic can be written as the sum of all the service traffic flow volumes:

$$R_{Ntwk}(t) = R_{ave,Ntwk} + \Delta R_{Ntwk} \cos\left((2\pi t/T) - \phi_{Ntwk}\right)$$
$$= \sum_{k=1}^{N^{(svc)}} \left(R_{avg}^{(k)} + \Delta R^{(k)} \cos\left(((2\pi t)/T) - \phi^{(k)}\right) \right) \tag{11.32}$$

Similarly, from (11.4), the total traffic for the j th network element, $R_j(t)$, will have the form:

$$R_j(t) = R_{avg,j} + \Delta R_j \cos\left((2\pi t/T) - \phi_j\right)$$
$$= \sum_{k=1}^{N^{(svc)}} \left(R_{avg,j}^{(k)} + \Delta R_j^{(k)} \cos\left(((2\pi t)/T) - \phi^{(k)}\right) \right) \tag{11.33}$$
$$= \sum_{k=1}^{N^{(svc)}} \alpha_j^{(k)} \left(R_{avg}^{(k)} + \Delta R^{(k)} \cos\left(((2\pi t)/T) - \phi^{(k)}\right) \right)$$

Again, the ϕ terms correspond to the time of the peak load within the diurnal cycle relative to a fixed origin. Setting the time of day relative to time t_0 (in hours), then, if the peak traffic occurs at time t_{peak}, we have $\phi = 2\pi(t_{peak} - t_0)/24$.

The service flows, $R^{(k)}(t)$, are referred to as "synchronised" when all their phases are approximately the same, i.e. $\phi^{(k)} \approx \phi^{(l)}$ for all services k and l. The flows are unsynchronised if the $\phi^{(k)}$ are widely distributed over the range 0 to 2π.

The degree of synchronisation is important for several reasons. For example, network design principles are based on the network being dimensioned to accommodate peak traffic [61–63]. As shown below, the more synchronised the different service flows, the greater the peak traffic is relative to average traffic. In legacy networks, equipment remains fully energised 24/7; therefore, dimensioning a network for peak load means that during off-peak times equipment is lowly utilised, which is less energy efficient (i.e. higher energy per bit) than at peak time. Depending upon the shape of the diurnal cycle, the network may well be operating well away from peak traffic for a significant proportion of the diurnal cycle. This situation is expected with the dominance of video streaming traffic causing the peak to average traffic ratio to increase [64].

Further, in new generation networks, the use of low energy (sleep) states during "off-peak" times is a widely proposed strategy for improving energy efficiency in networks [44, 65] The depth of the diurnal cycle, which is dependent upon the degree of traffic flow synchronisation, is an important consideration because it indicates how much equipment can be powered down during off-peak times.

The depth of the diurnal cycle strongly influences the energy efficiency metrics that include terms of the form $\langle 1/R \rangle_T$ and $\langle R^{(k)}/R \rangle_T$. Consequently, any strategies developed to improve energy efficiency (i.e. reduce the energy per bit value provided by the metric) are likely to be influenced by the depth of the diurnal cycle.

For a network and a network element in which the flows are highly synchronised (i.e. all flows peak at roughly the same time) we have, respectively:

$$R_{avg,Ntwk} \approx \sum_{k=1}^{N^{(svc)}} R_{avg}^{(k)} \qquad R_{max,Ntwk} \approx \sum_{k=1}^{N^{(svc)}} R_{max}^{(k)}$$
$$R_{avg,j} \approx \sum_{k=1}^{N^{(svc)}} R_{avg,j}^{(k)} \qquad R_{max,j} \approx \sum_{k=1}^{N^{(svc)}} R_{max,j}^{(k)} \tag{11.34}$$

When the flows are not synchronised, we get (see Section 11.10.1 in the Appendix):

$$R_{Ntwk}(t) = \sum_{k=1}^{N^{(svc)}} R_{mean}^{(k)}$$
$$+ \left\{ \sum_{k=1}^{N^{(svc)}} \left(\Delta R^{(k)} \right)^2 + \sum_{k \neq l}^{N^{(svc)}} \Delta R^{(k)} \Delta R^{(l)} \cos\left(\phi^{(k)} - \phi^{(l)} \right) \right\}^{1/2}$$
$$\times \cos\left((2\pi t/T) - \phi_{Ntwk} \right) \tag{11.35}$$

In the summation over $k \neq l$ is from 1 to $N^{(svc)}$ and the k-sum over all "off-diagonal" terms, i.e. all the $\Delta R^{(k)} \Delta R^{(l)} \cos(\phi^{(k)} - \phi^{(l)})$ terms with k and l ranging from 1 to $N^{(svc)}$ but excluding the "diagonal" terms which have $k = l$. Also,

$$\phi_{Ntwk} = \tan^{-1} \left(\sum_{k=1}^{N^{(svc)}} \Delta R^{(k)} \sin \phi^{(k)} \middle/ \sum_{k=1}^{N^{(svc)}} \Delta R^{(k)} \cos \phi^{(k)} \right) \tag{11.36}$$

A corresponding form can be written for $R_j(t)$ by using (11.4) with the k-sum over the $R_j^{(k)}$.

If the collection of $N^{(svc)}$ service flows are very unsynchronised, the differences $(\phi^{(k)} - \phi^{(l)})$ are distributed over the range of values 0 to 2π. Comparing a network with highly synchronised flows to the same network with unsynchronised flows, we find:

$$\left. \frac{\Delta R_{Ntwk}}{R_{avg,Ntwk}} \right|_{synch} > \left. \frac{\Delta R_{Ntwk}}{R_{avg,Ntwk}} \right|_{unsynch} \tag{11.37}$$

Taking this further, if the phases, $\phi^{(k)}$, are uniformly randomly distributed over the range 0 to 2π, then:

$$\left. \frac{\Delta R_{Ntwk}}{R_{avg,Ntwk}} \right|_{unsynch} \approx \frac{\left(\left\langle \left(\Delta R \right)^2 \right\rangle^{(svc)} \right)^{1/2}}{\sqrt{N^{(svc)}} \left\langle R_{avg} \right\rangle^{(svc)}} \tag{11.38}$$

where

$$\langle X \rangle^{(svc)} = \frac{1}{N^{(svc)}} \sum_{k=1}^{N^{(svx)}} X^{(k)}, \qquad \langle (X)^2 \rangle^{(svc)} = \frac{1}{N^{(svc)}} \sum_{k=1}^{N^{(svx)}} \left(X^{(k)} \right)^2 \tag{11.39}$$

From this we see that, as the number of unsynchronised flows increases, the network traffic maximum, given by $R_{avg,Ntwk} + \Delta R_{Ntwk}$, converges towards the average, $R_{avg,Ntwk}$:

$$
\begin{aligned}
\lim_{N^{(svc)} \to \infty} R_{max,Ntwk} &= \lim_{N^{(svc)} \to \infty} R_{avg,Ntwk} \left(1 + \frac{\Delta R_{Ntwk}}{R_{avg,Ntwk}} \right) \\
&\approx \lim_{N^{(svc)} \to \infty} R_{avg,Ntwk} \left(1 + \frac{\left(\left\langle (\Delta R)^2 \right\rangle^{(svc)} \right)^{1/2}}{\sqrt{N^{(svc)}} \left\langle R_{avg} \right\rangle^{(svc)}} \right) \approx R_{avg,Ntwk}
\end{aligned}
\tag{11.40}
$$

This means the depth of the overall diurnal cycle reduces when increasingly many unsynchronised traffic flows are brought together. This result applies to any single network element, network facility or overall network that deals with many unsynchronised service flows. These results tell us that facilities dealing with highly synchronised traffic (such as serving only a local time zone) are likely to experience a relatively greater variation between peak and minimum traffic levels than those dealing with many unsynchronised traffic flows. In a given geographical region or time zone, typically the vast majority of service flows have a diurnal cycle that aligns with local end-user activity habits [66–69]. Therefore, local time zone diurnal cycles are likely to display a greater relative variation.

In contrast, we would expect the diurnal cycle of a data centre that services users spread relatively uniformly around the globe (and hence covering all time zones) will experience a relatively shallow diurnal cycle.

11.5.2 Energy per Bit Metrics: ^1H, ^2H and ^3H

11.5.2.1 Energy Efficiency of a Network Element

Without a loss of generality, we can drop the phase term when applying the diurnal cycle to the traffic through a network element. For the j th network element, we have:

$$
\begin{aligned}
a) \; &^1H_j(t) = \frac{P_{idle,j}}{R_{avg,j} + \Delta R_j \cos(2\pi t/T)} + E_j \\
b) \; &^2H_j(T) = \frac{P_{idle,j}}{R_{avg,j}} + E_j \\
c) \; &^3H_j(T) = \frac{P_{idle,j}}{R_{min,j}^{1/2} R_{max,j}^{1/2}} + E_j
\end{aligned}
\tag{11.41}
$$

where $R_{min,j} = R_{avg,j} - \Delta R_j$. and $R_{max,j} = R_{avg,j} + \Delta R_j$. The details for evaluating $^3H_j(T)$ are in the Appendix to this chapter.

We note from its definition in (11.16) that $^2H(T)$ metric for a network element involves $1/\langle R \rangle_T = 1/R_{avg}$, where R_{avg} is the arithmetic mean of $R(t)$ over a diurnal cycle. This can be seen in (11.18)a) In contrast, the definition of the $^3H(T)$ metric in (11.24), when applied to a network element in (11.25) a), involves the harmonic mean ($\langle 1/R \rangle_T$) of the traffic over a diurnal cycle. In the case of a sinusoidal diurnal cycle, $R_{ave} + \Delta R\cos(\theta)$, (where $\theta = 2\pi t/T$) the harmonic mean has the form:

$$\int_0^\pi \frac{d\theta}{R_{avg} + \Delta R\cos(\theta)} = \frac{1}{R_{avg}}\int_0^\pi \frac{d\theta}{1 + (\Delta R/R_{avg})\cos(\theta)}$$

$$= \frac{\pi}{(R_{avg} + \Delta R)^{1/2}(R_{avg} - \Delta R)^{1/2}} = \frac{\pi}{R_{max}^{1/2}R_{mjn}^{1/2}} \tag{11.42}$$

where we have used 3.613.1 in Gradshteyn and Ryzhik [70].

That is, the harmonic mean of a sinusoidal results in the one over geometric mean of the maximum and minimum traffic values $(R_{max,j}R_{min,j})^{-1/2}$.

Applying the $^3H(T)$ metric to examples of actual diurnal cycle data, we found a correlation between the harmonic mean of the $R(t)$ data and $(R_{max}R_{min})^{1/2}$ of over 0.99. Therefore, the form of $^3H_j(T)$ in (11.41), rather than its definition in (11.24), could be a useful metric, even though (11.41) is based on a perfectly sinusoidal diurnal cycle.

Comparing $^2H_j(T)$ and $^3H_j(T)$ in (11.41) for more general diurnal cycle data, it was found that $^3H_j(T)$ reflects the impact of traffic variation over a diurnal cycle via its dependence on the minimum and maximum traffic throughput. In contrast, these details are averaged out in $^2H_j(T)$ and it gives a value unaffected by the depth of the diurnal cycle for a given average value. This is seen by calculating the ratio $^3H_j(T)/^2H_j(T)$ over a range of diurnal cycle depths (R_{min}/R_{max}) for the network element traffic, the results of which are shown in Figure 11.2. The plot in Figure 11.2 shows that, for a given average traffic, $^3H_j(T)$ will expose periods of low utilisation, which corresponds to R_{min} being much less than R_{max}, whereas $^2H_j(T)$ will not.

This outcome shows one of the advantages of the $^3H(T)$ metric, in that it can provide useful information on the impact of the shape of the diurnal cycle. For many network operators, continued high idle-power consumption during times of low traffic is considered to be energy inefficient. This is often addressed by encouraging some service providers to shift their traffic into off-peak times. For example, operating services such as financial institution reconciliations between sites, data back-up services and the like, during off-peak times will flatten the diurnal cycle. In these cases, the total amount of data transported through the network element or network over a diurnal cycle remains the

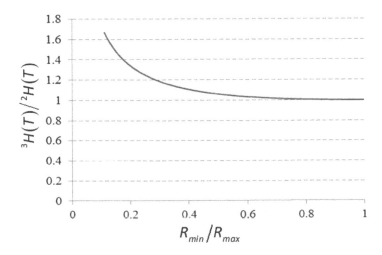

FIGURE 11.2 Plot of the ratio $^3H(T)/^2H(T)$ showing that the metric $^3H(T)$ reflects the impact of durations of low utilisation on network energy/bit metric for equipment in which idle power dominates.

same (hence $^2H(T)$ does not change); however, the ratio R_{min}/R_{max} will be closer to unity and so $^3H(T)$ will provide a measure for improvement of energy efficiency.

From this we can see that both $^2H(T)$ and $^3H(T)$ will reflect changes in energy efficiency attained by changing the equipment (i.e. P_{idle} and E). However, only $^3H(T)$ will reflect changes in energy efficiency attained by changing the shape of the diurnal cycle for a given total data throughput, $B(T)$.

11.5.2.2 Energy Efficiency of a Network

The network energy efficiency, $^1H_{Ntwk}(t)$, is given by merely substituting the form for $R_{Ntwk}(t)$ given by the middle term in (11.32) into (11.13)b). For $^2H_{Ntwk}(T)$ and $^3H_{Ntwk}(T)$ we use the forms (11.18)b) and (11.25)b), respectively. With this we get with the sinusoidal diurnal cycle approximation:

$$a)\ ^1H_{Ntwk}\left(t\right) \approx M_{OH}\left(\left\langle N_{hops}\right\rangle^{(svc)}+1\right)\left(\frac{\left\langle P_{idle}\right\rangle_E}{\mu_{max}\left\langle C_{max}\right\rangle_E}\frac{R_{peak}}{R_{Ntwk}\left(t\right)}+\left\langle E\right\rangle_E\right)$$

$$b1)\ ^2H_{Ntwk}\left(T\right) \approx M_{OH}\left(\frac{N_E\left\langle P_{idle}\right\rangle_E}{R_{avg,Ntwk}}+\left\langle N_E\right\rangle^{(svc)}\left\langle E\right\rangle_E\right)$$

$$b2)\ \approx M_{OH}\left(\left\langle N_{hops}\right\rangle^{(svc)}+1\right)\left(\frac{\left\langle P_{idle}\right\rangle_E}{\mu_{max}\left\langle C_{max}\right\rangle_E}\frac{R_{peak}}{R_{avg,Ntwk}}+\left\langle E\right\rangle_E\right) \qquad (11.43)$$

$$c1)\ ^3H_{Ntwk}\left(T\right) \approx M_{OH}\left(\frac{N_E\left\langle P_{idle}\right\rangle_E}{R_{max,Ntwk}^{1/2}R_{min,Ntwk}^{1/2}}+\left\langle N_E\right\rangle^{(svc)}\left\langle E\right\rangle_E\right)$$

$$c2)\ \approx M_{OH}\left(\left\langle N_{hops}\right\rangle^{(svc)}+1\right)\left(\frac{\left\langle P_{idle}\right\rangle_E}{\mu_{max}\left\langle C_{max}\right\rangle_E}\frac{R_{peak}}{R_{max,Ntwk}^{1/2}R_{min,Ntwk}^{1/2}}+\left\langle E\right\rangle_E\right)$$

where we have used $\langle N_E\rangle^{(svc)} = \langle N_{hops}\rangle^{(svc)} + 1$.

As with a network element, $^3H_{Ntwk}(T)$ is dependent upon the geometric mean of the maximum and minimum network traffic. For a network, the difference between the maximum and minimum network traffic is dependent on the degree of synchronisation of the service traffic flows through the network. To acquire an appreciation of the impact of flow synchronisation on the metrics, a network simulation was constructed. The simulated network consisted of 50 interconnected network elements ($N_E = 50$) each with a power profile given by (11.1) with a range of values for $P_{idle,j}$ (randomly selected in the range 1 kW to 1.5 kW) and E_j (randomly selected in the range 0.5 nJ/bit to 2 nJ/bit). These values are typical of current-generation router and switch technology [44]. The network carries 500 sinusoidal service flows ($N^{(svc)} = 500$), with average flow data rates randomly distributed over the range 0.5 Gbit/s $\leq R^{(k)}_{avg} \leq$ 2 Gbit/s. The network has mesh architecture and each flow travels through 10 network elements randomly selected from the 50 elements in the simulation. No flow travels through the same element more than once.

The synchronisation of the flows is parametrised by quantity b such that the phases of the $N^{(svc)}$ flows chosen randomly over the range $-b\pi \leq \phi^{(k)} \leq b\pi$. For highly synchronised flows, we set $b = 0.1$. For totally desynchronised flows, we set $b = 1$. The simulation was run for values of b from 0.1 to 1.0 in steps of 0.1. (i.e. $b = \{0.1, 0.2, 0.3, 0.4, 0.5, 0.6, 0.7, 0.8, 0.9, 1.0\}$) The b parameter is therefore a measure of the degree of synchronisation of the flows. For b close to zero, the flows are highly synchronised and, for b close to unity, the flows are very unsynchronised.

To parametrise the diurnal cycle depth, $\Delta R/R_{avg}$, the flows are distributed randomly over a range $\Delta R/R_{avg}$ to $\Delta R/R_{avg} + 0.2$, with values of $\Delta R/R_{avg}$ ranging from 0.1 to 0.8 in steps of 0.1 (i.e. the pairs ($\Delta R/R_{avg}$, $\Delta R/R_{avg} + 0.2$) range over the set of values {0.1,0.3), (0.2,0.4), (0.3,0.5), (0.4,0.6), (0.5,0.7), (0.6,0.8), (0.7,0.9), (0.8, 1.0)} representing increasing depth of the cycle).

Metric $^1H_{Ntwk}(t)$ varies over the diurnal cycle. For the case of synchronised flows and a deep diurnal cycle in the simulated network, the value of $^1H_{Ntwk}(t)$ can vary dramatically over the diurnal cycle. For the simulated network described above, at peak traffic time ($R_{max,Ntwk} = R(t_{peak})$), we get $^1H_{Ntwk}(t_{peak}) \approx 66$ nJ/bit. At the time of minimum traffic ($R_{min,Ntwk} = R(t_{min})$), $^1H_{Ntwk}(t_{min}) \approx 910$ nJ/bit. From the significant difference between these two values, we see that using this metric requires careful consideration of the time at which it is measured. Measuring at peak traffic time will give an underestimate for typical energy per bit.

The values of $^2H_{Ntwk}(T)$ and $^3H_{Ntwk}(T)$ ($T = 24$ hours) for ranges of cycle depth, $\Delta R/R_{avg}$, and synchronisation, b, are shown by the surface plots in Figure 11.3. The top-left region of the surface plots corresponds to highly synchronised service traffic flows with relatively deep diurnal cycles. We see that $^3H_{Ntwk}(T)$ reflects the impact of durations of low network utilisation that occur in networks with highly synchronised traffic and a deep diurnal cycle. In contrast, $^2H_{Ntwk}(T) \approx {}^3H_{Ntwk}(T)$ for networks that are desynchronised or have shallow diurnal cycles. This shows that, as with applying these metrics for quantifying energy efficiency of equipment, $^3H_{Ntwk}(T)$ will show a network energy efficiency improvement (i.e. reduction of network energy per bit) can be attained by changing the traffic behaviours of network users. In contrast, $^2H_{Ntwk}(T)$ will not.

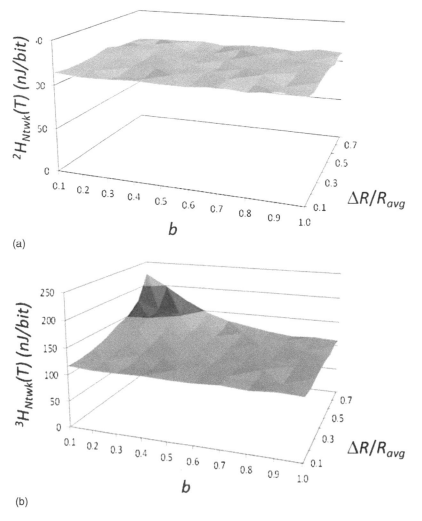

(a)

(b)

FIGURE 11.3 Surface plots of (a) $^2H_{Ntwk}(T)$ and (b) $^3H_{Ntwk}(T)$ for values of synchronicity parameter b and diurnal cycle depth $\Delta R/R_{ave}$.

11.5.2.3   Energy Efficiency of a Service

Using (11.13)c), (11.18)c1) and (11.25)c2), the three metrics for the k th service are as follows:

$$a)\ {}^1H(t)^{(k)} \approx M_{OH} N_E^{(k)} \left(\langle P_{idle} \rangle_E \left\langle \frac{1}{R(t)} \right\rangle_E + \langle E \rangle_E \right)$$

$$b)\ {}^2H^{(k)}(T) \approx M_{OH} N_E^{(k)} \langle P_{idle} \rangle_E \left(\left\langle \frac{1}{R_{max}^{1/2} R_{min}^{1/2}} \right\rangle_E + \right.$$

$$\left. + \frac{\Delta R^{(k)}}{R_{avg}^{(k)}} \left\langle \cos\left(\theta^{(k)}\right) \right\rangle_E \left\langle \frac{1}{\Delta R} \left(1 - \frac{R_{avg}}{R_{max}^{1/2} R_{min}^{1/2}} \right) \right\rangle_E \right) + \langle E \rangle_E \right)$$

$$c)\ {}^3H(T)^{(k)} \approx M_{OH} N_E^{(k)} \left(\langle P_{idle} \rangle_E \left\langle \frac{1}{R_{max}^{1/2} R_{min}^{1/2}} \right\rangle_E + \langle E \rangle_E \right)$$

$$(11.44)$$

where $\langle \cos(\phi^{(k)}) \rangle_E$ is the average (over network elements) of the phase offsets between the peak traffic time of the j th network element and the peak traffic time of the k th service.

$$\left\langle \cos\left(\phi^{(k)}\right) \right\rangle_E = \frac{1}{N_E} \sum_{j=1}^{N_E} \cos\left(\phi^{(k)} - \phi_j\right) \tag{11.45}$$

Where ϕ_j is the phase of the total traffic through the j th network element, given by (11.33)

Looking at (11.44)b), if the $N^{(svc)}$ service traffic flows are unsynchronised (i.e. $\phi_j^{(k)}$ are relatively uniformly spread over the range 0 to 2π), then $\langle \cos(\phi^{(k)}) \rangle_E \approx 0$, $R_{min,j} \approx R_{max,j} \approx R_{avg,j}$, in which case we can use (11.15), giving:

$$^2H^{(k)}(T) \approx M_{OH} N_E^{(k)} \left(\frac{\langle P_{idle} \rangle_E}{\langle R_{avg} \rangle_E} + \langle E \rangle_E \right) \tag{11.46}$$

With highly synchronised flows, the vast majority of the $\phi_j^{(k)}$ fall within a narrow range of values. For a highly synchronised network, $\langle \cos(\phi^{(k)}) \rangle_E \approx 1$ if the k th flow is "in-synch" with the vast majority of other flows and $\langle \cos(\phi^{(k)}) \rangle_E \approx -1$ if the k th flow is 180 degrees "out-of-synch" with the vast majority of flows. In this situation (11.44)b) indicates that the energy efficiency of the k th flow is determined by the sign of the $1 - R_{avg}/(R_{min}R_{max})^{1/2}$ terms for the network elements. We can determine the sign of this expression by using the fact that, for a sinusoidal diurnal cycle, $R_{avg,j} - \Delta R_j = R_{min,j}$ for every network element. Therefore:

$$R_{avg,j}^2 \geq R_{avg,j}^2 - \Delta R_j^2$$
$$\therefore R_{avg,j}^2 \geq \left(R_{avg,j} + \Delta R_j\right)\left(R_{avg,j} - \Delta R_j\right) = R_{min,j} R_{max,j}$$
$$\therefore R_{avg,j} \geq R_{min,j}^{1/2} R_{max,j}^{1/2}$$
$$\therefore \frac{R_{avg,j}}{R_{min,j}^{1/2} R_{max,j}^{1/2}} \geq 1$$
$$\therefore 0 \geq 1 - \frac{R_{avg,j}}{R_{min,j}^{1/2} R_{max,j}^{1/2}}$$

$$(11.47)$$

This inequality applies to all network elements and, because $\Delta R > 0$, the term $\langle (1 - R_{avg}/(R_{min}R_{max})^{1/2}/\Delta R \rangle_E$ will be negative. This means that for a network in which $R_{avg,j} \geq \Delta R_j$

the energy per bit of a service that is in-synch with the majority of the network flows will be lower than the energy per bit of a service that is significantly out-of-synch with the rest of the majority of network flows.

The reader may recall the discussion at the end of Section 10.7 in which it was noted that, depending upon the shape of the diurnal cycle, the energy consumption of a service that is out-of-synch with the overall traffic may be less than when it is in-synch. This does not conflict with the result above because the shape of the diurnal cycles for which an out-of-synch traffic consumes less energy than an in-synch traffic flow are well away from sinusoidal. In fact, if we insert values for a sinusoidal diurnal cycle into (10.97), we get:

$$
\begin{aligned}
Q^{(k)}_{out-of-synch} - Q^{(k)}_{in-synch} &= P_{idle}T\,\frac{\Delta R^{(k)}}{\Delta R}\left(\frac{1}{R_{min}^{1/2}R_{max}^{1/2}}\left(R_{max}+R_{min}\right)-2\right)\\
&= P_{idle}T\,\frac{\Delta R^{(k)}}{\Delta R}\left(\frac{R_{min}}{R_{max}}\right)^{1/2}\left(\frac{R_{max}}{R_{min}}-2\left(\frac{R_{max}}{R_{min}}\right)^{1/2}+1\right)\\
&= P_{idle}T\,\frac{\Delta R^{(k)}}{\Delta R}\left(\frac{R_{min}}{R_{max}}\right)^{1/2}\left(\left(\frac{R_{max}}{R_{min}}\right)^{1/2}-1\right)^{2}
\end{aligned}
\tag{11.48}
$$

From this, we see that $Q^{(k)}_{out-of-synch}$ will always be greater than $Q^{(k)}_{in-synch}$ for sinusoidal diurnal cycle profiles.

To graphically display the dependence of $^{2}H^{(k)}(T)$ and $^{3}H^{(k)}(T)$ on the parameters $\Delta R/R_{avg}$ and b, we average over the service index k. Plotting the simulation results for $\langle^{2}H(T)\rangle^{(svc)}$ and $\langle^{3}H(T)\rangle^{(svc)}$, we get identically shaped surface plots as those shown in Figure 11.3.

Turning our attention back to out-of-synch services, if the k th service is significantly out-of-synch with the vast majority of other services, its value of $^{2}H^{(k)}(T)$ can be significantly greater than in-synch services. For example, consider a service k for which $\phi^{(k)}$ is close to 180 degrees, whereas $\phi^{(m)} \approx 0$ for all $m \neq k$. In this case the network is highly synchronised with only the k th flow well out-of-synch. The surface plot of $^{2}H^{(k)}(T)$ for a service with $\phi^{(k)} = \pi$ for b ranging from 0.1 to 1, is shown in Figure 11.4. We see that an out-of-synch service has much higher energy per bit than the other (synchronised) services when the network is highly synchronised. As the degree of synchronisation reduces or the diurnal cycle depth reduces, $^{2}H^{(k)}(T)$ reduces to that of the other services.

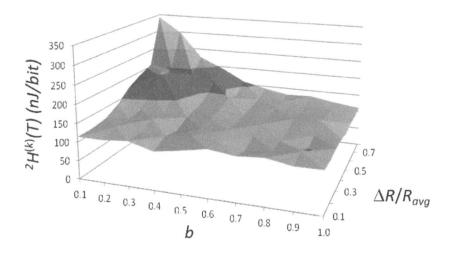

FIGURE 11.4 Surface plot for $^{2}H^{(k)}(T)$ for a service π out-of-synch with all other services in the network. Low values of b correspond to all other flows across the network being highly synchronised.

Comparing Figure 11.3a and 11.4, we see that service operators wishing to minimise the energy per bit of their service will want to avoid being significantly out-of-synch with the majority of services. This will lead to service providers trying to synchronise their services with everyone else. This, in turn, will lead to deeper diurnal cycles and the resulting dimensioning of the network will result in an increase in idle power consumption relative to a flatter diurnal cycle. To mitigate this, the network operator may consider a tariff structure that encourages service providers who can shift their time of service into off-peak times to do so there-by flattening the diurnal cycle.

11.5.3 POWER PER USER METRICS: ^1G, ^2G AND ^3G

We now repeat this analysis for the G metrics. We retain the sinusoidal approximation for the traffic diurnal cycles, $R(t)$, and additionally approximate the diurnal cycles for the number of users, $N^{(usr)}(t)$, with a sinusoid.

$$N^{(usr)}(t) = N_{avg}^{(usr)} + \Delta N^{(usr)} \cos\left(\frac{2\pi t}{T}\right) \tag{11.49}$$

and

$$N_{\max}^{(usr)} = N_{avg}^{(usr)} + \Delta N^{(usr)}, \qquad N_{\min}^{(usr)} = N_{avg}^{(usr)} - \Delta N^{(usr)} \tag{11.50}$$

To make the results as general as possible, we will include consideration of the situation when the traffic and user diurnal cycles have a phase offset. Where possible, without loss of generality, the phase offset will be placed in the numerator. This means the phase terms in the metrics represent the traffic peak times relative to the peak time for the number of users.

11.5.3.1 Power per User of a Network Element

For the j th network element we have

$$a)\ ^1G_j(t) = \frac{P_{idle,j} + E\left(R_{avg,j} + \Delta R_j\left(\cos\frac{2\pi t}{T} + \phi_j\right)\right)}{N_{avg,j}^{(usr)} + \Delta N_j^{(usr)} \cos\left(\frac{2\pi t}{T}\right)}$$

$$b)\ ^2G_j(T) = \frac{P_{idle,j} + ER_{avg,j}}{N_{avg,j}^{(usr)}} \tag{11.51}$$

$$c)\ ^3G_j(T) = P_{idle}\frac{1}{N_{max,j}^{(usr)1/2}N_{min,j}^{(usr)1/2}} +$$

$$+ E_j\left(\frac{R_{avg,j}}{N_{max,j}^{(usr)1/2}N_{min,j}^{(usr)1/2}} + \frac{\Delta R_j \cos\left(\phi_j^{(usr)}\right)}{\Delta N_j^{(usr)}}\left(1 - \frac{N_{avg,j}}{N_{max,j}^{(usr)1/2}N_{min,j}^{(usr)1/2}}\right)\right)$$

Where $N_{avg,j}^{(usr)}$, $N_{max,j}^{(usr)}$, $N_{min,j}^{(usr)}$ are the average, maximum and minimum number of users of the j th network element respectively. Also, $\Delta N_j^{(usr)} = N_{max,j}^{(usr)} - N_{min,j}^{(usr)}$ and the phase $\phi_j^{(usr)}$ is the phase of the peak time of the diurnal traffic R_j relative to the peak time of the number of users $N_j^{(usr)}$ of the j th network element.

11.5.3.2 Power per User of a Network

For a network, we have

$$
a)\ {}^1G_{Ntwk}(t) \approx \frac{M_{OH}\left(\langle N_{hops}\rangle^{(svc)}+1\right)}{N^{(usr)}_{avg,Ntwk}+\Delta N^{(usr)}_{Ntwk}\cos\left(\dfrac{2\pi t}{T}\right)}\times
$$

$$
\left(\frac{\langle P_{idle}\rangle_E\, R_{peak}}{\mu_{max}\langle C_{max}\rangle_E}+\langle E\rangle_E\left(R_{avg,Ntwk}+\Delta R_{Ntwk}\cos\left(\frac{2\pi t}{T}+\phi^{(usr)}_{Ntwk}\right)\right)\right)
$$

$$
b)\ {}^2G_{Ntwk}(T) \approx \frac{M_{OH}\left(\langle N_{hops}\rangle^{(svc)}+1\right)}{N^{(usr)}_{avg,Ntwk}}\left(\frac{\langle P_{idle}\rangle_E\, R_{peak}}{\mu_{max}\langle C_{max}\rangle_E}+\langle E\rangle_E\, R_{avg,Ntwk}\right)
$$ (11.52)

$$
c)\ {}^3G_{Ntwk}(T) \approx M_{OH}\left(\langle N_{hops}\rangle^{(svc)}+1\right)\times\left(\frac{\langle P_{idle}\rangle_E\, R_{peak}}{\mu_{max}\langle C_{max}\rangle_E\, N^{(usr)1/2}_{max,Ntwk}N^{(usr)1/2}_{min,Ntwk}}\right.
$$

$$
+\left(\frac{R_{avg,Ntwk}}{N^{(usr)1/2}_{max,Ntwk}N^{(usr)1/2}_{min,Ntwk}}+\right.\\
\left.+\Delta R_{Ntwk}\cos\left(\phi^{(usr)}_{Ntwk}\right)\left(\frac{1}{\Delta N^{(usr)}_{Ntwk}}\left(1-\frac{N^{(usr)}_{avg,Ntwk}}{N^{(usr)1/2}_{max,Ntwk}N^{(usr)1/2}_{min,Ntwk}}\right)\right)\right)\langle E\rangle_E
$$

In ${}^1G_{Ntwk}(t)$ and ${}^3G_{Ntwk}(T)$, the phase $\phi^{(usr)}_{Ntwk}$ is the phase of the peak time of the diurnal network traffic, R_{Ntwk}, relative to the peak time of the number or network users, $N^{(usr)}_{Ntwk}$.

11.5.3.3 Power per User of a Service

For the k th service, we have

$$
a)\ {}^1G^{(k)}(t) \approx M_{OH}N^{(k)}_E\left(\langle P_{idle}\rangle_E\left\langle\frac{1}{R_{avg}+\Delta R\cos\left(\frac{2\pi t}{T}+\phi\right)}\right\rangle_E+\langle E\rangle_E\right)\times
$$

$$
\frac{R^{(k)}_{avg}+\Delta R^{(j)}\cos\left(\frac{2\pi t}{T}+\phi^{(k)}\right)}{N^{(k)}_{avg}+\Delta N^{(k)}\cos\left(\frac{2\pi t}{T}\right)}
$$ (11.53)

$$
b)\ {}^2G^{(k)}(T) \approx \frac{M_{OH}N^{(k)}_E}{N^{(k)}_{avg}}\left(\langle P_{idle}\rangle_E\left(R^{(k)}_{avg}\left\langle\frac{1}{R^{1/2}_{max}R^{1/2}_{min}}\right\rangle_E+\right.\right.
$$

$$
\left.\left.+\Delta R^{(k)}\left\langle\cos\left(\phi^{(k)}\right)\right\rangle_E\left\langle\frac{1}{\Delta R}\left(1-\frac{R_{avg}}{R^{1/2}_{max}R^{1/2}_{min}}\right)\right\rangle_E\right)+\langle E\rangle_E\, R^{(k)}_{avg}\right)
$$

where $\langle\cos(\phi^{(k)})\rangle_E$ is the cosine of phase offset between the k th service flow and the total traffic, R_j, through the j th network element averaged over all the network elements. The ${}^3G^{(k)}(T)$ metric is not presented in (11.53) because, even with the simplification of approximating the traffic and user diurnal cycles with pure sinusoids, (11.30)c) does not enable analytical calculation of ${}^3G^{(k)}(T)$ due to the multiple averaging term $\langle\langle 1/R\rangle_E R^{(k)}/N^{(usr,k)}\rangle_T$, which results in a form that cannot be integrated into a closed form.

11.6   APPLICATIONS OF ENERGY EFFICIENCY METRICS

The application of energy efficiency metrics in real networks can be somewhat problematic due to the difficulty of attaining all the required diurnal cycle information. In particular, evaluating these metrics for a network or service can require collection of a significant amount of information typically not easily available. Approximations for the metrics can make their evaluation easier, although most likely at the cost of reduced accuracy. Also, the inter-relationships between the metrics may allow the data collected for one metric to be used to evaluate another.

The analytical expressions derived for the H metrics in Section 11.5.2 and G metrics in Section 11.5.3 are based on pure sinusoidal diurnal cycles. Although real networks' diurnal cycles are not pure sinusoids, there are advantages for taking the sinusoidal forms for these metrics as a valid metric in their own right. That is, we use the forms for the H and G metrics given in Sections 11.5.2 and 11.5.3 as definitions for energy efficiency metrics. Doing this enables us to reduce the amount of detailed data required to calculate the metrics. Using the forms for these metrics given by the sinusoidal cycles we avoid having to collect the hourly data for every diurnal cycle in the system under consideration; we only need averages, maxima, minima and the peak times for the cycles. This may significantly simplify the data collection process.

Referring back to the discussion in Section 10.7 and relating to (11.47) and (11.48), we recall that there can be situations in which a metric evaluated using the sinusoidal approximation has limitations. If the diurnal cycle profiles under consideration are very different from a typical diurnal cycle shape, it may be that using the sinusoidal forms for the metrics is not appropriate. In that case, the more direct forms should be used.

Adopting the sinusoidal diurnal cycle-based forms, the metrics $^{2}H_{j}(T)$ in (11.41)b and $^{2}H_{Ntwk}(T)$ in (11.43)b only involve $\langle R \rangle_{T} = R_{avg}$, and therefore these forms can be directly applied to any diurnal cycle profile.

From the equations above, we see that, with the sinusoidal diurnal cycle approximation, some of the H and G metrics involve quantities such as R_{max}, R_{min}, ΔR, N_{max}, N_{min}, ΔN and $\phi^{(k)}$. To generalise these metrics from a sinusoid to a real-world diurnal cycle, one possible approach is to replace these values with the corresponding values averaged over multiple diurnal cycles. For example, R_{max} will be replaced by $\langle R_{max} \rangle_{mT}$, R_{min} by $\langle R_{min} \rangle_{mT}$, ΔR by $\langle \Delta R \rangle_{mT}$ and so on, where $\langle X \rangle_{mT}$ represents the time average of those specific quantities taken over multiple, m, diurnal cycle durations, T. Where a quantity is raised to a power, a, we can replace X^{a} by $(\langle X \rangle_{mT})^{a}$. In these cases, the values for these quantities can be extracted from traffic data collected over multiple diurnal cycles, as indicated by the mT subscript.

If this approach is adopted, care must be taken when constructing the parameters ΔR and ΔN. For sinusoidal diurnal cycles, these parameters are easily evaluated. For sinusoids, we have $R_{max} = R_{avg} + \Delta R$ and $R_{min} = R_{avg} - \Delta R$. However, for a general diurnal cycle these forms may not accurately reflect the relationships between R_{max}, R_{min} and R_{avg}. Therefore, applying the metric forms based on sinusoidal approximations to real networks we need to consider how the ΔR and ΔN parameters are constructed.

To investigate the applicability of the sinusoidal forms to provide estimates for $^{2}H(T)$ and $^{3}H(T)$ when using the metric on real typical diurnal cycles, a set of simulations of a mesh telecommunications network was constructed. Using measured diurnal cycle data, a collection of 500 diurnal cycles was generated using the forecasting methods described in Section 6.4. This was done to produce cycles that are representative of real data and which can display significantly varying depths (i.e. varying peak to average ratios) to ensure the approximations can deal with flows that are forecast for the future [21, 64]. These cycles were used for the diurnal cycles of service flows, $R^{(k)}(t)$. The maximum values of the cycles $\left(R_{max}^{(k)} = R_{peak}^{(k)} = R^{(k)}\left(t_{peak}^{(k)} \right) \right)$ were randomly set over the range 2.5 Gbit/s to 40 Gbit/s. A total of $N^{(svc)} = 1500$ flows were generated for the simulation and were directed through between 8 and 16 network elements in a network consisting of $N_E = 500$ network elements. The simulated network was a mesh and the flow paths through the network were randomly generated so that no flow travelled through a network element more than once. Of the 1500 flows, half of the flows were split into two parallel paths and the $\alpha_j^{(k)}$ parameters for these flows were constructed accordingly.

The simulation was designed to allow control over the range of diurnal cycle depths and the degree of synchronisation of the peak traffic times, $R_{peak}^{(k)}$. The depth of the diurnal cycles was able to be set to values that ranged from $R_{max}^{(k)} / R_{min}^{(k)}$ averaging around 3 up to averaging around 47. The degree of synchronisation of the flows was set by allowing the peak traffic time of each flow, $t_{peak}^{(k)}$, to be uniformly random over a pre-set range of hours of the day. For highly synchronised flows, all the peak times were set to be within 2 hours of each other. For highly unsynchronised flows, the peak times were allowed to range over the full 24 hours of a diurnal cycle.

The network element traffic flows were generated using (11.4). The 500 network elements were represented by the power profile given in (11.1) with the values of $P_{idle,j}$ ranging between 1 kW and 1.5 kW and E_j ranging between 0.5 nJ/bit and 2 nJ/bit.

The service traffic flows were routed through the network and the simulation calculated network element, network and service power usages using (11.1) and the summations in (11.2) and (11.8), respectively. The network element traffic is given by (11.4) and the total network traffic is given by (11.3). Using these values, the $^2H(T)$ and $^3H(T)$ metrics were calculated in three ways:

1) **Using the original definitions**. That is, the $^2H(T)$ metrics were calculated using (11.16) and the $^3H(T)$ metrics using (11.24). These values are considered the "actual" values of the metrics because they do not involve any approximation or simplification.
2) **Using the forms given after averaging over one or more of service flows, network elements and the diurnal cycle**. The $^2H(T)$ metrics were calculated using (11.18)a), b1) and c2). The $^3H(T)$ metrics were calculated using (11.25)a), b1) and c2).
3) **Using the analytical sinusoidal forms after averaging**. The $^2H_j(T)$ and $^3H_j(T)$ metrics were calculated using (11.41)b) and c), respectively. The $^2H_{Ntwk}(T)$ and $^3H_{Ntwk}(T)$ metrics were calculated using (11.43)b1) and c1), respectively. The $^2H^{(k)}(T)$ and $^3H^{(k)}(T)$ metrics were calculated using (11.44)b) and c), respectively.

It was found that, with approach 3), the quantities $R_{avg}^{(k)}$, $\Delta R^{(k)}$, $R_{avg,j}$ and ΔR_j for the measured diurnal cycles had to be defined as follows:

$$R_{avg}^{(k)} = \frac{R_{max}^{(k)} + R_{min}^{(k)}}{2}, \quad R_{avg,j} = \frac{R_{max,j} + R_{min,j}}{2}$$
$$\Delta R^{(k)} = \frac{R_{max}^{(k)} - R_{min}^{(k)}}{2}, \quad \Delta R_j = \frac{R_{max,j} - R_{mjn,j}}{2} \tag{11.54}$$

These definitions ensure the corresponding sinusoidal traffic flow approximations for these parameters, that provide the closed-form expressions, are always positive.

The simulation generated 132 different network traffic flows with varying diurnal cycle depth and degree of synchronisation. To assess the accuracy of the approximations using the averaged values given in 2) above and the sinusoidal approximations given in 3) above, the "Normalised Root Mean Squared Deviation" (NRMSD) was calculated. The NRMSD is defined as follows [71]:

$$NRMSD(x,y) = \sqrt{\sum_{m=1}^{M} \frac{(x_m - y_m)^2}{M}} \Big/ mean(x) \tag{11.55}$$

TABLE 11.1

Average NRMSD Values for Approaches (2) and (3) for Approximating Energy Efficiency Metrics

NRMSD	Approach (2)	Approach (3)
$^2H_j(T)$	0	0
$^3H_j(T)$	0	0.024
$^2H^{(k)}(T)$	0.066	0.33
$^3H^{(k)}(T)$	0.066	0.07

where x_m are the actual values and the y_m are the estimations. The $NRMSD(x,y)$ provides a measure of the accuracy the approximations, y_m, provide for the actual values, x_m. In our use, the x_m are the "actual" values of the ^2H and ^3H metrics given by approach 1) above. The y_m are the approximations given by approaches 2) and 3) above. The closer the NRMSD is to zero, the better the approximation. The results are presented in Table 11.1

From Table 11.1 we see that approaches 2) and 3) above for calculating the energy efficiency metrics, on average, provide good estimates. Using these forms for the metrics can simplify the calculations and reduce the amount of data required.

Table 11.1 also shows that the sinusoidal approximations for $^2H^{(k)}(T)$, on average, are not as accurate as the other approximations. Closer inspection of the detailed results showed that this reduction in accuracy occurs for networks in which the services have deep diurnal cycles and are highly synchronised. This is due to the factor $1/R_{max}^{1/2}R_{min}^{1/2}$ resulting from the closed forms using the sinusoidal diurnal cycles. This sinusoidal approximation may not be applicable to systems with synchronised deep diurnal cycles.

The result in (11.44)b) is also reflected in the simulation results. This is seen by running the simulation with all services having a relatively deep diurnal cycle (average mean is approximately 40% of the peak), with the peak traffic time of the $k = 0$ service fixed at $t_{peak}^{(0)} = 12$ hours and all the other services given peak traffic times increasingly desynchronised, starting with the $k \neq 0$ peak times in the range 0 to 2 hours and increasing the range of the all $k \neq 0$ service peak hour times spread from 0 to 24 hours in steps of 2 hours. In other words, this set of simulations has the peak time of the $k = 0$ service always at $t_{peak}^{(0)} = 12$ hours. All the other services have peak times that are initially (randomly) clustered within the range of 0 to 2 hours meaning the $k = 0$ service is effectively totally out-of-synch with the all the other services. The peak times for all the $k \neq 0$ services is then step wise increased to (randomly) range over 0 to 4 hours, then over 0 to 6 hours and so on up to these services peak times being randomly spread over 0 to 24 hours. As the range of spread of the $k \neq 0$ increases, the phase difference between the $k = 0$ service and sum all the other services reduces until the peak time of the $k = 0$ service is within the spread of peak times of the other services. At this stage of the simulation, the $k = 0$ service is no longer "out-of-synch" with the other service traffic flows.

The results of these simulations are shown in Figure 11.5. In this figure we can see that, when the $k = 0$ service is totally out-of-synch with the other services (which corresponds to $t^{(k \neq 0)}_{peak} = 2$ on the horizontal axis in the figure), the energy per bit, $^2H^{(0)}(T)$, of the 0th service, shown as the light shaded dots, is significantly greater than the average for all the other services, shown as the dark shaded dots. As the spread of peak times of the $k \neq 0$ services increases, (which corresponds to $t^{(k \neq 0)}_{peak}$ increasing along the horizontal axis), the value of $^2H^{(0)}(T)$ reduces (because it is less "out-of-synch" with all the other services) and becomes aligned with the average over all the other services, $\langle ^2H(T) \rangle^{(svc)}$.

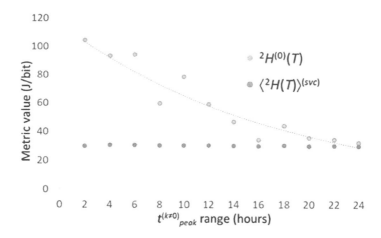

FIGURE 11.5 Values of $^2H^{(0)}(T)$ and $\langle ^2H(T)\rangle^{(svc)}$ for degrees of synchronisation of the $k = 0$ service flow relative to all other flows. The peak traffic time of the $k = 0$ service is fixed at 12 hours. The rest of the network flows are set with t_{peak} values covering from 0 hours to the number of hours on the horizontal axis. Therefore, the difference between phase of the $k = 0$ service traffic and overall phase of all the other service traffic flows decreases as $t_{peak}^{(k\neq0)}$ increases.

In general, when all the service traffic flows are similarly synchronised, such that only very few service traffic flows have a phase, $\phi^{(k)}$, well away from all other services, or the diurnal cycles are relatively flat, we find:

$$\langle ^2H(T)\rangle^{(svc)} = \sum_{k=1}^{N^{(svc)}} {}^2H^{(k)}(T) \bigg/ N^{(svc)} \approx {}^2H_{Ntwk}(T)$$

$$\langle ^3H(T)\rangle^{(svc)} = \sum_{k=1}^{N^{(svc)}} {}^3H^{(k)}(T) \bigg/ N^{(svc)} \approx {}^3H_{Ntwk}(T)$$

(11.56)

This means that, provided the vast majority of services are similarly synchronised, the average energy efficiency of the services is approximately equal to the energy efficiency of the network. With this proviso satisfied, we can approximate the energy efficiency of (typical) services by setting them equal to that of the network. (By "typical", we mean the service in question is not an outlier in terms of its synchronisation relative to the vast majority of other services.) Conversely, we can use the average energy efficiency of multiple typical services to provide an estimation of the energy efficiency of the network. This means we can use whichever of the two approaches is easier to calculate to provide an estimate of the other. (The justification of these approximations and the conditions that must be satisfied are given in detail in the Appendix).

Finally, we consider the approximate form for $^2H^{(k)}(T)$ given in (11.46). Using the simulation to assess the accuracy of (11.46), the average NRMSD for this form was around 0.08. Therefore, the approximation given in (11.46) can provide a reasonable estimate of service energy efficiency.

From the above, we can see that there are options available for estimating these energy efficiency metrics that do not require highly detailed hour-by-hour diurnal-cycle data. The metrics $^1H(t)$, $^2H(T)$, $^1G(t)$ and $^2G(T)$ are commonly used [3, 7–10, 21, 22, 26, 29, 30, 37, 38, 47], whereas the $^3H(T)$ and $^3G(T)$ metrics are not. The advantage provided by the $^3H(T)$ metric is that it quantifies the impact of the shape of the diurnal cycle and its relationship to other traffic flows (via R_{max}, R_{min} and $\phi^{(k)}$). This enables us to quantify the impact of changing diurnal traffic profiles on energy efficiency of networks and services.

11.6.1 Estimating Power Consumption

Although the energy efficiency metrics have primarily been created to provide a quantitative measure of "energy efficiency" [3, 7, 8, 12, 72, 73], they have also been used to estimate the power consumption of equipment, networks and services [9, 22–24, 26, 31, 40] or to allocate a power consumption footprint [22, 30, 54]. Both the H and G metrics have been used to estimate power consumption.

11.6.1.1 Power Consumption Estimation Using "Energy per Bit": H Metrics

Using energy per bit H metrics to estimate power or energy consumption is based on the principle that the power, P, consumed by a network element, network or service with energy efficiency H Joules/bit and with traffic load R bit/s is given by $P = HR$ [22, 23, 26, 32, 55]. Alternatively, the energy consumption is given by $Q = HB$, where B is the number of bits transferred [24, 31, 74, 75]. Although this appears to be intuitive, as we have seen above, there are multiple options for the H energy efficiency metric.

Many authors have used the definition $H' = P_{max}/C_{max}$, where the values of P_{max}, the maximum power consumption, and C_{max}, the maximum capacity, are both based on data provided in equipment specification sheets or some form of measurement [22, 23, 26, 31, 32]. In some cases, the utilisation $\mu(t)$ has been included to give metric $H''(t) = P_{max}/\mu(t)C_{max}$ [24, 75], where the "utilisation", $\mu(t)$, of a network element or network is defined as the ratio of carried traffic, $R(t)$, to its maximum capacity, C_{max}. That is:

$$\mu(t) = R(t)/C_{max} \qquad (11.57)$$

In many cases, the metric is used to calculate the power or energy consumption of a service or user, based upon a data rate for the service or user, $R^{(k)}(t)$ [9, 22, 24, 31, 32, 76]. With this approach to modelling, the appropriate equation is (11.8). Noting that $E_j = (P_{max,j} - P_{idle,j})/C_{max,j}$ and using (11.7), we get

$$P^{(k)}(t) = \sum_{j=1}^{N_E} \left(\frac{P_{max,j}}{C_{max,j}} + \frac{P_{idle,j}}{C_{max,j}} \left(\frac{1}{\mu_j(t)} - 1 \right) \right) \alpha_j^{(k)} R^{(k)}(t) \qquad (11.58)$$

From this we see that, for load proportional equipment ($P_{idle} \approx 0$), the H' metric is appropriate. For other equipment, including constant power equipment ($P_{idle} \approx P_{max}$), the H'' metric is appropriate.

Comparing $^2H(T)$, defined in (11.16), and $^3H(T)$, defined in (11.24), we see that it is more appropriate to use $^2H(T)$ to calculate the power or energy consumption of a service because it has the form $\langle P \rangle_T / \langle R \rangle_T$ and a service is typically parametrised with $\langle R \rangle_T$ or $B(T)$. Provided the conditions for (11.56) are satisfied and the k th services are in-synch with the vast majority of other service traffic flows, we have

$$P^{(k)}{}_T \approx {}^2H_{Ntwk}(T) R^{(k)}{}_T$$
$$Q^{(k)}(T) \approx {}^2H_{Ntwk}(T) B^{(k)}(T) \qquad (11.59)$$

This approach has been widely used to estimate power or energy consumption of a variety of Internet services [23, 26, 31, 32, 74]. As discussed with (11.46) and (11.47), this approximation is only reasonable if the service in question is not out-of-synch with the other network flows. Therefore, using (11.59) to estimate the energy consumption of out-of-synch services (such as off-peak backup or off-peak data transfer services) is inappropriate. This issue is likely to arise with data services that travel through time zones that are out-of-synch with their source.

11.6.1.2 Power Consumption Estimation Using "Power per User": G Metrics

Viewing the forms for $^3G_j(T)$, $^3G_{Ntwk}(T)$, $^2G^{(k)}(T)$ and $^3G^{(k)}(T)$, we can see there are two types of synchronisation that may occur when considering the power per user: firstly, the synchronisation between traffic flows, $R^{(k)}$; and secondly, the synchronisation between traffic and user number, $N^{(usr)}$. For totally unsynchronised flows, although the $R^{(k)}$ may exhibit significant diurnal variation, the diurnal cycles of R_j and R_{Ntwk} are likely to be relatively shallow. In this case, $R_{max,j} \approx R_{min,j} \approx R_{avg,j}$ and $R_{max,Ntwk} \approx R_{min,Ntwk} \approx R_{ave,Ntwk}$. Whether or not the flows are synchronised, the number of users, $N^{(usr)}$, almost always exhibits a significantly varying diurnal cycle.

The G metrics listed in the previous paragraph have a dependence upon the phase of the user diurnal cycle relative the traffic diurnal cycle. We will focus our considerations on the $^3G_{Ntwk}(T)$ and $^2G^{(k)}(T)$ metrics because they provide an instructive example.

The $^2G_{Ntwk}(T)$ metric does not account for the profile of the traffic or user diurnal cycles and so will not reflect any significant change in user behaviour, such as many users shifting their active times to traditionally off-peak hours.

The $^3G_{Ntwk}(T)$ reflects the phase relationship between users and traffic. This will result in the power per user changing if many users don't change the traffic they generate but change the times when they generate it. For example, if these users shift their active times from around the peak traffic time to an off-peak time, then these users may cause $R_{Ntwk}(t)$ to become out-of-phase with $N^{(usr)}(t)$. This changes the value of $^3G_{Ntwk}(T)$, thus possibly providing a mechanism for monitoring user behaviour.

Turning our attention to the $^2G^{(k)}(T)$ metric, we see that this metric accounts for both the phase and magnitude of service traffic. Therefore, this metric may be suitable for allocating a power footprint to services accounting for both the magnitude of the service traffic, the number of users of that service, and the phase of the service traffic diurnal cycle relative to overall network traffic.

11.7 NETWORK ENERGY EFFICIENCY TRENDS

Using the results from Section 8.5.2, we can construct energy efficiency forecasts. We shall focus on network-energy-per-bit efficiency metrics, $^1H_{Ntwk}(t)$ and $^2H_{Ntwk}(T)$. The three network evolution scenarios considered in Chapter 8 will be discussed here. These scenarios are as follows:

(i) No Equipment Improvement;
(ii) Latest Generation Equipment Across Entire Network;
(iii) Equipment Improvement and Replacement.

We recall that Scenario (i) provides an upper bound to network power, Scenario (ii) provides a lower bound, and Scenario (iii) represents a more realistic network evolution forecast. In the equations below, the superscript in square brackets, [N], refers to the year N into the future from year 0 for which the forecasts are made.

Scenario (i): No equipment improvements:
For $^1H_{Ntwk}^{[N]}\left(t_{peak}\right)$ we divide the network power consumption at t_{peak}, in (8.49), by $R_{peak}^{[N]}$ to give

$$^1H_{Ntwk}^{[N]}\left(t_{peak}\right) \approx M_{OH}\left(\langle N_{hops}\rangle^{(svc)}+1\right)\frac{\left\langle P_{max}^{[0]}\right\rangle}{\left\langle C_{max}^{[0]}\right\rangle}\left(y^{[N]}\left(\frac{1}{\mu_{max}}+1\right)-1\right) \tag{11.60}$$

For $^2H_{Ntwk}^{[N]}\left(T\right)$, we divide the network power consumption averaged over a diurnal cycle duration, T, in (8.48), by the average diurnal cycle traffic $\langle R^{[N]}\rangle_T$ given in (8.45). This gives

$$^2H_{Ntwk}^{[N]}\left(T\right) \approx M_{OH}\left(\langle N_{hops}\rangle^{(svc)}+1\right)\frac{\left\langle P_{max}^{[0]}\right\rangle}{\left\langle C_{max}^{[0]}\right\rangle}\left(y^{[N]}\left(\frac{\sigma^N}{\beta^N\mu_{max}\left\langle f^{[0]}\right\rangle_T}-1\right)+1\right) \tag{11.61}$$

Scenario (ii): Latest equipment across the entire network:
Dividing (8.52) by $R^{[N]}_{peak}$ gives

$$^1 H^{[N]}_{Ntwk}\left(t_{peak}\right) \approx M_{OH}\left(\left\langle N_{hops}\right\rangle^{(svc)} 1\right) \frac{\left\langle P^{[0]}_{max}\right\rangle}{\left\langle C^{[0]}_{max}\right\rangle} \gamma^N \left(y^{[N]}\left(\frac{1}{\mu_{max}}+1\right)-1\right)$$ (11.62)

And for the average over a diurnal cycle divide (8.51) by $\left\langle R^{[N]}\right\rangle_T$ to get

$$^2 H^{[N]}_{Ntwk}\left(T\right) \approx M_{OH}\left(\left\langle N_{hops}\right\rangle^{(svc)}+1\right) \frac{\left\langle P^{[0]}_{max}\right\rangle}{\left\langle C^{[0]}_{max}\right\rangle} \gamma^N \left(y^{[N]}\left(\frac{\sigma^N}{\beta^N \mu_{max}\left\langle f^{[0]}\right\rangle_T}-1\right)+1\right)$$ (11.63)

Scenario (iii): Deployment based on traffic forecasting with equipment replacement:
Again, energy efficiency at peak time we divide (8.57) by $R^{[N]}_{peak}$ to get

$$\begin{aligned}
^1 H^{[N]}_{Ntwk}\left(t_{peak}\right) = &\, M_{OH}\left(\left\langle N_{hops}\right\rangle^{(svc)}+1\right) \frac{\left\langle P^{(0)}_{max}\right\rangle_E}{\left\langle C^{(0)}_{max}\right\rangle_E} \sigma^{-N}\left(y^{[0]}\left(\frac{1}{\mu_{max}}-1\right)+\right. \\
&+\sum_{m=0}^N \gamma^{\lfloor m/q \rfloor q}\left(\frac{y^{[m]}}{\mu_{max}}\sigma^{\lfloor m/p \rfloor p}\left(\sigma^p-1\right)\left(\lfloor m/p \rfloor-\lfloor (m-1)/p \rfloor\right)\right. \\
&\qquad\left.+\left(1-y^{[m]}\right)\sigma^m\left(\sigma-1\right)\right) \\
&+\sum_{m=0}^N \theta\left(\lfloor m/p \rfloor-r\right)\left(\gamma^{\lfloor m/q \rfloor q}-\gamma^{\lfloor (m-rp)/q \rfloor q}\right)\left(\frac{y^{[m]}}{\mu_{max}}\sigma^{(\lfloor m/p \rfloor-r)p}\left(\sigma^p-1\right)\left(\lfloor m/p \rfloor-\lfloor (m-1)/p \rfloor\right)\right. \\
&\qquad\left.\left.+\sigma^{m-rp}\left(\sigma-1\right)\left(1-y^{[m]}\right)\right)\right)
\end{aligned}$$ (11.64)

For $^2 H^{[N]}_{Ntwk}\left(T\right)$ we divide (8.56) by $\left\langle R^{[N]}\right\rangle_T$ to get

$$\begin{aligned}
^2 H^{(N)}_{ntwk}\left(T\right) = &\, M_{OH}\left(\left\langle N_{hops}\right\rangle^{(svc)}+1\right) \frac{\left\langle P^{[0]}_{max}\right\rangle_E}{\left\langle C^{[0]}_{max}\right\rangle_E} \beta^{-N}\left(\frac{y^{[0]}}{\mu_{max}\left\langle f^{[0]}\right\rangle_T}+\left(1-y^{[0]}\right)+\right. \\
&+\sum_{m=0}^N \gamma^{\lfloor m/q \rfloor q}\left(\frac{y^{[m]}}{\mu_{max}\left\langle f^{[0]}\right\rangle_T}\sigma^{\lfloor m/p \rfloor p}\left(\sigma^p-1\right)\left(\lfloor m/p \rfloor-\lfloor (m-1)/p \rfloor\right)\right. \\
&\qquad\left.+\left(1-y^{[m]}\right)\beta^m\left(\beta-1\right)\right) \\
&+\sum_{m=0}^N \theta\left(\lfloor m/p \rfloor-r\right)\left(\gamma^{\lfloor m/q \rfloor q}-\gamma^{\lfloor (m-rp)/q \rfloor q}\right)\left(\frac{y^{[m]}}{\mu_{max}\left\langle f^{[0]}\right\rangle_T}\sigma^{(\lfloor m/p \rfloor-r)p}\left(\sigma^p-1\right)\left(\lfloor m/p \rfloor-\lfloor (m-1)/p \rfloor\right)\right. \\
&\qquad\left.\left.+\beta^{m-rp}\left(\beta-1\right)\left(1-y^{[m]}\right)\right)\right)
\end{aligned}$$ (11.65)

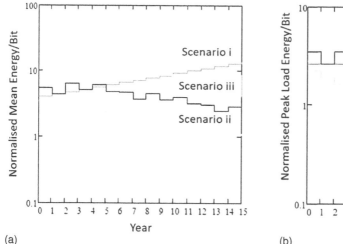
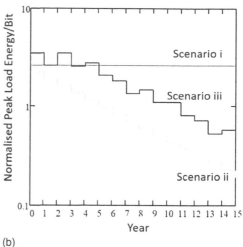

(a)   (b)

FIGURE 11.6   (a) Normalised mean and (b) peak load energy per bit evolution for Scenarios of; i "no equipment improvement", ii "latest generation equipment" and iii "traffic forecasting with equipment replacement".

These results can be normalised by dividing out the factor

$$M_{OH}\left(\left\langle N_{hops}\right\rangle^{\langle svc\rangle}+1\right)\frac{\left\langle P_{idle}^{[0]}\right\rangle_E}{\left\langle C_{idle}^{[0]}\right\rangle_E} \tag{11.66}$$

The resulting "normalised" values for the parameters given in Table 8.2 are plotted in Figure 11.6 for the values $p = 2$, $q = 3$ and $r = 2$. As expected, we can see that Scenario (iii) lies between (i) and (ii) for all except the first few years of deployment. This is because Scenario (iii) pre-deploys equipment for several years into the future, resulting in under-utilisation of the earliest generation of equipment for the first few years. This under-utilisation results in the mean and peak load energy per bit for Scenario (iii) being greater than the other two scenarios during the first few years. However, over the succeeding years the pre-deployment of latest-generation equipment still provides better network energy efficiency than not updating the equipment technology at all.

11.8   TECHNICAL SUMMARY

A variety of energy efficiency metrics have been proposed for use for network equipment. Examples are TEER, ECR, EER and TEEER. These metrics are based upon a generic form typified by the ECR given in (11.9). This type of metric has been designed primarily for use on network elements. It is not easily translated for use when considering the energy efficiency of networks and services.

The most commonly used energy efficiency metrics used in the literature for equipment, networks and services are based on either "energy per bit" or "power per user". With "energy per bit" metrics, some are calculated from "power per bit per second", which is an instantaneous measure, and others are calculated from "total energy per total bits throughput" over a pre-set duration, such as a diurnal cycle or year.

The "power per throughput" (in bits per second) metric, $^1H(t)$, can be defined for a network element, network and service, respectively, in (11.13). The metric, $^2H(T)$, based on "total energy per total bits throughput" over duration T is defined for a network element, network and service, respectively, in (11.18).

These forms, $^1H(t)$ and $^2H(T)$, for "energy per bit" have been widely used in publications on telecommunications energy efficiency. A third metric considered in this chapter is the $^3H(T)$ metric defined in (11.25). This metric is the time average of the $^1H(t)$ metric, as given in (11.24).

Although the $^3H(T)$ metric has not been adopted in the literature, it has the advantage that it can distinguish between situations in which the shape of the traffic diurnal cycle has changed even though the total data throughput has not. That is, it will provide different energy efficiency values for deep and shallow traffic diurnal cycles with a given network element or network. The corresponding $^2H(T)$ metrics will not provide this.

The instantaneous power per user metrics, $^1G(t)$, has the form for network equipment, networks and services as defined, respectively, in (11.28). When these metrics are used, they are typically evaluated at a specific time, such as the time of peak traffic. The average power per average number of users, $^2G(T)$, is also a widely used metric, defined in (11.29). Again, we can define a third power per user metric, $^3G(T)$, which is the time average of $^1G(t)$. The $^3G(T)$ metrics are defined in (11.30).

The collection of detailed diurnal-cycle data to calculate a metric can be problematic. The use of approximate forms can provide estimates for the metrics requiring fewer data. These forms are given in (11.41), (11.43) and (11.44) for energy-per-bit metrics. For power-per-user metrics, the approximations are (11.51), (11.52) and (11.53). When using these forms with real-world diurnal cycles, the definitions for R_{avg} and ΔR given in (11.54) must be applied. To apply these forms, the traffic and user diurnal cycles cannot be too far from an approximate sinusoidal form.

The $^2H(T)$ metric has been used to estimate the power consumption of networks and services. For a service that is in-phase with the vast majority of other services, its power or energy consumption can be estimated using the network $^2H_{Ntwk}$ energy-per-bit metric by using (11.59).

The synchronicity of a traffic flow with the rest of the network is important when estimating the power consumption of a service. Typically the energy per bit of a service that is "in-synch" with general network traffic will be lower than a service that is "out-of-synch". This can be seen by approximating the diurnal cycle of services with a sinusoid, as shown in (11.31). With this approximation, the impact of the phase difference between flows for the metrics is seen in (11.44)b).

Energy efficiency metrics are of value when planning network upgrading schedules. For this use, the $^1H(t)$ metric at peak traffic time and $^2H(T)$ metric can provide guidance on future network upgrading scenarios. Three upgrading scenarios are considered in detail. Scenario (i), no equipment upgrading, provides an upper bound to the energy per bit as network traffic grows. The equations for this scenario are given in (11.60) and (11.61). Scenario (ii), latest equipment across all the network, provides a lower bound on energy efficiency. The equations for this scenario are given in (11.62) and (11.63). Scenario (iii) provides a more realistic network planning scenario, with deployment based on traffic forecasting with equipment replacement. The equations for this scenario are given in (11.64) and (11.65).

11.9   EXECUTIVE SUMMARY

Metrics play an essential role in providing a quantitative description of a system, be it a piece of machinery, an organisation's output, or the energy consumption of a network or service. The primary purpose of any metric is to provide a quantitative measure for guidance when comparing and/or improving the system. This makes metrics integral to the management of any organisation, process or system.

Metrics are integral to the process of developing strategies for reducing the environmental footprint of telecommunications equipment, network and services. The need to mitigate the growth of the Internet's power consumption has produced a range of energy-efficiency metrics for this purpose. There are several alternatives for defining energy-efficiency metrics and the method by which they can be applied to network elements, to entire networks, and to services.

As a measure for energy per bit, although the metrics ^1H, ^2H and ^3H have the same dimensions, their numerical values can be significantly different; similarly, for the power-per-user metrics, ^1G, ^2G and ^3G. For example, in a network with somewhat synchronised traffic, value of the ^1H metric at peak traffic hour will be very different to that for the ^2H metric. Therefore, direct comparison of the numerical values of these metrics can be problematic [8–10].

When used to benchmark or improve energy efficiency, we see that desynchronising traffic flows can reduce the $^3H_j(T)$ and $^3H_{Ntwk}(T)$ metrics. Therefore, according to these metrics we can improve the energy efficiency (i.e. reduce the joules per bit) by desynchronising the network traffic flows. In contrast, desynchronising flows have no impact on $^2H_j(T)$ and $^2H_{Ntwk}(T)$.

The $^2H^{(k)}(T)$ metric indicates that service providers should endeavour to synchronise their service flows with any variation in the diurnal cycle. If all service providers do this, the diurnal cycle will increase in depth, which will impact network dimensioning. From this we see that the choice of metric can influence the choice of strategy for improvement; and even using the same metric in different situations may lead to different (and possibly conflicting) strategies. Therefore, the choice of metric must be given careful consideration and the use of different metrics in different circumstances may be appropriate.

11.10   APPENDIX

11.10.1   DERIVATION OF (11.35) AND (11.36)

The summation of sinusoids has the form

$$\sum_{k=1}^{N^{(usr)}} \Delta R^{(k)} \cos\left(\left(2\pi t/T\right)-\phi^{(k)}\right) = \sum_{k=1}^{N^{(usr)}} \Delta R^{(k)} \cos\left(2\pi t/T\right)\cos\left(\phi^{(k)}\right)+\sin\left(2\pi t/T\right)\sin\left(\phi^{(k)}\right)$$

$$= \cos\left(2\pi t/T\right)\sum_{k=1}^{N^{(usr)}} \Delta R^{(k)} \cos\left(\phi^{(k)}\right)+\sin\left(2\pi t/T\right)\sum_{k=1}^{N^{(usr)}} \Delta R^{(k)} \sin\left(\phi^{(k)}\right)$$

$$(11.67)$$

We express the summations in the form

$$\sum_{k-1}^{N^{(usr)}} \Delta R^{(k)} \cos\left(\phi^{(k)}\right) = \Delta R \cos\left(\varphi\right)$$

$$\sum_{k=1}^{N^{(usr)}} \Delta R^{(k)} \sin\left(\phi^{(k)}\right) = \Delta R \sin\left(\varphi\right)$$

$$(11.68)$$

Solving for ΔR and ϕ, we have

$$\Delta R = \sqrt{\Delta R^2 \cos^2\left(\varphi\right)+\Delta R^2 \sin^2\left(\varphi\right)}$$

$$= \left(\left(\sum_{k=1}^{N^{(usr)}} \Delta R^{(k)} \cos\left(\phi^{(k)}\right)\right)^2 + \left(\sum_{k=1}^{N^{(usr)}} \Delta R^{(k)} \sin\left(\phi^{(k)}\right)\right)^2\right)^{1/2}$$

$$= \left(\sum_{k=1}^{N^{(usr)}} \left(\Delta R^{(k)}\right)^2 + \sum_{k\neq l}^{N^{(usr)}} \Delta R^{(k)}\Delta R^{(l)}\left(\cos\left(\phi^{(k)}\right)\cos\left(\phi^{(l)}\right)+\sin\left(\phi^{(k)}\right)\sin\left(\phi^{(l)}\right)\right)\right)^{1/2} \quad (11.69)$$

$$= \left(\sum_{k=1}^{N^{(usr)}} \left(\Delta R^{(k)}\right)^2 + \sum_{k\neq l}^{N^{(usr)}} \Delta R^{(k)}\Delta R^{(l)} \cos\left(\phi^{(k)}-\phi^{(l)}\right)\right)^{1/2}$$

This gives the summation term in (11.35). For (11.36), we note that from (11.68)

$$\sum_{k=1}^{N^{(usr)}} \Delta R^{(k)} \cos\left(\left(2\pi t/T\right) - \phi^{(k)}\right) = \Delta R \cos\left(2\pi t/T\right)\cos\left(\varphi\right) + \Delta R \sin\left(2\pi t/T\right)\sin\left(\varphi\right) \qquad (11.70)$$

$$= \Delta R \cos\left(\left(2\pi t/T\right) - \phi_{Ntwk}\right)$$

where

$$\tan\left(\phi_{Ntwk}\right) = \frac{\Delta R \sin\left(\varphi\right)}{\Delta R \cos\left(\varphi\right)} = \frac{\sum_{k=1}^{N^{(svc)}} \Delta R^{(k)} \sin\left(\phi^{(k)}\right)}{\sum_{k=1}^{N^{(svc)}} \Delta R^{(k)} \cos\left(\phi^{(k)}\right)} \qquad (11.71)$$

which gives (11.36).

11.10.2   CALCULATION OF $^3H_j(T)$ IN (11.41)

To calculate $^3H_j(T)$ we have used Item 3.613.1 from Gradshteyn and Ryzhik [70]:

$$\left\langle \frac{1}{R} \right\rangle_T = \frac{1}{T} \int_0^T \frac{dt}{R_{avg} + \Delta R \cos \dfrac{2\pi t}{T}}$$

$$= \frac{1}{2\pi R_{ave}} \int_0^{2\pi} \frac{d\theta}{1 + \dfrac{\Delta R}{R_{avg}} \cos\theta} = \frac{1}{2\pi R_{avg}} \left(\int_0^{\pi} \frac{d\theta}{1 + \dfrac{\Delta R}{R_{avg}} \cos\theta} + \int_{\pi}^{2\pi} \frac{d\theta}{1 + \dfrac{\Delta R}{R_{avg}} \cos\theta} \right) \qquad (11.72)$$

$$= \frac{1}{2\pi} \left(\frac{\pi}{\left(R_{avg}^2 - \Delta R^2\right)^{1/2}} + \frac{1}{R_{ave}} \int_0^{\pi} \frac{d\theta}{1 - \dfrac{\Delta R}{R_{avg}} \cos\theta} \right) = \frac{1}{\left(R_{avg}^2 - \Delta R^2\right)^{1/2}}$$

$$= \frac{1}{\left(R_{avg} - \Delta R\right)^{1/2} \left(R_{avg} + \Delta R\right)^{1/2}} = \frac{1}{R_{min}^{1/2} R_{max}^{1/2}}$$

11.10.3   CALCULATION OF $^3H_j(T)$ IN (11.44)B

The term $\langle\langle R^{(k)}/R\rangle_E\rangle_T$ in (11.18)c can be written out in full as

$$\left\langle\left\langle \frac{R^{(k)}}{R} \right\rangle_E \right\rangle_T = \frac{1}{2\pi} \sum_{j=1}^{N_E} \int_0^{2\pi} \frac{R_{avg}^{(k)} + \Delta R^{(k)} \cos\left(\theta + \phi_j^{(k)}\right)}{R_{avg,j} + \Delta R_j \cos\left(\theta\right)} d\theta \qquad (11.73)$$

where $\theta = 2\pi t/T$ and $\phi(k)$ is the phase of the diurnal cycle of service k relative to the total traffic flow through network element j. That is, $\phi_j^{(k)} = 2\pi\left(t_{peak}^{(k)} - t_{peak,j}\right)/T$, where $t_{peak}^{(k)}$ is the peak traffic time for the k th service and $t_{peak,j}$ is the peak traffic time for the j th network element. Separating out the two terms in the numerator, we get:

$$\left\langle\left\langle\frac{R^{(k)}}{R}\right\rangle_T\right\rangle_E = \frac{R_{avg}^{(k)}}{2\pi}\sum_{j=1}^{N_E}\int_0^{2\pi}\frac{d\theta}{R_{avg,j}+\Delta R_j\cos(\theta)} + \frac{\Delta R^{(k)}}{2\pi}\sum_{j=1}^{N_E}\int_0^{2\pi}\frac{\cos\left(\theta+\phi_j^{(k)}\right)d\theta}{R_{avg,j}+\Delta R_j\cos(\theta)} \quad (11.74)$$

where we use (11.72) to evaluate the first integral on the right-hand side. To evaluate the second, we have:

$$\frac{\Delta R^{(k)}}{2\pi}\int_0^{2\pi}\frac{\cos\left(\theta+\phi_j^{(k)}\right)dt}{R_{avg,j}+\Delta R_j\cos(\theta)} = \frac{\Delta R^{(k)}}{2\pi}\int_0^{2\pi}\frac{\cos(\theta)\cos\left(\phi_j^{(k)}\right)-\sin(\theta)\sin\left(\phi_j^{(k)}\right)d\theta}{R_{avg,j}+\Delta R_j\cos(\theta)} \quad (11.75)$$

Looking at this, we note that the second term in the integration on the right-hand side of (11.75) has the integrated form $\ln(1+(\Delta R/R_{avg})\cos(\theta))$, which, when evaluated between 0 and 2π, gives zero. Therefore, that term can be discarded. Focussing on the first term in the integral on the right-hand side, using Gradshteyn and Ryzhik [70], 2.554.2, we have:

$$\frac{\Delta R^{(k)}}{2\pi}\int_0^{2\pi}\frac{\cos(\theta)\cos\left(\phi_j^{(k)}\right)d\theta}{R_{avg,j}+\Delta R_j\cos(\theta)} = \frac{\Delta R^{(k)}}{2\pi}\cos\left(\phi_j^{(k)}\right)\int_0^{2\pi}\frac{\cos(\theta)d\theta}{R_{avg,j}+\Delta R_j\cos(\theta)}$$
$$= \frac{\Delta R^{(k)}}{\Delta R_j}\cos\left(\phi_j^{(k)}\right)\left(1-\frac{R_{avg,j}}{\left(R_{avg,j}^2-\Delta R_j^2\right)^{1/2}}\right) \quad (11.76)$$

This gives for (11.74):

$$\left\langle\left\langle\frac{R^{(k)}}{R}\right\rangle_T\right\rangle_E = \frac{1}{N_E}\left(R_{avg}^{(k)}\sum_{j=1}^{N_E}\frac{1}{R_{max,j}^{1/2}R_{min,j}^{1/2}} + \Delta R^{(k)}\sum_{j=1}^{N_E}\frac{\cos\left(\phi_j^{(k)}\right)}{\Delta R_j}\left(1-\frac{R_{avg,j}}{R_{max,j}^{1/2}R_{min,j}^{1/2}}\right)\right)$$
$$= R_{avg}^{(k)}\left\langle\frac{1}{R_{max}^{1/2}R_{min}^{1/2}}\right\rangle_E + \Delta R^{(k)}\left\langle\frac{\cos\left(\phi^{(k)}\right)}{\Delta R}\left(1-\frac{R_{avg}}{R_{max}^{1/2}R_{min}^{1/2}}\right)\right\rangle_E \quad (11.77)$$

Assuming the phases $\phi_j^{(k)}$ are independent of the traffic flows, $R_{avg,j}$ and ΔR_j, we can split the $\cos(\phi^{(k)})$ term out as a separate average, giving:

$$\left\langle\left\langle\frac{R^{(k)}}{R}\right\rangle_T\right\rangle_E = R_{avg}^{(k)}\left\langle\frac{1}{R_{max}^{1/2}R_{min}^{1/2}}\right\rangle_E + \Delta R^{(k)}\left\langle\cos\left(\phi^{(k)}\right)\right\rangle_E\left\langle\frac{1}{\Delta R}\left(1-\frac{R_{avg}}{R_{max}^{1/2}R_{min}^{1/2}}\right)\right\rangle_E \quad (11.78)$$

which is the form that appears in (11.44)b).

A minor detail that needs to be addressed is whether or not the $1/\Delta R_j$ term on the right-hand side diverges in the limit as $\Delta R_j \to 0$. Because the variation in the diurnal cycle of the j th network element, ΔR_j, is dependent on the phases $\phi_j^{(k)}$ and variation $\Delta R^{(k)}$ of the flows through that element, we can have situations in which ΔR_j becomes vanishingly small. For example, the phases $\phi_j^{(k)}$ may be sufficiently widely spread over 0 to 2π that the flow through the j th network element is effectively flat over the duration of the diurnal cycle. To cover this eventuality, we consider the term:

$$
\lim_{\Delta R \to 0} \frac{1}{\Delta R}\left(1 - \frac{R_{avg}}{R_{max}^{1/2} R_{min}^{1/2}}\right) = \lim_{\Delta R \to 0} \frac{1}{\Delta R} \frac{\left(1+\dfrac{\Delta R}{R_{avg}}\right)^{1/2}\left(1-\dfrac{\Delta R}{R_{avg}}\right)^{1/2} - 1}{\left(1+\dfrac{\Delta R}{R_{avg}}\right)^{1/2}\left(1-\dfrac{\Delta R}{R_{avg}}\right)^{1/2}}
$$

$$
= \lim_{\Delta R \to 0} \frac{1}{\Delta R} \frac{\left(1+\dfrac{1}{2}\dfrac{\Delta R}{R_{avg}}\right)\left(1-\dfrac{1}{2}\dfrac{\Delta R}{R_{avg}}\right) - 1}{\left(1+\dfrac{1}{2}\dfrac{\Delta R}{R_{avg}}\right)\left(1-\dfrac{1}{2}\dfrac{\Delta R}{R_{avg}}\right)} \tag{11.79}
$$

$$
= \lim_{\Delta R \to 0} \frac{\Delta R}{\left(4R_{avg}^2 - \Delta R^2\right)} = 0
$$

where we have used the limit $(1 + x)^{1/2} \to 1 + (x/2)$ as $x \to 0$.

11.10.4   DERIVATION OF (11.56)

This equation can help simplify the task of estimating the energy efficiency of a network or service. Provided the services of interest are "typical" or representative of all the other services in the network and not significantly "out-of-synch" with the other services, we can use (11.56). To justify the ^2H approximation in (11.56), we start with (11.18)c). Justifications for the steps a) to d) in the derivation are provided in the following equation:

$$
\left\langle {}^2H(T)\right\rangle^{(svc)} \approx \frac{M_{OH}}{N^{(svc)}}\left(\left\langle P_{idle}\right\rangle_E \left\langle\left\langle\frac{1}{R}\right\rangle_T\right\rangle_E + \left\langle E\right\rangle_E\right)\sum_{k=1}^{N^{(svc)}} N_E^{(k)} \qquad a)
$$

$$
= \frac{M_{OH}}{N^{(svc)}}\left(\left\langle P_{idle}\right\rangle_E \left\langle\left\langle\frac{1}{R}\right\rangle_T\right\rangle_E + \left\langle E\right\rangle_E\right)\sum_{k=1}^{N^{(svc)}}\sum_{j=1}^{N_E} \alpha_j^{(k)} \qquad b)
$$

$$
\approx M_{OH}\left(N_E\left\langle P_{idle}\right\rangle_E \left\langle\left\langle\frac{1}{R}\right\rangle_T\right\rangle_E \left\langle\alpha\right\rangle + \left\langle N_E\right\rangle^{(svc)}\left\langle E\right\rangle_E\right) \qquad c) \tag{11.80}
$$

$$
\approx M_{OH}\left(N_E\left\langle P_{idle}\right\rangle_E \left\langle\left\langle\frac{1}{R}\right\rangle_T\right\rangle_E \frac{\left\langle\left\langle R\right\rangle_T\right\rangle_E}{\left\langle R_{Ntwk}\right\rangle_T} + \left\langle N_E\right\rangle^{(svc)}\left\langle E\right\rangle_E\right) \qquad d)
$$

$$
\approx M_{OH}\left(\frac{N_E\left\langle P_{idle}\right\rangle_E}{\left\langle R_{Ntwk}\right\rangle_T} + \left\langle N_E\right\rangle^{(svc)}\left\langle E\right\rangle_E\right) \approx {}^2H_{Ntwk}(T)
$$

Steps a) to b) used (8.23); b) to c) used two equalities in (8.24); c) to d) used (8.27) averaged over a diurnal cycle. The final step, which cancels $\left\langle\left\langle R\right\rangle_T\right\rangle_E$ and $\left\langle\left\langle 1/R\right\rangle_T\right\rangle_E$, uses the fact that the $\left\langle R_j\right\rangle_T$ satisfy (11.15) for the situation under consideration (that being all of the service flows are similarly synchronised).

To derive the ^3H approximation in (11.56) we start with (11.25)c1:

$$
\begin{aligned}
\left\langle {}^3H(T) \right\rangle^{(svc)} &= \frac{M_{OH}}{N^{(svc)}} \sum_{k=1}^{N^{(svc)}} \sum_{j=1}^{N_E} \left(P_{idle,j} \alpha_j^{(k)} \left\langle \frac{1}{R_j} \right\rangle_T + E_j \alpha_j^{(k)} \right) && a) \\[2ex]
&\approx \frac{M_{OH}}{N^{(svc)}} \left(\langle P_{idle} \rangle_E \sum_{k=1}^{N^{(svc)}} \sum_{j=1}^{N_E} \alpha_j^{(k)} \left\langle \frac{1}{\sum_{p=1}^{N^{(svc)}} \alpha_j^{(p)} R^{(p)}} \right\rangle_T + N^{(svc)} N_E \langle E \rangle_E \langle \alpha \rangle \right) && b) \\[2ex]
&\approx M_{OH} \left(\frac{\langle P_{idle} \rangle_E}{N^{(svc)}} \sum_{k=1}^{N^{(svc)}} \sum_{j=1}^{N_E} \frac{\alpha_j^{(k)}}{\langle \alpha_j \rangle^{(svc)}} \left\langle \frac{1}{R_{Ntwk}} \right\rangle_T + \langle N_E \rangle^{(svc)} \langle E \rangle_E \right) && c) \\[2ex]
&\approx M_{OH} \left(\langle P_{idle} \rangle_E \sum_{j=1}^{N_E} \frac{\langle \alpha_j \rangle^{(svc)}}{\langle \alpha_j \rangle^{(svc)}} \left\langle \frac{1}{R_{Ntwk}} \right\rangle_T + \langle N_E \rangle^{(svc)} \langle E \rangle_E \right) && d) \\[2ex]
&\approx M_{OH} \left(N_E \langle P_{idle} \rangle_E \left\langle \frac{1}{R_{Ntwk}} \right\rangle_T + \langle N_E \rangle^{(svc)} \langle E \rangle_E \right) \approx {}^3H_{Ntwk}(T)
\end{aligned}
$$

(11.81)

Steps from a) to b) use (8.15) and (8.24); steps from b) to c) use (8.14); steps from c) to d) use the definition of $\langle \alpha_j \rangle^{(svc)}$. The final step is a summation over the j-index.

REFERENCES

[1] L. Ryan and N. Campbell, "Spreading the Net: The Multiple Benefits of Energy Efficiency Improvements," *International Energy Agency - Insight Series*, 2012.

[2] IETF, "RFC 4984, "IAB Workshop on Routing & Addressing," IETF, 2007.

[3] ITU-T, "ITU-T L.1310 (11/2012), Energy Efficiency Metrics and Measurement Methods for Telecommunications Equipment," International Telecommunications Union, 2012.

[4] GreenTouch, "GreenTouch," 2015. [Online]. Available: https://s3-us-west-2.amazonaws.com/belllabs-microsite-greentouch/indcx.html. [Accessed 20 September 2020].

[5] The Green Grid, "The Green Grid," The Green Grid, 2020. [Online]. Available: https://www.thegreen-grid.org/. [Accessed 11 July 2020].

[6] World Resource Institute, "Greenhouse Gas Protocol," World Resource Institute, [Online]. Available: www.ghgprotocol.org. [Accessed 20 September 2020].

[7] S. Bianzoni *et al.*, "Apples-to-apples: A Framework Analysis for Energy Efficiency in Networks," *ACM Sigmetrics Performance Evaluation Review*, vol. 38, no. 3, p. 81, 2010.

[8] V. Coroama and L. Hilty, "Assessing Internet Energy Intensity: A Review of Methods and Results," *Environmental Impact Assessment Review*, vol. 45, p. 63, 2014.

[9] D. Schien *et al.*, "Modelling and Assessing Variability in Energy Consumption During the Use Stage of Online Multimedia Services," *Journal of Industrial Ecology*, vol. 17, no. 6, p. 800, 2013.

[10] D. Schien and C. Preist, "Approaches to the Energy Intensity of the Internet," *IEEE Communications Magazine*, vol. 52, no. 11, p. 130, 2014.

[11] J. Slutsky and A. El-Homsi, *Corporate Sigma: Optimizing the Health of Your Company with Systems Thinking*, CRC Press, 2009.

[12] R. S. Tucker, "Optical Communications—Part I: Energy Limitations in Transport," *IEEE Journal on Selected Topics in Communications*, vol. 17, no. 2, p. 245, 2011.

[13] M. Imran *et al.*, "Most Suitable Efficiency Metrics and Utility Functions," European Commission, 2012.

[14] S. Tombaz *et al.*, "*Energy Efficiency Assessment of Wireless Access Networks Utilizing Indoor Base Stations*," in *24th International Symposium Personal Indoor & Mobile Radio Communications, PIMRC 2013*, 2013.

[15] G. Auer *et al.*, "How Much Energy is Needed to Run a Wireless Network?," in *Green Radio Communication Networks*, E. Hossain *et al.*, Eds., Cambridge University Press, 2012, p. 360.

[16] A. Chand, "Potential of Energy Conservation of Femtocells in WCDMA Networks," *International Journal of Innovations in Engineering and Technology*, vol. 2, no. 2, p. 25, 2013.

[17] M. Alsharif *et al.*, "Survey of Green Radio Communications Networks: Techniques and Recent Advances," *Journal of Computer Networks and Communications*, vol. 2013, ID 453893, 2013.

[18] C. Fang *et al.*, "A Survey of Green Information-Centric Networking: Research Issues and Challenges," *IEEE Communication Surveys & Tutorials*, vol. 17, no. 3, p. 1455, 2015.

[19] A. Wu *et al.*, "*Intelligent Efficiency for Data Centres and Wide Area Networks*", in *Electronic Devices & Networks Annex*, May 2019

[20] D. Kilper, "Energy Efficiency in Optical Networks," in *Springer Handbook of Optical Networks*, B. Mukherjee *et al.*, Eds., Springer Nature, 2020, p. 631.

[21] GreenTouch, "GreenTouch Final Results from Green Meter Research Study: Reducing the Net Energy Consumption in Communications Networks by up to 98% by 2020," GreenTouch, 2015.

[22] J. Baliga *et al.*, "Energy consumption of Optical IP Networks," *Journal of Lightwave Technology*, vol. 27, no. 13, p. 2391, 2009.

[23] W. Van Heddeghem *et al.*, "Power Consumption Modeling in Optical Multilayer Networks," *Photonic Network Communications*, vol. 24, no. 2, p. 86, 2014.

[24] A. Taal *et al.*, "*Storage to Energy: Modelling the Carbon Emissions of Storage Task Offloading between Data Centres*," in *IEEE 11th Annual Consumer Communications and Networking Conference*, 2014.

[25] W. Van Heddeghem *et al.*, "*Evaluation of Power Rating of Core Network Equipment in Practical Deployments*," in *IEEE Online Conference on Green Communications*, 2012.

[26] C. Chan *et al.*, "Methodologies for Assessing the Use-Phase Power Consumption and Greenhouse Gas Emissions of Telecommunications Network Services," *Environmental Science and Technology*, vol. 47, no. 1, p. 485, 2013.

[27] J. Koomey *et al.*, "Network Electricity Use Associated with Wireless Personal Digital Assistants," *Journal of Infrastructure Systems*, vol. 10, no. 3, p. 131, 2004.

[28] S. Lanzisera *et al.*, "Data Network Equipment Energy Use and Savings Potential in Buildings," *Energy Efficiency*, vol. 5, p. 149, 2012.

[29] V. Coroama *et al.*, "*The Energy Intensity of the Internet: Home and Access Networks*," in *ICT Innovations for Sustainability, Advances in Intelligent Systems and Computing*, vol. 310, L. Hilty and B. Aebischer, Eds., Springer, 2015, p. 137.

[30] J. Baliga *et al.*, "Energy Consumption in Wired and Wireless Access Networks," *IEEE Communications Magazine*, vol. 49, no. 6, p. 70, 2011.

[31] J. Baliga *et al.*, "Green Cloud Computing: Balancing Energy in Processing, Storage and Transport," *Proceedings of the IEEE*, vol. 99, no. 1, p. 149, 2011.

[32] D. Kilper *et al.*, "Power Trends in Communication Networks," *IEEE Journal of Selected Topics In Quantum Electronics*, vol. 17, no. 2, p. 275, 2011.

[33] L. Barroso and U. Holze, "The Case for Energy-Proportional Computing," *IEEE Computer*, vol. 40, no. 12, p. 33, 2007.

[34] M. Ricca *et al.*, "*An Assessment of Power-Load Proportionality in Network Systems*," in *Sustainable Internet and ICT for Sustainability (SustainIT 2013)*, 2013.

[35] D. Abts *et al.*, "*Energy Proportional Data Center Networks*," in *ISCA '10: Proceedings of the 37th Annual International Symposium on Computer Architecture*, 2010.

[36] GeSI, Carbon Trust, "ICT Sector Guidance built on the GHG Protocol Product Life Cycle Accounting and Reporting Standard," 2017.

[37] D. Minoli, *Designing Green Networks and Network Operations*, CRC Press, Taylor & Francis Group, 2011.

[38] H. Hamdoun *et al.*, "Survey and Applications of Standardized Energy Metrics to Mobile Networks," *Annals of Telecommunications*, vol. 67, p. 113, 2012.

[39] Alliance for Telecommunications Industry Solutions (ATIS), "ATIS 0600015; Energy Efficiency for Telecommunication Equipment: Methodology for Measurement and Reporting – General Requirements," ATIS, 2018.

[40] L. Ceuppens *et al.*, "*Power Saving Strategies and Technologies in Network Equipment Opportunities and Challenges, Risk and Rewards*," in *International Symposium on Applications and the Internet*, 2008.

[41] Verizon, "Verizon NEBS TM Compliance: Energy Efficiency Requirements for Telecommunications Equipment, VZ.TPR.9205," Verizon, 2018.

[42] A. Riekstin et al., "A Survey on Metrics and Measurement Tools for Sustainable Distributed Cloud Networks," *IEEE Communications Surveys & Tutorials*, vol. 20, no. 2, p. 1244, 2018.

[43] M. Parker and S. Walker, "Roadmapping ICT: An Absolute Energy Efficiency Metric," *Optical Communications Networks*, vol. 3, no. 8, p. a49, 2011.

[44] J. Elmirghani et al., "GreenTouch GreenMeter Core Network Energy efficiency Improvement Measures and Optimization," *Journal of Optical Communications and Networking*, vol. 10, no. 2, p. A250, 2018.

[45] ITU-T "ITU-T L.1330 (03/2015) Energy efficiency measurements and metrics for telecommunications networks," International Telecommunications Union, 2015.

[46] ITU-T "ITU-T L.1332 (01/2018) Total network infrastructure energy efficiency metrics," International Telecommunications Union, 2015.

[47] ITU, "ITU-T Y.3022(08/2014) Global Information Infrastructure, Internet Protocol Aspects and Next-Generation Networks: Future Networks", Measuring energy in Networks," *International Telecommunications Union*, 2014.

[48] J. Franklin, "Quantity and Number," in *Neo-Aristotelian Perspectives in Metaphysics*, D. Novotný and L. Novák, Eds., Routledge, 2014, p. 221.

[49] S. Scheider and D. Huisjes, "Distinguishing Extensive and Intensive Properties for Meaningful Geocomputation and Mapping," *International Journal of Geographical Information Science*, vol. 33, no. 1, p. 28, 2019.

[50] B. Skubic et al., "Energy-Efficient Next Generation Optical Access Networks," *IEEE Communications Magazine*, vol. 50, no. 1, p. 122, 2012.

[51] S. Aleksić, and A. Lovrić, "*Power Consumption of Wired Access Network Technologies*," in *7th International Symposium on Communication Systems, Networks & Digital Signal Processing (CSNDSP 2010)*, 2010.

[52] S. Lambert et al., "Worldwide Electricity Consumption of Communications Networks," *Optics Express*, vol. 20, no. 26, p. 513, 2012.

[53] B. Skubic et al., "*Power Efficiency of Next-Generation Optical Access Architectures*," in *Optical Fiber Conference/National Fiber Optical Engineers Conference*, 2010.

[54] Google, "Google's Green Computing: Efficiency at Scale," Google, 2011.

[55] A. Vishwanath et al., "*Energy Consumption Comparison of Interactive Cloud-Based and Local Applications*," IEEE Journal on Selected Areas in Communications, vol. 33, no. 4, p. 616, 2015.

[56] A. Vishwanath et al., "*Energy Consumption of Interactive Cloud-Based Document Processing Applications*," in *IEEE International Conference on Communications (ICC)*, 2013.

[57] K. Hinton and F. Jalali, "*A Survey of Energy Efficiency Metrics*," in *5th International Conference on Smart Cities and Green ICT Systems (SMARTGREENS)*, 2015.

[58] DE-CIX, "DE-CIX New York Statistics," 22 November 2020. [Online]. Available: https://www.de-cix.net/en/locations/united-states/new-york/statistics. [Accessed 23 November 2020].

[59] Hong Kong Internet Exhange, "HKIX Switching Statistics," Hong Kong Internet Exhange, 12 July 2020. [Online]. Available: https://www.hkix.net/hkix/stat/aggt/hkix-aggregate.html. [Accessed 12 July 2020].

[60] Netnod, "Aggregate traffic graphs for Netnod," 12 July 2020. [Online]. Available: https://www.netnod.se/ix-stats/sums/. [Accessed 2020 July 2020].

[61] Cisco, "Best Practices in Core Network Capacity Planning," Cisco, 2013

[62] P. Loskot et al., "Long-Term Drivers of Broadband Traffic in Next-generation Networks," *Annals of Telecommunications*, vol. 70, p. 1, 2015.

[63] International Telecommunication Union, "Telecom Network Planning for Evolving Network Architectures - Reference Manual, V5.1," International Telecommunication Union, 2008.

[64] Cisco, "Cisco Visual Networking Index: Forecast and Trends, 2017–2022", Cisco White Paper, 2018.

[65] L. Chiaraviglio et al., "*Energy-Aware Backbone Networks: A Case Study*," in *2009 IEEE International Conference on Communications Workshop*, 2009.

[66] G. Maier et al., "*On Dominant Characteristics of Residential Broadband Internet Traffic*," in *Proceedings of 9th ACM SIGCOMM Conference on Internet Measurement*, 2009.

[67] A. Betker et al., "Comprehensive Topology and Traffic Model of a Nationwide Telecommunications Network," *Journal of Optical Communications Networks*, vol. 6, no. 11, p. 1038, 2014.

[68] J. Aslan *et al.*, "Electricity Intensity of Internet Data Transmission: Untangling the Estimates," *Journal of Industrial Ecology*, vol. 22, no. 4, p. 785, 2017.

[69] S. Strowes, *"Diurnal and Weekly Cycles in IPv6 Traffic,"* in *Proceedings of the 2016 Applied Networking Research Workshop (ANRW)*, 2016.

[70] L. Gradshteyn and I. Ryzhik, *Table of Integrals, Series and Products*, 6th Edition, A. Jeffrey and D. Zwillinger, Eds., Academic Press, 2000.

[71] A. Botchkarev, "A New Typology Design of Performance Metrics to Measure Errors in Machine Learning Regression Algorithms," *Interdisciplinary Journal of Information, Knowledge, and Management*, vol. 14, p. 45, 2019.

[72] A.-C. Orgerie *et al.*, "A Survey on Techniques for Improving the Energy Efficiency of Large Scale Distributed Systems," *ACM Computing Surveys*, vol. 46, no. 4, p. 1, 2014.

[73] S. Aleksic,, "Energy Efficiency of Electronic and Optical Network Elements," *IEEE Journal Of Selected Topics In Quantum Electronics*, vol. 70, no. 2, p. 296, 2011.

[74] F. Jalali *et al.*, "Energy Consumption of Content Distribution from Nano Data Centers versus Centralized Data Centers," *ACM SIGMETRICS Performance Evaluation Review*, vol. 42, no. 3, p. 49, 2014.

[75] M. Makkes *et al.*, "A decision framework for placement of applications in clouds that minimizes their carbon footprint," *Journal of Cloud Computing*, vol. 2, p. Article 21, 2013.

[76] A. Vishwanath *et al.*, "Modeling Energy Consumption in High-Capacity Routers and Switches," *Journal of Selected Areas in Communications*, vol. 32, no. 8, p. 1524, 2014.

Notation Index

This index provides a listing of the notation used in this book, chapter by chapter. Because there are very many mathematical symbols used in this book, the interpretation of some symbols is chapter specific. This Index should provide the reader with definitions for the majority of mathematical terms used in the book.

CHAPTER 3

Symbol	Definition	Symbol	Definition
$\langle E_{active} \rangle_{active}$	Average incremental energy per bit of active line cards (J/bit)	$P_{active,k}(t)$	k th active line card sle power consumption (W)
$\langle P_{active\_idle} \rangle_{active}$	Average idle power of active line cards (W)	P_{base}	Power consumption of network element base configuration (W)
$\langle P_{sleep} \rangle_{active}$	Sleep power averaged over active line cards (W)	P_{Core}	Total core network power (W)
$\langle P_{sleep} \rangle_{sleep}$	Sleep power averaged over line cards in sleep state (W)	$P_{Core,j}$	Power consumption of j th core network element (W)
$\langle R_{active} \rangle_{active}$	Average throughput of active line cards (bit/s)	P_{CPE}	Power consumption of CPE (W)
$C_{LC,max,v}$	Maximum equipment throughput capacity of type v line card (bit/s)	$P_{CPE,j}$	Power consumption of j th CPE (W)
C_{max}	Maximum equipment throughput capacity (bit/s)	P_{Edge}	Total edge network power (W)
ΔP	Increase in network element power consumption (W)	$P_{Edge,j}$	Power consumption of j th edge network element (W)
ΔR	Increase in network element throughput (bit/s)	P_{idle}	Network element idle power consumption (W)
$\Delta R^{[n]}$	Traffic increase in year n (bit/s)	P_{LA}	Power consumption of HFC line amplifiers (W)
E	Incremental energy per bit (J/bit)	$P_{LC,v}$	Power consumption of type v line card (W)
$E_{active,k}$	Incremental energy per bit for k th line card (J/bit)	P_{max}	Maximum equipment power consumption (W)
$E_{LC,v}$	Incremental energy per bit of type v line card (J/bit)	$P_{Network}$	Total network power (W)
E_P	Per packet processing energy (J)	P_{RN}	Power consumption of remote nodes (W)
$E_{S\&F}$	Per byte store and forward energy (J)	$P_{RN,k}$	Power consumption of k th remote node (W)
L_{pkt}	Input packet length (bytes)	P_{sleep}	Sleep mode power consumption (W)
$M_{X,LA}$	Average number of customers sharing the line amplifier	$P_{sleep,k}$	k th line card sleep mode power consumption (W)
$M_{X,RN}$	Average number of customers sharing the remote node	P_{TU}	Power consumption of terminal units (W)
$M_{X,TU}$	Average number of customers sharing the terminal unit	$P_{TU,m}$	Power consumption of m th terminal unit (W)
N_{Access}	Number of Customer Premises Equipment (CPE) equipment	P_{Wired_Access}	Wired Access network power (W)
N_{active}	Number of active line cards	$P_{Wireless\_Access}$	Wireless Access network power (W)

Symbol	Definition	Symbol	Definition
N_{Core}	Number of core network elements	$R(t)$	Network element throughput (bit/s)
N_{Edge}	Number of edge network elements	$R_{active,k}(t)$	Traffic throughput of k th line card (bit/s)
N_{LA}	Number of Hybrid Fibre Coax (HFC) line amplifiers	R_{bit}	Router input bit rate (bit/s)
$n_{LC,v}$	Number of line cards of type v	R_{byte}	Router input byte rate (byte/s)
N_{RN}	Number of remote nodes	R_{pkt}	Router input packet rate (packets/s)
N_{sleep}	Number of line cards in sleep mode	$R_{threshold}$	Threshold bit rate for sleep to active transition (bit/s)
N_{TU}	Number of terminal units	$v^{[n]}$	Type v line card deployed in year n
N_Z	Number of customers connected by technology Z	V_{LC}	Number of different types of line cards in the element
$N_{Z,LA}$	Number of line amplifiers using technology Z	W_X	Multiplier to account for additional power overheads; cooling, power supplies, etc.
$N_{Z,RN}$	Number of remote nodes using technology Z	X_{DSL}	Proportion of Digital Subscriber Line connections
$N_{Z,TU}$	Number of terminal units using technology Z	X_{FTTN}	Proportion of FTTN connections
$P(t)$	Network element power consumption (W) at time t	X_{FTTR}	Proportion of FTTP connections
$P^{[N]}$	Network element power consumption in year N (W)	X_{HFC}	Proportion of HFC connections
P_{Access}	Access network power		

CHAPTER 4

Symbol	Definition	Symbol	Definition
A_{BS}	Area of base station service area or cell (m²)	$N_{TRX,Sector}$	Number of transmit/receive antennas per sector of the base station
$\langle SE(t_u)\rangle^{(usr)}$	Average Spectral Efficiency over users for the u th Radio Frame (bit/s/Hz)	$P(t)$	Network element power consumption at time t (W)
$\langle SE(t_u)\rangle_{RF}$	Average Spectral Efficiency over PDSCH REs in the u th Radio Frame (bit/s/Hz)	P_{BB}	Baseband interface power consumption (W)
$\langle SE_p\rangle_{RF}$	Average Spectral Efficiency over a Radio Frame for the p th base station sector (bit/s/Hz)	$P_{BB,j}$	Baseband interface power consumption of j th base station (W)
$B^{(k)}(t_u)$	Data bits delivered to the k th user in u th Radio Frame	P_{BS}	Base station power consumption (W)
$B_i(t_u)$	Number of bits contained in the I th PDSCH RE transmitted in the u th Radio Frame	$P_{BS,j}$	Power consumption of j th base station (W)
$B_{RF}(t_u)$	Number of user data bits transmitted in the u th Radio Frame	P_{idle}	Network element idle power consumption (W)
BW	Base station bandwidth (Hz)	$P_{idle,BS}$	Base station idle power consumption (W)
C	Speed of light (m/s)	$P_{interf}(z^{(k)},t_u)$	Interference power at the location of user k, for the u th Radio Frame (W)

Symbol	Definition	Symbol	Definition
C_{max}	Wireline element maximum possible throughput (bit/s)	$P_{max,\,BS}$	Base station maximum power consumption (W)
$C_1(t_u)$	Base station throughput for 100% utilisation in u th Radio Frame	$P_{max,\,out}$	Maximum antenna radiated output power (W)
Δf	Spectral width of a Physical Resource Block (Hz)	P_{Mobile_Access}	Mobile network power consumption (W)
ΔL_f	Frequency correction factor for path loss	P_{out}	Transmit power (W)
ΔL_{rh}	Receiver antenna height correction factor	$P_{out}(t_u)$	Power provided to the transmit antenna for u th Radio Frame(W)
Δ_P	Slope factor relating transmit power to base station incremental power	$P_{out,\,j}$	Transmit power of j th base station (W)
$\Delta_{P,\,m}$	Slope factor relating transmit power to m th base station incremental power	P_{RF}	Base station transceiver power consumption (W)
δt_{sf}	Time duration of a subframe (s)	$P_{RF,\,j}$	Base station transceiver power consumption of j th base station (W)
E	Incremental energy per bit (J/bit)	P_{RS}	Reference Signal Resource Element power (W)
f	Carrier frequency used to transmit data (Hz)	$P_{Rx,\,m,\,p}(z^{(l)},t_u)$	Signal at $z^{(k)}$ transmitted from the p th sector antenna of the m th base station
F	Loss factor to account for local environment	$P_{Rx,\,Sig}(z^{(k)},t_u)$	Received signal power at the location of user k, for the u th Radio Frame (W)
$f^{(usr)}(z)$	User location probability density function	$P_{RxN}^{(k)}$	Receiver Noise power of user device for k th user(W)
$\Phi_m(z^{(k)})$	Radiation pattern value for the position $z^{(k)}$ relative to the vertical radiation pattern peak of the m th base station	P_{small_BS}	Small cell base station power consumption (W)
γ	Path loss exponent	$P_{TypeA,\,RE}$	Type A Resource Element power (W)
G_{RX}	User handset receiver antenna gain factor	$P_{TypeB,\,RE}$	Type B Resource Element power (W)
G_{TX}	Base station transmission antenna gain factor	P_{WiFi_Access}	Mobile network power consumption (W)
h_b, h_r	Base station transmitter, user receiver heights above ground, respectively (m)	$P_{WiFi\_AP,\,j}$	Power consumption of j th Wi-Fi Access Point (W)
η_{PA}	Power amplifier efficiency	$P_{Wireless\_Access}$	Total power consumption of wireless access (W)
$\eta_{PA,\,j}$	Power amplifier efficiency of j th base station	$\theta^{(k)}$	Horizontal angular location of k th user relative to transmitters of the p th sector of m th base station
$\varphi^{(k)}$	Vertical angular location of k th user relative to transmitters of m th base station	θ_{3dB}	Half power point for horizontal transmit antenna pattern
φ_{3dB}	Half power point for vertical transmit antenna pattern	$\Phi_{m,\,p}(z^{(k)})$	Radiation pattern value for the angle subtended by the position $z^{(k)}$ relative to the horizontal antenna pattern peak of sector p of base station m
λ	Sub-carrier wavelength for frequency range $\leq$ 2GHz (m)	$R^{(k)}(t_u)$	Data rate to the k th user by the u th Radio Frame (bit/s)
$L(r_m^{(k)},f,h_b,h_r,F,\upsilon)$	Path loss function	$R(t)$	Network element throughput (bit/s)
L_{free_space}	Free space propagation loss	$R(t_u)$	Base station throughput for u th Radio Frame (bit/s)

Symbol	Definition	Symbol	Definition
$\mu_{BS}^{(k)}(t_u)$	Base station utilisation by the k th user of the u th Radio Frame	ρ_A	Sets power of Type A RE relative to RS RE power
$P_{max,out,m}$	Maximum antenna radiated output power of the m th base station (W)	ρ_B	Sets power of Type B RE relative to RS RE power
$\mu_{BS,1}^{(k)}(t_u), \mu_{BS,2}^{(k)}(t_u)$	Utilisation of layers 1 and 2 transmitters for k th user with 2x2 SU-MIMO	$r_m^{(k)}$	Radial distance from m th base station to location $z^{(k)}$ (m)
μ_{BS}	Base station utilisation	r_{norm}	Normalising distance parameter for path loss function (m)
$\mu_{BS,m,p}$	Utilisation of p th sector of the m th base station	R_p	Download data rate for the p th base station sector (bit/s)
$\mu_{BS,m,p,q}$	Utilisation of q th transmitter in the p th sector of the m th base station	υ	Statistical "fading loss"
$\mu_{RS,p}$	Utilisation of p th base station transmitter	σ_{cool}	Loss factor representing cooling of the base station
$\mu_{PDCCH}(t_u)$	Utilisation of PDCCH REs for the u th Radio Frame	$\sigma_{cool,j}$	Loss factor representing cooling of the base station of j th base station
$\mu_{PDSCH}(t)$	Utilisation of PDSCH REs	σ_{DC}	Loss factor for DC to DC power supply
$N^{(usr)}(t_u)$	Number of active users at time of u th Radio Frame	$\sigma_{DC,j}$	Loss factor for DC to DC power supply of j th base station
$N^{(usr)}(z,t_u)$	Number of users in area dA at location z at time of u th Radio Frame	$SE_1\left(z^{(k)},t_u\right),$	Loss factor for DC to DC power supply of j th base station
$N_{BS}^{(usr)}(t_u)$	Number of active users in the base station service area at time of u th Radio Frame	$SE_2\left(z^{(k)},t_u\right)$	Spectral Efficiencies of layers 1 and 2 transmitters for k th user with 2x2 SU-MIMO (bit/s/Hz)
$N_{max,PDSCH}$	Maximum number of available PDSCH REs	$SE_i(t_u)$	Spectral Efficiency of the i th PDSCH RE in the u th Radio Frame
$N_{max,RF,PDCCH}$	Maximum number of available PDCCH REs in a Radio Frame	σ_{feed}	Loss factor for feed cable from power amplifier to antenna
$N_{max,RF,PDSCH}$	Maximum number of available PDSCH REs in a Radio Frame	$\sigma_{feed,j}$	Loss factor for feed cable from power amplifier to antenna of j th base station
$N_{RF,RE}$	Number Resource Elements in a Radio Frame	$SINR(z^{(k)},t_u)$	SINR at the location of user k, at time of u th Radio Frame
$N_{max,sd,RE}$	Number Resource Elements in a symbol duration	σ_{MS}	Loss factor for AC to DC power supply
N_{mobile}	Number of base stations in mobile network	$\sigma_{MS,j}$	Loss factor for AC to DC power supply of j th base station
$N_{PDSCH}(t)$	Number of PDSCH REs transmitted at time t	σ_{PDSCH}	Proportion of PDSCH REs to total REs in a Radio Frame
$N_{RF,PDCCH}(t_u)$	Number of PDCCH REs in the Radio Frame being transmitted at time of u th Radio Frame	T_{RF}	Duration of a Radio Frame (s)
$N_{RF,PDSCH}(t_u)$	Number of PDSCH REs in the Radio Frame being transmitted at time of u th Radio Frame	t_u	Time corresponding to the centre of u th Radio Frame (sec)
$N_{sf,PDSCH}(t_u)$	Number of PDSCH REs in the subframe being transmitted at time of u th Radio Frame	$z_q^{(k)}$	Location of k th user in the q th repeat of the simulation
N_{sim}	Number of times the simulation is repeated	$z^{(k)}=(x^{(k)},y^{(k)})$	Location of k th user
$\Pi_{m,p}(z^{(k)})$	Three dimensional antenna transmit radiation pattern		
N_{TRX}	Number of transmit/receive antennas per base station		

CHAPTER 5

Symbol	Definition	Symbol	Definition
$(x^{(k)}, y^{(k)}) = z^{(k)}$	Location of the k th user in (x,y) coordinates	$\mu_{BS,m,p}(t)$	Utilisation of p th sector of m th base station at time t
a) $N_j^{(usr)-}$	Number of users going off-line in the j th epoch with a successful download	$N^{(usr)}$	Target number of users for which data is to be collected in the simulation
A_{BS}	Service area of a base station (km^2)	$N_{grid}^{(usr)}$	Number of user locations on location grid
$\langle \mu_{BS,m,p} \rangle_T$	Average utilisation p th sector of m th base station over duration T	$N_{j,m,p}^{(usr)}$	Number of active users being served by the p th sector of the m th base station during the j th epoch
$\langle \mu_{BS,m} \rangle_T$	Average utilisation of m th base station over duration T	$N_{j,m,p,PDSCH}^{(usr)}$	Number of PDSCH REs allocated to each user in the p th sector of m th base station during the j th epoch
$\langle \mu \rangle_{BS,D}$	Average base station utilisation over a diurnal cycle	$N_{Tot}^{(usr)}$	Actual number of users for which data is collected in the simulation
$\langle R \rangle_{BS,D}$	Average base station throughput over a diurnal cycle (bit/s)	N_{BS}	Number of base stations
$\langle R \rangle_{BS,T}$	Average base station throughput over duration, T of the simulation (bit/s)	N_{ISD}	Number of inter base station distances for length and width of service area
$\langle R \rangle_D^{(usr)}$	Average user throughput over a diurnal cycle (bit/s)	$N_j^{(usr)}$	Number of on-line users in the j th epoch
A_{SA}	Simulation service area (km^2)	$N_j^{(usr)-}$	Number of users going off-line in the j th epoch
$B_{Tot}^{(k)}$	Total data downloaded by the k th user (bits)	$N_j^{(usr)+}$	Number of new user requests that occur in the j th epoch
$B^{(usr)}$	Size of file downloaded by users (bits)	N_{tot}	Total number of epochs in simulation
b) $N_j^{(usr)-}$	Number of users going off-line in the j th epoch with a failed download	$N_{TTI,PDSCH}$	Average number of PDSCH REs in a TTI
BS_m	m th base station	$N_{TTI,RE}$	Average number of REs in a TTI
B_{Tot}	Total data downloaded to all users in the simulation (bits)	$p^{(k)}$	Index of the base station sector serving the k th user
$B_{Tot,S}$	Total downloaded data during stage S of the diurnal cycle	$P_{int\,erf,j}^{(k)}$	Interference power for k th user in j th epoch (W)
δ	Fraction of ISD between user location grid points	P_{BB}	Baseband interface power consumption (W)
$D^{(usr)}$	Average download traffic demand per user (kbit/s)	P_{idle}	Network element idle power consumption (W)
$\Delta B_j^{(k)}$	Amount of data downloaded by k th user during the j th epoch (bits)	$P_{idle,out}$	Base station Idle transmit power (W)
Δ_P	Slope factor relating transmit power to base station incremental power	$P_{idle,out,m,p}$	Idle transmit output power of p th sector of m th base station (W)
ΔP_{out}	Transmit power for each sector antenna of each base station (W)	$P_{max,out}$	Maximum transmit power for all sectors of all base stations (W)

Symbol	Definition	Symbol	Definition
Δt	Simulation epoch duration (s)	$P_{out,m,p}(\mu_{BS,m,p})$	Transmit power of p th sector of m th base station as a function of its utilisation
$\Delta t_{on}^{(k)}$	Time duration k th user is on-line (s)	P_{RF}	Base station transceiver power consumption (W)
$\Delta t_{max,dl}$	Maximum allowed download time (s)	$\text{Prob}(\Delta N_j^{(usr)+} \leq 2)$	Probability no more than 2 user requests are added to the service area in the j th epoch
δt_{TTI}	Duration of a Transmission Time Interval, TTI (s)	$\text{Prob}(z^{(k)} = z)$	Probability location of k th user will be z
f	Carrier frequency used to transmit data (Hz)	$P_{Rx,m,p}(z^{(k)})$	Signal at $z^{(k)}$ transmitted from the p th sector antenna of the m th base station
F	Loss factor to account for local environment	$P_{RxN}^{(k)}$	Receiver Noise power of k th user's device (W)
$\Phi_m(z^{(k)})$	Radiation pattern value for the position $z^{(k)}$ relative to the vertical radiation pattern peak of the m th base station	$Q_{BS,m}$	Energy consumed by the m th base station (J)
F_{Prop}	Proportion of downloads that fail (are terminated before $B^{(usr)}$ bits downloaded)	$Q_{BS,S,m}$	Energy consumption of m th base station during stage S of the diurnal cycle
$F_{Prop,D}$	Proportion of downloads that fail over a diurnal cycle	R_j	Total download data throughput in the j th epoch (bit/s)
$GPM^{(usr)}$	User download volume per month (GB/month/user)	ρ_s	Proportion of 24 hours for stage S of the diurnal cycle
G_{Rx}	User handset receiver antenna gain factor	υ	Statistical "fading loss"
G_{Tx}	Base station transmission antenna gain factor	σ_{cool}	Loss factor representing cooling of the base station
h_b	Base station transmitter height above ground (m)	σ_{DC}	Loss factor for DC to DC power supply
h_r	User receiver height above ground (m)	σ_{feed}	Loss factor for feed cable from power amplifier to antenna
H	Base station energy efficiency (J/bit)	$SINR^{(k)}$	SINR for the k th user
H_D	Base station energy efficiency over a diurnal cycle (J/bit)	σ_{MS}	Loss factor for AC to DC power supply
η_{PA}	Power amplifier efficiency	σ_{PDSCH}	Average number of PDSCH REs in a Radio Frame
H_S	Base station energy efficiency during stage S of the diurnal cycle (J/bit)	T	Total time duration of the simulation (s)
ISD	Inter base station distance (m)	$t_{off,j}^{(k)}$	Time at which k th user goes off line
$j_{off}^{(k)}$	Epoch in which k th user goes off-line	$t_{on,j}^{(k)}$	Time at which k th user goes on line
$j_{on}^{(k)}$	Epoch in which k th user comes on-line	$W(m,p,z^{(k)})$	Channel function defined in (5.22)
$j_{max,dl}$	Number of epochs corresponding to maximum allowed download time	X_S	Multiplier to rescale λ for the stage S of the diurnal cycle
λ	Download request rate (requests/s)	Y, Z	Generic simulation output result
$L(r_m^{(k)}, f, h_b, h_r, F, \upsilon)$	Path loss function	$z^{(k)} = (x^{(k)}, y^{(k)})$	Location of the k th user
$m^{(k)}$	Index of base station serving the k th user		
$M_{BS,m,p}$	Cumulative number of epochs during which the p th sector of the m th base station downloaded data		

CHAPTER 6

Symbol	Definition	Symbol	Definition
$(x_{m,p,w(1)}, y_{m,p,w(1)})$	Location of first positioned small cell in the process for positioning small cells in a macro cell sector	MD_{SC}	Minimum distance of user from a small cell base station (m)
$(x_{m,p,w(2)}, y_{m,p,w(2)})$	Location of second positioned small cell in the process for positioning small cells in a macro cell sector	μ_{SC}	Small cell base station utilisation
$(x_{m,p,w(q)}, y_{m,p,w(q)})$	Location of q th positioned small cell in the process for positioning small cells in a macro cell sector	$\mu_{SC,w,j}$	Utilisation of w th small cell base station in j th epoch
$(x_{m,p,w,v}, y_{m,p,w,v})$	(x,y) location of point with quadruplet index $\{m,p,w,v\}$ relative to BS_0	ν	Randomly generated number to locate $\{nSC\}$ users in a small cell in $\{nSC\}$, $0 \le \nu \le 1$
$(x_{m,p,w}, y_{m,p,w})$	(x,y) location of point with triplex index $\{m,p,w\}$	$N^{(usr)}_{\{X\},grid}$	Number of user location grid points available in the $\{X\}$ network
$(x_{p,w}, y_{p,w})$	(x,y) location of point with duplex index $\{p,w\}$	$N^{(usr)}_{\{X\},data}$	Target number of users served by the $\{X\}$ network for which data is collected
(x_w, y_w)	(x,y) location of point with index $\{w\}$	$N^{usr}_{\{X\}}$	Number of user location grid points available in the $\{SC\}$ network
$\{m,p,w,v\}$	Quadruplet index for locations of small cell grid points for user locations, v, in small cell, w, in p th sector of m th macro cell	$N^{(usr)}_{MC}$	Number of macro cell users in the service area
$\{m,p,w\}$	Triplet index for locations of small cell grid points, w, in p th sector of m th macro cell	$N^{(usr)}_{MC,grid}$	Number of user location grid points available in the macro cells
$\{nSC\}$	Not the small cell only sub-network	$N^{(usr)}_{SA}$	Number of uses in the service area
$\{p,w\}$	Duplet index for locations of small cell grid points, w, in p th macro cell sector	$N^{(usr)}_{SC}$	Number of small cell users in the service area
$\{SC\}$	Small cell only sub-network	$N^{(usr)}_{SCArea}$	Total number of users located within a small cell
$\{w\}$	Index for locations of small cell grid points, (x_w, y_w)	$N^{(usr)}_{\{X\},data}$	Target number of users for which data is collected in the $\{X\}$ simulation
$\langle B_{\{x\}}(D^{(usr)}) \rangle_T$	Downloaded bit per epoch for user demand $D^{(usr)}$ of $\{X\}$ network averaged over time T (bits)	$N^{(usr)}_{Tot,\{X\},data}$	Actual number of users for which data is collected in the $\{X\}$ simulation
$\langle B_{Het} \rangle_D$	Average downloaded bits per epoch of heterogeneous network over a diurnal cycle (bits)	$N^{(usr)}_{Tot,\{x\},u}$	Actual number of users for which data is collected in the u th run of the $\{X\}$ simulation
A_{BS}	Service area of a base station (km²)	$N_{\{nSC\},SC}$	Number of small cells in $\{nSC\}$
$\langle \mu_{SC,q,u} \rangle_T$	q th small cell base station utilisation for the u th run of the small cell simulation averaged over simulation duration T	$N_{\{SC\}}$	Number of small cells in $\{SC\}$

Symbol	Definition	Symbol	Definition
$\langle u_{SC,u}\rangle_{SC}$	Average small cell base station utilisation for the u th run of the small cell simulation	$N_{\{X\},fail}(D^{(usr)})$	Number of failed downloads in $\{X\}$ for user demand $D^{(usr)}$
$\langle u_{SC,u}\rangle_{SC,sim,T}$	Small cell base station utilisation averaged over all of small cell simulations, small cells and simulation duration T	$N_{\{X\},fail,S}$	Number of failed downloads in $\{X\}$ during diurnal cycle stage S
$\langle u_{\{X\}}(D^{(usr)})\rangle_T$	Average utilisation of $\{X\}$ base stations for user demand $D^{(usr)}$ over duration T	N_{grid}	Number of available grid points in (x,y) location grid
$\langle u_{\{X\}}\rangle_D$	Average utilisation of $\{X\}$ network base stations over a diurnal cycle	N_{ISD}	Number of inter macro cell site distances used for the length and width of the service area
$\langle u_{Het}(D^{(usr)})\rangle^{(usr)}$	Average base station utilisation for user demand $D^{(usr)}$ averaged over all base stations in the heterogeneous network	N_{MC}	Number of macro cells in the service area
$\mu_{Het\,D}^{(usr)}$	User base station utilisation averaged over all base stations in the heterogeneous network and the diurnal cycle	$N_{MC,SC}$	Number of macro base stations with small cells located within them
$\langle P_{\{X\}}(D^{(usr)})\rangle_T$	Network power consumption for user demand $D^{(usr)}$ averaged over time T of $\{X\}$ network (W)	N_{SC}	Number of small cells per macro cell sector
$\langle P_{Het}\rangle_D$	Average power consumption of heterogeneous network over a diurnal cycle (W)	$N_{SC,grid}$	Number of grid points for user location within a small cell
$\langle Q_{\{X\}}(D^{(usr)})\rangle_T$	Energy consumption per epoch for user demand $D^{(usr)}$ of $\{X\}$ network averaged over time T (J)	$N_{SC,Tot}$	Total number of small cells in the network
$\langle Q_{Het}\rangle_D$	Average energy consumption per epoch of heterogeneous network over a diurnal cycle (J)	$N_{Sctr,grid}$	Number of small cell location grid points in the sector
$\left\langle R_{\{X\}}\right\rangle_D^{(usr)}$	User throughput of $\{X\}$ network averaged over the diurnal cycle (bit/s)	N_{sim}	Number of small cell simulation runs
$\langle R_{\{X\}}(D^{(usr)})\rangle^{(usr)}$	Average user throughput of $\{X\}$ network for use demand $D^{(usr)}$ (bit/s)	$N_{Tot,\{SC\},u}$	Number of epochs for which data was collected in the u th run of the $\{SC\}$ simulation
$\langle R_{\{X\}}(D^{(usr)})\rangle_T$	Download data rate for user demand $D^{(usr)}$ averaged over time T of $\{X\}$ network (bit/s)	$N_{Tot,SC,S,u}$	Number of epochs for which data was collected in the u th run of the small cell simulation corresponding to diurnal stage S
$\langle R_{Het}(D^{(usr)})\rangle^{(usr)}$	Average user throughput for use demand $D^{(usr)}$ averaged over all users in the heterogeneous network	$N_{Tot,u}$	Number of epochs in the u th simulation run
$\langle R_{Het}\rangle_D$	Heterogeneous network throughput averaged over the diurnal cycle (bit/s)	$P_{Interf\{X\},j}^{(k)}$	Interference power for k th user in network $\{X\}$ during j th epoch (W)
$\left\langle R_{Het}\right\rangle_D^{(usr)}$	User throughput averaged over all users in the heterogeneous network and the diurnal cycle	$P_{\{nSC\},m,u}$	Power consumption of m th macro cell in $\{nSC\}$ in u th simulation run (W)

Symbol	Definition	Symbol	Definition
A_{SA}	Total service area (m²)	$P_{\{SC\},q,u}$	Power consumption of q th base station in $\{SC\}$ for u th simulation run (W)
$\langle SINR_{SC,u}\rangle^{(usr)}$	Average user SINR for the u th run of the small cell simulation	P_{BB}	Baseband interface power consumption (W)
$\langle SINR_{SC}\rangle^{(usr)}$	Average small cell user SINR	$P_{idle,out,SC}$	Transmit idle power for small cell transmitters (W)
$\langle SINR_{Het}(D^{(usr)})\rangle^{(usr)}$	Average user SINR for user demand $D^{(usr)}$ averaged over all users in the heterogeneous network	$P_{max,out,SC}$	Maximum transmit power for small cells (W)
$\langle SINR_{Het}\rangle_{D}^{(usr)}$	User SINR averaged over all users in the heterogeneous network and the diurnal cycle	$P_{MC,m}$	Power consumption of m th macro cell base station (W)
$B^{(usr)}$	Size of file downloaded by users (bits)	$P_{MC,m,u}$	Power consumption of m th macro cell base station in for u th simulation run (W)
$B_{\{X\},q,u}$	Total download of q th base station in $\{X\}$ network for u th simulation run (bit)	P_{RF}	Base station transceiver power consumption (W)
$B_{MC,m,u}$	Total download of m th macro cell base station in for u th simulation run (bit)	$\mathrm{Prob}((x^{(k)},y^{(k)})=(x,y))$	Probability that user will be located at grid point (x,y) in macro cell network
BS_m	m th base station	$\mathrm{Prob}\big((x^{(k)},y^{(k)})=(x_{m,p,w,v},y_{m,p,w,v})\big)$	Probability that user will be located at grid point $(x_{m,p,w,v},y_{m,p,w,v})$ in small cell network
δ	Proportion relating Δx and Δy to SCD	$\mathrm{Prob}\big(\Delta N_{j}^{(usr)+}\leq 2\big)$	Probability no more than 2 user requests are added to the service area in the j th epoch
$D^{(usr)}$	Average download rate per user (kbit/s)	$P_{Rx,w}(z^{(k)})$	Received power by k th user from w th small cell
$D'^{(usr)}_{\{X\}}$	Rescaled average user demand data rate for network $\{X\}$	$P_{SC}(\mu_{SC})$	Power consumption of small cell base station for utilisation μ_{SC} (W)
δ_{MC}	Grid spacing factor, $\Delta x=\Delta y=\delta_{MC}ISD$	$Q_{\{nSC\},m,u}$	Energy consumption of m th macro cell the u th $\{nSC\}$ simulation run (J)
P_{out}	Transmit power for each sector antenna of each base station (W)	$Q_{\{SC\},q,u}$	Energy consumption of q th small cell in u th $\{SC\}$ simulation run (J)
$\Delta t_{\{X\}}$	Epoch duration for $\{X\}$ network simulation (s)	$R_{\{X\},u}$	$\{X\}$ network throughput for u th simulation run (bit/s)
$\Delta t_{max,dl}$	Maximum allowed download time (s)	ρ_S	Proportion of 24 hours for stage S of the diurnal cycle
$\Delta x,\Delta y$	Step for grid separation in x and y directions (m)	SCD	Small cell diameter (m)
ε	Proportion relating MD_{SC} to SCD	σ_{DC}	Loss factor for DC to DC power supply
$F_{\{X\},Prop,D}$	Proportion of failed downloads in the $\{X\}$ network averaged over the diurnal cycle	$SINR_{\{nSC\}}^{(k)}$	SINR for k th user in $\{nSC\}$

Symbol	Definition	Symbol	Definition
$F_{Het,Prop}(D^{(usr)})$	Proportion of heterogeneous network failed downloads for user demand $D^{(usr)}$	$SINR_{\{SC\}}^{(k)}$	SINR for k th user in $\{SC\}$
$F_{Het,Prop,D}$	Proportion of failed downloads in the heterogeneous network averaged over the diurnal cycle	σ_{MS}	Loss factor for AC to DC power supply
$H_{\{X\},S}$	Energy efficiency of $\{X\}$ network for stage S of the diurnal cycle (J/bit)	$U^{(usr)}$	Average user density across the service area (users/km²)
$H_{\{X\},u}$	Energy efficiency of $\{X\}$ network for u th small cell simulation run (J/bit)	$U_{MC}^{(usr)}$	Density of macro cell users (users/m²)
$H_{Het}(D^{(usr)})$	Energy efficiency of heterogeneous network for user demand $D^{(usr)}$ (J/bit)	$U_{SC}^{(usr)}$	Density of small cell users (users/m²)
$H_{Het,D}$	Energy efficiency of the heterogeneous network averaged over the diurnal cycle (J/bit)	$w^{(k)}$	w-index of small cell base station serving k th user
ISCD	Inter small cell base station distance (m)	$W_{MC}(m,p,z^{(k)})$	Channel function for location $z^{(k)}$ for transmission from p th of the m th macro base station
ISD	Inter site distance for macro cells (m)	$W_{SC}(w,z^{(k)})$	Channel function for location $z^{(k)}$ for transmission from w th small cell base station
φ_{SC}	Offload ratio: proportion of users offloaded to small cells	X_S	Multiplier to rescale λ for the stage S of the diurnal cycle
φ'_{SC}	Changed offload ratio: proportion of users offloaded to small cells	Y, Z	Generic simulation output result
λ	Download request rate (requests/s)	$z^{(k)}$	Location of k th user
$\lambda_{\{X\}}$	Service request rate for $\{X\}$ network (request/s)		
$\mu_{\{X\}}$	Utilisation of base station in $\{X\}$ network		
MD	Minimum allowed distance from a small cell base station to every macro cell base station (m)		

CHAPTER 7

Symbol	Definition	Symbol	Definition
$A(t,n)$	Mean access speed per user (bit/s)	$N_{b,tot}$	Total number of Type b) Access Network users
A_a	Average access speed of Type a) users	N_E	Number of network elements
$A_a(n)$	Typical access speed of Type a) users over the diurnal cycle in year n	N_{min}	Minimum number of Access Network active users
A_b	Average access speed of Type b) users	N_{peak}	Maximum number of Access Network active users
$A_b(n)$	Typical access speed of Type b) users over the diurnal cycle in year n	N_{tot}	Total number of access network users = number of CPE
$\langle f^{[n]} \rangle_T$	Average of $f(t,\tau)$ over duration T in year n	$P(t)$	Network element power consumption (W)
$A^{(k)}(t,n)$	Access speed of k th user at time t in year n (bit/s)	P_{idle}	Network element idle power (W)
$\langle R(n) \rangle_T$	Mean Access Network traffic in year n averaged over duration T	$P_{idle,j}$	Idle power consumption of j th network element (W)
β	Annual growth rate of mean traffic	$P_{Ntwk}(t)$	Total network power consumption (W)

Symbol	Definition	Symbol	Definition
χ	Proportion of Type a) users; $N_{a,tot} = \chi N_{tot}$	$P_{save}(t)$	Power saved at time t by use of sleep state
$\Delta N_a(t)$	Increase in number of Type a) users over $N_{a,min}$ at time t	P_{sleep}	Sleep state power consumption
$\Delta N_b(t)$	Increase in number of Type b) users over $N_{b,min}$ at time t	$Q_{Ntwk}(t,T)$	Network energy consumption over time t to $t+T$ (J)
$\Delta R(t)$	Increase in access traffic over R_{min} at time t	$R(N)$	Access Network traffic as a function of number of active users, N
E	Network element incremental energy per bit (J/bit)	$R(t)$	Network element throughput (bit/s)
$\varepsilon(t,\tau)$	Defined in (7.21)	$R(t,n)$	User dependent Access Network traffic = $R'(t,n) - R_{oh}$ (bit/s)
E_j	Incremental energy per bit of j th network element (J/bit)	$R'(t,n)$	Total Access Network traffic at time t in the diurnal cycle in year n (bit/s)
$f(t,\tau)$	Diurnal cycle traffic normalised relative to peak traffic	$R_{peak}^{[m]}$	Peak traffic in year m (bit/s)
$\gamma, \delta, \eta_a, \eta_b$	Annual exponential growth factors	$R_2(t)$	Diurnal cycle traffic averaged over minutes to hours (bit/s)
H	Incremental traffic per user = $(R_{peak} - R_{min})/(N_{peak} - N_{min})$	$R_3(t)$	Diurnal cycle traffic averaged over days to weeks (bit/s)
$\kappa(\tau)$	Defined in (7.24)	$R_j(t)$	Throughput of j th network element (bit/s)
N	Number of active Access Network users	R_{oh}	Network overhead traffic (bit/s)
$N(t,n)$	Number of Access Network active users at time t in year n	R_{peak}	Peak time traffic (bit/s)
$N_a(t)$	Number of Type a) users active at time t	σ	Annual growth rate of peak traffic
$N_{a,min}$	Minimum number of active Type a) users over the diurnal cycle	T_1	Averaging time of minutes to hours
$N_{a,tot}$	Total number of Type a) Access Network users	T_2	Averaging time of days to weeks
$N_b(t)$	Number of Type b) users active at time t	t_{peak}	Time of peak traffic in the diurnal cycle (hours)
$N_{b,min}$	Minimum number of active Type b) users over the diurnal cycle	ζ	Constant defined by $\Delta N_b(t) \approx \zeta \Delta N_a(t)$

CHAPTER 8

Symbol	Definition	Symbol	Definition
$\left\langle C_{max}^{[n]} \right\rangle_E$	Average maximum network element throughput capacity in year n (bit/s)	N_E	Number of network elements
$\langle E \rangle_E$	Average incremental energy per bit across all network elements (J/bit)	$N_E^{(k)}$	Number of network elements through which the k th service flows
$\alpha_j^{(k)}$	Portion of the traffic flow of the k th service that is goes through the j th network element	$N_j^{(svc)}$	Number of services flowing through the j th network element
$\langle N_{hops} \rangle^{(svc)}$	Average number of hops for a service	$P_{Ntwk}^{[n]}(t)$	Forecast network power consumption at time t in the diurnal cycle in year n (W)
$\left\langle P_{max}^{[n]} \right\rangle_E$	Average maximum network element power consumption in year n (W)	$P_{Ntwk}^{[n]}(t_{peak})$	Forecast network power consumption at peak traffic time in the diurnal cycle in year n (W)
$\left\langle P_{Ntwk}^{[n]} \right\rangle_T$	Forecast network power consumption averaged over a duration T for year n (W)	$P_{idle,j}$	Idle power of j th network element (W)

Symbol	Definition	Symbol	Definition
$\langle P_{idle} \rangle_E$	Average idle power across all network elements (W)	$P_{max,j}$	Maximum power consumption of j th network element (W)
$\langle x^{(k)} \rangle_j$	$x_j^{(k)}$ averaged over the j index for N_E network elements	$P_{Ntwk}(t)$	Total network power consumption at time t (W)
$\langle x_j \rangle^{(svc)}$	$x_j^{(k)}$ averaged over the k index for $N^{(svc)}$ services	$P_{Ntwk}(t, \tau)$	Network power consumption at time t in the diurnal cycle at time τ on the time scale of months to years
$\langle x \rangle$	$x_j^{(k)}$ averaged over both j and k indices	$\theta(t)$	Unit step function, $\theta(x) = 0$ for $x < 0$ and $\theta(x) = 1$ for $x \geq 0$.
β	Annual proportional increase in traffic averaged over a diurnal cycle	$R^{(k)}(t)$	Data rate of k th service (bit/s)
$C_{max,j}$	Maximum throughput capacity of j th network element (bit/s)	$R^{[n]}(t)$	Forecast traffic at time t in year n (bit/s)
$\Delta R^{[m,p]}(t)$	Increase in traffic over the years between two successive deployments, $\lfloor m/p \rfloor p$ and $(\lfloor m/p \rfloor + 1)p$ (bit/s)	$R^{[n]}_{peak}$	Forecast peak traffic in year n (bit/s)
$\Delta R^{[n]}(t)$	Forecast increase traffic in year n at time t (bit/s)	$R_j^{(k)}(t)$	Traffic flow of the k th service that is goes through the j th network element (bit/s)
E_j	Incremental energy per bit of j th network element (J/bit)	$R_j(t)$	Throughput of j th network element (bit/s)
$f(t, \tau)$	Diurnal cycle traffic at time t in the diurnal cycle at time τ on the time scale of months to years normalised relative to peak traffic $R_{peak}(\tau)$	$R_{Ntwk}(t)$	Total network traffic (bit/s)
$f^{[n]}(t)$	Forecast normalised diurnal cycle traffic at time t in year n relative to peak traffic $R^{[n]}_{peak}(\tau)$	R_{peak}	Peak network traffic (bit/s)
γ	Annual improvement factor for $\langle P_{max} \rangle / \langle C_{max} \rangle$ (W/bit/s)	σ	Annual proportional increase in peak traffic
μ_{max}	Maximum allowed utilisation of network elements	t_{peak}	Time of peak network traffic (hour)
M_{OH}	Multiplier used to represent the proportionate increase in power consumption due to overheads	y	Ratio of P_{idle}/P_{max}
$N_{hops}^{(k)}$	Number of hops in the path of the k th service	$y^{[n]}$	Forecast ratio of P_{idle}/P_{max} in year n
$N^{(svc)}$	Number of services	$\lfloor x \rfloor$	The largest integer less than or equal to x

CHAPTER 9

Symbol	Definition	Symbol	Definition
$a_1, b_1, a_2,$ $b_2, c_2, a_3,$ b_3, c_c, d_3	Server power consumption fitting parameters	$P_{PDU}(U)$	PDU power consumption for utilisation U (W)
$\langle u(t) \rangle_{svr}$	Average utilisation over servers at time t	$P_{PDU, idle}$	Idle power consumption of PDU unit (W)
$\langle u_c \rangle_{svr}$	Average utilisation per server for consolidation c	P_{pumps}	Water circulation pump power consumption (W)
β_{CPU}	Improvement rate per year for the maximum number of operations per second for the CPU	P_{rem}	Power required to remove heat at rate dQ_{heat}/dt (W)
B_d	Transferred file size (bits)	P_{sleep}	Server sleep state power consumption (W)

Symbol	Definition	Symbol	Definition
β_{NIC}	Maximum capacity improvement rate per year for the NIC	P_{st}	Storage server power consumption (W)
$B_{storage}$	Capacity of the data storage system (bits)	$P_{st,\,max}$	Maximum power consumption of the storage server (W)
c	Consolidation parameter	$P_{storage}$	Power consumption of the data storage system (W)
$C_{NIC}^{[n]}$	NIC maximum capacity in year n	$P_{storage}(u_{CPU})$	Netflix "storage appliance" server power consumption (W)
C_{air}	Specific heat of air(1.0 kJ/kg°C)	$P_{svr}(u(t))$	Server power consumption for utilisation u (W)
CoP	Ratio of heat removed per second (dQ_{heat}/dt) and the power required to remove that heat	$P_{svr,\,CPU}$	Server CPU power consumption (W)
$CoP_{condenser}$	CoP of the condenser in the CRAC unit	$P_{svr,\,idle}$	Idle server power consumption (W)
$C_{st,\,max}$	Maximum throughput capacity of the storage server (bit/s)	$P_{svr,\,max}$	Maximum server power consumption (W)
ΔP_{CPU}	Incremental CPU power defined as $P_{CPU,max} - P_{CPU,idle}$ (W)	$P_{svr,\,mem/disk}$	Power consumption of server memory & disk access (W)
ΔP_{CRAH}	defined as $P_{CRAH,max} - P_{CRAH,idle}$ (W)	$P_{svr,\,NIC}$	Power consumption of sever NIC (W)
ΔP_{CRAH}	Incremental CRAH power defined as $P_{CRAC,max} - P_{CRAC,idle}$ (W)	$P_{svr,\,tot}(1)$	Maximum total server power consumption given (W)
ΔP_{disk}	Incremental disk power defined as $P_{disk,max} - P_{disk,idle}$ (W)	PUE	Power Use Effectiveness of the data centre in which the server is located
ΔP_{mem}	Incremental memory power defined as $P_{mem,max} - P_{mem,idle}$ (W)	$P_{UPS}(U)$	UPS power consumption for utilisation U (W)
ΔP_{NIC}	Incremental UPD power defined as $P_{NIC,max} - P_{NIC,idle}$ (W)	$P_{UPS,\,idle}$	UPS idle power consumption (W)
ΔP_{UPS}	Incremental UPD power defined as $P_{UPS,max} - P_{UPS,idle}$ (W)	$R_{CPU}(t)$	CPU operations per second (operations/s)
dQ_{heat}/dt	Rate of heat energy removal (J/s)	R_D	File transfers per hour (transfers/hour)
ΔT	Temperature difference between exit and entry air flow (°C)	$R_{disk}(t)$	Disk read/writes per second (read/s, write/s)
E_{CPU}	CPU energy per operation (J/operation)	$R_{disk\_B/s}$	Total number of bytes sent to and retrieved from the disk per second (bit/s)
E_{disk}	Disk energy per disk read or write (J/read, J/write)	$R_{mem}(t)$	Memory accesses per second (accesses/s)
E_{mem}	Memory energy per access (J/access event)	$R_{NIC}(t)$	NIC data rate (bit/s)
E_{NIC}	NIC energy per bit (J/bit)	R_{page_fault}	Number of memory page faults per second (fault/s)
E_{proc}	Server energy consumption (J)	ρ_{PDU}	PDU power increase coefficient (1/W)
f_{air}	Air volume flow rate (m³/s)	S_{air}	Specific gravity of air (1.2 kg/m³)
$f_{air,\,norm}$	Normalised air flow rate $= f_{air}/f_{CRAH,max}$	T_{inlet}	Temperature of the air inlet into the server housing (°C)
$f_{CRAH,\,max}$	Maximum CRAH air volume flow rate (m³/s)	T_{outlet}	Temperature of the air outlet from the server housing (°C)
η_{heat}	Heat transfer efficiency	t_{proc}	Time taken to complete processing (s)
κ_{air}	Air flow containment factor	T_{supply}	Temperature of the cold air emerging from the CRAH (°C)
m_{CRAH}	Mass flow rate of the CRAH unit (kg/s)	T_{water}	Supply water temperature to the CRAC (°C)
$N_{chiller}$	Number of chiller units deployed	$U(t)$	Data centre utilisation at time t
N_{CRAH}	Number of CRAH units deployed	$u(t)$	Server utilisation at time t
N_{svr}	Number of servers	$u_{CPU}^{[n]}$	CPU utilisation in year n

Symbol	Definition	Symbol	Definition
$N_{svr,\,act}$	Number of active servers	$u_{CPU,\,flash}^{[n]}$	CPU utilisation of flash application server in year n
$ops_{CPU}^{[n]}$	Maximum number of CPU operations per second in year n	$u_{CPU,\,global}^{[n]}$	CPU utilisation of global application server in year n
$P_{chiller}(1)$	Maximum power consumption of the chiller (W)	$u_{CPU,\,storage}^{[n]}$	CPU utilisation of storage application server in year n
$P_{chillter}(U)$	Chiller power consumption for utilisation U (W)	$u_{NIC}^{[n]}$	NIC utilisation in year n
$P_{cooling}(1)$	Maximum power consumption of the server cooling system (W)	$u_{CPU}(t)$	CPU utilisation at time t
$P_{CPU,\,idle}$	CPU idle power consumption (W)	$u_{CPU}(u_{NIC})$	CPU utilisation as a function of NIC utilisation
$P_{CRAC}(U)$	CRAC power consumption for utilisation U (W)	$u_{disk}(t)$	Disk read/writes utilisation at time t
$P_{CRAC,\,idle}$	Idle power consumption of CRAC unit (W)	u_{disk_time}	Proportion of disk time the server disk was busy servicing the read and write requests
$P_{CRAH,\,m}$	Power consumption of the m th CRAC unit (W)	u_{mem}	Proportion of memory used by the application
$P_{(CRAC,\,tot)}(t)$	Maximum power consumption of the CRAC (W)	$u_{mem}(t)$	Memory utilisation
$P_{CRAH}(U)$	CRAH power consumption for utilisation U (W)	$u_{NIC}(t)$	NIC utilisation at time t
$P_{CRAH,\,idle}$	Idle power consumption of CRAH unit (W)	$u_{NIC,\,flash}$	NIC utilisation for flash appliance server
$P_{CRAH,\,tot}(1)$	Maximum power consumption of the CRAH (W)	$u_{NIC,\,global}$	NIC utilisation for global appliance server
$P_{DC}(1)$	Total data centre power consumption at maximum utilisation (W)	$u_{NIC,\,storage}$	NIC utilisation for storage appliance server
$P_{disk,\,idle}$	Disk idle power consumption (W)	X_{red}	Redundancy used in the system in case of failure
$P_{flash}(u_{CPU})$	Netflix "flash appliance" server power consumption (W)	y	Ratio $P_{svr,idle}/P_{svr,max}$
$P_{global}(u_{CPU})$	Netflix "global appliance" server power consumption (W)	y_{sleep}	Ratio $P_{sleep}/P_{svr,max}$
$P_{mem,\,idle}$	Memory idle power consumption (W)		
P_{misc}	Miscellaneous power consumption (W)		
$P_{NIC,\,idle}$	Network Interface Card idle power consumption (W)		

CHAPTER 10

Symbol	Definition	Symbol	Definition
$\langle \alpha \rangle$	$\alpha_j^{(k)}$ averaged over all service flows and network elements	$N_{max,\,PDSCH}$	Maximum number of PDSCH REs
$\langle N_{hops} \rangle^{(svc)}$	Average number of hops in the path of a service	$N_{PDSCH}(t)$	Number of PDSCH REs allocated at time t
$\langle P^{(k)} \rangle_T$	Average power consumption of k th service over duration T (W)	$P^{(k)}(t)$	Network element power consumption of k th traffic flow (W)
$\langle P^{(svc)} \rangle$	Average power consumption of the equipment (W)	$P_{Ntwk}^{(k)}(t)$	Total power consumption of k th service across the network (W)
$\langle P_{idle} \rangle_E$	Idle power averaged over network elements (W)	$P^{(k,svc)}$	Base station power attributed to the k th service (W)

Symbol	Definition	Symbol	Definition
$\langle R \rangle$	Average data rate for service (bit/s)	$P^{(op)}(t)$	Idle network element power allocated to the network operator (W)
$\langle T \rangle$	Average inactive times for single service (s)	$P^{(svc)}$	Network element power consumption of a given service (W)
$\langle t \rangle$	Average active times for single service (s)	$P(t)$	Network element power at time t (W)
$B^{(k)}(t)$	Number of bits for the k th service over time duration t (bits)	$P_{BS}(\mu_{BS})$	Base station power for utilisation μ_{BS} (W)
$B^{(svc)}$	Number of bits transferred in time T_{tot} (bits)	$P_{BS}^{(op)}(t)$	Base station idle power allocated to mobile network operator (W)
B_m	Number of bits transferred in time t_m (bits)	P_{idle}	Network element idle power (W)
$BW\sigma_{PDSCH}$	base station bandwidth available for data transmission to active users	P_{max}	Maximum network element power (W)
C_{line}	Fixed capacity line rate of the device (bit/s)	$P_{max,\,out}$	Maximum base station transmit power (W)
C_{max}	Maximum throughput capacity of the network element (bit/s)	$P_{Ntwk}(t)$	Total network power consumption at time t(W)
$C_{Tot,\,Node}$	Total network node throughput (bit/s)	$P_{Tot,\,Node}$	Total network node power consumption (W)
Δ_P	Scaling factor relating transmit power to base station consumed power	$Q^{(k)}(T)$	Energy consumption of k th service over duration T (W)
$\delta P^{(k)}$	Change in power consumption of k th service due to traffic change $\delta R^{(k)}$ (W)	$Q_{in-phase}^{(k)}$	Energy consumption of k th service when it is in-phase with network traffic (J)
$\delta P^{(k,p)}$	Change in power consumption of k th service due to change $\delta R^{(p)}$ of the p th service (W)	$Q_{out-of-phase}^{(k)}$	Energy consumption of k th service when it is out-of-phase with network traffic (J)
ΔR	defined as $= R_{max} - R_{min}$ (bit/s)	Q_m	Energy consumption over duration t_m (J)
$\delta R^{(k)}$	Small change in traffic flow of k th service (bit/s)	Q_{tot}	Total energy consumption of the equipment (J)
$\Delta R^{(k)}$	defined as $= R^{(k)}{}_{max} - R^{(k)}{}_{min}$	ρ	Portion of time T_{tot} during which the user device is active
E	Incremental power per throughput $= \Delta P/\Delta R$ (Watts/bit/s)	$R^{(k)}(t)$	Bit rate of k th traffic flow through a network element (bit/s)
γ	Constant relating service traffic, $R^{(k)}(t)$, to total traffic, $R(t)$	$R_{max}^{(k)}$	Maximum data rate of the k th service (bit/s)
$h^{(k)}(t)$	Allocation of idle power to k th traffic flow	$R_{min}^{(k)}$	Minimum data rate of the k th service (bit/s)
$H(t)$	Energy efficiency metric (J/bit)	$R^{(k,\,svc)}$	Total base station throughput of the k th service (bit/s)
M	Number of active and inactive times in T_{tot}	$R^{(op)}(t)$	Network element throughput traffic allocated to the network operator (bit/s)
μ	Network operator policy for network utilisation ($\mu < 1$) of network elements	$R^{(svc)}$	Traffic due to the service of interest (bit/s)
$\mu_{BS}^{(k)}(t)$	Utilisation of base station by the k th user	$R(t)$	Network element throughput at time t (bit/s)
$\mu(t)$	Network element utilisation at time t	R_{fixed}	Pre-set network element throughput for allocation of element idle power (bit/s)
$\mu_{BS}^{(k,p)}$	Utilisation of the k th service by the p th user	R_{max}	Maximum network element throughput (bit/s)
$\mu_{BS}^{(k,svc)}$	Base station utilisation for the k th service	R_{min}	Minimum network element throughput (bit/s)

Symbol	Definition	Symbol	Definition
$\mu_{fixed,\,BS}$	Pre-set base station utilisation for allocation of element idle power to network operator	R_{peak}	Peak network traffic (bit/s)
M_{OH}	multiplier used to represent the proportionate increase in power consumption due to overheads	$SE(z^{(k)},t)$	spectral efficiency at the location of the k th user (bit/s/Hz)
μ_{op}	Average operational utilisation of network elements	$SE(z^{(k,p)},t)$	Spectral efficiency for the p th user of the k th service (bit/s/Hz)
$N_{hops}^{(k)}$	Number of hops in the path of the k th service	t_m	Times when user is accessing the service for single service (s)
$N_{PDSCH}^{(k)}(t)$	Number of PDSCH REs allocated to transmit data to the k th user	T_m	Times when user is not accessing the service (s)
$N^{(k,svc)}$	Number of users of the k th service that are active	$T_{n,m}$	Time within time duration T_m during which the n th service for multi-service (s)
$N^{(svc)}$	Number of service traffic flows	T_{tot}	Total time the user equipment is operational for single service (s)
N_E	Number of network elements	ξ	Proportion of idle power allocated to that service
$N_E^{(svc)}$	Number of network elements through which a given service traffic flows		

CHAPTER 11

Symbol	Definition	Symbol	Definition
$^1G^{(k)}(t)$	Instantaneous Power per user energy efficiency metric for k th service at time t (Watts/user)	ECR_j	Energy Consumption Ratio-Variable Load for the j th network element (W/bit/s)
$^1G_j(t)$	Instantaneous Power per user energy efficiency metric for j th network element at time t (Watts/user)	E_j	Incremental energy per bit of the j th network element (J/bit)
$^1G_{Ntwk}(t)$	Instantaneous Power per user energy efficiency metric for network at time t (Watts/user)	$\phi^{(k)}$	Phase of k th flow to account for the fact that the diurnal cycles of the individual services may have a different time of peak traffic
$^1H^{(k)}(t)$	Instantaneous power per throughput energy efficiency metric for k th service at time I (Watts/bit/s)	ϕ_{Ntwk}	Phase of total network traffic
$^1H_{Ntwk}^{[N]}(t_{peak})$	^1H network energy efficiency at peak traffic time in year N (W/bit/s)	$\mu(t)$	Utilisation at time t
$^1H_j(t)$	Instantaneous power per throughput energy efficiency metric for j th network element at time t (Watts/bit/s)	μ_{max}	Maximum network element utilisation
$^1H_{Ntwk}(t)$	Instantaneous power per throughput energy efficiency metric for network at time t (Watts/bit/s)	M_{OH}	multiplier used to represent the proportionate increase in power consumption due to overheads
$^2G^{(k)}(T)$	Average power per average number of users energy efficiency metric for k th service (Watts/user)	$N^{(type)}$	Number of user "types"
$^2G_j(t)$	Average power per average number of users energy efficiency metric for j th network element (Watts/user)	$N^{(type,p)}(t)$	Number of users of the p th user type online at time t

Symbol	Definition	Symbol	Definition
$^2G_{Ntwk}(T)$	Average power per average number of users energy efficiency metric for network (Watts/user)	$N^{(usr)}(t)$	Number of users at time t
$^2H^{(k)}(T)$	Energy per bit energy efficiency metric for k th service (J/bit)	$N^{(usr)}_{avg}$	Average number of users
$^2H^{[N]}_{Ntwk}(T)$	^2H network energy efficiency in year N (J/bit)	$N^{(usr,k)}$	Number of users of k th service
$^2H_j(T)$	Energy per bit energy efficiency metric for j th network element (J/bit)	N_E	Number of network elements
$^2H_{Ntwk}(T)$	Energy per bit energy efficiency metric for a network (J/bit)	$N^{(usr)}_{max}$	Maximum number of users
$^2G^{(k)}(T)$	Average power per user energy efficiency metric for k th service (Watts/user)	$N^{(usr)}_{min}$	Minimum number of users
$^3G_j(T)$	Average power per user energy efficiency metric for j th network element (Watts/user)	$NRMSD$	Normalised Root Mean Squared Deviation
$^3G_{Ntwk}(T)$	Average power per user energy efficiency metric for network (Watts/user)	$P_{idle,j}$	Idle power consumption of j th network element throughput (W).
$^3H^{(k)}(T)$	Average power per throughput energy efficiency metric for k th service (Watts/bit/s)	$P^{(k)}_j(t)$	Power consumption of the j th network element allocated to the k th service flow (W)
$^3H_j(T)$	Average power per throughput energy efficiency metric for j th network element (Watts/bit/s)	$P_{max,j}$	Maximum power consumption of j th network element throughput (W).
$^3H_{Ntwk}(T)$	Average power per throughput energy efficiency metric for network (Watts/bit/s)	$P_{Ntwk}(t)$	Network power consumption at time t (W)
$\langle ^2H(T)\rangle^{(svc)}$	Average of energy per bit energy efficiency metric over services (J/bit)	$Q^{(k)}_{in-synch}$	Energy of k th service if it is in-synch with network traffic (J)
$\langle ^3H(T)\rangle^{(svc)}$	Average power per throughput energy efficiency metric over services (Watts/bit/s)	$Q^{(k)}_{out-of-synch}$	Energy of k th service if it is out-of-synch with network traffic (J)
$\langle C_{max}\rangle_E$	Average maximum throughput capacity of network elements (bit/s)	$Q^{(svc)}$	Energy consumption of a service (J)
$\langle \cos(\phi^{(k)})\rangle_E$	Average of the phase offsets between the peak traffic time of the j th network element and the peak traffic time of the k th service	$\theta(x)$	Unit step function, $\theta(x) = 0$ for $x < 0$ and $\theta(x) = 1$ for $x \geq 0$.
$\langle E\rangle_E$	Average incremental energy per bit of network elements (J/bit)	$R^{(k)}(t)$	Data rate of k th service (bit/s)
$\alpha^{(k)}_j$	Portion of the traffic flow of the k th service that is goes through the j th network element	$R^{(k)}_{avg}$	Average traffic of the k th service flow (bit/s)
a_m, b_m	Parameters used in definition of ECR	$R^{(svc)}$	Data rate of a service (bit/s)
$\langle N_E\rangle^{(svc)}$	Average number of services through a network element	$R(t)$	Network element throughput at time t (bit/s).
$\langle N_{hops}\rangle^{(svc)}$	Average number of hops per service	$R^{(usr,p)}(t)$	Date rate for the p th user at time t (bit/s)
$\langle P_{idle}\rangle_E$	Average idle power of network elements (W)	$R^{[N]}_{peak}$	Peak traffic in year N (bit/s)
$\langle R(t)\rangle^{(usr)}$	Average bandwidth per user at time t (bit/s)	$R_{avg,j}$	Average traffic of j th network element (bit/s)
$\langle R^{(usr)}(t)\rangle^{(type,p)}$	Average data rate for the p th type of user (bit/s)	$R_{avg,Ntwk}$	Average network traffic (bit/s)

Symbol	Definition	Symbol	Definition
$\langle R^{[N]} \rangle_T$	Average traffic data over time T in year N (bit/s)	$R_j^{(k)}(t)$	Traffic flow of the k th service that is goes through the j th network element (bit/s)
$B^{(svc)}$	Total data transfer of a service (bits)	$R_{max,j}$	Maximum traffic of j th network element (bit/s)
C_{line}	CPE access line rate (bit/s)	$R_{max,Ntwk}$	Maximum network traffic (bit/s)
$C_{max,j}$	Maximum throughput capacity of j th network element throughput (bit/s).	$R_{min,j}$	Minimum traffic of j th network element (bit/s)
$\Delta N^{(usr)}$	Amplitude of the variation away from the average number of users	$R_{min,Ntwk}$	Minimum network traffic (bit/s)
$\Delta R^{(k)}$	Amplitude of the variation away from the average for the k th flow (bit/s)	$R_{Ntwk}(t)$	Total network throughput (bit/s)
$\Delta R^{(k)}(t)$	Variation away from the average for the k th flow at time t (bit/s)	$R_{Tot}(t)$	Total (network element, network or service) traffic at time t (bit/s)
ECR	Energy Consumption Ratio-Variable Load (Wat/bit/s)	T_{tot}	Total active time of a service (s)

Index

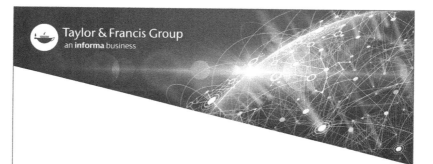